Bibliographies on
Regional Geography and Area Studies

Bibliographien zur
Regionalen Geographie und Landeskunde

Edited by / Herausgegeben von
Walter Sperling · Lothar Zögner

4

Hans-Jürgen Philipp

Saudi Arabia

Bibliography on
Society · Politics · Economics

Saudi-Arabien

Bibliographie zu
Gesellschaft · Politik · Wirtschaft

Literatur seit dem 18. Jahrhundert
in westeuropäischen Sprachen
mit Standortnachweisen

K·G·Saur München·New York·London·Paris 1984

HERAUSGEBER DER REIHE

Dr. Walter Sperling
o. Professor der Geographie und ihrer Didaktik
an der Universität Trier

Dr. Lothar Zögner
Direktor der Kartenabteilung der Staatsbibliothek
Preußischer Kulturbesitz, Berlin

CIP-Kurztitelaufnahme der Deutschen Bibliothek

Philipp, Hans-Jürgen:
Saudi Arabia : bibliogr. on society, politics,
economics ; Literatur seit d. 18. Jh. in west-
europ. Sprachen mit Standortnachweisen = Saudi-
Arabien / Hans-Jürgen Philipp. — München : Saur,
1984.
  (Bibliographien zur regionalen Geographie und
  Landeskunde ; 4)
  ISBN 3-598-21134-1

NE: GT

All rights reserved. No part of this publication may be reproduced, stored in a retrieval system, or transmitted in any
form or by any means, electronic, mechanical, photocopying,
recording, or otherwise, without permission in writing from
the publisher.

©1984 by K. G. Saur Verlag KG, München
Printed in the Federal Republic of Germany
Druck: WS-Druckerei, Mainz
Binden: Buchbinderei Karl Schaumann, Darmstadt
ISBN 3-598-21134-1

VORWORT

Diese Bibliographie verdankt ihre Entstehung zu einem Gutteil der freundlichen und kompetenten Hilfe einer Reihe von Personen und Organisationen. Danken muß ich vorrangig dem Personal der Universitätsbibliotheken Hohenheim und Tübingen; dort trug Frau Dipl.-Bibl. C. Harifi bzw. Herr Dr. W. Werkmeister in besonderem Maße zur Literatursuche und -beschaffung bei. Die erbetene Hilfe wurde mir auch in (weiteren) Stuttgarter, Tübinger, Hamburger, Münchner und Göttinger Bibliotheken zuteil; Erwähnung verdient, wie Herr Dr. U. Steinbach, der Direktor des Deutschen Orient-Instituts in Hamburg, und Herr Dr. E. Franz, der Leiter der Dokumentations-Leitstelle Moderner Orient ebenfalls in Hamburg, mir die Bestände ihrer Einrichtungen zugänglich machten. Für seine wohlwollende Tolerierung meiner jahrelangen Literaturstudien sei Herrn Prof. Dr. U. Planck, dem geschäftsführenden Direktor des Instituts für Agrarsoziologie, landwirtschaftliche Beratung und angewandte Psychologie der Universität Hohenheim, und für ihr sorgfältiges Schreiben des schwierigen Typoskripts Frau K. Maulbetsch im selben Institut herzlich gedankt.

Daß eine Bibliographie dieses Umfangs - trotz großer Bemühungen - nicht fehler- und lückenlos ist, versteht sich für Kenner von selbst. Allen Benutzern, die mich direkt oder über den Saur Verlag auf Unrichtigkeiten und Ergänzungen (und damit auf die Möglichkeit ihrer Beseitigung bzw. Berücksichtigung in einer zweiten Auflage dieser Bibliographie) aufmerksam machen, sei hiermit schon im voraus vielmals gedankt.

Dr. Hans-Jürgen Philipp

Institut für Agrarsoziologie, landwirtschaftliche Beratung und angewandte Psychologie
Universität Hohenheim
D 7ooo Stuttgart 7o

FOREWORD

This bibliography owes its compilation to a considerable extent to the friendly and qualified help of a number of persons and organizations. First of all, I would like to express my gratitude to the staffs of the University Libraries of Hohenheim and Tübingen. Mrs. C. Harifi, Librarian, and Dr. W. Werkmeister respectively particularly contributed to the search and procurement of the relevant literature. As well, libraries other than the abovementioned in Stuttgart, Tübingen, Hamburg, Munich, and Göttingen gave me the support I had asked for; worth mentioning is the way in which Dr. U. Steinbach, Director of the Deutsches Orient-Institut (German Orient Institute) in Hamburg, and Dr. E. Franz, Head of the Dokumentations-Leitstelle Moderner Orient (Middle East Documentation Center), again in Hamburg, facilitated access to the holdings of their institutions. Special thanks go to Prof. Dr. U. Planck, executive Director of the Institut für Agrarsoziologie, landwirtschaftliche Beratung und angewandte Psychologie der Universität Hohenheim (Institute of Rural Sociology, Agricultural Extension and Applied Psychology of the University of Hohenheim) for kindly tolerating my search for literature which lasted for several years. Mrs. K. Maulbetsch in the same institute deserves gratitude for her diligent typing of a complicated manuscript.

Thanks in advance to those users who, directly or through the publisher, K.G. Saur Verlag KG, draw my attention to mistakes or omissions (thereby enabling their correction or deletion in a second edition of this bibliography).

Dr. Hans-Jürgen Philipp

Institut für Agrarsoziologie, landwirtschaft-
liche Beratung und angewandte Psychologie
Universität Hohenheim
D 7ooo Stuttgart 7o

INHALT / CONTENTS

Vorwort / Foreword		V/VI
EINFÜHRUNG / INTRODUCTION		IX
I	Saudi-Arabien in Vergangenheit und Gegenwart / Saudi Arabia: Past and Present	IX/XXVIII
II	Beschränkungen und Zusammensetzung der Literaturauswahl / Limiting the Literature Selection and Composition of the Selected Literature	XLVII/LVIII
III	Hinweise zur Zitierweise und Benutzung / Citation and Use Instructions	LXVI/LXXIV
IV	Abkürzungen / Abbreviations	LXXXI
V	Bibliothekssigel und -adressen / Library Symbols and Addresses	LXXXIV
BIBLIOGRAPHISCHER TEIL / BIBLIOGRAPHIC SECTION		1
I	Bibliographien / Bibliographies	1
II	Literatur / Literature	7
III	Nachträge / Additions	3oo
REGISTERTEIL / INDEX SECTION		317
I	Schlagwörter / Subjects	317/343
II	Persönliche Verfasser / Authors	369
III	Persönliche Herausgeber / Editors	397
IV	Korporative Verfasser und Herausgeber / Corporate Bodies	4oo

EINFÜHRUNG / INTRODUCTION

I SAUDI-ARABIEN IN VERGANGENHEIT UND GEGENWART

Das Königreich Saudi-Arabien hat im 2o. Jahrhundert wie kaum ein zweiter Flächenstaat 'Karriere gemacht'. Zu Beginn dieses Jahrhunderts zählte sein mehr als 2 Mill. km^2 umfassendes Territorium noch zu den wirtschaftlich ärmsten, gesellschaftlich traditionalsten, politisch instabilsten, religiös fanatischsten, militärisch unbefriedetsten, infrastrukturell unerschlossensten, technologisch rückständigsten und am dünnsten besiedelten Großräumen der Erde. Hieraus sowie aus der klimatischen Ungunst, der vermuteten Ressourcenarmut und der außerordentlichen Fremdenfeindlichkeit der Bewohner jenes Raums gegenüber Nicht-Moslems erklärte sich auch der höchst unbefriedigende Forschungs- und Literaturstand zu jener Zeit - trotz Arabiens Lage zwischen zwei vielbefahrenen Schifffahrtswegen, dem Roten Meer und dem Persisch-Arabischen Golf. Dieser langen Negativliste stand insbesondere die überregionale Bedeutung eines kleinen Teilraums - des südlichen Hedschas - als Wiege und Zentrum des Islam gegenüber; zu den Heiligen Stätten in und bei Mekka und Medina pilgerten jährlich Zehn- bis Hunderttausende von Gläubigen. Positiv muß man gewiß auch die weitgehende ethnische und kulturelle Homogenität der nomadischen, bäuerlichen und städtischen Bevölkerung außerhalb des südlichen Hedschas bewerten; dieser Teilraum wirkte seinerseits als Schmelztiegel der vielen Zuwanderer (vor allem Pilger und Sklaven) aus verschiedenen Rassen und Kulturen.

Das obige Bild begann sich während des langsamen Machtaufstiegs und der langen Regierungszeit des legendären Staatsgründers, König Abdalasis ibn Abdarrachman ibn Faisal Ahl Saud (populärer westlicher Name: Ibn Saud), grundlegend zu wandeln. Am 15. Januar 1902 gelang es ihm, erst 21jährig, mit seinen nur 40 Getreuen, Riad, die Hauptstadt des zweiten saudischen Staates (s.u.), durch einen kühnen Handstreich dem in Innerarabien herrschenden Fürstenhaus Ibn Raschid zu entreißen. Drei Jahrzehnte lang hatte er dann mit der Ausweitung, Befriedung und Verwaltung seines aufstrebenden Herrschaftsgebiets - nicht ohne Rückschläge - alle Hände voll zu tun. Erst nach dem siegreichen Blitzkrieg mit dem Jemen im Frühjahr 1934 begann in diesem am 23. September 1932 zum Königreich Saudi-Arabien ausgerufenen Gebiet eine dauerhafte friedliche Entwicklung. Zuvor hatte es auch an Kapital und Personal für eine staatlich gelenkte wirtschaftliche und technische Modernisierung gemangelt. Diese setzte eigentlich erst mit den ersten, noch bescheidenen Erdölfunden ab 1935 durch eine US-amerikanische Gesellschaft, die spätere Arabian American Oil Company (ARAMCO), ein, wurde aber schnell durch den Zweiten Weltkrieg unterbrochen. Danach konnten Ölförderung und -ausfuhr zwar schnell gesteigert werden, dennoch wurde Saudi-Arabiens Chance, binnen kurzem zu einer wirtschaftlichen und damit verbunden auch zu einer politischen Führungsmacht aufzusteigen, noch bis Ende der 50er Jahre weithin verkannt.

Wohl noch der Großteil der Nicht-Saudis (und vielleicht auch der Saudis selbst) hegt teils ungenaue, teils falsche Vorstellungen davon, welcher Entwicklungsstand und welche Bedeutung das jetzige Saudi-Arabien kennzeichnen, weshalb diese im folgenden wenigstens schlaglichtartig beleuchtet werden sollen.* Ihr 'Rückgrat' (und

---
*
Die nachstehenden Angaben stützen sich größtenteils auf die folgenden Quellen:

Blume, H. (Hg.): Saudi-Arabien. Natur, Geschichte, Mensch und Wirtschaft. Ländermonographien 7; Tübingen, Basel: Erdmann, 1976.

Ingenieurgemeinschaft Lässer-Feizlmayr Consulting Engineers: Water for Riyadh. Innsbruck: Tyrolia, 1983.

Kingdom of Saudi Arabia, Ministry of Planning: Saudi Arabia: achievements of the First and Second Development Plans, 1390-1400 (1970-1980). Facts & Figures; Jeddah: Tihama Publications, 1982.

Kingdom of Saudi Arabia, Saudi Arabian Monetary Agency, Research and Statistics Department: Annual report, 1401 (1981). Riyadh: Saudi Arabian Printing Company Ltd. Ammariyah, n.d.

selbstverständlich das des phänomenalen Aufstiegs in den 6oer und insbesondere in den 7oer Jahren) ist die Ausbeutung des fast unermeßlichen Erdöl- und Erdgasreichtums der Ostprovinz am und des vorgelagerten Hoheitsgebiets im Persisch-Arabischen Golf. In diesem Raum lagern - mit Ende 1981 mindestens 116,7 Mrd. nachgewiesenen und 177,2 Mrd. vermuteten Faß Rohöl (à 159 l) - rd. ein Viertel der Weltölreserven und auch ein beachtlicher Anteil der Weltgasreserven; bezeichnenderweise befinden sich dort das größte bekannte Onshore- (Ghawar) und Offshoreölvorkommen (Safanija) sowie Gasvorkommen (Haradh). Saudi-Arabiens Rohölfördermengen von 3,48-3,62 Mrd. Faß (464-483 Mill. t) in den Jahren 1979-81 entsprachen rd. einem Sechstel der Weltförderung und wurden nur noch von denjenigen der Sowjetunion übertroffen. 97-98 % jener Mengen stammten allein aus den verbliebenen Konzessionsgebieten des größten Ölproduzenten in der Welt, der ARAMCO, die 198o von der Regierung - nach einer Mehrheitsbeteiligung seit 1974 - vollständig verstaatlicht wurde. Die restlichen 2-3 % wurden von der überwiegend japanischen Arabian Oil Company und der US-amerikanischen Getty Oil Company gefördert. Die Situation auf dem Weltölmarkt und in der Organization of Petroleum Exporting Countries (OPEC) veranlaßten Saudi-Arabien allerdings, seine durchschnittliche Tagesförderung zwischen Mitte 1981 und Herbst 1983 von 1o,2 auf 5,5 Mill. Faß zu senken.

Ende 198o wurden die sicheren und die wahrscheinlichen Erdgasreserven allein in den Konzessionsgebieten der ARAMCO auf 1,95 bzw. 3,25 Billionen Kubikmeter geschätzt. Das bei der Ölförderung anfallende (sog. assoziierte) Erdgas belief sich in jenem Jahr auf 53,3 Mrd. m^3 (gegenüber 2o,6 Mrd. m^3 197o). Dieser anfänglich ausschließlich abgefackelte und reinjizierte Energieträger wird zunehmend wirtschaftlich genutzt: der genutzte Anteil stieg zwischen 197o und 198o von 11,2 auf 27,5 %.

Fortsetzung von Anmerkung *

Koszinowski, T. (Hg.): Saudi-Arabien: Ölmacht und Entwicklungsland. Beiträge zur Geschichte, Politik, Wirtschaft und Gesellschaft. Mitteilungen des Deutschen Orient-Instituts 2o; Hamburg: Deutsches Orient-Institut, 1983.

Niblock, T. (ed.): State, society and economy in Saudi Arabia. London: Croom Helm; Exeter: Centre for Arab Gulf Studies, 1982.

Statistisches Bundesamt: Saudi-Arabien. Statistik des Auslandes, Länderkurzberichte; Stuttgart, Mainz: Kohlhammer, 1982.

Die vorstehenden Angaben zu den Erdöl- und Erdgasreserven und -fördermengen lassen vermuten, daß sich Saudi-Arabien mindestens noch bis Mitte des nächsten Jahrhunderts günstige wirtschaftliche Aussichten bieten.

Aufgrund des geringen Eigenbedarfs wurden die geförderten Ölmengen um 1980 noch zu über 97 % exportiert, und zwar fast ausschließlich in unbearbeitetem Zustand. Dank jenes Umstands steht Saudi-Arabien als Ölexportland in der Welt konkurrenzlos da. Aus dem Export von Erdöl und Erdölprodukten wurden 1978 39,7, 1979 61,7 und 1980 105,9 Mrd. $ erlöst, aus dem Export von verflüssigtem Erdgas dagegen 1978 0,66, 1979 1,11 und 1980 2,36 Mrd. $. Die Erdöl- und Erdgasexporte machten in diesen Jahren 99,1-99,2 % des gesamten Ausfuhrwertes aus; ihnen allein verdankte Saudi-Arabien damit seine hohen Ausfuhrüberschüsse (1979: 39,0; 1980: 79,2; 1981: 85,0 Mrd. $).

Wie die seit 1970, insbesondere seit der sog. Ölkrise im Herbst 1973 stark gestiegenen Ölfördermengen und -preise einerseits sowie Royalties und Steuern der Fördergesellschaften andererseits Saudi-Arabiens Einnahmen aus dem 'Ölgeschäft' in die Höhe schnellen ließen, verdeutlicht folgende Zahlenreihe: 1970: 1,21; 1972: 2,74; 1974: 22,57; 1976: 30,75; 1978: 32,23; 1980: 84,47 Mrd. $; 1981 wurden erstmals über 100 'Petrodollarmilliarden' eingenommen (1940, 1950 und 1960 demgegenüber erst 1,5, 56,7 bzw. 333,7 Mill. $). In diesem Zeitraum machten die Öleinnahmen zwischen 86,2 % und 97,3 % der gesamten Staatseinnahmen aus. Trotz ebenfalls rapide gewachsener Staatsausgaben wurden 1980/81 und 1981/82 Mehreinnahmen in Höhe von 4,8 bzw. 12,2 Mrd. $ erzielt. Angesichts dieser Größenordnungen muß daran erinnert werden, daß Ibn Sauds jahrzehntelanger Finanzminister Abdallah Sulaiman al-Hamdan noch Anfang der 30er Jahre die chronisch fast leere Staatskasse unter seinem Bett aufbewahren konnte und daß Ibn Sauds Nachfolger, der verschwenderische König Saud (1953-64), sein Land binnen fünf Jahren an den Rand des Bankrotts brachte.

Weitere wirtschaftliche Kennzahlen seien angeschlossen. Das Bruttosozialprodukt zu Marktpreisen stieg zwischen 1970 und 1979 - in jeweiligen Preisen gerechnet - von 4,0 auf 94,6 Mrd. $ und zwischen 1970 und 1982 - zur Zeit großer Wirtschaftskrisen in der Welt - durchschnittlich um real zehn Prozent jährlich. Im letztgenannten

Zeitraum wuchs das Pro-Kopf-Einkommen der schätzungsweise 8 Mill. Landesbewohner von rd. 800 auf rd. 16 800 $ an, so daß es sich mehr als verzwanzigfacht hat und heutzutage zu den höchsten in der Welt zählt. Da die Lebenshaltungskosten im Jahrzehnt 1970-80 um 16,5 % jährlich und seither deutlich weniger gestiegen sind, ist das Realeinkommen und damit verbunden der Wohlstand des 'durchschnittlichen Saudis' beträchtlich gestiegen. Zur Entstehung des Bruttoinlandsprodukts trug der Ölsektor Ende der 7oer Jahre - unter Zugrundelegung der Preise von 1969 - erstmals mit weniger als 5o % bei, und die Anteilswerte für den Agrarsektor, in dem immer noch relativ die meisten Saudis (wahrscheinlich nur mehr 35-45 %) beschäftigt sind, beliefen sich in der zweiten Hälfte der 7oer Jahre auf 3-4 %; in den 7oer Jahren nahm das relative Gewicht beider Sektoren ab, wogegen das des Baugewerbes deutlich zunahm. Der Devisenbestand wurde im Mai 1978 und Mai 1979 mit rd. 17 Mrd. $ und jeweils zwei Jahre später mit rd. 28 Mrd. $, der Goldbestand in diesen Jahren gleichmäßig mit rd. 4,5 Mrd. $ deklariert. Die Auslandsvermögen und die Währungsreserven bezifferten sich Ende 1978 auf ca. 6o bzw. 2o Mrd. $; erstere dürften zwischenzeitlich beträchtlich angewachsen sein.

Aus den angeführten Zahlen muß gefolgert werden, daß sich Saudi-Arabien von einem der ärmsten Entwicklungsländer z.Z. seiner Gründung binnen 5o Jahren zu einem der reichsten Länder in der Welt entwickelt hat. Kann man es auch noch nicht zu den Industrieländern zählen, muß man ihm doch den Status eines fortgeschrittenen Landes der Dritten Welt zubilligen. Seine Öl-, Finanz- und Wirtschaftspolitik sind für das Wohl und Wehe der übrigen Welt heutzutage alles andere als eine quantité négligeable.

Wie der saudiarabische Staat seine enormen finanziellen Ressourcen nutzt, weisen allein zwei Zahlen aus: die Gesamtausgaben (einschließlich Verteidigungs- und Entwicklungshilfeausgaben) während der Zweiten Fünfjahresplanperiode (1975/76-1979/80) erreichten 133,o Mrd. $, und die im Dritten Fünfjahresplan (für die Jahre 198o/81-1984/85) eingeplanten Gesamtausgaben belaufen sich auf 363,o Mrd. $.

In diesen beiden Entwicklungsplänen haben u.a. Projekte zur Weiterleitung, Verarbeitung und Nutzung von gefördertem Erdöl und Erdgas hohe Priorität. 1975 beauftragte die Regierung die ARAMCO damit, das sog. Master Gas System zu planen, zu bauen und zu betreiben. Mit Hilfe dieses schätzungsweise 15 Mrd. $ kostenden Großprojektes soll der Großteil (rd. 85 Mill. m^3 pro Tag) des von der ARAMCO in Onshore-Ölfeldern geförderten und zuvor abgefackelten assoziierten Erdgases durch ein Netz von Sammelleitungen, Öl-Gastrenn-, Fraktionierungs- und Verflüssigungsanlagen, Verschiffungseinrichtungen usw. wirtschaftlicher Nutzung zugeführt werden; konkret heißt dies, daß der größere Teil des bearbeiteten Gases den neuen petrochemischen Komplexen, kombinierten Meerwasserentsalzungsanlagen und Kraftwerken, Stahlwerken usw. von Dschubail und Janbo an der Ost- bzw. Westküste als Rohstoff und Energieträger und der kleinere Teil für die Ausfuhr - von den dortigen neuen Ausfuhrhäfen aus - dienen soll. Das imposanteste Teilprojekt ist die 117o km lange, 66-76 cm dicke und computergesteuerte sog. East-West Natural Gas Liquids/Ethane Transpeninsular Pipeline (Kapazität: 27o ooo Faß pro Tag) von Shedgum unweit des Persisch-Arabischen Golfs nach Janbo, die bisher längste und modernste Hochdruck-Gasleitung überhaupt. Die während der ersten Bauphase fertiggestellten Einrichtungen des Master Gas System sind seit Herbst 1982 in Betrieb.

Überwiegend parallel zur obigen Flüssiggaspipeline verläuft die 12oo km lange sog. East-West Crude Oil Pipeline, die bis Mitte 1981 von der staatlichen General Petroleum and Mineral Organization (PETROMIN) mit einem Kostenaufwand von 1,6 Mrd. $ fertiggestellt wurde. Ihre bisherige Durchsatzkapazität von 1,85 Mill. Faß Rohöl pro Tag kann verdoppelt werden. Diese Pipeline, die als die sicherste und fortschrittlichste ihrer Art in der Welt gilt, erfüllt zwei Aufgaben: sie verkürzt den Reiseweg von Öltankern aus westlichen Ländern um Tausende von Seemeilen und versorgt die Industriestadt und den Industriekomplex von Janbo mit Rohöl.

Die angesprochenen Industriekomplexe von Dschubail und Janbo sind das Herzstück der Industriepolitik im Zweiten und Dritten Fünfjahresplan. Diese noch vor zehn Jahren völlig bedeutungslosen Küstenstädtchen sind inzwischen zu infrastrukturell supermodernen Industriestädten ausgebaut worden, in denen im Jahre 2ooo zusammengenommen ungefähr 5oo ooo Menschen, darunter 144 ooo Erwerbstätige, leben sollen. Durch die dortige Industriekonzentration soll die Wirtschaft des Landes beträchtlich diversifiziert und stabilisiert werden. Dies soll durch die Ansiedlung von wenigen energieintensiven Primärindustrien auf Erdöl- und Erdgasbasis sowie - teilweise als Abnehmer ihrer Produkte - eine Vielzahl von Sekundär- und Tertiärindustrien gelingen. 1980 war die Einrichtung von 17 primär-, 136 sekundär- und über 100 tertiärindustriellen Betrieben bis 2000 geplant, darunter mehrere Ölraffinerien, Fraktionierungsanlagen, Eisen- und Stahlwerke, Zement- und Düngemittelfabriken sowie eine Aluminiumhütte.

Als Saudi-Arabien Ende der 40er Jahre seinen ersten Bauboom - damals noch beschränkt auf Dschidda, Riad und andere mittelgroße Städte - erlebte, stand seine Bauwirtschaft noch auf ganz schwachen Füßen. U.a. Zement mußte zu 100 % importiert werden. Die anhaltende Baukonjunktur regte ab Anfang der 50er Jahre zur Planung von kleinen Portland-Zementfabriken an. Drei solche Fabriken wurden zwischen 1958 und 1966 bei Dschidda, Hofuf und Riad in Betrieb genommen. Da sie mit ihrer Anfangskapazität von jeweils 300 t pro Tag und durch unvorhergesehene Anfangsschwierigkeiten weit hinter dem Landesbedarf zurückblieben, wurden sie sukzessive ausgebaut. Ihre gesamte Jahresproduktion lag Mitte der 70er Jahre zwar bei 1,1 Mill. t, erreichte z.Z. der Inangriffnahme der großen Industrievorhaben im Rahmen des Zweiten Fünfjahresplans aber nur noch einen Bedarfsdeckungsgrad von 40 % und weniger. Deshalb wurde eine Steigerung der Gesamtkapazität von damals 1,4 auf mehr als 10 Mill. t 1984 - durch den weiteren Ausbau der drei bestehenden und den Bau von sieben weiteren Zementfabriken - mit dem Ziel der längerfristigen Sicherstellung einer 100%igen Selbstversorgung beschlossen. Daß dieses Programm umgehend verwirklicht wurde und noch wird, beweist die effektive Produktionssteigerung von 1,3 Mill. t (1977) über 2,6 Mill. t (1979) auf 4,3 Mill. t (1981).

Der akuten oberirdischen Wasserknappheit sucht die Staatsführung insbesondere seit Anfang der 60er Jahre mit allen möglichen Mitteln abzuhelfen. Eine unter mehreren Problemlösungen sieht folgendermaßen aus: Zwischen 1969 und 1981 wurden unter der Leitung der Saline Water Conversion Corporation in acht Siedlungen am Roten Meer und Persisch-Arabischen Golf 16 Meerwasserentsalzungsanlagen mit einer Gesamtkapazität von 179 Mill. Litern pro Tag zur Trinkwasserversorgung küstennaher Gebiete eingerichtet. 1981 befanden sich sieben weitere Anlagen mit einer Tageskapazität von insgesamt 1 253 Mill. l im Durchführungs- und nochmals sechs mit einer Tageskapazität von insgesamt 413 Mill. l im Planungsstadium. 1981 wurde mit der Fertigstellung der Wasserpipeline Janbo-Medina erstmals entsalztes Meerwasser bis zu 12o km weit ins Landesinnere transportiert. Mitte 1983 stand - nach dreijähriger Bauzeit und Kosten von rd. 2 Mrd. $ - das größte Hochdruckwasserleitungssystem der Erde vor dem Abschluß. Seine wichtigsten Bestandteile sind zwei 466 km lange parallele Hauptleitungen mit einem Durchmesser von jeweils 1,425 m, durch die täglich bis zu 83o Mill. l entsalztes Golfwasser vom zur größten Anlage ihrer Art ausgebauten Al Jubayl Desalination Plant bis nahe Riad gepumpt werden können. Dieses von Dschubail aus mittels Telekontrolle gesteuerte sog. Riyadh Water Transmission System soll den schnell wachsenden Trinkwasserbedarf der Millionenstadt Riad bis zum Jahre 2o12 decken helfen. Weitere großdimensionierte Wasserpipelines - zur Trinkwasserversorgung von Mekka, Taif, Abha, Tebuk usw. - sind von der Rotmeerküste ins Landesinnere im Bau.

Neben konventionellen Lösungen, wie der Anlage von Tiefbrunnen, dem Bau von Stau- und Rückhaltedämmen sowie der Aufbereitung von Schmutzwässern, liebäugelten einzelne saudische Politiker und Planer in letzter Zeit mit zwei revolutionären Ideen: der Einfuhr von Frischwasser zum einen mittels Nutzung von aus der Antarktis herangeschleppten Eisbergen und zum anderen aus dem wasserreichen Nordjapan in bislang leer anreisenden japanischen Öltankern.

Trotz der rapide gestiegenen städtischen und industriellen Wasserbedarfe ist der weitaus größte Wasserverbraucher - mit 198o 8o - 85 % des Gesamtverbrauchs - immer noch der Bewässerungslandbau. Letzterer ist außerhalb vom südlichen Hedschas und von Asir, wo ertragreicher Regenfeldbau möglich ist, überall im Lande traditionell die wichtigste Voraussetzung für sichere und gute Ernten der angebauten Baum-, Getreide-, Gemüse- und Futterkulturen. Der landwirtschaftliche Wasserverbrauch ist nun gerade im letzten Jahrzehnt durch zwei gegenläufige Entwicklungen beeinflußt worden. Auf der einen Seite gaben viele tausend Oasenbauern ihre kleinbetriebliche Bodennutzung auf, um außerhalb des Agrarsektors ein weniger mühseliges und besser bezahltes Auskommen zu finden; hiermit waren vielfach Sozialbrache und Landflucht verbunden. Auf der anderen Seite drängten mehrere hundert Oberschichtszugehörige - Prinzen, Minister, Großhändler, Industrielle usw. - in diesen Sektor. Sie erwarben - z.T. durch Schenkungen des Königshauses - mehr oder weniger große Ödlandflächen in natürlichen Gunsträumen und im Umland großer Städte. Als Anreize, diese Flächen in Kultur zu nehmen und intensiv zu bewirtschaften, wirkten und wirken noch

(1) die starke Abhängigkeit des Landes von Nahrungsgüterimporten verschiedenster Art;
(2) die hohe staatliche Subventionierung bestimmter landwirtschaftlicher Produktionsmittel und Erträge;
(3) die großzügige Gewährung von Modernisierungskrediten zu günstigen Konditionen durch die Saudi Arabian Agricultural Bank und den Saudi Industrial Development Fund;
(4) der beträchtliche Nachfrageüberhang einer wachsenden Anzahl von In- und Ausländern im Hinblick auf höherwertige landwirtschaftliche Erzeugnisse und - als Konsequenz aus dieser Konstellation -
(5) die günstigen Profitchancen von landwirtschaftlichen Investitionen.

Planung, Aufbau und Bewirtschaftung ihrer Großbetriebe lassen jene 'nebenberuflichen Landwirte' von westlichen und arabischen Consultants, Managern, Informatikern, Mechanikern, Tierärzten und anderen Spezialisten vornehmen. Diese beschaffen und nutzen auf Wunsch oder mit Billigung ihrer vielfach steinreichen Arbeitgeber die neuesten Modelle der westlichen Agrartechnik (großkalibrige Planierraupen, Scheibenpflüge, Rundregner, Mähdrescher, Vollerntemaschinen, Plastiktreibhäuser, vollautomatisierte Rinder- und Hühnerställe, Maschinenhallen, Silos usw. weit überwiegend 'made in USA'). Daß die meisten dieser äußerst kapitalintensiven, eng spezialisierten und hoch produktiven Agribusinessunternehmen aufgrund ihrer Flächenausdehnung einen enormen Wasserverbrauch haben, läßt sich denken.

Der schnell wachsende moderne Agrarsektor fördert absehbar den weiteren Niedergang des traditionellen Agrarsektors. Ersterem ist andererseits hauptsächlich die in den letzten Jahren deutlich gestiegene Agrarproduktion (z.B. wurden 1977 12o ooo und 1983 mehr als 5oo ooo t Weizen sowie im erstgenannten Jahr insgesamt 283 ooo und zwei Jahre später 377 ooo t Gemüse geerntet, weiter 1975 14 Mill. und 198o 4o Mill. Brathühner sowie 1978 9 2oo und 198o 32 ooo t Milch erzeugt) zu verdanken, auf die Saudi-Arabiens Agrarpolitiker und Wirtschaftsplaner heutzutage stolz sind. So berechtigt dieser Stolz ist, muß angesichts der bisherigen Entwicklung von Umfang und Zusammensetzung der Agrarimporte und ihrer Einflußfaktoren (Bevölkerungswachstum, Nachfrageverschiebungen, Wasserdargebot, Vermarktungs- und Verarbeitungseinrichtungen usw.) doch bezweifelt werden, ob die fast ohne Rücksicht auf natürliche Ressourcen und wirtschaftliche Kalküle angestrebte Autarkie tatsächlich bis 1990 oder überhaupt erreicht werden kann.

Mehrere Städte Saudi-Arabiens sind in den letzten Jahrzehnten geradezu explosionsartig gewachsen. Noch zu Beginn dieses Jahrhunderts dürfte jeweils nur ein halbes Dutzend Siedlungen mehr als 10 000 bzw. zwischen 5 000 und 10 000 Einwohner gezählt haben. Mekka war damals die größte Stadt (sogar ganz Arabiens) mit wahrscheinlich mehr als 50 000 ständigen Einwohnern, gefolgt von Dschidda und Hofuf mit 20 000-30 000. Riad, bereits Hauptstadt des zweiten saudischen Staates (s.u.), zählte noch weniger als 10 000 und selbst 1920 noch weniger als 20 000 Einwohner. Seinen Charakter als abgelegenes Landstädtchen (mit ausgedehntem Palmenhain) begann es erst in den 20er und 30er Jahren zu verlieren. Daß seine Einwohnerzahl seit dem Zweiten Weltkrieg exponential angestiegen ist, zeigen folgende Zahlen: 1945: 62 000, 1955: 106 000, 1965: 225 000, 1974: 667 000, 1980: ca. 1 Mill. Ähnlich spektakulär wuchs Dschidda: 1940: 30 000, 1965: 194 000, 1974: 561 000, 1981: ca. 1,3 Mill. Hinter der politischen Metropole Riad und der wirtschaftlichen Metropole Dschidda treten heutzutage einerseits die traditionellen religiösen Zentren Mekka und Medina (mit 1965 185 000 bzw. 1963 73 000 und 1980 schätzungsweise 550 000 bzw. 290 000 Einwohnern) und andererseits die traditionellen landwirtschaftlichen Zentren Hofuf, Katif, Nedschran, Taif usw. deutlich zurück, doch wurde auch ihr Stadtbild in den letzten 30-40 Jahren durch starke Zuwanderung von In- und Ausländern, moderne Verkehrs- und sonstige Stadtplanung, westliche Bau- und Lebensstile u.a.m. völlig auf den Kopf gestellt. Verstädterung und Urbanisierung gehören zweifelsohne zu den wichtigsten Kennzeichen des 'neuen' Saudi-Arabiens.

Die heutigen Saudis sind regelrecht von einer 'Auto-Manie' besessen, was um so verwunderlicher ist, als Kraftfahrzeuge lange als Teufelswerk verfemt waren (im ländlichen Raum kamen Fälle von 'Auto-Stürmerei' vor). In nennenswerter Zahl wurden Personenwagen

und Kleintransporter erst nach der Eroberung des Hedschas 1924-25
durch Ibn Saud eingeführt. Fast zwei Jahrzehnte lang wurden sie
vorrangig zum Transport von Pilgern und von Angehörigen der königlichen Familie benutzt; mangels befestigter Straßen - die erste
Asphaltstraße wurde erst Ende der 3oer Jahre zwischen Dschidda und
Mekka (72 km) gebaut, und zwar von Ägyptern - mußte vor allem auf
die alten Karawanenwege und mangels einheimischer Fahrer auf ausländische, insbesondere indische Moslems zurückgegriffen werden.
Wie grundlegend sich dieses Bild gerade in der letzten Zeit gewandelt hat, zeigen folgende Zahlen: Der gesamte zugelassene Kfz-Bestand wuchs zwischen 197o und 198o von 59 9oo auf 1 998 ooo an
(Steigerung um 3336 %); die jährlichen Zuwachsraten betrugen 21-
65 %. 198o gab es 1 o2o ooo Pkw (darunter 1o5 5oo Taxis), so daß
im Mittel acht Landesbewohner, in den größten Städten sogar jeder
zweite Einwohner über ein Auto verfügten bzw. verfügte. Fast ebenso
hoch lag die Lkw-Zahl und -Relation, was sich z.T. aus dem Fehlen
eines - allerdings seit kurzem geplanten - ausgedehnten Eisenbahnnetzes erklärt. Jene Motorisierungswelle steht mit dem präferentiellen Ausbau des Straßennetzes in den beiden Entwicklungsplänen
für die Jahre 197o-8o in engem Zusammenhang: die Gesamtlänge der
Asphaltstraßen wuchs von mehr als 8 ooo auf mehr als 2o ooo und
die der Erdstraßen von rd. 3 5oo auf mehr als 24 ooo km an. Daß
das jetzige Straßennetz trotz dieses imposanten Ausbaus noch zu
wünschen übrig läßt, machen einerseits die Landesgröße (Nord-Süd-
und West-Ost-Ausdehnung von rd. 2 2oo bzw. 1 2oo km) und andererseits der teilweise chaotische innerstädtische Straßenverkehr
(stundenlange Verstopfungen und Hupkonzerte mit entsprechender Umweltbelastung) deutlich.

Das große Pilgerfest in der ersten Hälfte des islamischen Monats
Dhu al-Hidscha stellt seit seiner Entstehung eine religiöse Massenveranstaltung dar. Nie zuvor erreichte es jedoch ähnliche Ausmaße
wie gerade in den letzten Jahren: die Gesamtzahl der in- und ausländischen Pilger überstieg wahrscheinlich 197o erstmals die Millionengrenze und 1979 erstmals die Zweimillionengrenze; knapp unter oder über zwei Millionen - darunter 8oo ooo-9oo ooo Ausländer,
von denen 6o-75 % das Flugzeug benutzen - sind seit 1978 die Regel.
Daß solcher Massenandrang den Ausbau fast der gesamten materiellen
Infrastruktur im Raum Mekka-Medina-Dschidda forciert hat, liegt
auf der Hand. Ein herausragendes Beispiel hierfür ist der King Ab-

dul Aziz International Airport nördlich von Dschidda, dessen erster Bauabschnitt im Frühjahr 1981 und dessen zweiter Bauabschnitt planmäßig 1985 abgeschlossen wurde bzw. wird. Dieser Flughafen wird mit 1o5 km^2 Ausdehnung, einem besonderen Pilger-Terminal, Unterbringungsmöglichkeiten für 8o ooo Pilger und 6 3oo 'Normalpassagiere', Parkraum für 22 ooo Autos, einer eigenen Moschee und Wasserentsalzungsanlage usw. einer der größten und modernsten Flughäfen der Erde sein; dort werden 1985 schätzungsweise 8,6 Mill. und im Jahre 2 ooo schätzungsweise 1o Mill. Passagiere abgefertigt. Noch gewaltigere Dimensionen weist der im Herbst 1983 eingeweihte King Khalid International Airport nördlich von Riad auf (24o km^2 Ausdehnung, Jahreskapazität von 15 Mill. Passagieren). Ein dritter internationaler Großflughafen entsteht im Agglomerationsraum Dammam-Dhahran-al-Khobar.

Um 19oo gab es im jetzigen Staatsgebiet nur eine Handvoll von öffentlichen und privaten türkischen und arabischen Schulen und nicht viel mehr einfachste Koran- und Moscheeschulen. Außerhalb von Mekka und Medina, traditionellen Zentren islamischer Gelehrsamkeit, waren infolgedessen fast alle Landesbewohner Analphabeten. Mit dem Ausbau jenes rudimentären Bildungswesens wurde in den drei ersten Jahrzehnten dieses Jahrhunderts sowohl im damals noch haschemitisch beherrschten Hedschas als auch in Ibn Sauds Herrschaftsgebiet begonnen. Letzterer richtete 1926 das sog. Directorate of Education ein, das sich – mit bescheidenen Mitteln ausgestattet – bis Anfang der 5oer Jahre einerseits um die Förderung von Schulbau und -besuch, andererseits um die langsame Abkehr von religiös bestimmten Lehrplänen verdient gemacht hat. Bis 196o blieb der Besuch öffentlicher Bildungseinrichtungen Jungen vorbehalten. Nur 2o Jahre später waren die Verhältnisse von einst wie umgekrempelt: 198o bestanden mehr als 1o 7oo Bildungseinrichtungen (ohne Vorschulen), und zwar 5 4oo Grundschulen, 1 8oo Mittel- und höhere Schulen, 25 Sonderschulen, 3o berufsbildende Schulen, 1o7 Lehrerausbildungsstätten, sechs Universitäten sowie 3 4oo Erwachsenenbildungsanstalten (für Alphabetisierungskurse). Davon waren für Mädchen eingerichtet – Koedukation ist verpönt – 1 7oo Grundschulen, 5oo Mittel- und höhere Schulen, neun Sonderschulen, 64 Lehrerausbildungsstätten sowie sechs fakultätsähnliche Colleges und Institute; 2 9oo der 9 3oo Erwachsenenbildungsklassen waren für Teilnehmerinnen reserviert. Welche Anstrengungen in den letzten Jahren unternommen wurden, städtische gleichermaßen wie ländliche Siedlungen mit Schulen auszustatten, erhellt daraus, daß im Rahmen des Ersten und Zweiten Fünfjahresplans (1970-80) 2,2 Schulen pro Tag fertiggestellt und angemietet wurden. Die aufgezählten Einrichtungen wurden 1980 von 862 000 Grundschülern, 339 000 Mittel- und Oberschülern, 2 000 Sonderschülern, 6 000 Berufsschülern, 22 000 pädagogischen und 48 000 sonstigen Studenten beiderlei Geschlechts besucht, und die Anzahl der Teilnehmer und Teilnehmerinnen an Alphabetisierungskursen lag bei 142 000. Die Gesamtzahl der Vollzeit-Lehrkräfte erreichte 78 000, von denen jede zweite Ausländer war (der Anteil der ausländischen Lehrkräfte an den Universitäten lag bei 66 %, derjenige an den Mittel- und höheren Schulen bei 79 % und derjenige an den sog. Teacher Training Institutes sogar bei 86 %). Diesem und anderen Schwachpunkten im Bildungssystem (unbefriedigende Einschulungsrate insbesondere der Nomadenkinder, hohe Raten der Schul- und Studienabbrecher, Defizit an berufsbildenden Einrichtungen, ungleiche regionale Bildungsversorgung usw.) sucht man in der laufenden Dritten Fünfjahresplanperiode mit einem Aufwand von fast 36 Mrd. $ zu Leibe zu rücken.

Seit 1963, als es die royalistische Seite im jemenitischen Bürgerkrieg zu unterstützen begann, hat sich Saudi-Arabien militärisch von einem Zwerg zu einem Riesen gemausert. Im Rahmen eines defensiven Militärkonzepts, das Größe und Natur des Landes aufzwingen, wurde ab Mitte der 60er Jahre zuerst die reguläre Armee und in den letzten Jahren auch die beduinische Nationalgarde hinsichtlich Ausrüstung und Ausbildung großzügig modernisiert. In den 70er Jahren stiegen die jährlichen Militärausgaben von 0,4 auf 14,2 Mrd. $ an, wovon allerdings rd. die Hälfte für Infrastrukturprojekte (Krankenhaus-, Straßen- und Hausbau) verwendet wurde. Im Zeitraum 1971-80 avancierte Saudi-Arabien durch den Kauf von Rüstungsgütern im Gesamtwert von mehr als 34 Mrd. $ allein in den USA zu deren wichtigstem Abnehmer. Zu seinem Arsenal zählen heutzutage modernste Waffensysteme, wie F-5- und F-15-Kampfflugzeuge sowie mit AWACS ausgerüstete Boing-Jets. Milliarden verschlingt auch jede der vier im Bau befindlichen supermodernen Militärstädte, deren bekannteste die King Khalid Military City im Nordosten des Landes ist.

Die extreme Modernisierung in einigen Lebensbereichen hat bislang vor den politischen Strukturen haltgemacht. Saudi-Arabiens Staats-

und Regierungsform trägt in westlichen Veröffentlichungen überwiegend das Etikett absolute Monarchie, weil alle Macht letztlich in den Händen des Königs und Regierungschefs (Premierminister) - in einer Person - liegt und sich dieser bei seiner Herrschaft nur auf ein von ihm ernanntes politisches Kabinett (Ministerrat), eine Versammlung von Religionsgelehrten (Ulema) und einen Familienrat (Prinzen) stützt. Diese Etikettierung stimmt jedoch nur teilweise mit der Wirklichkeit überein. Saudi-Arabien versteht sich nämlich als islamischer Staat, dessen Grundlage, praktisch die Verfassung, der Inhalt insbesondere von Koran und Sunna, damit das aus letzteren abgeleitete islamische Recht (Scharia) bildet. Durch seine strenge Befolgung und Anwendung dieses Rechts sieht sich der König hinsichtlich seiner Position und Rolle legitimiert. Dieses religiöse Fundament erlaubt nicht uneingeschränkte Machtfülle und schrankenlose Machtausübung, sondern verlangt die Beratung des Herrschers durch Repräsentanten seiner Untertanen (Religionsgelehrte, Stammesführer, Berater, Minister und andere Notabeln) mit dem Ziel, im Hinblick auf anstehende Entscheidungen zu einem Konsensus zu gelangen. Dies sollten und sollen die in diesem Jahrhundert geschaffenen politischen Institutionen (Bürgerrat von Mekka, Stellvertreterrat, Ministerrat usw.) bewerkstelligen; viel älteren Datums ist die durch den Mordanschlag auf König Faisal im Frühjahr 1975 weithin bekannt gewordene Institution des Medschlis: zu bestimmten Zeiten steht jedem Untertan, der etwas erbitten oder beklagen möchte, der direkte Zugang zu seinem Herrscher frei, der sich seinerseits um Abhilfe bemüht.

Im skizzierten konservativen politischen System ist die Gründung von Parteien und Gewerkschaften, selbst politische Agitation (als Verbreitung von Propaganda) verboten; da es kein Wahlrecht kennt, unterbleiben allgemeine, freie und geheime Wahlen jeder Art. Die Tagespolitik wird demgegenüber von königlichen Dekreten, Befehlen und Anweisungen bestimmt. Hiergegen wurde in den letzten Jahrzehnten manchmal unorganisiert opponiert. Das Königshaus scheint der Forderung nach einer Demokratisierung - in Richtung auf eine konstitutionelle Monarchie - entsprechen zu wollen, wurde doch schon mehrmals - zuletzt Ende März 1982 vom damaligen Kronprinzen und jetzigen König Fahd - die Ausarbeitung einer eigentlichen Verfassung, die Einsetzung einer Beratenden Versammlung und die Beteiligung breiterer Volksschichten am politischen Entscheidungsprozeß

versprochen. Einzelne andere Gruppen kritisierten in letzter Zeit
die zu schnelle Modernisierung außerhalb des politischen Sektors.
Zu ihnen zählten die rd. 5oo religiös motivierten in- und ausländischen Besetzer der Großen Moschee von Mekka im November 1979.
Wenngleich es Oppositionsgruppen und -aktivitäten fast zu allen
Zeiten seit Mitte der 2oer Jahre gegeben hat, muß dem Regime der
Ahl Saud bislang eine für ein Entwicklungsland wie auch ein arabisches Land ganz außergewöhnliche politische Stabilität und Kontinuität bescheinigt werden.

Sein jetziges politisches Ansehen in der Welt verdankt Saudi-Arabien weniger jener Stabilität und Kontinuität als seinem Aufstieg
zu einer auch politischen Führungsmacht seit dem Zweiten Weltkrieg.
Als Hüter der Heiligen Stätten in und bei Mekka und Medina fühlt es
sich zur Förderung der islamischen Solidarität und Einheit ver-
pflichtet; damit sind großzügige Finanz- und Entwicklungshilfelei-
stungen zugunsten islamischer Staaten und Organisationen verbunden.
Daneben hat es sich durch seine geschickte Vermittlung und seinen
mäßigenden Einfluß bei Konflikten zwischen Staaten und sonstigen
Parteien im arabischen Raum, zwischen den Ölförder- und westlichen
Ölverbraucherländern sowie in internationalen Organisationen her-
vorgetan; selbst im israelisch-arabischen Konflikt tritt es aktiv
für eine friedliche Lösung ein. Waren auch seine diplomatischen
Initiativen, wie der Fahd-Friedensplan von August 1981, teilweise
erfolglos, steht es unstreitig auf der Arabischen Halbinsel als
die Ordnungsmacht und darüber hinaus als der führende arabische
Staat da.

Die vorstehenden Streiflichter sollen und dürfen keineswegs den
Eindruck erwecken, als wäre in Saudi-Arabien bereits alles zum be-
sten bestellt oder wenigstens auf dem Wege dahin. Zusammen mit
einigen hier außer acht gelassenen Aspekten (z.B. Gesundheitswesen,
Kriminalitätsniveau, Entwicklungsplanung und Hafenbau) repräsentie-
ren sie nur die 'Schokoladenseite' einer Medaille, die auch eine
Kehrseite mit viel Schatten hat.

Die aufgezählten umwälzenden Veränderungen binnen kürzester Zeit -
gleichbedeutend mit dem Überspringen von Entwicklungsstadien, die
in den meisten Industrieländern mindestens ein Jahrhundert gedauert
haben - wurden schon, bevor es in Saudi-Arabien 'richtig losging',
als Sprung vom Rücken eines Kamels ans Steuer eines Cadillacs be-

schrieben (und belächelt). Einen solchen Sprung haben am Anfang der Modernisierungslawine erstaunlich viele Saudis tatsächlich mit Leichtigkeit geschafft. Was jedoch in den letzten Jahren - auf Veranlassung und mit Billigung der schmalen gebildeten und kosmopolitischen Oberschicht - über die Landesbewohner hereingebrochen ist, hat den Großteil derselben, insbesondere die älteren, maßlos überfordert und zutiefst verunsichert, regelrecht überrollt. Ein Volk, dessen über 14jährigen Männer und Frauen 1980 noch zu 70,1 % bzw. 97,7 % und insgesamt zu 83,8 % Analphabeten gewesen sein sollen und zu dessen religiösem und politischem Credo jahrhundertelang die strikte Ablehnung fremder Kulturelemente und -repräsentanten gehörte, befindet sich manifest und latent in einem Kulturkonflikt, der starke Überfremdung und Entwurzelung zeitigt. Viele Saudis dürften heutzutage 'hin- und hergerissen' sein - hier Loyalität gegenüber dem kulturellen einschließlich religiösen Erbe, da Imitation westlicher Lebens- und Arbeitsstile. Diese Zerreißprobe wird bislang noch teils kaschiert, teils gemildert durch einerseits großzügige staatliche Sozialleistungen, Subventionen, Entschädigungen, Steuervergünstigungen usw. und andererseits die Präsenz von 1,5-2 Mill. Gastarbeitern (Jemeniten, Ägypter, Pakistanis, Südkoreaner, Filipinos, US-Amerikaner, Westeuropäer), die in ausführenden und leitenden Positionen den Großteil der Arbeit des wirtschaftlichen und infrastrukturellen Ausbaus erledigen, aber wie Menschen zweiter Klasse behandelt werden. Ohne ihre tatkräftige Mitarbeit wären fast alle milliarden- und millionenschweren petrochemischen Industriebetriebe, Kfz-Montagewerke, Klinikzentren, Universitäten, Telekommunikationssysteme, Be- und Entwässerungsprojekte usw. weder entstanden noch funktionsfähig. Wird der Großteil von ihnen im Zuge der gegenwärtigen und zukünftigen 'Saudization'-Politik ersetzt, dann steht die Gesellschaft Saudi-Arabiens vor ihrer eigentlichen Bewährung.

Eine zweite Bewährungsprobe ist mit der jüngsten und absehbaren sozialen Dynamik verbunden: Können die beträchtlichen sozialen Ungleichheiten und Inter-Generationen-Konflikte vermindert werden? Wird die gesellschaftliche Gleichstellung und berufliche Chancengleichheit der Frauen von religiösen und politischen Instanzen weiterhin verhindert oder aber durchgesetzt? Wo enden die jetzige Geld-, Technik- und Wachstumsgläubigkeit vieler Saudis? Welche Politik soll zukünftig gegenüber Andersdenkenden, -gläubigen und

-handelnden verfolgt werden? Wie können die weitere Verschwendung einerseits und Nutzenminderung andererseits von wertvollen natürlichen Ressourcen (fossiles Grundwasser, Umweltverschmutzung, Landaufgabe usw.) gebremst werden? Erst wenn diese Bewährungsproben bestanden sind, dürfte Saudi-Arabien seiner in Wirtschaft und Politik erlangten Führungsposition voll gerecht werden und sich sein immenser Ressourcenreichtum als wahrer Gottessegen erweisen.

Die bisherigen Ausführungen geben nicht Auskunft über Ursprünge und Prägungen des jetzigen Königreichs Saudi-Arabien. Jene Ursprünge und Prägungen sind nicht im Geschehen der ersten Jahrzehnte dieses Jahrhunderts, sondern im Geschehen der beiden vorausgegangenen Jahrhunderte auf der Arabischen Halbinsel zu suchen.

Als Begründer der Ahl Saud-Dynastie gilt Mohammed ibn Saud Ahl Mukrin aus der mächtigen Anese-Stammesgruppe, der 1726/27 seinem verstorbenen Vater als Herrscher über die zentralarabische Oase Dirija (bei Riad) nachfolgte. Bei diesem Duodezfürsten suchte 1744/45 der andernorts in Nedschd (Zentralarabien) vertriebene Religionslehrer Mohammed ibn Abdalwahhab Zuflucht, der die Rückkehr zum ursprünglichen Islam des Korans und Propheten Mohammed und folglich die Abkehr von allen seitherigen Neuerungen der Glaubenslehre predigte; er stand damit in der Tradition der konservativsten der vier islamischen Rechtsschulen, derjenigen der Hanbaliten. Nachdem sich die Bevölkerung Dirijas dieser Reformlehre angeschlossen hatte - ihre Anhänger nennen sich selbst seither Unitarier (aufgrund ihres Bekenntnisses zur Einheit Gottes), werden aber von anderen allgemein Wahhabiten genannt -, schlossen Emir (Fürst) Mohammed und Alim (Religions-, Rechtsgelehrter) Mohammed ein Bündnis, in dem sie sich dem Kampf für die Sache Gottes und die Eroberung Arabiens verschrieben. Letztere kam trotz zahlloser Kriegszüge bis zum Tode des Fürsten 1766 und auch des Reformators 1792 nur langsam voran. Große Landgewinne in allen Himmelsrichtungen wurden erst Ausgang des 18. und Anfang des 19. Jahrhunderts erzielt. Auf dem Höhepunkt seiner Ausdehnung und Macht um 1810 umfaßte dieser erste saudische Staat ein größeres Territorium als der jetzige dritte und war durch seine fanatischen wahhabitischen Glaubenskämpfer der Schrecken der Bewohner einiger Randgebiete der Arabischen Halbinsel.

Diese saudisch-wahhabitische Expansion, insonderheit die Besetzung und Verwüstung von Mekka 1803 und Medina 1805, bewog das wiederer-

starkte Osmanische Reich, das zuvor fast zwei Jahrhunderte lang
die Oberhoheit über den Hedschas und damit die Vorrangstellung in
der islamischen Welt innegehabt hatte, zum Eingreifen. Vom Pascha
und späteren Vizekönig von Ägypten, Mechmed Ali, aufgestellte und
teilweise auch angeführte Heere wurden mit der Rückeroberung des
Hedschas und der Zerschlagung des saudischen Staates beauftragt.
Ersteres gelang bis Anfang 1813, letzteres erst im Spätsommer 1818
mit der Eroberung und anschließenden völligen Zerstörung von Dirija. Damit verbunden wurde an der saudischen Herrscherfamilie und
ihren wahhabitischen Anhängern blutig und grausam Rache genommen.

Der von jener Tyrannei ausgelöste Türkenhaß sowie der Existenz-
wille der restlichen Familie Saud und übrigen Wahhabiten waren die
Keime des zweiten saudischen Staates. Als seine eigentliche Ge-
burtsstunde gilt die Vertreibung der damals letzten türkischen Gar-
nison in Nedschd, derjenigen von Riad, im Jahre 1824 durch Turki
ibn Abdallah ibn Mohammed Ahl Saud; Riad wurde nun zur neuen Hauptstadt erklärt. Diesem Herrscher fielen bis zu seiner Ermordung 1834
meist kampflos große Teile des ehemaligen Staatsgebietes wieder zu.
Dieser Aufstieg wurde - nach einer weiteren erfolgreichen ägyptischen Intervention in den Jahren 1837-4o - von seinem Sohn Faisal
fortgesetzt, dessen zweite und wichtigste Regierungszeit von Anfang 1843 bis zu seinem Tode Ende 1865 dauerte und die politische
wie auch wirtschaftliche Blütezeit des zweiten saudischen Staates
bildete. Die nachfolgenden dynastischen Auseinandersetzungen und
militärischen Schwächen endeten mit einem neuerlichen Niedergang.
Dieser wurde von zwei Seiten gefördert und ausgenutzt. In der zweiten Hälfte des 19. Jahrhunderts, besonders nach der Eröffnung des
Suezkanals 1869, maß das Osmanische Reich Rotem Meer, Persisch-
Arabischem Golf und ihren Anrainerländern erhöhte strategische Bedeutung zu. 1870-72 gelang ihm die Wiedergewinnung einerseits des
Jemens und des Hedschas und andererseits des schon in der zweiten
Hälfte des 16. und während fast des gesamten 17. Jahrhunderts beherrschten nordöstlichen Arabiens, der Region al-Hasa; zum letzt-
genannten Schritt ermutigte sie der Hilferuf eines der Kontrahen-
ten im Hause Saud. Von der Hohen Pforte wurde nun auch die 1835
begründete wahhabitische Dynastie der Ibn Raschid unterstützt, die
im nördlichen Zentralarabien, vom Gebiet des Dschebel (Berg)
Schammar mit dem Zentrum Hail aus, langsam an Macht und Einfluß
gewonnen hatte. Unter ihrem bedeutendsten Emir, Mohammed ibn Ra-

schid (1872-97), erlangte diese Dynastie nicht nur die Unabhängigkeit, sondern auch die Herrschaft vom bzw. über das Reich der Ahl Saud. Jener Emir 'schluckte' im Frühjahr 1891 die Reste des zweiten saudischen Staates, nachdem dessen letzter Herrscher, der jüngste Sohn Abdarrachman des großen Faisal, mit seiner Familie, der der damals zehnjährige Ibn Saud zugehörte, vor ihm aus Riad geflohen war. Später fanden die Flüchtlinge bei einem Feind der Ibn Raschids, dem berühmten Scheich Mubarak ibn Sabbah, in Kuwait Zuflucht, und Riad wurde von Statthaltern der Ibn Raschids verwaltet. Von Kuwait aus unternahm der junge Ibn Saud seinen eingangs erwähnten Überraschungsangriff auf Riad, mit dem die Geschichte des dritten, bis heute bestehenden saudischen Staates begann.

Die bisherigen Ausführungen stecken den zeitlichen, räumlichen und sachlichen (oder fachlichen) Rahmen dieser Bibliographie ab; diese erstreckt sich auf
- den Zeitraum seit dem ersten gemeinsamen Aufstieg der saudischen Herrschaft und der wahhabitischen Bewegung, der 1745 begann und sich seither durch eine beträchtliche politische und religiöse Kontinuität ausgezeichnet hat;
- das Gebiet des Königreichs Saudi-Arabien, des dritten saudischen Staates, das den Großteil der Arabischen Halbinsel einnimmt, jedoch noch einer anerkannten Grenzziehung mit seinen südlichen Nachbarstaaten ermangelt;
- die Gesellschaft, Politik und Wirtschaft (einschließlich Infrastruktur) in diesem Zeitraum und Gebiet, also auf die zentralen Erkenntnisgegenstände der Gesamtheit der Sozialwissenschaften, so daß dies ausschließlich eine sozialwissenschaftliche Bibliographie ist.

Die Titel der auf den Seiten 1-6 und 3oo angeführten nützlichen Bibliographien deuten schon darauf hin, daß es an einer solchen Bibliographie höchstwahrscheinlich weltweit mangelte. Diese Lücke wird hiermit zu schließen versucht.

I SAUDI ARABIA: PAST AND PRESENT

Unlike almost any other large state of the 2oth century the Kingdom of Saudi Arabia has made a 'brilliant career' for itself. At the beginning of the century its territory of more than 2 million km^2 still belonged to those sparsely populated extensive areas of the world that were economically very poor, socially extremely traditional, politically highly unstable, fanatically religious, constantly involved in armed conflicts, lacking a developed infrastructure, and technologically very backward. All this, as well as the climatic inclemency, the suspected lack of natural resources, and the extremely hostile attitude of the locals towards non-Moslem strangers, accounts for the highly dissatisfactory state of research and literature on Saudi Arabia at that time—in spite of Arabia's geographical position between two much-used shipping routes, the Red Sea and the Persian or Arabian Gulf. But, putting aside this long list of negative factors, one has to primarily consider the far-reaching importance of a small region—southern Hijaz—as the cradle and center of Islam. Every year tens or hundreds of thousands of believers went on a pilgrimage to the Holy Places in and around Mecca and Medina. Another surely positive fact is the high degree of ethnical and cultural homogeneity of the nomadic, rural, and urban population outside of the southern Hijaz, whereas this region formed the melting pot for the many immigrants (especially pilgrims and slaves) of different races and cultures.

With King Abd al-Aziz ibn Abd ar-Rahman ibn Faisal Ahl Saud (known as Ibn Saud to the western world) gradually taking over power and the subsequent long reign of this legendary founder of state, the situation described above began to change fundamentally. On Janu-

ary 15th, 1902, with the help of merely 40 loyal followers, the
21 year old Ibn Saud succeeded in taking Riyadh, the capital of
the second Saudi state (see below), which up to then had been in
the hands of the Ibn Rashid dynasty which ruled Central Arabia,
by surprise. For three decades he was busy extending, pacifying,
and administering—not without setbacks—his expanding territory.
Only after he had won the blitzkrieg with Yemen in spring 1934,
did a lasting peaceful development begin in the region that had
been proclaimed the Kingdom of Saudi Arabia on September 23rd,
1932. Previously, Saudi Arabia had lacked both the capital and
personnel to initiate a policy of planned economic and technological modernization. In fact, it only started after a North American company, later to become the Arabian American Oil Company
(ARAMCO), discovered oil on and after 1935, at first only in small
quantities, and was quickly interrupted by World War II. Although
Saudi Arabia's oil production and oil export increased enormously
after the war, its chances of playing a leading economic and consequently political role in the world were generally underestimated
until the end of the '50's.

Since perhaps the majority of non-Saudis (and possibly even the
Saudi Arabs themselves) still have an either vague or mistaken
conception of the present state of affairs, the following is intended to highlight a few important facts concerning the state of
development and the importance of today's Saudi Arabia.* Its

---

* Most of the data given in this introduction is derived from the
following sources:
Blume, H. (Hg.): Saudi-Arabien. Natur, Geschichte, Mensch und Wirtschaft. Ländermonographien 7; Tübingen, Basel: Erdmann, 1976.
Ingenieurgemeinschaft Lässer-Feizlmayr Consulting Engineers: Water for Riyadh. Innsbruck: Tyrolia, 1983.
Kingdom of Saudi Arabia, Ministry of Planning: Saudi Arabia:
  achievements of the First and Second Development Plans, 1390-
  1400 (1970-1980). Facts & Figures; Jeddah: Tihama Publications, 1982.
Kingdom of Saudi Arabia, Saudi Arabian Monetary Agency, Research
  and Statistics Department: Annual report, 1401 (1981). Riyadh:
  Saudi Arabian Printing Company Ltd. Ammariyah, n.d.
Koszinowski, T. (Hg.): Saudi-Arabien: Ölmacht und Entwicklungsland.
  Beiträge zur Geschichte, Politik, Wirtschaft und Gesellschaft.
  Mitteilungen des Deutschen Orient-Instituts 20; Hamburg: Deut-
  sches Orient-Institut, 1983.
Niblock, T. (ed.): State, society and economy in Saudi Arabia. Lon-
  don: Croom Helm; Exeter: Centre for Arab Gulf Studies, 1982.
Statistisches Bundesamt: Saudi-Arabien. Statistik des Auslandes,
  Länderkurzberichte; Stuttgart, Mainz: Kohlhammer, 1982.

'backbone' (and needless to say also that of the spectacular boom of the '6o's and the '7o's) is the exploitation of the immense oil and natural gas deposits of the Eastern Province and of the offshore national territory in the Persian Gulf. A quarter of the world's proved oil resources and a considerable amount of its natural gas is deposited in this area: by the end of 1981 at least 116.7 billion barrels of proved and 177.2 billion barrels of probable (including proved) crude oil reserves (159 l per barrel) were calculated. It is therefore not surprising to learn that here we find the largest known onshore (Ghawar) and offshore oil fields (Safaniya) as well as natural gas field (Haradh). Saudi Arabia's output of crude oil between 1979 and 1981, namely 3.48 to 3.62 billion barrels, represents approximately a sixth of the global output, and the only country producing more crude oil was the Soviet Union. 97 to 98 % of this quantity was mined on oil fields where the largest oil-producing company in the world, ARAMCO, had a concession. The government, which had already held the majority of shares since 1974, completely nationalized ARAMCO in 198o. The remaining 2 to 3 % of the Saudi Arabian output of oil was produced by the predominantly Japanese Arabian Oil Company and the U.S. Getty Oil Company. However, the world oil market and the situation of the Organization of Petroleum Exporting Countries (OPEC), caused Saudi Arabia to reduce its average daily output between mid-1981 and the fall of 1983 from 1o.2 to 5.5 million barrels.

By the end of 198o the proved and the probable natural gas reserves, exclusively for the areas where ARAMCO was licensed to produce, were estimated at 1.95 respectively 3.25 trillion m^3. In this year 53.3 billion m^3 (compared to 2o.6 billion m^3 in 197o) of natural gas were gained as a by-product of the oil production. This source of energy, which in the beginning was simply flared and reinjected, is now, to a greater and greater degree, being economically exploited: the percentage utilized increased from 11.2 to 27.5 % between 197o and 198o. The data on crude oil and natural gas reserves and production given above lead one to assume that Saudi Arabia's economic prospects, at least until the middle of the coming century, are very favourable.

Round about 198o, because of the low at-home requirements, more than 97 % of the oil produced was still exported, most of it crude. Due to this, Saudi Arabia as an oil-exporting country is virtually without competition in the world. The proceeds from the export of oil and oil products were $ 39.7 billion in 1978, $ 61.7 billion in 1979 and $ 1o5.9 billion in 198o, whereas those from liquefied natural gas came to $ o.66 billion in 1978, $ 1.11 billion in 1979 and $ 2.36 billion in 198o. In these years the exports in oil and gas represented 99.1 to 99.2 % of the total export value; only this can explain Saudi Arabia's high export surplus (1979: $ 39.o; 198o: $ 79.2; 1981: $ 85.o billion).

The following figures show how quickly rising production and prices, especially since the so-called oil crisis of autumn 1973, on the one hand, and royalties and taxes for the oil producing companies on the other, increased Saudi Arabia's revenue from the oil business: 197o: $ 1.21; 1972: $ 2.74; 1974: $ 22.57; 1976: $ 3o.75; 1978: $ 32.23; 198o: $ 84.47 billion; 1981 was the first year the revenue in 'petro dollars' exceeded 1oo billions (compared to not more than $ 1.5, $ 56.7 respectively $ 333.7 million in 194o, 195o and 196o). During this period, the income from oil represented between 86.2 and 97.3 % of the total revenue of the state. Despite the also rapidly growing government expenditures, surplus receipts of $ 4.8 respectively $ 12.2 billion were gained in 198o/81 and 1981/82. In view of these figures, one should keep in mind that Sheikh Abdulla Sulaiman al-Hamdan, Ibn Saud's Minister of Finance for several decades, could still keep the chronically nearly empty coffers of the state underneath his bed at the beginning of the '3o's, and that Ibn Saud's successor, the prodigal King Saud (1953-64), had brought the country near bankruptcy in only five years.

Some other important figures concerning the economy: between 197o and 1979, the gross national product based on the then valid market prices rose from $ 4.o to $ 94.6 billion, and between 197o and 1982, during the time of serious economic crises all over the world, the average annual increase in real terms was 1o %. During this latter period, the per-capita-earnings of the approximately 8 million inhabitants rose from about $ 8oo to about $ 16.8oo, thus multiplying more than 2o times so that Saudi Arabs today enjoy among the highest incomes in the world. Since the annual increase in the cost-of-living amounted to 16.5 % in the decade between 197o and 198o, but since then has risen much less drastically, the real income, and at the same time the wealth of the average

Saudi Arab, improved considerably. In the late '7o's, the oil sector contributed for the first time less than 5o %—based on the prices in 1969—to the gross national product, and the figures for the agricultural sector, that still employs a relative majority (probably now no more than 35 to 45 %) of the Saudis, amounted to 3 to 4 % in the last five years of the '7o's; during the '7o's the relative influence of both sectors sank, whereas the building trade became more and more important. In May 1978 and May 1979, the foreign currency reserves amounted to about $ 17 billion and, two years later, to about $ 28 billion, and the gold holdings for all these years were uniformly declared at about $ 4.5 billion. The external assets and the monetary reserves amounted to roughly $ 6o billion and $ 2o billion respectively at the end of 1978; the former should have grown enormously in the mean time.

These figures show that from the time of its foundation, when it was one of the poorest developing countries, to 5o years later, Saudi Arabia developed into one of the world's richest countries. Even though one could not rightly call it an industrialized nation, one has to see Saudi Arabia as an advanced third world country. Its fiscal, economic and oil policy is anything other than a quantité négligeable to the welfare of the rest of the world.

Just two figures show how the Saudi Arabian state makes use of its enormous financial resources: the total government expenditure allocated for the second five-year plan (1975/76-1979/8o) (including spendings for defense and development aid) amounted to $ 133.o billion, and that scheduled for the third five-year plan (198o/81-1984/85) to $ 363.o billion.

In these two development plans priority is given, among other things, to projects dealing with petroleum and natural gas collection, transportation, processing and utilization. In 1975, the government ordered ARAMCO to plan, construct, and run the so-called Master Gas System. With the help of this large-scale project, costing approximately $ 15 billion, the major part of the presently flared associated gas (about 85 million m^3 per day) from ARAMCO onshore oil fields is to be conducted through a network of pipelines, gas-oil separator, fractionation and liquefaction plants, as well as shipment facilities, thus being made easily available for economic exploitation. This means that the greater part of processed gas is intended to supply the petrochemical complexes, the combined desalination and power plants, the steelworks, etc., of Jubail and Yanbu at the east respectively west coast with raw material and energy, whereas the smaller part is to be exported by way of the recently built marine terminals at that places. The most impressive part of the project is the 1,17o km long, 66-67 cm wide, computer-controlled "East-West Natural Gas Liquids/Ethane Transpeninsular Pipeline" (capacity: 27o,oo barrels daily), which will be the longest and most modern high-pressure natural gas pipeline known, and runs from Shedgum near the Persian Gulf to Yanbu. Since fall 1982, the Master Gas System installations of the first stage of construction have been in operation.

Parallel to much of the above-described liquefied-gas pipeline runs the 1,2oo km long so-called East-West Crude Oil Pipeline finished in mid-1981 which was built by the national General Petroleum and Mineral Organization (PETROMIN) at a cost of $ 1.6 billion. Its flow capacity of 1.85 million barrels of crude oil per day can be doubled. This pipeline, considered both the safest and the most modern construction of its type in the world, fulfills two tasks: it reduces the journey of oil tankers from western countries by thousands of sea miles and provides crude oil for the industrial city and the industrial complex of Yanbu.

The already mentioned industrial complexes of Jubail and Yanbu form the very core of the industrial policy pursued in the second and third five-year plan. In the meantime these coastal towns, insignificant ten years ago, have been developed into industrial cities with an ultramodern infrastructure, planned to house 5oo,ooo altogether, including a working force of 144,ooo, by the year 2ooo. The concentration of industry in those areas is expected to effect both a considerable diversification and consolidation of the national economy. By establishing a few primary industries, which being based on oil and natural gas use up a lot of energy, as well as a multitude of secondary and tertiary industries, some of them buying the products of the former, one hopes for success. In 198o, plans existed to set up 17 primary, 136 secondary, and over one hundred tertiary industrial plants, among them several oil refineries, fractionation plants, iron and steel works, concrete and fertilizer factories, as well as an aluminium mill.

By the end of the '4o's, when Saudi Arabia—limited though to the cities of Jeddah and Riyadh—registered its first boom in the building trade, this branch of the economy still rested on a very weak foundation. Among other products, 1oo % of the cement needed had to be imported. The constant upward trend in the building trade at the beginning of the '5o's gave rise to the idea of planning small Portland cement factories. Between 1958 and 1966, three factories of this type were set to work near Jeddah, Hofuf, and Riyadh. Because of initial difficulties and a starting capacity of only 3oo tons each per day, they could not meet the country's demand by far, and, consequently, were gradually expanded. But even with an overall production output of 1.1 million tons per year in the middle of the '7o's, only 4o % or less of the general requirements could be supplied at the time the important industrial projects of the second five-year plan were to be realized. It was decided to raise the total capacity of the then 1.4 million tons to more than 1o million tons in 1984—by means of expanding the existing three and constructing another seven cement factories— thus trying to ensure complete self-sufficiency in the long run. The actual increase in productivity from 1.3 million tons (1977) to 2.6 million tons (1979) and then 4.3 million tons (1981) shows that this idea was immediately put into practice.

Especially since the beginning of the '6o's, the government has striven to overcome the grave shortage of aboveground water by various means. One of several solutions is the following: Between 1969 and 1981 16 seawater desalination plants with a total capacity of 179 million litres per day were built under the supervision of the Saline Water Conversion Corporation in 8 settlements on the Red Sea and the Persian Gulf to supply coastal areas with drinking water. In 1981, another 7 plants with a daily capacity of 1.253 million litres were under construction, and, on top of that, 6 with a daily capacity of 14.3 million litres in the planning stage. With the completion of the water pipeline Yanbu-Medina in 1981, desalinated seawater was transported as far as 12o km into the interior of the country for the first time. By the middle of 1983, the largest high-pressure water pipeline system in the world was nearly completed, after three years' construction and at a cost of $ 2 billion. Its most important elements are two parallel 466 km long main pipes with a diameter of 1.425 m through which it is possible to pump up to 830 million litres per day of desalinated Gulf water from the largest plant of its type, the Al Jubayl Desalination Plant, to the vicinity of Riyadh. This so-called Riyadh Water Transmission System, remote-controlled from Jubail, is projected to meet the growing demand for drinking water of the city of Riyadh with its more than 1 million inhabitants until the year 2012. Other big-sized water pipelines, planned to supply drinking water for the cities of Mecca, Taif, Abha, Tabuk, etc., and running from the Red Sea shore into the interior of the country, are under construction.

Besides conventional solutions, e.g., the installation of deep wells, the construction of storage and retainer dams and the recycling of sewage, some Saudi Arabian politicians and planners have recently been flirting with two revolutionary ideas for importing fresh water: utilizing icebergs towed from the Antarctic, or filling hitherto empty oil tankers coming from water abundant north Japan.

Despite the rapidly growing urban and industrial demand for water, the largest water consumer by far—with a share of 80 to 85 % of the total consumption in 1980—is still irrigated agriculture. Traditionally, irrigation remains the most important precondition to secure and favourable fruit tree, grain, vegetable and fodder crops in all those parts of the country where, unlike in southern Hijaz and Asir, productive rainfed cultivation is impossible. Over the last decade, the agricultural requirements for water have been influenced by two developments running in opposite directions. On the one hand, thousands of oasis farmers gave up small-scale agricultural production in order to earn a better living in a less demanding job outside the agricultural sector; very often the consequences were fallow ground and rural exodus. On the other hand, a few hundred members of the upper class—princes, ministers, businessmen, industrialists, etc.—made efforts to break into this sector. They acquired—some of them through donations from the Royal Family—more or less extensive tracts of fallow land, situated in naturally favourable regions or in the neighbourhood of big cities. Incentives to take up intensive cultivation of these areas were and still are:

(1) the country's great dependence on the import of all kinds of food

(2) the high government subsidization for certain means of production and yields
(3) the generous granting of loans for modernization purposes with very favourable conditions by the Saudi Arabian Agricultural Bank and the Saudi Industrial Development Fund
(4) the hitherto unsatisfied demand for high-quality agricultural products by both natives and foreigners—and as a consequence of this
(5) the prospect of large profits from agricultural investments.

This new type of "part-time farmer" leaves his large enterprise to be planned, established and run by western and Arab consultants, managers, computer operators, mechanics, veterinarians, and other specialists. These are the people who buy and apply, according to their often immensely rich employers' orders, or at least with their approval, the latest western farm machinery (large bulldozers, disk ploughs, center pivot systems, combines, harvesting machines, plastic greenhouses, fully automated stables for cattle and poultry, machine garages, silos, etc., most of them "made in USA"). One can imagine that most of these extremely capital-intensive, highly specialized, immensely productive agribusiness enterprises have an enormous demand for water.

The rapid growth of the modern agricultural sector visibly accelerates the decline of the traditional one. The former, however, is mainly responsible for the noticeable increase in agricultural production over the last few years, a fact Saudi agricultural politicians and economic planners are very proud of (e.g., in 1977 12o,ooo tons and in 1983 more than 5oo,ooo tons of wheat, as well as 283,ooo tons in 1977 and two years later 377,ooo tons of vegetables were harvested; 14 million broilers were produced in 1975 and 4o million in 198o, 9,2oo tons of milk were produced in 1978 and 32,ooo tons in 198o). Although there are reasons to be proud, the development of the quantity and structure of the food imports so far, and the factors influencing it (growth of population, shifts of demand, water supply, marketing and processing institutions, etc.) give rise to doubts as to whether or not the goal of self-sufficiency pursued without much regard for natural resources and economic calculations can be reached up to 199o or at all.

During the last decades, several Saudi Arabian cities have grown, as it were, explosively. At the beginning of this century there existed scarcely more than half a dozen settlements of more than 1o,ooo or of 5,ooo to 1o,ooo inhabitants. At that time Mecca was the largest city (even in the whole of Arabia), counting probably more than 5o,ooo permanent residents, followed by Jeddah and Hofuf with 2o,ooo to 3o,ooo each. Riyadh, already the capital of the second Saudi state (see below), had less than 1o,ooo and even in 192o still less than 2o,ooo inhabitants. Only during the '2o's and '3o's did it start to lose its character as a quite out-of-the-way provincial town (with an extensive palm grove). The following figures prove its exponential increase in inhabitants since World War II: 1945: 62,ooo; 1955: 1o6,ooo; 1965: 225,ooo; 1974: 667,ooo; 198o: c. 1 million. Similarly spectacular was the growth of Jeddah: 194o: 3o,ooo; 1965: 194,ooo; 1974: 561,ooo; 1981: c. 1.3 million. Compared to the political metropolis Riyadh, and the economic metropolis Jeddah, the traditional religious centers Mecca (with 185,ooo inhabitants in 1965 and estimately 55o,ooo in 198o) and Medina (with 73,ooo inhabitants in 1963 and estimately 29o,ooo in 198o) on the one hand, and, on the other hand, the traditional agricultural centers Hofuf, Qatif, Najran, Taif, etc., have lost much of their importance. However, during the last 3o to 4o years, their townscapes were turned upside down by the heavy immigration of natives and foreigners, modern traffic and town planning, and western ways of building and living, etc. Urbanization is doubtlessly one of the most important features of the 'new' Saudi Arabia.

The Saudis of today are virtually 'automobilemaniacs', surprisingly enough, in view of the fact that for a long time cars were considered devilish things (in the country there were cases of car-iconoclasm). Only after the conquest of Hijaz in 1924-25 were cars and small transport vehicles introduced by Ibn Saud in numbers worth mentioning. For nearly 2 decades they were used primarily for the transport of pilgrims and of members of the Royal Family. Due to a lack of paved roads—the first asphalt road between Jeddah and Mecca (72 km) was not built before the end of the '3o's, and that by Egyptians—and a lack of native drivers, one had to use mainly the old caravan tracks, and employ foreign driv-

ers, in most cases Indian Moslems. The following figures show the
fundamental change in this situation, especially in recent times:
between 1970 and 1980 the total number of registered motor cars
rose from 59,900 to 1,998,000 (an increase of 3,336 %); the annual
increase rates were 21 to 65 %. In 1980, the number of passenger
cars was 1,020,000 (including 105,000 taxis), meaning that on the
average every eighth, and in the big cities nearly every second,
citizen had a car at his disposal. Nearly the same numbers and
proportions apply to trucks, a fact that can be partly explained
by the non-existence of an ample railroad system (now being planned). The motorization boom is closely related to the preferential
expansion of the road-network as fixed in the two development
plans for the years 1970-80: the total length of paved roads increased from more than 8,000 to more than 20,000 km and that of
the earth-surfaced roads from about 3,500 to more than 24,000 km.
That the present road-network, despite its enormous expansion, is
still insufficient is seen by looking at the one hand at the huge-
ness of the country (north-south and east-west extension of more
than 2,200 respectively 1,200 km) and on the other hand at the
often chaotic traffic situation in the cities (which leads to
traffic jams and general honking for several hours, with the side-
effect of environmental pollution).

Since its beginnings, the great pilgrim celebration in the first
half of the Islamic month of Dhu al-Hijja has always been a religious mass meeting. However, never before has it reached dimensions similar to those of the last few years especially: The total
number of native and foreign pilgrims came, probably for the first
time, to more than a million in 1970 and to over 2 million in 1979;
since 1978 the normal number of pilgrims is a little under or over
2 million—among them 800,000 to 900,000 foreigners with 60 to 75 %
of them coming by plane. It is obvious that such a rush of people
furthered the expansion of nearly the whole material infrastructure
in the area of Mecca-Medina-Jeddah. An outstanding example of this
kind is the King Abdul Aziz International Airport north of Jeddah
whose first stage of construction ended in 1981 and whose second,
according to the plans, is due to be completed in 1985. With an
expansion of 105 km^2, a special pilgrims' terminal, accomodation
for 80,000 pilgrims and 6,300 'normal' passengers, parking space
for 22,000 cars, a mosque and a seawater desalination plant of its

own, etc., this will be one of the largest and most modern airports in the world. It will be possible to check in an estimated 8.6 million passengers in 1985, and 1o million in the year 2ooo. The King Khalid International Airport north of Riyadh (24o km^2 expansion, annual capacity of 15 million passengers), which was inaugurated in autumn 1983, has even huger dimensions. A third international major airport is under construction in the agglomeration area of Dammam-Dhahran-al-Khobar.

There were only a few public and private Turkish and Arab schools and not many more Koran and mosque schools in today's national territory round about 19oo. Consequently, nearly all natives outside Mecca and Medina, two traditional centers of Islamic learning, were illiterate. The development of the existing rudimentary educational system began both in the still Hashimite-governed Hijaz and in Ibn Saud's territory during the three first decades of our century. In 1926, Ibn Saud established the so-called Directorate of Education which—provided with very moderate means only—did a good job in promoting, on the one hand, the construction of and the attendance at schools, and, on the other hand, the gradual departure from religiously-oriented curricula, until the beginning of the '5o's. Up to 196o, only boys were allowed to attend public educational institutions. Just 2o years later, the former situation was completely reversed; in 198o, Saudi Arabia had more than 1o,7oo educational institutions (not included pre-schools), 5,4oo of which were primary schools, 1,8oo intermediate and secondary schools, 25 special schools, 3o vocational schools, 1o7 teacher training institutions, 6 universities, and 3,4oo institutions for adult education (established to promote literacy). Of these, given the fact that coeducation is disapproved of, 1,7oo primary schools, 5oo intermediate and secondary schools, 9 special schools, 64 teacher training institutions, as well as 6 colleges and institutes comparable to faculties were for girls only. 2,9oo of the 9,3oo classes for adult education were reserved for female participants. One can imagine the efforts made during the last years to provide both urban and rural settlements with schools, when one learns that under the first and the second five-year plan (197o to 198o) an average of 2.2 schools a day were either finished or rented. In 198o, the institutions listed above were attended by 862,ooo pupils of primary schools, 339,ooo pupils of intermediate and secondary schools, 2,000 pupils of special schools, 6,000 pupils of vocational schools, 22,000 students of teacher training institutions, and 48,000 other students of both sexes; the number of men and women attending courses in reading und writing came to 142,000. The total number of full-time teachers was 78,000; half of them were foreigners (66 % of the university teachers, 79 % of the teachers at intermediate and secondary schools, 86 % of those teaching at the so-called Teacher Training Institutes). A solution for this problem, and other deficiencies of the educational system (dissatisfactory enrollment in elementary schools, especially for the nomad children, high rates of school and university drop-outs, deficit of vocational schools, regional differences in the supply of educational facilities) is being attempted during the present five-year plan with an expenditure of nearly $ 36 billion.

Since 1963, when Saudi Arabia began to support the royalists in the Yemenite civil war, it has developed from a military dwarf into a military giant. Within the scope of a defensive military policy, made necessary by the large area and the harsh nature of the country, the government began, from the middle of the '60's onwards, to generously modernize the equipment and training of first the regular army and, during the last few years, the Bedouin national guard as well. The annual military expenditure rose in the '70's from $ 0.4 to $ 14.2 billion, half of which was however spent for infrastructural projects (hospitals, road construction, and housing). Through its purchase of armaments from the USA of a total value of more than $ 34 billion, between 1971 and 1980, Saudi Arabia became their most important customer. Today, its arsenal includes ultramodern weapon systems, e.g., F-5 and F-15 combat aircrafts and Boeing jets equipped with AWACS. Of the 4 ultramodern military cities under construction, each of which swallows up billions of dollars, the best known is the King Khalid Military City in the North East of the country.

Thus far the extreme modernization in some spheres of life has had no impact on political structures. Most western publications label Saudi Arabia's form of state and government an absolute monarchy, because all power is concentrated in the hands of one person—the King/Prime Minister—whose rule is unchallenged by a designated political cabinet (Council of Ministers), a council of religious scholars (Ulema), and a family council (princes). However, this label only partly corresponds to reality. Saudi Arabia's conception of itself is that of an Islamic state with the Koran and the Sunna (and the Islamic law, Sharia, derived from them) as its basis, comparable to its constitution. By keeping strictly to and applying this law, the King feels his position and role justified. Besides, these religious foundations do not concede unlimited power or unchecked reign but claim that the sovereign has to consult representatives of his people (religious scholars, chiefs of the tribes, advisers, ministers and other important personalities) in order to facilitate general agreement for decisions to be made. This was and still is the task of the political institutions created in this century (Council of Citizens of Mecca, Deputy Council, Council of Ministers, etc.). Much older is the institution called Majlis, widely known since spring 1975 for the attempt on King Faisal's life: at certain times, each subject who wants to either plead for something or complain about something is free to see the ruler who for his part will try to settle the matter.

In a conservative political system like this one, the foundation of parties and trade unions, and even political agitation (under the name of propagandism) is forbidden. Since universal suffrage does not exist, neither general, free, nor secret elections of any kind take place. Daily politics consist of royal decrees, orders, and instructions. During the last few years there were a few unorganized protests against this kind of reign. It seems that the royal dynasty is considering more democracy, e.g., the installation of a constitutional monarchy, because several times—for the last time at the end of March 1982 by Fahd ibn Abd al-Aziz (then crown prince, now King)—the drawing-up of a proper constitution, creation of a consultative council, and the organization of a broader participation of the different social strata in political decision-making was promised. Some other groups recently criticized the too rapid modernization outside the political sector. Among them were the 5oo religiously motivated native and foreign occupiers of the Great Mosque of Mecca in November 1979. Although, since the middle of the '2o's, opposition groups and activities have nearly always existed, one must admit that one distinctive feature—exceptional for a developing and an Arab country—of the Ahl Saud regime is its extraordinary political stability and continuity.

Saudi Arabia owes its present political prestige less to this continuity and stability than to its ascent to a leading power in politics since World War II. As keeper of the Holy Places in and round Mecca and Medina, it feels obliged to promote Islamic solidarity and unity. This attitude manifests itself in generous financial and development aids for Islamic states and organizations. And too, Saudi Arabia has distinguished itself by its clever mediation and its moderating influence in conflicts between states and other parties in the Arab world, between the oil producing and the western oil consuming countries and in international organizations; it even actively supports a peaceful solution of the Arab-Israeli conflict. Although diplomatic initiatives such as the Fahd peace plan of August 1981 are not always completely successful, it is indisputable that Saudi Arabia is the keeper of order on the Arabian Peninsula and the leading Arab power.

The above details are not intended to and must not convey the impression that everything in Saudi Arabia is perfect or at least near to it. Together with some aspects not dealt with here (e.g., public health service, crime rate, development planning, and port construction) they only represent the best side of a coin, the reverse of which is not so bright.

The above-described fundamental changes, which occurred within a very short period of time—whereby stages of development were skipped over that lasted for at least a century in most developed countries—were called (and ridiculed) the "jump from a camel's back into a Cadillac", even before things really began to move in Saudi Arabia. At the beginning of the modernization-avalanche surprisingly many Saudis 'jumped' without problem. But what has befallen the natives during the last years—initiated and agreed upon by the small and cosmopolitan upper class—was too much for the majority of them, especially for the older ones, in that it made them thoroughly insecure and virtually overran them. A people of whose men and women over 14 years old in 1980 still 70.1 % respectively 97.7 % and altogether 83.8 % were said to be illiterate, and whose religious and political creed was for centuries one of strict rejection of foreign cultural elements and representatives, is driven into a cultural conflict which produces the feeling of being uprooted and infiltrated by foreign elements. Today, supposedly many Saudis are torn between loyalty to their cultural including religious heritage and the imitation of western life- and work-style. So far, this severe trial was partly camouflaged partly alleviated

by generous government benefits, subsidies, compensations, tax
reductions, etc., on the one hand, and, on the other, by the presence of 1.5 to 2 million foreign workers (Yemenites, Egyptians,
Pakistanis, South-Koreans, Filipinos, Americans, Western Europeans,
etc.), who although filling leading and executive positions and
carrying out most of the work concerning economic and infrastructural development, are nevertheless treated as second-rate. Without their efficient help nearly all petrochemical plants, car-assembly factories, hospital centers, universities, telecommunication systems, irrigation and drainage projects, etc., costing millions and millions of dollars, would neither have been established
nor be in operation. If, in the course of the present 'Saudization'
policy, most of these foreigners are replaced, Saudi Arabian society will face its test.

Another impending test is tied to the latest and continuing social
dynamics: is it possible to diminish the considerable social inequality and the inter-generational conflicts? Will religious and
political authorities go on preventing social equality and equal
professional opportunities for women, or will women fight it out?
What are the limits to the present belief of many Saudis in money,
technology and growth? What kind of politics shall be applied to
those of different attitude, religion or action? How to stop on
the one hand the further waste and on the other hand the sinking
utility of valuable natural resources (fossil groundwater, environmental pollution, rural exodus, etc.)? Only when Saudi Arabia has
passed these tests, might it be able to fully master the task of
its leading economic and political position, so that its abundance
of resources will prove a real blessing.

The comments made so far do not clarify either origin or patterns
of the present Kingdom of Saudi Arabia. But these cannot be found
in the history of the first decades of this century, but in that
of the Arabian Peninsula of the two preliminary centuries.

Muhammad ibn Saud Ahl Muqrin of the powerful Anaza tribal confederation, who in 1726/27 followed his father as ruler over the Central Arabian oasis Diriya (near Riyadh), is considered to be the
founder of the Ahl Saud dynasty. In 1744/45, the religious scholar
Muhammad ibn Abd al-Wahhab, who had been chased away from other
places in Najd (Central Arabia) and who preached the return to the

original Islam of the Koran and the Prophet Muhammad, and consequently the abandonment of all new doctrines since then, sought refuge with this petty prince; thus, this scholar followed the tradition of the most conservative of the four Islamic schools of law, that of the Hanbalites. After the population of Diriya embraced these reformistic teachings—since then their supporters call themselves Unitarians (because they declare the unity of God), but are generally called Wahhabis by the others—Emir (prince) Muhammad and Alim (religous and legal scholar) Muhammad signed a pact saying that they would devote themselves to the fight for God and the conquest of Arabia. Despite numerous military campaigns the latter proceeded only very slowly until the Prince's death in 1766 and that of the reformer in 1792. Conquest of extensive areas of land in all directions did not take place until the end of the 18th and the beginning of the 19th century. At the height of its territorial expansion and power, round about 1810, this first Saudi state had a larger territory than the present third state, and because of its fanatic Wahhabi religious fighters it spread fear and terror among the inhabitants of some adjoining areas on the Arabian Peninsula.

This Saudi-Wahhabi expansion, especially the occupation and demolition of Mecca in 1803 and Medina in 1805, caused the Ottoman Empire, which before that, for nearly 200 years, had had sovereignty over Hijaz and with it a superior rank in the Islamic world, to intervene. Muhammad Ali Pasha, the later viceroy of Egypt, raised and sometimes also commanded armies which were ordered to reconquer Hijaz and to destroy the Saudi state. He had succeeded in the former goal by the beginning of 1813, in the latter only by late summer 1818 when Diriya was conquered and then completely devastated. He also seized this opportunity to take bloody and cruel revenge upon the Saud family and its Wahhabi supporters.

Hatred for the tyrannous Ottomans, and the will of the remaining Wahhabis and members of the Saud family to survive formed the germ of the second Saudi state. The expulsion of the at that time last Turkish garrison in Najd, namely that of Riyadh, by Turki ibn Abdulla ibn Muhammad Ahl Saud in 1824 is considered to be its real hour of birth. Riyadh was declared the new capital. Until he was assassinated in 1834, this sovereign regained, mostly without fighting at all, vast parts of the former territory. His son Faisal—after a further successful Egyptian intervention from 1837 to 1840—resumed his father's successful policy. Faisal's second and most important term of office from the beginning of 1843 until his death late in 1865 represented the political and economic prime of the second Saudi state. The ensuing conflicts within the dynasty and military weaknesses resulted in another decline, furthered and exploited by two sides. In the second half of the 19th century, especially after the Suez Canal was opened in 1869, the Ottoman Empire ascribed special strategic importance to the Red Sea, the Persian Gulf and the adjoining countries. From 1870 to 1872, it succeeded in regaining on the one hand Yemen and Hijaz, and, on the other, north eastern Arabia, that is the region of al-Hasa, which had already been under Ottoman rule during the second half of the 16th and during almost the whole of the 17th century. It was encouraged to do the latter by one of the adversaries in the Saud dynasty, who had pleaded for help. Now the Porte also supported the Wahhabi dynasty of the Ibn Rashid founded in 1835, that starting from the region of the Jabal (mountain) Shammar with Hail as its center, had gradually gained power and influence in northern Central Arabia. Under the reign of their most important Emir, Muhammad ibn Rashid (1872 to 1897), this dynasty not only achieved independence from but also eventually rule over the Ahl Saud empire. It was again this Emir who in spring 1891 'swallowed' the remainder of the second Saudi state, after its last sovereign, the great Faisal's youngest son Abd ar-Rahman, and his family, including the then 10 year old Ibn Saud, had left Riyadh fleeing from him. Later the refugees were given shelter in Kuwait by an enemy of the Ibn Rashids, the famous Sheikh Mubarak ibn Sabbah. Riyadh remained in the hands of governors installed by the Ibn Rashids. From Kuwait, the young Ibn Saud started his above-mentioned surprise attack on Riyadh which marks the beginning of the third, still existing Saudi state.

The remarks made so far describe the temporal, spatial and factual (or professional) framework of this bibliography. It covers:
- the period of time commencing with the first common ascent of Saudi power and the Wahhabi movement, which began in 1745 and since then has been characterized by a considerable political and religious continuity

- the territory of the Kingdom of Saudi Arabia, the third Saudi state, comprising the major part of the Arabian Peninsula, but which still lacks acknowledged boundaries with the neighbouring states in the south
- society, politics and economy (including the infrastructure) during this period of time and in this region, i.e., the central objects of interest in all the social sciences which means that this is exclusively a social science bibliography.

The titles of the useful bibliographies listed on pages 1-6 and 3oo indicate by themselves that such a bibliography has been lacking worldwide. With this contribution we will try to fill this gap.

II BESCHRÄNKUNGEN UND ZUSAMMENSETZUNG DER LITERATURAUSWAHL

In bezug auf die zeitlich, räumlich und fachlich eingegrenzte Thematik liegt ein wahrer 'Literaturberg' vor, der allerdings erst in den letzten Jahren - parallel zum phänomenalen Entwicklungstempo und Bedeutungszuwachs Saudi-Arabiens - seine jetzige 'Höhe' erreicht hat (s.u.). Dies legte den Verzicht auf einen Teil der relevanten und interessanten Titel nahe; andernfalls hätte diese Bibliographie einen 'unhandlichen Umfang' erreicht. Vorwiegend aus diesem quantitativen Grund blieben unberücksichtigt eigentlich in Betracht kommende
- Artikel in und Beilagen von Tages- und Wochenzeitungen,
- Beiträge in Zeitschriften mit einer kürzeren als vierzehntäglichen Erscheinungsweise,
- lexikalische Artikel mit Ausnahme von wichtigen Artikeln vor allem in den beiden Auflagen der Enzyklopädie des Islam,
- Rezensionen im engeren Sinne,
- vertrauliche Anträge, Berichte, Gutachten und Studien,
- unveröffentlichte Vortragsmanuskripte, die dem Bearbeiter nicht vorlagen,
- Atlanten, Landkarten, Luftbilder, Stadtpläne u.dgl.

Dazu einige Erläuterungen:
Informative Saudi-Arabien-Artikel und -Beilagen in Tageszeitungen und Wochenschriften (wie Le Commerce du Levant, The Economist, Financial Times, Informazioni per il Commercio Estero, Life, Middle East Economic Digest, Le Monde, Neue Zürcher Zeitung, Newsweek, New York Times, The Oil and Gas Journal, Der Spiegel, Time und The Times) vor allem aus den letzten zehn Jahren sind Legion und lohnen somit eine eigenständige Bibliographie. Vergleichbare Veröffentlichungen liegen auch aus früherer Zeit vor; hingewiesen sei bei-

spielhaft auf H.St.J.B. Philbys und L. Weiss' vielen Korrespondentenberichte aus den 2oer und 3oer Jahren in der Wochenschrift The Near East and India und ihrer Nachfolgerin Great Britain and the East bzw. in den Jahrgängen 1927-32 der Tageszeitung Neue Zürcher Zeitung, weiter auf die im 19. Jahrhundert in den Wochenschriften Globus und Ausland erschienenen Beiträge von C.M. Doughty, I. Goldziher, S. Langer, A. Zehme u.a.

Mehr oder weniger lange Artikel über Landschaften, Siedlungen, Persönlichkeiten und Ereignisse in Saudi-Arabien sind in einer Vielzahl von Enzyklopädien und Lexika enthalten (Britannica Book of the Year; Collier's Yearbook; Encyclopaedia Universalis; The Encyclopedia Americana; Encyclopedia Britannica; Der Große Brockhaus; Grote, Winkler, Prins; Larousse; Meyers Enzyklopädisches Lexikon; M. Heravi (ed.): Concise encyclopedia of the Middle East. Washington: Public Affairs Press, 1973; S. & N. Ronart: Lexikon der Arabischen Welt. Ein historisch-politisches Nachschlagewerk. Zürich, München: Artemis, 1972; W. Tietze (Hg.): Westermann Lexikon der Geographie. 5 Bde. Braunschweig: Westermann, 1968-72 usw.). Auf einzelne davon kann wohl in jeder größeren öffentlichen Bibliothek zurückgegriffen werden. Für die Zitierung von wichtigen Artikeln in der ersten und zweiten Auflage der Enzyklopädie des Islam stand nur die deutschsprachige bzw. die englischsprachige Ausgabe zur Verfügung; mit Hilfe dieser Bibliographie dürften die entsprechenden Artikel in den englisch- und französischsprachigen Ausgaben der ersten Auflage bzw. in der französischsprachigen Ausgabe der zweiten Auflage problemlos zu finden sein.

Unberücksichtigt blieben alle Rezensionen, die sich auf Inhalt und Gehalt von hier angeführten Veröffentlichungen beschränken; gegenteilig wurde nur entschieden, wenn eine Besprechung umfangreiche sozialwissenschaftliche Berichtigungen und Ergänzungen enthält (Beispiel: Nr. 2552).

Den Verzicht auf die Aufnahme vertraulicher Anträge, Berichte, Gutachten und Studien erklären deren eingeschränkten Benutzungsmöglichkeiten. Meist läßt sich nicht absehen, ob überhaupt und, wenn ja, wann solche Arbeiten zu allgemeiner Benutzung freigegeben werden.

Als vertraulich werden auch viele Vortragsmanuskripte bis zu ihrer eventuellen Veröffentlichung in Zeitschriften, Konferenzberichten und sonstigen Sammelwerken 'gehandelt'; manche Manuskripte liegen zudem nicht völlig ausgearbeitet vor.

Im Verlauf seiner mehrjährigen Beschäftigung mit Saudi-Arabien sah der Bearbeiter davon ab, Karten, Pläne, Bilder u.dgl. systematisch zu erfassen und zu sammeln.

Die (weitgehende) Nicht-Berücksichtigung der umrissenen Arbeiten wird durch die Berücksichtigung von (sonstiger) 'grauer Literatur' mehr als ausgeglichen, liegen doch aus den letzten 4o Jahren fast unübersehbar viele Veröffentlichungen von Regierungsstellen, Ölgesellschaften, internationalen Organisationen, wissenschaftlichen Instituten, Verbänden, Banken, Bauunternehmen, Consultingfirmen usw. innerhalb und außerhalb Saudi-Arabiens vor, die nicht über den Buchhandel erhältlich sind. Berge an grauer Literatur scheinen geradezu ein Kennzeichen der neureichen Ölländer im Vorderen Orient zu sein! Diese Bibliographie will den Zugang zu dieser Literaturgattung fördern, weil deren Informationswert aus zwei Gründen nicht unterschätzt werden darf: die eingeschränkte Forschungsfreiheit auf vielen sozialwissenschaftlichen Gebieten in Saudi-Arabien zwingt zum Rückgriff auf graue Literatur, und die im Auftrag und mit Unterstützung von saudiarabischen Regierungsstellen, Ölgesellschaften, Consultingfirmen usw. verfaßte Literatur ist in einem beachtlichen Umfang das Produkt modernster Forschungsarbeit - deren Qualität hatte fast immer Vorrang vor ihren Kosten. Der Großteil der von dortigen Regierungsstellen (insbesondere einzelnen Ministerien) in letzter Zeit herausgegebenen (Jahr-)Bücher und Broschüren zeichnet sich einerseits durch eine aufwendige Ausstattung (fester Einband, Vierfarbdruck, hervorragende Fotos usw.), andererseits durch vielfältiges statistisches Material (in ein- und mehrdimensionalen Tabellen und Graphiken) aus; der propagandistische Gehalt ist meist bemerkenswert gering.

Sozialwissenschaftlich relevante graue Saudi-Arabien-Literatur ohne eine spezielle Beschaffungsreise und einen 'großen Apparat' möglichst vollständig nachzuweisen, ist mit mehr Aufwand verbunden und mehr von Glücksfällen abhängig als der Nachweis mittels Buchhandelskatalogen, Verlagsprospekten usw. Einzelne solche Arbeiten besitzt heutzutage eine große Zahl der Staats-, Landes-, Universitäts- und Institutsbibliotheken in der Bundesrepublik. Die umfangreichste Sammlung liegt zweifellos im Deutschen Orient-Institut in Hamburg vor, das seit Ende 1969 im Auftrag der Deutschen Forschungsgemeinschaft zentral graue Orientliteratur beschafft und in seiner Präsenzbibliothek zur Benutzung anbietet. Erwähnenswert sind auch - nach eigenen Erfahrungen - die Bestände der Universitätsbibliothek Tübingen, der Bibliothek des Instituts für Weltwirtschaft an der Universität Kiel, der Bibliothek des HWWA-Instituts für Wirtschaftsforschung in Hamburg sowie der Bibliothek des Instituts für Auslandsbeziehungen in Stuttgart.

Aus den angedeuteten Gründen mußte der Nachweis 'grauer' Literatur unvollständiger bleiben als derjenige 'käuflicher' Literatur. Letztere überwiegt in dieser Bibliographie eindeutig. Da gerade viele käufliche Veröffentlichungen nicht ausschließlich Saudi-Arabien, sondern den islamischen Kulturkreis, die Länder der Arabischen Halbinsel, einzelne internationale Organisationen mit saudiarabischer Mitgliedschaft u.dgl. betreffen, war eine Abgrenzung notwendig. Als für diese Bibliographie relevant wurden auch alle diejenigen Arbeiten erachtet, in denen lediglich einzelne Kapitel oder längere Abschnitte Aspekten der Gesellschaft, Politik und/oder Wirtschaft im Gebiet des jetzigen Saudi-Arabiens seit Mitte des 18. Jahrhunderts gewidmet sind, sowie diejenigen, in denen zwar nur an verstreuten Stellen, aber insgesamt doch in nennenswertem Umfang deutlich auf solche Aspekte eingegangen wird. Was noch bzw. nicht mehr als 'nennenswerter Umfang' gelten sollte, konnte nur von Fall zu Fall entschieden werden und findet möglicherweise nicht die Zustimmung anderer Kenner. Neben dieser quantitativen Abgrenzung wurde von qualitativen Maßstäben abgesehen, weil allgemein verschiedene Leser ein und dieselbe Arbeit unterschiedlich bewerten und damit jede Arbeit potentiell wertvoll ist.

Mit der vorgenommenen Abgrenzung ist vereinbar, daß zum einen biographische Veröffentlichungen in bezug auf bekannte westliche Arabienreisende und zum anderen Veröffentlichungen zu Wasserbau und -wirtschaft jeweils im Betrachtungszeitraum berücksichtigt worden sind. Hierfür gibt es einleuchtende Gründe. Die Reiseberichte von J.L. Burckhardt, R.F. Burton, C.M. Doughty, Ch. Huber, H. von Maltzan, A. Musil, W.G. Palgrave, L. Pelly, H.St.J.B. Philby, G.F. Sadlier, C. Snouck Hurgronje, G.A. Wallin u.a. waren bis weit in das 2o. Jahrhundert die wichtigsten Informationsquellen bezüglich des neuzeitlichen Arabiens; diese meist durch Archivmaterial und Sekundärliteratur gestützte Einschätzung ist in fast allen berücksichtigten Biographien enthalten. Daß Wasserbau und -wirtschaft, die nur beschränkt sozialwissenschaftliche Erkenntnisgegenstände sind, angesichts des extrem bis semiariden Klimas in Saudi-Arabien von existentieller Bedeutung sowohl im ländlich-landwirtschaftlichen als auch im städtisch-industriellen Sektor sind, bedarf kaum der Erwähnung; der letzte Abschnitt läßt erkennen, daß Saudi-Arabien diese natürliche Herausforderung seit Beginn seines Ölzeitalters wie kaum ein zweites subtropisches Land erfolgreich angenommen hat.

Eine größere Unvollständigkeit wurde in bezug auf drei disparate Themen in Kauf genommen: die Lehre und Bedeutung des Islam, die Person und Taten von T.E. Lawrence ('Lawrence von Arabien') während des Arabischen Aufstands (1916-18) sowie die Ölpolitik von OPEC und OAPEC (seit 1960); diesbezüglich können bei Bedarf Spezialbibliographien konsultiert werden.

Ersichtlich kann diese Bibliographie - trotz jahrelanger intensiver Literatursuche und -auswertung - nicht Anspruch auf Vollständigkeit erheben. Diese - auch in umfangreicheren Bibliographien übliche - Feststellung ist erläuterungsbedürftig. Eine Saudi-Arabien-Bibliographie, die sich auch auf die graue Literatur erstreckt, ist nolens volens unvollständig, weil letztere nirgendwo zentral oder speziell registriert worden und deshalb teilweise unbekannt ist; bekannt ist lediglich, daß die Titeleinträge z.B. im National Union Catalog der Library of Congress diejenigen im British Museum (Library) General Catalogue of Printed Books zahlenmäßig weit übersteigen. Nur noch Sprachgenies können das weite Spektrum an westeuropäischen Sprachen, in denen inzwischen Saudi-Arabien-Veröffentlichungen erschienen sind, (einigermaßen) beherrschen. Noch wichtiger ist schließlich, daß eine Abgrenzung wie die obige willkürlich und außerdem kaum operationalisierbar ist. Mit gutem Gewissen könnte man beispielsweise dafür plädieren, auch diejenigen Arbeiten z.B. zu den drei traditionellen Lebens- und Wirtschaftsformengruppen (Nomaden, Bauern und Städter) im Vorderen Orient und - viel spezieller - zur dortigen Versorgungslage während des Zweiten Weltkriegs aufzunehmen, deren Aussagen für saudiarabische Verhältnisse (überwiegend) gültig sind, ohne dies jedoch explizite zu sagen. Durch eine solche Ausweitung wären sicherlich noch Tausende von Veröffentlichungen in Betracht gekommen.

Diese Bibliographie umfaßt 3680 Titeleinträge, und zwar 51 in bezug auf nützliche Bibliographien und 3629 in bezug auf sachlich relevante Veröffentlichungen. Dies entspricht einem Mehrfachen der Titeleinträge mit sozialwissenschaftlichem Saudi-Arabien-Bezug in jenen 51 Bibliographien. Angesichts der skizzierten Entwicklung und Bedeutung Saudi-Arabiens (und der infolgedessen absehbar weiterrollenden 'Literaturlawine') ist abzusehen, daß einzelne künftige Bibliographien wiederum umfangreicher ausfallen werden...

Die im Bibliographischen Teil verzeichneten Arbeiten (ohne die 51 Bibliographien und die Nachträge 3653-3680) wurden anhand der Merkmale Erscheinungsjahr und -sprache einer einfachen statistischen Analyse unterzogen. Davon blieben lediglich die wenigen Arbeiten mit gänzlich unbekanntem Erscheinungsjahr, aber nicht diejenigen mit nur ungenau oder unsicher angegebenem Erscheinungsjahr (Beispiele: ca. 1981, 1973?) ausgenommen. Die aufgenommenen zwei- und mehrbändigen Werke fanden entsprechend häufig Berücksichtigung. Die Titel mit Jahresangaben wie 1969/73 und 1974/75 wurden lediglich unter dem jeweils erstgenannten Jahr in Rechnung gestellt. Sofern mehrere Ausgaben und/oder Auflagen nachgewiesen werden konnten, wurde nur auf die erstgenannte Ausgabe bzw. Auflage zurückgegriffen. Unter diesen Bedingungen wurden 3595 Literaturtitel in die Untersuchung einbezogen (s. Tab. 1).

Von den 3595 Titeln liegen 2565 (71,3 %) englischsprachig, 620 (17,2 %) deutschsprachig, 295 (8,2 %) französischsprachig, 71 (2,0 %) italienischsprachig und die restlichen 44 (1,2 %) in sonstigen westeuropäischen Sprachen (22 in holländischer, zehn in spanischer sowie jeweils 2-5 in dänischer, finnischer, flämischer und schwedischer Sprache) vor. Die sozialwissenschaftliche Saudi-Arabien-Literatur ist somit - dies gilt auch unter Einschluß der hier ausgeklammerten Sprachen wie Arabisch, Russisch und Türkisch - weit überwiegend englischsprachig abgefaßt worden, wofür es mehrere Gründe gibt: die jahrhundertelange britische Vorherrschaft am Rande der Arabischen Halbinsel, die führende Rolle US-amerikanischer Öl-, Bergbau-, Baugesellschaften usw. bei der Ressourcenausbeutung und Infrastrukturentwicklung seit den 30er Jahren, Englisch als erste Fremdsprache der relativ meisten Saudis und anderen Araber sowie als weltweit führende Wissenschaftssprache usw. Die spaltenweise Betrachtung offenbart, daß diese Dominanz erst jüngeren Datums ist. Bis zum Ersten Weltkrieg erschienen - unter Berücksichtigung

Tab. 1 - Verteilung der Literaturtitel auf Erscheinungsjahre und -sprachen 1750-1983[+] (Absolutzahlen)

Ersch.- jahre	Sprache					Summe
	Englisch	Deutsch	Französ.	Italien.	sonst.	
1750-99	1	3	-	-	-	4
1800-09	1	3	1	-	-	5
1810-19	2	2	5	-	-	9
1820-29	1	3	3	-	-	7
1830-39	3	3	2	-	-	8
1840-49	-	3	13	-	-	16
1850-59	5	3	4	-	-	12
1860-69	6	5	3	-	1	15
1870-79	3	3	5	-	-	11
1880-89	8	10	5	-	1	24
1890-99	4	2	3	-	-	9
1900-04	5	2	4	-	-	11
1905-09	16	17	11	-	1	45
1910-14	10	16	4	1	1	32
1915-19	13	16	7	1	-	37
1920-24	31	7	9	1	5	53
1925-29	43	22	6	1	4	76
1930-34	56	22	12	9	-	99
1935-39	38	32	11	16	2	99
1940-44	28	19	2	8	1	58
1945-49	76	5	8	-	-	89
1950-54	108	16	4	7	4	139
1955-59	130	30	10	2	5	177
1960-64	182	32	8	5	1	228
1965-69	273	33	28	2	2	338
1970-74	358	49	31	5	6	449
1975-79	649	152	49	11	5	866
1980-83	515	110	47	2	5	679
Summe	2565	620	295	71	44	3595

[+] Bis Herbst 1983.

einzelner Übersetzungen - ungefähr gleich viele englisch-, deutsch- und französischsprachige Werke, und bis dahin genossen die deutsche und auch die französische Orientalistik insgesamt ein größeres Ansehen als die englische (die amerikanische steckte noch in den Kinderschuhen). Der Vormarsch der englischsprachigen Literatur begann ersichtlich in den 2oer Jahren. In den 3oer Jahren machten die deutsch- und die französischsprachige Literatur verlorenen Boden teilweise wieder gut. Während des Zweiten Weltkriegs ging die Publikationszahl in allen ausgewiesenen Sprachräumen stark zurück. Erst seit seinem Ende sind die englischsprachigen Veröffentlichungen numerisch drückend überlegen; deren Anteilswerte reichen von 73,4 % (1955-59) bis 85,4 % (1945-49) und erreichen einen Durchschnitt von 78,4 %.

Der Summenspalte ist u.a. folgendes zu entnehmen: In der zweiten Hälfte des 18. Jahrhunderts erschienen ganze vier (o,1 %), im gesamten 19. Jahrhundert nur 116 (3,2 %) und im bisherigen 2o. Jahrhundert demgegenüber 3475 (96,7 %) Veröffentlichungen. Die Publikationszahlen stagnierten grosso modo in den beiden Hälften des 19. und in den beiden ersten Vierteln des 2o. Jahrhunderts auf jeweils etwas höherem Niveau. Erst Anfang der 5oer Jahre setzte sowohl außerhalb als auch innerhalb Saudi-Arabiens eine rege Publikationstätigkeit ein, die bis Anfang der 8oer Jahre nicht nur angedauert, sondern noch stark zugenommen hat: Zuwachsraten von 27,3-92,9 % zwischen aufeinanderfolgenden Jahrfünften sprechen mehr für eine kurvilineare Entwicklung ('Literaturlawine') als für eine lineare.

Tab. 2 weist aus, wie sich die Publikationstätigkeit seit 1973, als Saudi-Arabien durch die sog. Ölkrise in aller Munde war, entwickelt hat. Saudi-Arabiens damalige (und seitherige) Öl- und Außenpolitik, Petrodollarflut, Entwicklungsplanung usw. wirkten sich offensichtlich erst mit ein- bis zweijähriger Verzögerung auf dem Buchmarkt richtig aus. Im bisherigen 'Rekordjahr' 198o erschienen fast ebensoviele Titel wie zwischen 175o und 1919 (241 gegenüber 261); diese 17o Jahre machen fast drei Viertel des Betrachtungszeitraums aus. Der leichte Rückgang 1981 und 1982 dürfte nur eine Folge kurzfristiger Umstände sein: Neuerscheinungen werden erst im Abstand von mehreren Monaten bis Jahren in Indizes angezeigt, in Zeitschriften besprochen, von Bibliotheken erworben, in Literaturverzeichnisse aufgenommen usw. Hieraus und aus dem Zeit-

Tab. 2 - Verteilung der Literaturtitel auf Erscheinungsjahre und -sprachen 1973-1983 (Absolutzahlen)

Ersch.-jahr	Sprache					Summe
	Englisch	Deutsch	Französ.	Italien.	sonst.	
1973	71	11	11	1	2	96
1974	72	17	11	2	3	1o5
1975	116	23	12	3	-	154
1976	1o9	3o	4	-	1	144
1977	134	34	11	-	4	183
1978	151	26	11	4	-	192
1979	139	39	11	4	-	193
198o	193	33	13	-	2	241
1981	154	35	11	-	1	2o1
1982	152	22	23	2	1	2oo
1983	16	2o	-	-	1	37
Summe	13o7	29o	118	16	15	1746

punkt des Redaktionsschlusses erklärt sich insbesondere die niedrige Angabe für 1983. Unter der Voraussetzung einer ähnlich vollständigen Titelaufnahme dürfte sich der bis 198o feststellbare Trend auch in den letzten und vermutlich noch in den nächsten Jahren fortgesetzt haben bzw. fortsetzen.

Im Jahrzehnt 1973-82, in dem mit 17o9 (47,5 %) die relativ meisten der berücksichtigten 3595 Veröffentlichungen erschienen sind, schwankte der Anteil englischsprachiger Titel zwischen 68,6 % (1974) und 8o,1 % (198o), der Anteil deutschsprachiger zwischen 11,o % (1982) und 2o,8 % (1976) sowie der Anteil französischsprachiger zwischen 2,8 % (1976) und 11,5 % (1973 und 1982). Die englischsprachige Literatur hatte damit durchgängig und unangefochten den ersten Rangplatz inne, und die Anteilswerte für die deutschsprachige Literatur wurden nur 1973 und 1982 von denjenigen für die französischsprachige Literatur erreicht.

Diese Bibliographie bestätigt den Engländer H.St.J.B. Philby (1885-196o) als den weitaus fruchtbarsten persönlichen Verfasser in bezug auf Saudi-Arabien (mit 14 Buch- und 28 Zeitschriftenveröffentlichungen). Philby hielt sich zwischen 1917 und 196o ungefähr 25 Jahre lang als Forschungsreisender, Historiker, Geschäftsmann, Journalist,

Moslem und Vertrauter Ibn Sauds im damals noch weitgehend abgeschlossenen Saudi-Arabien auf und beschäftigte sich ungefähr 4o Jahre lang wissenschaftlich mit diesem Land; auf seinem Grabstein steht zurecht, daß er der größte aller Arabien-Forschungsreisenden gewesen ist. Sein umfangreiches Gesamtwerk umfaßt neben den berücksichtigten 42 Arbeiten noch eine große Zahl einerseits von Artikeln in Tageszeitungen, Wochenschriften und Enzyklopädien und andererseits von unveröffentlichten Manuskripten, Reports, Tagebüchern, Korrespondenzen usw.

Interessierte Benutzer seien noch darauf aufmerksam gemacht, daß in bezug auf Saudi-Arabiens Gesellschaft, Politik und Wirtschaft im Betrachtungszeitraum nicht nur vielfältige Veröffentlichungen, sondern auch vielfältige Archivalien (Konsulats- und Handelsberichte, Pilger- und Schiffahrtsstatistiken, Verträge, Memoranden, Korrespondenzen, Tagebücher, Fotos usw.) vorliegen. Solche historischen Quellen befinden sich in den Staats- und einzelnen anderen Archiven
- der Kolonialmächte Großbritannien (British Library (Reference Division), India Office Library and Records, Public Record Office, Royal Geographical Society usw. in London, Middle East Centre, St. Antony's College in Oxford, Sudan Archive in Durham usw.), Frankreich (Archives Nationales und Archives du Ministère des Affaires Étrangères in Paris), Rußland (Zentrale Staatsarchive der UdSSR in Leningrad und Moskau), Niederlande (Algemeen Rijksarchief in Den Haag) und Italien (Reale Archivio di Stato und Istituto per Oriente in Rom); diese Mächte waren zusammen mit Österreich-Ungarn schon im 19. oder beginnenden 2o. Jahrhundert in Dschidda und teilweise auch anderswo am Rande der Arabischen Halbinsel konsularisch und später diplomatisch repräsentiert;
- des Türkischen Reichs (Başbakanlık Arşivi und Topkapi Sarayi Muzesi Arşivi in Istanbul), unter dessen Oberhoheit große Teile der Arabischen Halbinsel bis ins 2o. Jahrhundert standen;
- der politischen und wirtschaftlichen Großmächte USA (National Archives in Washington) und Deutschland (Bundesarchiv in Koblenz, Politisches Archiv des Auswärtigen Amtes in Bonn, Deutsches Zentralarchiv in Potsdam), die beide erst 1939 diplomatische Beziehungen zu Saudi-Arabien aufgenommen, aber dort schon vorher - vergleichsweise unbedeutende - wirtschaftliche Interessen verfolgt haben;

- Ägyptens (Dar al-Mahfuzat in Kairo), Indiens (National Archives of India in Neu-Delhi) und Österreichs (Österreichisches Staatsarchiv, Abteilung Haus-, Hof- und Staatsarchiv in Wien).

Angesichts von Großbritanniens jahrhundertelanger wirtschaftlicher und politischer Hegemonie im Nahen und Mittleren Osten sowie der durch den Public Records Act von 1967 auf 3o Jahre verkürzten Sperrfrist für die Benutzung fast aller Staatsdokumente sind Arbeitsaufenthalte in englischen Archiven besonders zu empfehlen.

II LIMITING THE LITERATURE SELECTION AND COMPOSITION OF THE SELECTED LITERATURE

This bibliography contains a total of 3,680 title entries, that is several times the number in comparable bibliographies on the topic Saudi Arabia. Given the flood of works published on this country especially in the last few years, it was nevertheless necessary to be selective and to limit the entries listed.

Listed are chiefly English-language publications, and, to a lesser extent, German, French, Italian, Dutch, Spanish, Danish, Finnish, Flemish, and Swedish titles (s. Tables 1 and 2).

Not only monographs and periodical articles are listed, but also government publications, business reports (from oil companies, consulting firms, and banks, to name a few), as well as research reports from international organizations and research institutions. Given the known difficulties in finding out about this so-called "grey" literature, even this bibliography, despite long years of intensive research and evaluation, can lay no claim to being complete. The worth of this type of literature is not to be underestimated for two reasons: the limited freedom of research in many social science areas in Saudi Arabia forces a return to "grey" literature, and the literature published by or with the support of government offices, oil companies, consulting firms, etc., in Saudi Arabia is largely the product of modern research, the quality of which is usually put before the cost. A large number of government offices, in particular the different ministries, have recently produced books, yearbooks, and brochures, which are characterized by expensive production and extensive statistical material. The propaganda content of these publications is surprisingly low.

Articles in and supplements to daily and weekly newspapers have not been included in this bibliography. Informative articles and supplements about Saudi Arabia have appeared, especially during the last ten years, in dailies and weeklies such as Le Commerce du Levant, The Economist, Financial Times, Informazioni per il Commercio Estero, Life, Middle East Economic Digest, Le Monde, Neue Zürcher Zeitung, Newsweek, New York Times, The Oil and Gas Journal, Der Spiegel, Time, and The Times; they are legion and therefore call for their own bibliography. Comparable publications are also available from earlier times, for example H.St.J.B. Philby's and L. Weiss' numerous news reports which appeared in the weekly The Near East and India and in its successor Great Britain and the East in the '2o's and '3o's and in the daily Neue Zürcher Zeitung in the years 1927-32 respectively, and further, the articles written by C.M. Doughty, I. Goldziher, S. Langer, A. Zehme, etc., appearing in the weeklies Globus and Ausland during the 19th century.

Furthermore, encyclopedia articles have not been taken into account, with the exception particularly of important articles in The Encyclopaedia of Islam; articles are taken from the first (German) and the second (English) edition. Other encyclopedia and lexicons which contain more-or-less in-depth articles about regions, settlements, personalities and events in Saudi Arabia can only be referred to here. They are: Britannica Book of the Year; Collier's Yearbook; Encyclopaedia Universalis; The Encyclopedia Americana; Encyclopedia Britannica; Der Große Brockhaus; Grote, Winkler, Prins; Larousse; Meyers Enzyklopädisches Lexikon; M. Heravi (ed.): Concise Encyclopedia of the Middle East. (Washington: Public Affairs Press, 1973); S. & N. Ronart: Lexikon der Arabischen Welt. Ein historisch-politisches Nachschlagewerk. (Zurich, Munich: Artemis, 1972); W. Tietze (ed.): Westermann Lexikon der Geographie. 5 vols. (Brunswick: Westermann, 1968-72).

Atlasses, maps, aerial photos, city maps, and similar publications have not been listed in this bibliography, nor have book reviews—with the exception of reviews which themselves contain a significant amount of social science material (e.g.: no. 2552).

It should be noted that many publications in book form available through book dealers do not deal exclusively with Saudi Arabia, but with the entire Islamic culture area, the countries of the Arabian

Peninsula, or international organizations in which Saudi Arabia is a member. Included also in this bibliography are works or collections of works which contain only a single chapter or long section on (Saudi) Arabia's society, politics and economy since the mid-18th century, as well as those works which—although perhaps only in scattered places—provide in total significant information about these aspects.

Since the travel reports of J.L. Burckhardt, R.F. Burton, C.M. Doughty, Ch. Huber, H. von Maltzan, A. Musil, W.G. Palgrave, L. Pelly, H.St.J.B. Philby, G.F. Sadlier, C. Snouck Hurgronje, G.A. Wallin and others were the most important sources of information well into the 2oth century, particular attention has been given to biographical and autobiographical works of well-known western explorers of Arabia.

Although literature dealing with water resources development and water usage is only of limited interest for social scientists, it has been taken into account in this bibliography, given the fact that problems related to Saudi Arabia's extreme to semi-arid climate are of existential importance in the agricultural, industrial and domestic sectors.

Three disparate themes were only partially and selectively treated: the teachings and importance of Islam; the personality and deeds of T.E. Lawrence ('Lawrence of Arabia') during the Arab Revolt of 1916-18, as well as the oil politics of OPEC and OAPEC since 196o. Special bibliographies can be consulted for information about literature on these subjects.

Of 3,595 titles listed (not counting the 51 bibliographies, the supplementary entries nos. 3653-368o, and the titles with unknown publication dates) 2,565 (71.3 %) are in English, 62o (17.2 %) in German, 295 (8.2 %) in French, 71 (2.o %) in Italian, and the remaining 44 (1.2 %) in various West European languages (22 in Dutch, 1o in Spanish, and from 2 to 5 in Danish, Finnish, Flemish, and Swedish). Thus, the social science-oriented Saudi Arabia literature is written largely in English, even when one takes into consideration languages excluded here, such as Arabic, Russian, and Turkish. There are several reasons for this: the long British dominance on the borders of the Arabian Peninsula; the leading role of U.S. oil, construction, mining and other companies in the exploitation of resources and in the development of the infrastructure since the '3o's; the fact that English is the first foreign language of most Saudis and other Arabs and is also the leading scientific language worldwide, etc. A study of Table 1 shows that the dominance of English is quite recent. Including a few translations, approximately the same numbers of English, German, and French works were published up to World War I. At that point, German and French oriental studies enjoyed a greater respect than the English (American oriental studies were still in their infancy). The advance of English-language literature began to be observable in the '2o's. In the '3o's, however, German- and French-speaking literature began to regain lost ground. During World War II, publications in all the above-named languages declined sharply. Only since its end have English-language publications begun to numerically tip the balance. Their percentages ranged from 73.4 % (1955 to 1959) to 85.4 % (1945 to 1949), and have reached an average of 78.4 %.

Among other conclusions which can be drawn from the totals column is the following: in the second half of the 18th century only 4 (o.1 %) and in the whole 19th century only 116 (3.2 %) works were published, in comparison to 3,475 (96.7 %) in the 2oth century up to the present time. The publication figures roughly stagnated in both halves of the 19th and in the first two quarters of the 2oth century. Only since the beginning of the '5o's can a consistent publishing activity be observed, both within and outside of Saudi Arabia, which has not just simply lasted up to the '8o's, but has also greatly increased. Rates of increase of 27.3 % to 92.9 % in five-year periods speak more for a curvilinear (a "literature avalanche") than for a linear development.

Table 2 shows how publication activity increased since 1973, when thanks to the oil crisis Saudi Arabia was on everyone's lips. Saudi Arabia's oil and foreign policy, the flood of petrodollars, development planning and so forth at that time (and since then) have had an obvious effect on the book market one or two years later. In the "record year" of 198o, almost as many titles appeared as had been published between 175o and 1919 (241 as opposed to 261); these 17o years take up nearly 3/4 of the considered time period. The mild recession of 1981 and 1982 was probably only short-term

Table 1. Distribution of titles according to date of publication and language 1750-1983[+] (Absolute figures)

Date of Publication	Language					Total
	English	German	French	Italian	Other	
1750-99	1	3	–	–	–	4
1800-09	1	3	1	–	–	5
1810-19	2	2	5	–	–	9
1820-29	1	3	3	–	–	7
1830-39	3	3	2	–	–	8
1840-49	–	3	13	–	–	16
1850-59	5	3	4	–	–	12
1860-69	6	5	3	–	1	15
1870-79	3	3	5	–	–	11
1880-89	8	1o	5	–	1	24
1890-99	4	2	3	–	–	9
1900-04	5	2	4	–	–	11
1905-09	16	17	11	–	1	45
1910-14	1o	16	4	1	1	32
1915-19	13	16	7	1	–	37
1920-24	31	7	9	1	5	53
1925-29	43	22	6	1	4	76
1930-34	56	22	12	9	–	99
1935-39	38	32	11	16	2	99
1940-44	28	19	2	8	1	58
1945-49	76	5	8	–	–	89
1950-54	1o8	16	4	7	4	139
1955-59	13o	3o	1o	2	5	177
1960-64	182	32	8	5	1	228
1965-69	273	33	28	2	2	338
1970-74	358	49	31	5	6	449
1975-79	649	152	49	11	5	866
1980-83	515	11o	47	2	5	679
Total	2,565	62o	295	71	44	3,595

[+] Up to fall 1983

and circumstantial: in a time period of between several months to years, new appearances would be first advertised and reviewed in periodicals, acquired by libraries, listed in bibliographies, etc. Due to these factors, and taking the editorial deadline of this book into account, the low figures for 1983 in particular are explained. Given a future similar listing of titles, the trend observed up to 1980 should be seen also in the last few years as well as in the years to come.

Table 2. Distribution of titles according to date of publication and language 1973-1983 (Absolute figures)

Date of Publication	Language					Total
	English	German	French	Italian	Other	
1973	71	11	11	1	2	96
1974	72	17	11	2	3	105
1975	116	23	12	3	-	154
1976	109	30	4	-	1	144
1977	134	34	11	-	4	183
1978	151	26	11	4	-	192
1979	139	39	11	4	-	193
1980	193	33	13	-	2	241
1981	154	35	11	-	1	201
1982	152	22	23	2	1	200
1983	16	20	-	-	1	37
Total	1,307	290	118	16	15	1,746

In the decade from 1973 to 1982, in which with 1,709 (47.5 %) the largest amount of the 3,595 evaluated titles were published, the percent of English-language titles swayed between 68.6 % (1974) and 80.1 % (1980), the percent of German-language titles between 11.0 % (1982) and 20.8 % (1976), and the percent of French titles between 2.8 % (1976) and 11.5 % (1973 and 1982). English-language literature has thus achieved first place without a serious challenger, and the percentage of German-language literature is equalled by the French-language literature only in 1973 and 1982.

This bibliography confirms that the Englishman H.St.J.B. Philby (1885-1960) is by far the most productive author of works relating to Saudi Arabia. For approximately 25 years between 1917 and 1960,

Philby was a travelling researcher, historian, businessman, journalist, Moslem and intimate of Ibn Saud in the then largely closed Saudi Arabia, and he busied himself for c. 4o years with research on this country. On his gravestone, appropriately, it is written that he was the greatest of all the Arabian explorers. His extensive works—aside from the 14 books and 28 periodical articles listed here—include a large number of articles in dailies, weeklies, and encyclopedia, and also of unpublished manuscripts, reports, diaries, correspondence, etc.

Interested users should note that with regard to Saudi Arabia's society, politics and economy in the time period under consideration, not only various publications, but also various archival materials (consulate and trade reports, pilgrim and port statistics, treaties, memoranda, correspondence, diaries, photos, etc.) are available. Such historical sources are to be found in governmental and other archives of:
- the colonial powers Great Britain (British Library (Reference Division), India Office Library and Records, Public Record Office, Royal Geographical Society, etc., in London, Middle East Centre, St. Antony's College in Oxford, Sudan Archive in Durham, etc.), France (Archives Nationales and Archives du Ministère des Affaires Étrangères in Paris), Russia (Central State Archives of the USSR in Leningrad and Moscow), the Netherlands (Algemeen Rijksarchief in The Hague), and Italy (Reale Archivio di Stato and Istituto per Oriente in Rome). These powers, together with Austria-Hungary, were already represented in Jeddah and partly in other centers on the Arabian Peninsula on a consular and later on an ambassadorial level partially in the 19th or the beginning of the 2oth century
- the Ottoman Empire (Başbakanlık Arşivi and Topkapi Sarayi Muzesi Arşivi in Istanbul), under whose rule were large parts of the Arabian Peninsula up into the 2oth century
- the political and economic world-powers USA (National Archives in Washington) and Germany (Bundesarchiv in Koblenz, Politisches Archiv des Auswärtigen Amtes in Bonn, Deutsches Zentralarchiv in Potsdam), which first took up diplomatic relations with Saudi Arabia in 1939, but had already pursued—comparatively unimportant—economic interests there

- Egypt (Dar al-Mahfuzat in Cairo), India (National Archives of India in New Delhi), and Austria (Österreichisches Staatsarchiv, Abteilung Haus-, Hof- und Staatsarchiv in Vienna).

Given Great Britain's long economic and political dominance in the Middle East, as well as the Public Records Act of 1967 which opened up access to almost all documents dating back thirty years or more, a research trip to England is highly recommended.

III HINWEISE ZUR ZITIERWEISE UND BENUTZUNG

Mit dieser Bibliographie soll Wissenschaftlern gleichermaßen wie Nicht-Wissenschaftlern ein einfaches und nützliches Hilfsmittel bereitgestellt werden. Jenem Ziel sollen vorrangig die gewählte Zitierweise und Titelordnung dienen. Diese orientierten sich an den folgenden Grundsätzen:

(1) Berücksichtigung fanden nur die zur zweifelsfreien Identifizierung einer Arbeit notwendigen und die hinsichtlich ihrer Art und ihres Inhalts auf einen Blick aufschlußreichen bibliographischen Angaben:
  - Name(n) des Verfassers/der Verfasser und/oder des Herausgebers/der Herausgeber (persönlich und korporativ),
  - Ober- und eventuell Untertitel der Veröffentlichung,
  - Name der Zeitschrift oder des fortlaufenden Sammelwerks (Schriftenreihe, Jahrbuch, periodische Statistik o.ä.) mit Band-, Jahrgangs- und/oder Heftnummer(n),
  - Art und Ort einer Hochschulschrift,
  - Band- und Auflagenzahl bei mehreren Bänden bzw. Auflagen eines monographischen oder nicht-periodischen Sammelwerks,
  - Erscheinungsort(e) und Erscheinungsjahr(e),
  - Verlagsname(n) in verkürzter Form,
  - Seitenzahl gemäß arabischer und eventuell römischer Paginierung.

  Ungenannt blieben damit z.B. das Format, die gesonderten Verfasser von Vorworten und Einleitungen, die Übersetzer, Fotographen und Illustratoren sowie die Anzahl der Karten, Abbildungen, Tabellen u.dgl. Die gewählte unvollständige Zitierweise ist den im Bibliotheks- und Dokumentationswesen üblichen Zitierweisen sehr ähnlich.

(2) Die berücksichtigten bibliographischen Angaben wurden je Literaturart (Monographien, Beiträge in periodischen und nichtperiodischen Sammelwerken, Hochschulschriften usw.) in gleicher Abfolge verzeichnet.

(3) Im Anschluß an die bibliographischen Angaben wurden beim Großteil der aufgenommenen Arbeiten 1-3 Standortnachweise in deutschen, in relativ wenigen Fällen alternativ dazu in ausländischen Bibliotheken angeführt. Diese Nachweise sollen zum 'direkten Zugriff' und zu Fernleihbestellungen anregen und damit in besonderem Maße zur 'Benutzerfreundlichkeit' dieser Bibliographie beitragen.

(4) Es wurde weitgehend auf die Verwendung von Abkürzungen verzichtet. Die verwendeten Abkürzungen (s. Abschnitt IV dieser Einführung) sind im deutschen, englischen und französischen Sprachraum fast ausnahmslos gebräuchlich und damit bekannt.

(5) Die berücksichtigten Arbeiten wurden alphabetisch nach Verfasser- und Herausgebernamen und je Verfasser und Herausgeber - bei mehreren Arbeiten - nach ihrem Erscheinungsjahr geordnet aufgeführt. Der Wert dieser primär alphabetischen und sekundär chronologischen Ordnung wird durch die Nachträge in Abschnitt III des Bibliographischen Teils nur wenig beeinträchtigt.

Bei der Zitierung wurde im einzelnen folgendermaßen verfahren:

Bei arabischen Familiennamen wurde der bestimmte Artikel in den Schreibungen Al-, Al, El- und El entsprechend einer bibliothekarischen Gepflogenheit nach dem oder den abgekürzten Vornamen mittels uniformer Kleinschreibung al- bzw. -el angeführt (Beispiel: Abdalwahed, A.M. al-) und damit bei der alphabetischen Titelordnung ignoriert; davon wurde abgesehen, wenn der Artikel 'fester Bestandteil' der Namensschreibung ist (Beispiel: Almana, M.). In den wenigen Fällen, in denen arabische Vornamen den bestimmten Artikel aufweisen, wurde letzterer bei ihrer Abkürzung übergangen. Die Zusätze de, le, van, von usw. von westlichen Verfassernamen wurden nachgestellt (Beispiel: Gaury, G. de). Die im Original bei der Schreibung von arabischen Namen und Begriffen benutzten diakritischen Zeichen wurden übernommen; vereinfachend steht das Apostroph (') sowohl für den Buchstaben 'ain als auch für die hamza. Alle diakritischen Zeichen blieben bei der alphabetischen Ordnung unberücksichtigt. Ebenso wurden die Titel (Baron, Doktor, Lady, Pascha,

Sheikh, Sir usw.) von Verfassern und Herausgebern außer acht gelassen. Alle Vornamen wurden mittels ihres ersten Buchstabens abgekürzt; hiervon wurde abgesehen, wenn dies der alphabetischen Katalogisierung dienlich war.

Die Arbeiten ohne bekannt gewordene Verfasser und Herausgeber wurden entsprechend der Buchstabenabfolge ihres Titels - unter Einschluß der bestimmten und unbestimmten Artikel - eingeordnet. Jeder Untertitel fand Berücksichtigung. Die im Original fett oder kursiv gedruckten Begriffe wurden unterstrichen. Zur besseren Unterscheidung der Titel der nicht-deutschsprachigen Arbeiten von den Namen der nicht-deutschsprachigen Zeitschriften, Schriftenreihen, Jahrbücher u.dgl. wurden alle Substantive und Adjektive in letzteren groß geschrieben.

Durchgängig wurde darauf verzichtet, die Namen von Zeitschriften und fortlaufenden Sammelwerken abzukürzen. Nur bei gleichnamigen Zeitschriften mit verschiedenen Erscheinungsorten wurde der jeweils in Betracht kommende Erscheinungsort angeführt. Es gilt zu beachten, daß der Name einzelner Zeitschriften im Laufe ihres Erscheinens geändert worden ist (aus 'The Moslem World' wurde 'The Muslim World', aus 'Aramco World' 'Aramco World Magazine' usw.).

Bei fortlaufender Seitenzählung in Zeitschriftenbänden und -jahrgängen wurde auf die Angabe der Heftnummer von Beiträgen verzichtet; bei separater Seitenzählung in den Einzelheften wurden die betreffenden Heftnummern in runden Klammern unmittelbar hinter den Band- und Jahrgangsangaben angeführt. Letzteres geschah vorsichtshalber auch dann, wenn die Paginierung der Hefte von Zeitschriftenbänden und -jahrgängen nicht festgestellt werden konnte. Die relevanten Heftnummern wurden selbstverständlich immer dann angegeben, wenn die Erscheinungsweise von Zeitschriften nicht band- oder jahrgangsweise gegliedert worden ist. Römische Ziffern und Großbuchstaben kennzeichnen einzelne (Zeit-)Schriftenreihen.

Schreibfehler in Verfassernamen sowie Ober- und Untertiteln wurden 'kommentarlos' berichtigt. Unkorrigiert blieben demgegenüber veraltete Schreibungen in Titeln (wie "rothes Meer" und "Hedjaz"). Die Umlaute ä, ö und ü wurden solchermaßen geschrieben und wie ae, oe und ue in Rechnung gestellt; das 'ß' wurde durch 'ss' ersetzt.

In Einzelfällen wurden die zwei oder mehr - zumeist unmittelbar aufeinanderfolgenden - Beiträge eines Verfassers in demselben Sammelwerk unter einer Nummer angeführt (Beispiel: Nr. 884). Gedankenstriche dienen zur leichteren Unterscheidung dieser Beiträge. Nur wenn (fast) alle Beiträge eines Sammelwerks inhaltlich relevant waren, wurde auch das Sammelwerk (unter seinem Herausgeber oder seinen Herausgebern) als Einzelnummer aufgenommen (Beispiel: Nr. 2266); um Doppelnennungen zu vermeiden, wurden solche Nummern im Sachregister (s. Abschnitt I des Registerteils) übergangen.

Bezüglich eines kleinen Teils der Arbeiten konnte mehr als eine Auflage und/oder Ausgabe nachgewiesen werden; die diesbezüglichen Angaben beanspruchen aber nicht, vollständig zu sein.

Einzelne der von saudiarabischen Regierungsstellen verfaßten oder herausgegebenen Arbeiten liegen mehrsprachig vor; ihr arabischsprachiger Teil wurde bibliographisch außer acht gelassen.

Die teilweise unterschiedliche Schreibung von Erscheinungsorten innerhalb Saudi-Arabiens (z.B. von Dschidda und Riad) in verschiedenen Arbeiten, die in ein und derselben Sprache vorliegen, wurde vereinheitlicht; so fanden in bezug auf englischsprachige Arbeiten einheitlich die Schreibungen Jeddah und Riyadh Verwendung. Der Eigenname University of Riyad wurde nicht an die gewählte Schreibung von Riyadh als Erscheinungsort angepaßt. Auf die Nennung des Erscheinungsortes wurde verzichtet, wenn diese Information redundant ist, was für einen Teil der Hochschulschriften zutrifft, sind doch Angaben wie Universität Bonn und Claremont Graduate School eindeutig. Wenn die Bände eines mehrbändigen Werks an verschiedenen Orten und/oder von verschiedenen Verlagen veröffentlicht worden sind, wurden diese Orte bzw. Verlage durch einen Schrägstrich voneinander abgesetzt (Beispiel: Nr. 1451).

Die bibliographischen Angaben zum Großteil der US-amerikanischen Hochschulschriften aus den letzten zehn Jahren entstammen den unregelmäßigen "Catalogues of Doctoral Dissertations" und den monatlichen "Dissertation Abstracts International" von University Microfilms International. Bezüglich dieser Schriften wurde stets das Jahr ihrer Annahme durch eine Universität, nicht das Jahr ihres Nachdrucks (und damit ihrer Veröffentlichung) durch 'UMI' angegeben. Die ausschließlich arabischen Seitenangaben dieses Unternehmens erstrecken sich auf die römische und arabische Seitenzahl im Original.

Bei den Seitenangaben wurde nicht zwischen Seiten und Blättern unterschieden. Die ungezählten Seiten und Blätter blieben außer Betracht.

Für die Standortnachweise in deutschen Bibliotheken wurden deren Sigelnummern - in Übereinstimmung mit dem "Sigelverzeichnis für die Bibliotheken der Bundesrepublik Deutschland einschließlich Berlin (West)" und einem Verzeichnis des Deutschen Orient-Instituts - benutzt; auf die Angabe von Buchsignaturen wurde verzichtet. Alle diejenigen deutschen Bibliotheken, in denen Titel nachgewiesen werden konnten, sind in Abschnitt V dieser Einführung mit ihrem Sigel und ihrer Adresse aufgelistet. Gedankenstriche rahmen die maximal drei Nachweise je Titel ein. Diese Obergrenze wurde willkürlich gewählt (einige der aufgenommenen Arbeiten sind in einer Vielzahl von öffentlichen Bibliotheken vorhanden). Allgemein wurden in dieser Bibliographie diejenigen wissenschaftlichen Bibliotheken bevorzugt aufgeführt, die bekanntermaßen viel Orientliteratur besitzen und dazuerwerben. Zwei weitere Gesichtspunkte fanden Berücksichtigung: nach Möglichkeit wurde je Veröffentlichung (mindestens) ein persönlich kontrollierter Nachweis in einer dem nationalen und internationalen (oder Auswärtigen) Leihverkehr angeschlossenen Bibliothek angeführt. Welche Bibliotheken diesem Leihverkehr angeschlossen und welche demgegenüber Präsenzbibliotheken sind, ist Abschnitt V dieser Einführung zu entnehmen.

Bedauerlicherweise konnte nicht für alle Titel zumindest ein Standort nachgewiesen werden. Das Fehlen solcher Nachweise ist jedoch nicht unbedingt gleichbedeutend damit, daß eine zitierte Veröffentlichung in keiner deutschen wissenschaftlichen Bibliothek vorhanden ist, wenngleich in einigen Fällen Fernleihbestellungen erfolglos waren.

Eine ausgewogene räumliche Verteilung von 1-3 Standortnachweisen ist natürlich unmöglich. Die verzeichneten Nachweise lassen vielmehr zwei Schwergewichte erkennen. Relativ die meisten Nachweise beziehen sich auf Bibliotheken im Raum Stuttgart-Tübingen. Dies kam teils wohnortbedingt, teils durch die folgenden fünf Umstände zustande:
(a) Die Württembergische Landesbibliothek in Stuttgart verfügt
 u.a. über viel Reiseliteratur aus den letzten Jahrhunderten
 und hat als Zentralbibliothek von der Verlagskonzentration im

Raum Stuttgart seit dem Zweiten Weltkrieg profitiert. Unter einem Dach mit dieser Bibliothek befindet sich die Bibliothek für Zeitgeschichte (ehemalige Weltkriegsbücherei), die einige relevante - insbesondere ältere - Bestände besitzt.

(b) Die Bibliothek des Instituts für Auslandsbeziehungen in Stuttgart ist eine Spezialbibliothek für auslandskundliche Literatur mit z.Z. rd. 3oo ooo Bänden sowie rd. 3ooo Zeitungen und Zeitschriften aus dem Ausland.

(c) Die Universitätsbibliotheken der Universität (ehemaligen Landwirtschaftlichen Hochschule) Hohenheim und der Universität (ehemaligen Technischen Hochschule) Stuttgart sind zwei leistungsstarke wissenschaftliche Bibliotheken vor allem in bezug auf das dortige Fächerangebot. Die erstgenannte Bibliothek fungiert zudem als landwirtschaftliche Zentralbibliothek in Baden-Württemberg, und die letztgenannte als technische Zentralbibliothek des württembergischen Landesteils.

(d) In der Universitätsbibliothek Tübingen wird orientalistische Literatur bereits seit dem 16. Jahrhundert gesammelt. Aufgrund ihrer umfangreichen und wertvollen alten Bestände wurde diese Bibliothek zwischen den beiden Weltkriegen von der Notgemeinschaft der Deutschen Wissenschaft und wird sie seit 1952 von der Deutschen Forschungsgemeinschaft im Fach Orientalistik großzügig finanziell unterstützt. Dank dieser Unterstützung dürfte die Universitätsbibliothek Tübingen heutzutage in der Bundesrepublik mit Abstand u.a. über die meiste Saudi-Arabien-Literatur verfügen.

(e) Im Geographischen Institut der Universität Tübingen liegt seit 1972 die Koordination des Sonderforschungsbereichs 19 'Tübinger Atlas des Vorderen Orients', der ebenfalls weitgehend von der Deutschen Forschungsgemeinschaft finanziert wird. Seither stellt der Vordere Orient ein Spezialgebiet der Institutsbibliothek dar.

Das zweite Schwergewicht ist punktueller Natur. Es liegt auf räumlich verstreuten Bibliotheken mit großen relevanten Buchbeständen. Dazu zählen insbesondere die Staatsbibliothek Preußischer Kulturbesitz in Berlin (West), die Bayerische Staatsbibliothek in München, die Niedersächsische Staats- und Universitätsbibliothek in Göttingen, die Bibliothek des Instituts für Weltwirtschaft in Kiel sowie die Bibliothek des Deutschen Orient-Instituts in Hamburg.

Die meisten Standortnachweise in bezug auf Beiträge in Zeitschriften und fortlaufenden Sammelwerken wurden mit Hilfe des 9. Gesamtausdrucks der Zeitschriftendatenbank (Redaktionsschluß: April 1983) des Deutschen Bibliotheksinstituts, Staatsbibliothek Preußischer Kulturbesitz in Berlin, recherchiert. Da die in der Zeitschriftendatenbank verzeichneten Bestandsangaben der beteiligten Bibliotheken unterschiedlich 'up-to-date' sind, fanden bei den Nachweisen vorzugsweise diejenigen Bibliotheken Berücksichtigung, deren Angaben als besonders zuverlässig gelten; es sind dies u.a. die Bibliotheken mit den folgenden Sigeln: 1a, 5, 21, 24, 25, 9o, 93, 18o, 188, 289, 291 und 352.

Vom Großteil der amerikanischen Hochschulschriften ohne Standortnachweise können bei University Microfilms International vollständige Nachdrucke in Form von Mikrofilmen, Mikrofiches und xerographischen Kopien bezogen werden. Wesentlich ungünstiger ist es allgemein um die Erhältlichkeit - auch im Rahmen des internationalen Leihverkehrs - von unveröffentlichten britischen und französischen Hochschulschriften bestellt.

Die relativ wenigen Standortnachweise im Ausland haben drei Quellen: briefliche Mitteilungen einzelner Staatsbibliotheken, Auslandsbestellungen sowie kurze Arbeitsaufenthalte am Ort. Die meisten dieser Nachweise sollen anzeigen, daß die betreffenden Arbeiten in der Bundesrepublik vergeblich auszuleihen versucht wurden.

Lediglich im Hinblick auf den Registerteil wurden die aufgeführten Titel fortlaufend numeriert.

Mangels Annotationen und Klassifikationen im Bibliographischen Teil (abgesehen von der Unterscheidung Bibliographien vs. Literatur) kommt dem Registerteil für die Benutzung eine große Bedeutung zu. Von den vier Registern dieses Teils ist das erste, das Schlagwort- oder Sachregister, zweifellos das wichtigste. Es wird von 171 mehr oder weniger abstrakten Ober- und Unterkategorien (Deskriptoren) gebildet, die größtenteils in Anlehnung an die inhaltliche Palette der relevanten Literatur frei gewählt und zum kleineren Teil dem "Thesaurus für wirtschaftliche und soziale Entwicklung" der Deutschen Stiftung für internationale Entwicklung, Zentrale Dokumentation, Bonn, und dem ergänzenden "Orient-Thesaurus" der Dokumentations-Leitstelle Moderner Orient, Hamburg, entnommen worden sind.

Wie bei anderen Bibliographien auch, enthält sowohl die Formulierung des Kategorienschemas als auch die mit dessen Hilfe vorgenommene Verschlagwortung der einzelnen Arbeiten gewisse subjektive Elemente, was von den Benutzern in Rechnung zu stellen ist.

Weit mehr als die Hälfte der 3629 Literaturtitel im Bibliographischen Teil konnte vom Bearbeiter anhand von Druckexemplaren verschlagwortet werden. Die Verschlagwortung der meisten übrigen Titel stützte sich auf vorliegende Zusammenfassungen (z.B. Dissertation abstracts), Rezensionen, Auszüge sowie Kodierungen durch einzelne Stellen (Dokumentations-Leitstelle Moderner Orient, Institut für Weltwirtschaft an der Universität Kiel usw.). Lediglich dann, wenn in bezug auf einen Titel nicht mehr als bibliographische Angaben vorlagen, wurde von einer Verschlagwortung abgesehen; dies war nur bei wenigen Titeln der Fall. Soweit möglich wurde der mehrfachen gegenüber der einfachen Verschlagwortung einzelner Arbeiten der Vorzug gegeben. Landeskunden wurden dagegen nur als solche, damit einfach verschlagwortet; ihr Inhalt hätte sonst unter eine Vielzahl von Kategorien subsumiert werden müssen.

Die Register II und III enthalten die alphabetisch geordneten Namen der persönlichen Verfasser bzw. Herausgeber zusammen mit den Nummern der von ihnen verfaßten bzw. herausgegebenen Arbeiten. In bezug auf die korporativen Verfasser und Herausgeber - sie sind z.T. schwer zu unterscheiden - erschien nur das teilweise vereinfachende Register IV lohnend. Nutzungsweise und Nützlichkeit dieser drei Register liegen auf der Hand.

Alle Register lassen die Bibliographien des Bibliographischen Teils mit Ausnahme von Nr. 18 außer acht, da letztere neben einer zweiseitigen eigentlichen Bibliographie auch sachliche Ausführungen zur Geschichte des Druck- und Pressewesens in Saudi-Arabien enthält.

III CITATION AND USE INSTRUCTIONS

The purpose of this bibliography is to provide scientists and laymen alike a simple and useful research tool. Form of citation and arrangement of titles have been determined accordingly and are based on the following principles:

(1) Only useful and conclusive bibliographic data are cited:
- Author(s), editor(s) (personal and corporative)
- Title proper and subtitle of publication
- Title of journal or other serial publication, with volume number, year of publication, and/or issue number
- Nature and originating institution of an academic publication
- Number of volumes and print run in the case of more than one volume or more than one edition of a monographic publication
- Place(s) and year(s) of publication
- Publisher(s), abbreviated
- Pagination, arabic and roman

Elements not cited therefore include format, authors of forewords and introductions, translators, photographers and illustrators, as well as number of maps, illustrations, tables, etc. This short citation form is very similar to the form generally practiced in library and documentation work.

(2) The order of citation is the same for each kind of publication listed (monograph, article in periodical or collection, academic publication, etc.).

(3) Most of the entries include references to 1 to 3 German libraries in which the material listed may be located. In a few cases, foreign libraries are listed as an alternative. These

(4) Use of abbreviations has in general been avoided. Those forms adopted (s. Section IV of this Introduction) are found almost without exception in German, English, and French practice and are therefore familiar.

(5) Entries are arranged by author and editor. Multiple entries under a single name are subarranged by year of publication. The value of this arrangement is only slightly impaired by the additions at the end of the Bibliographic Section.

The following special practices are followed:

In accordance with common library practice, the definite articles Al-, Al, El-, and El in Arabic family names are inverted in lower case letters and ignored in filing (e.g.: Abdalwahed, A.M. al-), except when the prefix is an integral part of the name (e.g.: Almana, M.). In the few cases in which it is attached to a first name, the definite article is ignored in filing. The prefixes de, le, van, von, etc., in Western names are inverted (e.g.: Gaury, G. de). Arabic diacritical marks have been adopted as used. For the sake of simplicity, the apostrophy (') is used both for the letter 'ain and the hamza. Diacritical marks are ignored in filing. Likewise ignored are honorific titles such as Baron, Doctor, Lady, Pasha, Sheikh, Sir, etc., in names of authors and editors. First names are reduced to the initial letter except when detrimental to the alphabetization.

Anonymous publications are arranged by title. All subtitles are reproduced. Words which appeared in italics or boldface in the original are underlined. For the sake of easier differentiation between non-German-language titles of articles and serials, all nouns and adjectives in the titles of serial publications are reproduced in capital letters.

As a matter of principle, the titles of periodicals and other serial publications are given in full. The place of publication is cited to distinguish differing publications with the same title. It should be noted that some periodicals have changed their title during the course of publication. Thus, The Moslem World has become The Muslim World, Aramco World is now Aramco World Magazine, etc.

The issue number is omitted from the entry for articles in publications which are numbered continuously. Otherwise, the issue number appears in parentheses directly following volume number and year. This practice was followed, as well, when the pagination of an issue could not be determined. Issue numbers are also always cited when the frequency of publication is not specified in volumes or years. Roman numerals and capital letters identify individual series of one and the same serial publication.

Misspellings in names of authors, titles proper, and subtitles have been corrected without comment. Outdated spellings, on the other hand, have been retained (e.g.: "rothes Meer" and "Hedjaz"). The umlauts ä, ö, and ü are reproduced accordingly but filed as ae, oe and ue. The letter "ß" is given as "ss".

In individual cases, more than one contribution by the same author in the same publication—usually in immediate proximity to one another—are entered under a single number (e.g.: no. 884). Such articles are separated by dashes. A composite book is cited under a single number only when (almost) all the articles therein contained are relevant (e.g.: no. 2266). To avoid double entry, these numbers are omitted from the subject index (s. the first index of the Index Section).

More than one printing and/or edition could be determined for a small number of publications, but this list may not claim to be complete.

Some official publications of the Saudi Arabian government are available in multilingual editions. The Arabic portion of such publications has not been evaluated in this bibliography.

Differing forms of a Saudi Arabian place name used as place of publication in various publications in one and the same language have been standardized. Thus, in English-language publications the forms "Jeddah" and "Riyadh" are used. This rule, however, was not applied to "Riyadh" in the proper name, "University of Riyad". The place of publication has been omitted if this information is redundant, as in the case of many academic publications, where forms such as "Universität Bonn" and "Claremont Graduate School" are unambiguous. If the individual volumes of a multivolume work have appeared in various places and/or with various publishers, the places or publishers have been separated from one another by slashes (e.g.: no. 1451).

The bibliographic information on most of the U.S. academic publications from the past ten years was taken from the irregular catalogues of doctoral dissertations and the monthly <u>Dissertation Abstracts International</u> of University Microfilms International. The entry for these publications always shows the year of their acceptance by a university, not the year of the reprint edition (and, therefore, publication) by UMI. The exclusively Arabic form of the pagination note used by UMI covers both the Roman and Arabic paging of the original.

Pages and leaves are not distinguished in the pagination note. Unnumbered pages and leaves are not counted.

The union codes (library symbols) for German libraries were taken from the <u>Sigelverzeichnis für die Bibliotheken der Bundesrepublik Deutschland einschließlich Berlin (West)</u> (Union Codes of the Libraries of the Federal Republic of Germany Including West Berlin) and from a list of the Deutsches Orient-Institut (German Orient Institute). Call (press) numbers have not been used. All the German libraries from which titles have been registered are listed in Section V of this Introduction together with their union code and address. A maximum of three holding libraries is given, the codes enclosed in dashes. This limitation was chosen arbitrarily (some of the titles listed may be found in numerous public libraries). Generally speaking, research libraries have been preferred which are known to hold much Oriental literature and to collect currently in this field. Two further aspects were honored: insofar as possible, at least one personally verified reference to a national or foreign library active in interlibrary lending is listed per title. Libraries offering interlibrary loan services as opposed to reference libraries are indicated in Section V of this Introduction.

Unfortunately it was not possible to list at least one holding library for each title included. However, the fact that such a reference is missing does not mean that a cited title is not held by a German library, even when in certain cases interlibrary loan requests were without success.

A balanced geographical distribution of one to three holding libraries is, of course, not possible. Thus, two priorities are evident in the choice of libraries cited. Firstly, most of them are located in the Stuttgart-Tübingen area, determined in part by the residence of the compiler, in part by the following five circumstances:

(a) The Württembergische Landesbibliothek (Wurttemberg State Library) in Stuttgart has, among others, a large collection of 18th and 19th centuries travel literature and, being a main library, could profit from the concentration of publishing firms in the Stuttgart area following World War II. Housed in the same building as this library is the Bibliothek für Zeitgeschichte (Library of Modern History), with relevant, particularly older, holdings.

(b) The library of the Institut für Auslandsbeziehungen (Institute of Foreign Cultural Relations) in Stuttgart is a special library with currently some 3oo,ooo volumes and 3,ooo foreign serials in the field of area studies.

(c) The libraries of the Universität Hohenheim (formerly Landwirtschaftliche Hochschule Hohenheim) and the Universität Stuttgart (formerly Technische Hochschule Stuttgart) are two research libraries with substantial resources, particularly in the subjects offered in these schools. The library in Hohenheim functions as the central agricultural library in Baden-Wurttemberg. The Stuttgart institution is the central technical library for Wurttemberg.

(d) Literature in the field of Oriental studies has been collected by Tübingen University Library since the 16th century. In view of its comprehensive and valuable holdings in this field, this library received substantial financial support in the period between the two world wars from the Notgemeinschaft der Deutschen Wissenschaft (Emergency Association of German Research) and since 1952 from the Deutsche Forschungsgemeinschaft (German Research Association). Thanks to this support, Tübingen University Library today maintains surely the largest single collection of Saudi Arabian literature in the German Federal Republic.

(e) Since 1972 the Geographical Institute of Tübingen University has housed the Coordinating Committee of Special Research Area 19, "Tübingen Atlas of the Near and Middle East", which is likewise financed by the German Research Association. The Near and Middle East therefore represent an area of specialization for the Institute Library.

The second priority is dispersive in nature and lies in the geographical distribution of libraries with substantial relevant holdings. To these may be counted, in particular, the Staatsbibliothek Preussischer Kulturbesitz in West Berlin, the Bayerische Staatsbibliothek in Munich, the Niedersächsische Staats- und Universitätsbibliothek in Göttingen, the Library of the Institut für Weltwirtschaft in Kiel, as well as the Library of the Deutsches Orient-Institut in Hamburg.

Most of the references to libraries holding serial publications were researched with the aid of the 9th edition of the <u>Zeitschriften-Datenbank</u> (German Serials Data Base) (editorial deadline April 1983) of the Deutsches Bibliotheksinstitut, Staatsbibliothek Preussischer Kulturbesitz in Berlin. Since the information in the data base is not uniformly up-to-date, preference was given to those libraries known to be especially reliable. These are, among others, libraries with the following union codes: 1a, 5, 21, 24, 25, 9o, 93, 18o, 188, 289, 291, and 352.

Complete copies of most of the American academic publications lacking union codes may be obtained in microfilm, microfiche, and xerox form from University Microfilms International. The situation, however, is considerably less fortunate—within the framework of international borrowing, as well—when it comes to unpublished manuscripts from British and French schools.

The relatively few references to holdings in foreign countries originated in three sources: written communication from individual state libraries, lending requests, and the author's working visits in one or another foreign library. Most of these references are an indication that the material in question could not be located in the Federal Republic of Germany.

The titles listed have been numbered continuously merely to facilitate use of the indexes.

Since this bibliography is neither annotated nor classified (apart from the distinction between bibliographies and literature), the four indexes are particularly useful. Of these, the Subject Index is certainly the most important. It is comprised of 171 more or less abstract descriptors, suggested in the most part by literary warrant, but also borrowed from the <u>Thesaurus for economic and social development</u> of the Deutsche Stiftung für internationale Entwicklung, Zentrale Dokumentation in Bonn and the supplementary <u>Orient Thesaurus</u> of the Dokumentations-Leitstelle Moderner Orient in Hamburg. As is common in other bibliographies, as well, both classification scheme and actual indexing bear certain subjective elements which the user must take into account.

More than half of the 3,629 literature titles in the bibliography were indexed by the author directly from the piece itself. For most of the remaining titles, abstracts, reviews, excerpts, or the code marks of some documentation center (Dokumentations-Leitstelle Moderner Orient, Kiel University's Institut für Weltwirtschaft, etc.) were used as an indexing aid. Only a very few titles, for which nothing more than the bibliographic data could be ascertained, were not indexed. As far as possible, a given title was assigned more than one index term. Area studies, on the other hand, were indexed only once (as geography: in depth).

In addition to the Subject Index, the Bibliography includes an Index of Authors, Index of Editors, and Index of Corporate Bodies.

With the exception of item no. 18, which includes, in addition to a two-page bibliography, textual material on the history of printing and the press in Saudi Arabia, none of the bibliographies are indexed.

IV ABKÜRZUNGEN / ABBREVIATIONS

Apr.	April
ARAMCO	Arabian American Oil Company
Arch.D.	Doctor of Architecture dissertation
Aug.	August
B.A.	Bachelor of Arts thesis
Bd(e).	Band/Bände
bearb.	bearbeitet
Beibd.	Beiband
Beih.	Beiheft
B.Litt.	Bachelor of Litterature (Letters) thesis
bzw.	beziehungsweise
c(a).	circa
D.B.A.	Doctor of Business Administration dissertation
Dec.	December
Déc.	Décembre
d.i.	das ist
Diss.	Dissertation
D.L.S.	Doctor of Library Science dissertation
D.P.A.	Doctor of Public Administration dissertation
ed(s).	edition(s), editor(s), edited
éd.	édition, éditeur
Ed.D.	Doctor of Education dissertation
EI^1	Enzyklopädie des Islām (1. Auflage)
EI^2	The Encyclopaedia of Islam, New Edition
enl.	enlarged
env.	environs
Ergänzungsbd.	Ergänzungsband
erw.	erweitert

et al.	et alter
etc.	et cetera
FAO	Food and Agriculture Organization
fasc.	fascicle(s), fascicule(s)
Feb.	Februar, February
GATT	General Agreement on Tariffs and Trade
getr.Pag.	getrennte Paginierung
GTZ	Deutsche Gesellschaft für Technische Zusammenarbeit (GTZ) GmbH
H.	Heft
Habil.-Schrift	Habilitationsschrift
Halbbd.	Halbband
Hg(g).	Herausgeber (Singular bzw. Plural), herausgegeben (von einer Person oder Stelle bzw. von zwei oder mehr Personen oder Stellen)
HMSO	His (Her) Majesty's Stationery Office
HRAF	Human Relations Area Files (Press)
i.e.	that is
ILO	International Labor Office (Organization)
impr.	improved
Jan.	January
Jg(e).	Jahrgang/Jahrgänge
M.A.	Master of Arts thesis
M.P.A.	Master of Public Administration (Affairs) thesis
M.Phil.	Master of Philosophy thesis
M.Sc.	Master of Science thesis
n.d.	no date
neubearb.	neubearbeitet
N.F.	Neue Folge
no(s).	number(s)
Nov.	November, Novembre
n.p.	no place
n.pag.	no paging
Nr.	Nummer
N.S.	Neue Serie, New Series, Nouvelle Série
Oct.	October
OECD	Organisation for Economic Co-operation and Development
o.J.	ohne Jahresangabe
o.O.	ohne Ortsangabe

o.Pag.	ohne Paginierung
OPEC	Organization of Petroleum Exporting Countries
p.	pages
Ph.D.	Doctor of Philosophy dissertation
Pseud.	Pseudonym
pseud.	pseudonym(e)
pt.	part
resp.	respectively, respectivement
rev.	revised
s.	siehe, see
S.	Seiten
s.d.	sans date
Sept.	September, Septembre
s.l.	sans lieu
Sbd.	Sonderband
Sh.	Sonderheft
S.J.D.	Doctor of Juridical Science dissertation
Sp.	Spalten
Suppl.	Supplement, supplément
t.	tome
umgearb.	umgearbeitet
UN(O)	United Nations (Organization)
UNCTAD	United Nations Conference on Trade and Development
UNDP	United Nations Development Programme
UNESCO	United Nations Educational, Scientific and Cultural Organization
UNIDO	United Nations Industrial Development Organization
US(A)	United States (of America)
var.pag.	various pagings
vol(s).	volume(s)
WHO	World Health Organization

V   BIBLIOTHEKSSIGEL UND -ADRESSEN / LIBRARY SYMBOLS AND
     ADDRESSES

+ dem Auswärtigen Leihverkehr angeschlossen / participating
  in the system of interlibrary loans

-1a- Staatsbibliothek Preussischer Kulturbesitz +
      Potsdamer Str. 33
      1ooo Berlin 3o

-4-   Universitätsbibliothek Marburg +
      Wilhelm-Röpke-Str. 4
      355o Marburg/Lahn 1

-5-   Universitätsbibliothek Bonn +
      Adenauerallee 39-41
      53oo Bonn 1

-6-   Universitätsbibliothek Münster +
      Krummer Timpen 3-5
      44oe Münster/Westfalen

-7-   Niedersächsische Staats- und Universitätsbibliothek +
      Prinzenstr. 1
      34oo Göttingen

-12- Bayerische Staatsbibliothek +
      Ludwigstr. 16
      8ooo München 34

-15- Universitätsbibliothek Leipzig +
      Beethovenstr. 6
      DDR 7o1o Leipzig

-16- Universitätsbibliothek Heidelberg +
Plöck 1o7-1o9
69oo Heidelberg

-17- Hessische Landes- und Hochschulbibliothek +
Schloss
61oo Darmstadt

-18- Staats- und Universitätsbibliothek Hamburg +
Moorweidenstr. 4o
2ooo Hamburg 13

-19- Universitätsbibliothek München +
Geschwister-Scholl-Platz 1
8ooo München 22

-2o- Universitätsbibliothek Würzburg +
Am Hubland
87oo Würzburg

-21- Universitätsbibliothek Tübingen +
Wilhelmstr. 32
74oo Tübingen

-24- Württembergische Landesbibliothek +
Konrad-Adenauer-Str. 8
7ooo Stuttgart 1

-24/213- Bibliothek für Zeitgeschichte, ehemalige Weltkriegsbücherei +
Konrad-Adenauer-Str. 8
7ooo Stuttgart 1

-25- Universitätsbibliothek Freiburg +
Werthmannplatz 2
78oo Freiburg/Breisgau

-26- Universitätsbibliothek Giessen +
Bismarckstr. 37
63oo Giessen

-29- Universitätsbibliothek Erlangen-Nürnberg, Hauptbibliothek +
Universitätsstr. 4
852o Erlangen

-3o- Stadt- und Universitätsbibliothek Frankfurt +
Bockenheimer Landstr. 134-138
6ooo Frankfurt/Main 1

-31- Badische Landesbibliothek +
       Lammstr. 16
       7500 Karlsruhe

-35- Niedersächsische Landesbibliothek +
       Waterloostr. 8
       3000 Hannover 1

-38- Universitäts- und Stadtbibliothek Köln +
       Universitätsstr. 33
       5000 Köln 41

-38 M- Zentralbibliothek der Medizin
          Joseph-Stelzmannstr. 9
          5000 Köln 41

-46- Staats- und Universitätsbibliothek Bremen +
       Bibliothekstrasse
       2800 Bremen 33

-61- Universitätsbibliothek Düsseldorf +
       Universitätsstr. 1
       4000 Düsseldorf 1

-77- Universitätsbibliothek Mainz +
       Saarstr. 21
       6500 Mainz 1

-83- Universitätsbibliothek der Technischen Universität Berlin +
       Strasse des 17. Juni 135
       1000 Berlin 12

-84- Universitätsbibliothek der Technischen Universität Braun-
       schweig +
       Pockelsstr. 13
       3300 Braunschweig

-89- Universitätsbibliothek Hannover +
       Welfengarten 1 b
       3000 Hannover 1

-90- Universitätsbibliothek Karlsruhe +
       Kaiserstr. 12
       7500 Karlsruhe

-93- Universitätsbibliothek Stuttgart +
       Holzgartenstr. 16
       7000 Stuttgart 1

-98-   Universitätsbibliothek Bonn, Abteilung Zentralbibliothek
       der Landbauwissenschaft +
       Meckenheimer Allee 172
       5300 Bonn 1

-100- Universitätsbibliothek Hohenheim +
       Garbenstr. 15
       7000 Stuttgart 70

-104- Universitätsbibliothek Clausthal-Zellerfeld +
       Leibnizstr. 2
       3392 Clausthal-Zellerfeld

-107- Pfälzische Landesbibliothek +
       Johannesstr. 22 a
       6720 Speyer

-156- Fürstlich Hohenzollernsche Hofbibliothek Sigmaringen +
       Schloss
       7080 Sigmaringen

-180- Universitätsbibliothek Mannheim +
       Schloss/Ostflügel
       6800 Mannheim 1

-188- Universitätsbibliothek der Freien Universität Berlin +
       Garystr. 39
       1000 Berlin 33

-188/812- Institut für Islamwissenschaft der Freien Universität
           Berlin, Bibliothek
           Boltzmannstr. 4
           1000 Berlin 33

-201- Deutsches Patentamt, Bibliothek
       Zweibrückenstr. 12
       8000 München 2

-204- Ibero-Amerikanisches Institut, Preussischer Kulturbesitz,
       Bibliothek +
       Potsdamer Str. 37
       1000 Berlin 30

-206- Institut für Weltwirtschaft an der Universität Kiel, Zen-
       tralbibliothek der Wirtschaftswissenschaften in der Bundes-
       republik Deutschland +
       Düsternbrooker Weg 120
       2300 Kiel 1

-21o- Deutsches Museum, Bibliothek +
Museumsinsel 1
8ooo München 26

-212- Institut für Auslandsbeziehungen, Bibliothek +
Charlottenplatz 17
7ooo Stuttgart 1

-281- Deutscher Bundestag, Bibliothek
Görresstr. 15 (Bundeshaus)
53oo Bonn 1

-282- Statistisches Bundesamt, Bibliothek, Dokumentation, Archiv +
Gustav-Stresemann-Ring 11
62oo Wiesbaden 1

-289- Universitätsbibliothek Ulm +
Schlossbau 38
79oo Ulm/Donau

-291- Universitätsbibliothek der Universität des Saarlandes +
St. Johanner Stadtwald
66oo Saarbrücken

-294- Universitätsbibliothek Bochum +
Universitätsstr. 15o
463o Bochum-Querenburg

-352- Universitätsbibliothek Konstanz +
Universitätsstr. 1o
775o Konstanz

-385- Universitätsbibliothek Trier +
Kohlenstr. 68
55oo Trier-Tarforst

-464- Universitätsbibliothek Duisburg +
Koloniestr. 55
41oo Duisburg

-465- Gesamthochschulbibliothek Essen +
Universitätsstr. 9
43oo Essen 1

-467- Universitätsbibliothek Siegen +
Adolf-Reichwein-Strasse
59oo Siegen 21

-468-   Gesamthochschulbibliothek Wuppertal +
       Gaussstr. 2o
       56oo Wuppertal 1

-7o8- Fernuniversität, Gesamthochschule Hagen, Hochschulbibliothek +
       Feithstr. 14o
       58oo Hagen

-B 19- Auswärtiges Amt, Bibliothek
       Adenauerallee 99-1o3
       53oo Bonn 1

-B 2o8- Max-Planck-Institut für ausländisches öffentliches
        Recht und Völkerrecht, Bibliothek
        Berliner Str. 48
        69oo Heidelberg 1

-B 211- Politische Wissenschaft (Otto-Suhr-Institut), Fachbereichsbibliothek
        Ihnestr. 21
        1ooo Berlin 33

-B 212- Max-Planck-Institut für ausländisches und internationales Privatrecht, Bibliothek
        Mittelweg 187
        2ooo Hamburg 13

-B 851- Institut für Internationales und Ausländisches Recht
        und Rechtsvergleichung, Bibliothek
        Van't-Hoff-Str. 8
        1ooo Berlin 33

-Bm 25- Institut für Entwicklungsforschung und Entwicklungspolitik der Universität Bochum, Bibliothek
        Gebäude GB, Im Lottental
        463o Bochum-Querenburg

-Bo 133- Bibliothek des Archivs der sozialen Demokratie, Bibliothek der Friedrich-Ebert-Stiftung
         Godesberger Allee 149
         53oo Bonn 2

-Bo 139- Seminar für Orientalische Sprachen der Universität Bonn,
         Bibliothek
         Adenauerallee 1o2
         53oo Bonn 1

XC

-Bo 149-	Deutsche Stiftung für Internationale Entwicklung (DSE), Zentrale Dokumentation, Bibliothek Endenicher Str. 41 53oo Bonn 1
-Esb 2ooo-	Deutsche Gesellschaft für Technische Zusammenarbeit (GTZ) GmbH, Bibliothek Dag-Hammarskjöld-Weg 1 6236 Eschborn
-F 1-	Senckenbergische Bibliothek, Universitätsbibliothek für Naturwissenschaften und Alte Medizin, Archivbibliothek für Alte Medizin, Spezialbibliothek für Biowissenschaften + Bockenheimer Landstr. 134-138 6ooo Frankfurt/Main 1
-F 2oo3-	Kreditanstalt für Wiederaufbau, Bibliothek Palmengartenstr. 5 6ooo Frankfurt/Main 1
-Frei 119-	Arnold-Bergstraesser-Institut für kulturwissenschaftliche Forschung e.V., Bibliothek Windausstr. 16 78oo Freiburg/Breisgau
-Gö 153-	Institut für Rurale Entwicklung der Universität Göttingen, Bibliothek Büsgenweg 2 34oo Göttingen
-H 2-	Deutsches Hydrographisches Institut, Bibliothek Bernhard-Nocht-Str. 78 2ooo Hamburg 4
-H 3-	HWWA-Institut für Wirtschaftsforschung, Bibliothek + Neuer Jungfernstieg 21 2ooo Hamburg 36
-H 1o8-	Unesco-Institut für Pädagogik, Bibliothek Feldbrunnenstr. 58 2ooo Hamburg 13
-H 221-	Institut für Afrika-Kunde, Bibliothek Neuer Jungfernstieg 21 2ooo Hamburg 36

-H 223-   Deutsches Orient-Institut, Bibliothek
          Mittelweg 15o
          2ooo Hamburg 13

-Kn 28-   Erzbischöfliche Diözesan- und Dombibliothek +
          Gereonstr. 2-4
          5ooo Köln 1

-Tü 17-   Geographisches Institut der Universität Tübingen,
          Bibliothek
          Hölderlinstr. 12
          74oo Tübingen

BIBLIOGRAPHISCHER TEIL / BIBLIOGRAPHIC SECTION

I BIBLIOGRAPHIEN / BIBLIOGRAPHIES

1 Abstracta Islamica. Supplément à la Revue des Études Islamiques. Paris, 1927-.

2 Altoma, S.J., 1975: Modern Arabic literature: a bibliography of articles, books, dissertations, and translations in English. Asian Studies Research Institute, Occasional Papers 3; Bloomington: Indiana University. 73 p.

3 American University of Beirut, Economic Research Institute, 1954: A selected and annotated bibliography of economic literature on the Arabic speaking countries of the Middle East, 1938-1952. Beirut. IX, 199 p.

4 ——, ——, 1967: A selected and annotated bibliography of economic literature on the Arabic speaking countries of the Middle East, 1953-1965. Beirut. XVII, 458 p.

5 Anthony, J.D., 1973: The states of the Arabian Peninsula and the Gulf littoral: a selected bibliography. Washington: The Middle East Institute. 21 p.

6 Atiyeh, G.N., 1975: The contemporary Middle East, 1948-1973: a selective and annotated bibliography. Boston: Hall. XXVI, 664 p.

7 Bloomfield, B.C., 1967: Theses on Asia accepted by universities in the United Kingdom and Ireland, 1877-1964. London: Cass. XI, 127 p.

8   Bonnenfant, P., 1980: Bibliographie de la péninsule Arabique: sciences de l'homme. Fasc. 1: Titres concernant toute la péninsule, 1979. Université de Provence, Groupement d'intérêt scientifique: 'Sciences humaines sur l'aire méditerranéenne', Centre d'Études et de Recherches sur l'Orient Arabe Contemporain; Paris: Centre National de la Recherche Scientifique. 153 p.

9   Burke, J.T., 1956: An annotated bibliography of books and periodicals in English dealing with human relations in the Arab states of the Middle East with special emphasis on modern times (1945-1954). Beirut: American University of Beirut. XIV, 117 p.

10 Clements, F.A., 1976: The emergence of Arab nationalism, from the nineteenth century to 1921. London: Diploma Press. X, 290 p.

11 ——, 1979: Saudi Arabia. World Bibliographical Series 5; Oxford, Santa Barbara: Clio Press. XIV, 195 p.

12 Couland, J. (éd.), 1981: Bibliographie de la culture arabe contemporaine. Sous la direction de J. Berques. Paris: Sindbad, Les Presses de l'Unesco. 483 p.

13 Deutsches Orient-Institut, 1967: Bibliographie zum Erziehungs- und Bildungswesen in den Ländern des muslimischen Orients. Hamburg. VI, 52 S.

14 ——, Dokumentations-Leitstelle Moderner Orient, 1.1970-: Dokumentationsdienst Moderner Orient. Ausgewählte neuere Literatur. A selected bibliography of recent literature. Hamburg.

15 ——, ——, 1979: Saudi-Arabien. Einführende Literatur. Dokumentationsdienst Moderner Orient, Kurzbibliographie; Hamburg. 9 S.

16 Dodgeon, H. & A.M. Findlay, 1979: Ports of the Arabian Peninsula: a guide to the literature. Occasional Paper Series 7; Durham: Centre for Middle Eastern and Islamic Studies, University of Durham. IV, 49 p.

17 Dotan, U., 1970: A bibliography of articles on the Middle East 1959-1967. Ed. by A. Levy. The Shiloah Center Teaching and Research Aids 2; Tel Aviv: Tel Aviv University. 227 p.

18 Ende, W., 1975: Bibliographie zur Geschichte des Druckwesens und der Presse in Saudi-Arabien. In: Dokumentationsdienst Moderner Orient, Mitteilungen 4(1): 29-37.

19 Ettinghausen, R. (ed.), 1953: A selected and annotated bibliography of books and periodicals in Western languages dealing with the Near and Middle East, with special emphasis on medieval and modern times; with supplement. 2nd ed. Washington: The Middle East Institute. VIII, 137 p.

2o Finke, D., G. Hansen & R.-D. Preisberg, 1973: Deutsche Hochschulschriften über den modernen islamischen Orient. German thesis on the Islamic Middle East. Dokumentationsdienst Moderner Orient A, 1; Hamburg: Deutsches Orient-Institut. VIII, 177 S.

21 Gaury, G. de, 1944: An Arabian bibliography. In: Journal of the Royal Central Asian Society 31: 315-32o.

22 Gay, J., 1875: Bibliographie des ouvrages relatifs à l'Afrique et à l'Arabie. Catalogue méthodique de tous les ouvrages français & des principaux en langues étrangères traitant de la géographie, de l'histoire, du commerce, des lettres & des arts de l'Afrique & de l'Arabie. San Remo, Paris, Turin (reproduction: Amsterdam: Meridian, 1961). XI, 312 p.

23 Geddes, C.L., 1974: Analytical guide to the bibliographies on the Arabian Peninsula. Bibliographic Series 4; Denver: The American Institute of Islamic Studies. 5o p.

24 Grimwood-Jones, D., D. Hopwood & J.D. Pearson, with the assistance of J.P.C. Auchterlonie, J.D. Latham & Y. Safadi (eds.), 1977: Arab Islamic bibliography. The Middle East Library Committee Guide, based on G. Gabrieli's Manuale di bibliografia musulmana. Hassocks: The Harvester Press; Atlantic Highland: Humanities Press. XVII, 292 p.

25 Grimwood-Jones, D. (ed.), 1979: Middle East and Islam: a bibliographical introduction. Rev. & enl. ed. ([1]1972). Bibliotheca Asiatica 15; Zug: Inter Documentation. IX, 429 p.

26 Hansen, G., I. Otto & R.-D. Preisberg, 1976: Wirtschaft, Gesellschaft und Politik der Staaten der Arabischen Halbinsel. Eine bibliographische Einführung. Economy, society and politics of the countries of the Arabian Peninsula.

A bibliographic introduction. Dokumentationsdienst Moderner Orient A, 7; Hamburg: Deutsches Orient-Institut. 271 S.

27 Hazard, H.W., R.W. Crawford et al., 1956: Bibliography of the Arabian Peninsula. Prepared by the American Geographical Society for the Human Relations Area Files. Behavior Science Bibliographies; New Haven: HRAF. 256 p.

28 Hebrew University, Economic Research Institute, 1954: A selected bibliography of articles dealing with the Middle East 1939-1950. Jerusalem. VIII, 95, VI p.

29 Heyworth-Dunne, G.-E., 1952: Bibliography and reading guide to Arabia. The Muslim World Series 1; Cairo: The Renaissance Bookshop. 16 p.

30 Hopwood, D., 1972: Some Western studies of Saudi Arabia, Yemen and Aden. In: D. Hopwood (ed.): The Arabian Peninsula: society and politics. Studies on Modern Asia and Africa 8; London: Allen & Unwin, 1972: 13-27.

31 Library of Congress, Reference Department, Division of Orientalia, Near East Section, 1951: The Arabian Peninsula: a selected, annotated list of periodicals, books, and articles in English. Washington: US Government Printing Office (reprint: New York: Greenwood Press, 1969). XI, 111 p.

32 Lowenstein, A.C. & R.M. Weinthal (eds.), 1.1978-: The Middle East: abstracts and index. Pittsburgh: Library Information and Research Service.

33 Macro, E., 1958: Bibliography of the Arabian Peninsula. Coral Gables: University of Miami Press. XIV, 80 p.

34 Meghdessian, S.R., 1980: The status of the Arab woman: a select bibliography. Westport: Greenwood Press. 176 p.

35 Mekeirle, J.O., 1980: The Arab world: a guide to business, economic and industrial information sources. Dallas: Inter-Crescent. XXV, 492 p.

36 Otto, I. & M. Schmidt-Dumont, 1982: Frauenfragen im modernen Orient. Eine Auswahlbibliographie. Women in the Middle East and North Africa. Dokumentationsdienst Moderner Orient A, 12; Hamburg: Deutsches Orient-Institut. XVI, 248 S.

37 Pantelidis, V.S., 1982: Arab education 1956-1978. A bibliography. London: Mansell. XVII, 552 p.

38 Pearson, J.D., 1958-77: Index Islamicus 1906-1975. A catalogue of articles on Islamic subjects in periodicals and other collective publications. 5 vols. Cambridge/England: Heffer / London: Mansell.

39 —— (ed.), 1.1977-: The Quarterly Index Islamicus: current books, articles and papers on Islamic studies. London: Mansell.

40 Preisberg, R.-D., 1978: Bevölkerung und Beschäftigung im Vorderen Orient. Eine bibliographische Einführung. Population and labour in the Middle East. A bibliographic introduction. Dokumentationsdienst Moderner Orient A, 9; Hamburg: Deutsches Orient-Institut. XXVIII, 160 S.

41 Raccagni, M., 1978: The modern Arab woman: a bibliography. Metuchen, London: The Scarecrow Press. X, 262 p.

42 Rentz, G., 1950: Literature on the Kingdom of Saudi Arabia. In: The Middle East Journal 4: 244-249.

43 Rossi, P.M., W.E. White & A.E. Goldschmidt, Jr. (eds.), 1980: Articles on the Middle East, 1947-1971; a cumulation of the bibliographies from The Middle East Journal. 4 vols. Cumulated Bibliography Series 7; Ann Arbor: Pierian Press.

44 Schwarz, K., 1980: Der Vordere Orient in den Hochschulschriften Deutschlands, Österreichs und der Schweiz. Eine Bibliographie von Dissertationen und Habilitationsschriften (1885-1978). Islamkundliche Materialien 5; Freiburg/Br.: Schwarz ([1]1971). XXIII, 721 S.

45 Selim, G.D., 1976: American doctoral dissertations on the Arab world, 1883-1974. 2nd ed. ([1]1970). Washington: Library of Congress. XVIII, 173 p.

46 Sluglett, P., 1983: Theses on Islam, the Middle East and North-West Africa 1880-1978. Accepted by universities in the United Kingdom and Ireland. London: Mansell. XII, 147 p.

47 Stevens, J.H. & R. King, 1973: A bibliography of Saudi Arabia. Occasional Papers Series 3; Durham: Centre for Middle Eastern and Islamic Studies, University of Durham. 81 p.

48 University of Saint Joseph, Centre d'Études pour le Monde Arabe Moderne, 1973: Arab culture and society in change; a partially annotated bibliography of books and articles in English, French, German and Italian. Beirut: Dar el Mashreq. XLI, 318 p.

49 US Department of State, Division of Library and Reference Services, 1951: Point Four: Near East and Africa; a selected bibliography of studies on economically under-developed countries. Washington: US Government Printing Office (reprint: New York: Greenwood Press, 1969). 136 p.

II LITERATUR / LITERATURE

A

5o Abanami, A.A., 1982: Readability analysis of the 11th and 12th grade earth science textbooks used in the public schools in Saudi Arabia. Ed.D., University of Houston. 155 p.

51 Abbas, H.A., 1982: A plan for public library system development in Saudi Arabia. Ph.D., University of Pittsburgh. 147 p.

52 Abbasi, M.Y., 1979: Arabia in the accounts of the South Asian travellers. In: Islamic Studies 18: 49-63.   -1a, 21, 3o-

53 Abbassi, A. el- (pseud. de Badia y Leblich, D.), 1814: Voyages d'Ali Bey el Abbassi en Afrique et en Asie, pendant les années 18o3, 18o4, 18o5, 18o6 et 18o7. T. 2. 3. Paris: Didot (englische Ausgabe: Travels of Ali Bey in Morocco, Tripoli, Cyprus, Egypt, Arabia, Syria, and Turkey, between the years 18o3 and 18o7. 2 vols. London: Longman, 1816. Spanische Ausgaben: Viajes de Ali Bey el Abbassi - Domingo Badia y Leblich - por Africa y Asia, durante los años 18o3, 18o4, 18o5, 18o6 y 18o7. Bd. 2. 3. Valencia, 1836; Barcelona, 1888-89). 464 resp. 41o p.   -24-

54 Abdalwahed, A.M. al-, 1981: Human resource development and manpower planning in Saudi Arabia. Ph.D., Claremont Graduate School. 13o p.

55 Abdel Azim, M. & A. Gismann, 1956: Bilharziasis survey in south-western Asia, covering Iraq, Israel, Jordan, Lebanon, Sa'udi Arabia, and Syria: 195o-51. In: Bulletin of the World Health Organization / Bulletin de l'Organisation Mondiale de la Santé 14: 4o3-456.   -7, 24, 26-

56 Abd-el Wassie, A.W., 1965: Saudi Arabia: educational developments in 1964-1965. In: International Yearbook of Education 27: 3oo-3o5.   -1a, 7, 18-

57 ——, 197o: Education in Saudi Arabia. A history of fifteen years effort to spread education in a developing country, an orthodox diagnosis, and some proposals for a better future. Houndmills Basingstoke: Macmillan. XV, 73 p. -21, 2o6-

58 Abdo, Assad S., 1968/69: The evolution of modern roads in Saudi Arabia. In: Bulletin de la Société de Géographie d'Égypte 41/42: 23-41.   -1a, 3o, 18o-

59 ——, 1969: A geographical study of transport in Saudi Arabia, with special reference to road transport. Ph.D., Department of Geography, University of Durham.   -Main Library, University of Durham-

6o ——, 197o: Domestic passenger air transport in Saudi Arabia. In: Bulletin of the Faculty of Arts, University of Riyad 1: 21-39.

61 ——, 1971: Land and air transport in Saudi Arabia. Middle East Centre Monograph 3; University of Durham.

62 ——, 1971/72: Road traffic in Saudi Arabia. In: Bulletin of the Faculty of Arts, University of Riyad 2: 77-95.

63 Abdul Bari, M., 1955: The early Wahhābīs and the Sharīfs of Makkah. In: Journal of the Pakistan Historical Society 3: 91-1o4.   -1a, 21-

64 ——, 1971: The early Wahhābīs, some contemporary assessments. In: D. Sinor, with the assistance of T. Jacques, R. Larson & M.E. Meek (eds.): Proceedings of the twenty-seventh International Congress of Orientalists, Ann Arbor, Michigan, 13th-19th August 1967. Wiesbaden: Harrassowitz, 1971: 264-266.   -21-

65 Abdul-Fattah, A.-R.A.-F., 1978: Engineering and safety analysis of dual-purpose nuclear desalination plants: a case study (Saudi Arabia). Ph.D., Iowa State University, Ames. XII, 44o p.   -H 223-

66 Abdulfattah, K., 1981: Mountain farmer and fellah in 'Asīr, southwest Saudi Arabia. The conditions of agriculture in a traditional society. Erlanger Geographische Arbeiten Sbd.

12; Erlangen: Fränkische Geographische Gesellschaft. 123 p.
-21, 29, Tü 17-

67 Abdulkader, A.A. al-, 1978: A survey of the contribution of higher education to the development of human resources in the Kingdom of Saudi Arabia. Ph.D., University of Kansas, Lawrence. XV, 279 p. -H 223-

68 Abdul Majid, H., 1926: A Malay's pilgrimage to Mecca. In: Journal of the Malayan Branch of the Royal Asiatic Society 4: 269-287. -1a-

69 Abdul-Rauf, M., 1978: Pilgrimage to Mecca. In: National Geographic 154: 581-6o7. -21, 25, 212-

7o Abdulwahid, Y., 1979: Sand stabilization in al-Hasa. Riyadh: Forestry Department, Ministry of Agriculture and Water.

71 Abercrombie, T.J., 1966: Saudi Arabia: beyond the sands of Mecca. In: National Geographic 129: 1-53. -21, 25, 93-

72 A billion barrels ago... In: Aramco World 13.1962(5): 3-6. -H 223-

73 Abir, M., 1971: The 'Arab Rebellion' of Amīr Ghālib of Mecca (1788-1813). In: Middle Eastern Studies 7: 185-2oo. -1a, 21, 3o-

74 ——, 1977: Modernisation, reaction and Muhammad Ali's 'Empire'. In: Middle Eastern Studies 13: 295-313. -1a, 21, 3o-

75 Abir, Mordechai, 1974: Oil, power and politics: conflict in Arabia, the Red Sea and the Gulf. London: Cass. XIII, 221 p. -21, 24, 212-

76 Abo Ali, S.A., 1975: A study of educational goals for secondary education as determined by principals and teachers of selected districts in Saudi Arabia. Ed.D., University of Northern Colorado, Greeley. 172 p.

77 Abohassan, A.A., 1976: Sand stabilization by afforestation in Al Hassa Oasis, Saudi Arabia. Ph.D., Department of Forestry, Michigan State University, East Lansing. 14o p. -5-

78 Abo-Laban, M.A., 1978: A study of teachers perceptions of the principal's performance in selected high schools in Mecca and Riyadh, Saudi Arabia. Ph.D., University of Oklahoma, Norman. 139 p.

79 Abolkhair, Y.M.S., 1981: Sand encroachment by wind in al-Hasa of Saudi Arabia. Ph.D., Indiana University, Bloomington. 214 p.

80 Abou el-Nasr, K., 1962: Architecture in Asir. In: Middle East Forum 38(4): 29-32.   -188, 212-

81 Abraham, N.A., ^{2}1980: Doing business in Saudi Arabia. Boston: Tradeship Publishing (^{1}1979).

82 Abrahams, Anthony, 1977: Doing business in Saudi Arabia. In: Droit et Pratique du Commerce International 3: 579-602. -1a, 2o6, H 3-

83 Abrahams, A.M., 1978: Company formation and structure in Saudi Arabia. - Contract law in Saudi Arabia. In: R.M. Nelson (ed.): Corporate development in the Middle East. London: Oyez Publishing, 1978: 1-13 resp. 71-85.   -21-

84 Abu-Bakr, A.S., 1976: Cost-size relationship and efficiency of palm groves in the Al-Hasa Oasis, Saudi Arabia. Ph.D., Department of Geography, College of Arts and Sciences, University of Northern Colorado, Greeley. X, 99 p.   -2o6-

85 Abu Hakima, A.M., 1965: History of Eastern Arabia, 175o-18oo: the rise and development of Bahrain and Kuwait. Beirut: Khayats. XIX, 213 p.   -21, 24/213-

86 ——, 1969: Wahhabi religio-political movement of Arabia and its impact on India in the nineteenth century. In: M. Ahmad (ed.): India and the Arab World. Proceedings of the Seminar on India and the Arab World. New Delhi: Indian Council for Cultural Relations, 1969: 17-23.   -21-

87 Abu-Ihya, S., 1982: Religionsunterricht an den öffentlichen Knabenschulen des Königreichs Saudi-Arabien. Schriften zur Islamkunde 1; Frankfurt/M.: R.G. Fischer (zugleich: Diss., Philosophische Fakultät, Universität Mainz, 1981). 219 S. -21-

88 Abu-Laban, B. & S. McIrwin Abu-Laban, 1976: Education and development in the Arab world. In: The Journal of Developing Areas 1o: 285-3o4.   -16, 21, 24-

89 Abul-Ela, M.T., 1959: A geographical study of man and his environment in al-Ahsa province (Saudi Arabia). Ph.D., Department of Geography, Trinity College, Dublin.

90   Abul-Ela, M.T., 1965: Some geographical aspects of Al Riyadh
     (Saudi Arabia). In: Bulletin de la Société de Géographie
     d'Égypte 38: 31-72.   -1a, 3o, 18o-

91   Abul-Ezz, S., 1962: Report on the projected agricultural bank
     of the Kingdom of Saudi Arabia. Riyadh: The Supreme Plan-
     ning Board. 37 p.   -H 223-

92   Abul-Haggag, Y., 1963: Remarks on the artesian water of Nejd,
     Saudi Arabia. In: Annals of the Faculty of Arts, Ain Shams
     University 8: 1o3-111 (reprinted in: Bulletin de la Société
     de Géographie d'Égypte 37.1964: 57-65).   -21-

93   Abu-Manneh, B., 1973: Sultan Abdülhamid II and the Sharifs of
     Mecca (188o-19oo). In: Asian and African Studies 9: 1-21.
     -21, 3o, 2o6-

94   Abunabaa, A.M., 1981: An analysis of marketing in Saudi Arabia
     and American marketing executives' knowledge about the Sau-
     di Arabian market. Ph.D., North Texas State University,
     Denton. 252 p.

95   Abu Ras, A.S., 1979: Factors affecting teachers' utilization
     of elements of educational technology in Saudi Arabia.
     Ph.D., Indiana University, Bloomington. 178 p.

96   Accord commercial et économique entre le Gouvernement de la
     République Arabe Syrienne et le Gouvernement du Royaume de
     l'Arabie Séoudite. In: Syrie et Monde Arabe 19.1972(219):
     1o7-122.   -2o6, H 3-

97   Accordo 18 novembre 1936 fra l'Egitto e l'Arabia Sa'ūdiana
     per il 'maḥmal', i 'waqf al-haramain' e la cittadinanza.
     In: Oriente Moderno 16.1936: 666-668.   -1a, 5, 21-

98   Accordo fra Gran Bretagna e Arabia Saudiana per la 'sospensi-
     va' circa la questione di el-Bureimi (25 ottobre 1952).
     In: Oriente Moderno 33.1953: 23o-233.   -5, 18, 21-

99   Accordo generale fra il Regno Arabo Sa'ūdiana e il Regno Ye-
     menita per la soluzione delle questioni che sorgono fra i
     sudditi dei due Regni. In: Oriente Moderno 18.1938: 4-6.
     -1a, 5, 21-

1oo Achtnich, W. & B. Homeyer, 198o: Protective measures against
     desertification in oasis farming, as demonstrated by the exam-
     ple of the oasis Al Hassa, Saudi Arabia. In: W. Meckelein (ed.):
     Desertification in extremely arid environments. Stuttgarter
     Geographische Studien 95; Stuttgart: Geographisches Institut
     der Universität Stuttgart, 198o: 93-1o5.   -21, 24, 93-

1o1 A dam in Saudi Arabia. In: Aramco World Magazine 25.1974(2): 4-9.   -24-

1o2 A day in the life of a Saudi Arab doctor. In: Aramco World 14.1963(4): 2-7.   -H 223-

1o3 Adelman, M.B. & M.W. Lustig, 1981: Intercultural communication problems as perceived by Saudi Arabian and American managers. In: International Journal of Intercultural Relations 5: 349-362.   -1a, 12, 212-

1o4 Adler, G., A. Alkazaz & S. Scholtyssek, 1977: Poultry production in the Kingdom of Saudi Arabia: a study of the present situation and future prospects. Submitted to Saudi Arabian Agricultural Bank by German Agency for Technical Cooperation, Ltd. Eschborn. 22o p.   -Esb 2ooo-

1o5 Ageel, H.A., 1982: Job satisfaction of staff members of Umm Al-Qura University in Makkah, Saudi Arabia. Ph.D., Michigan State University, East Lansing. 162 p.

1o6 Agreement for the determination of boundaries between the Hashimite Kingdom of Jordan and the Kingdom of Saudi Arabia. In: The Middle East Journal 22.1968: 346-348.   -12, 21, 2o6-

1o7 Agricultural sector shows significant improvement: Saudi Arabia. In: The Arab Economist 12.198o(13o): 21-23.   -2o6, H 3-

1o8 Ahmad, A., 1967: Zum Finanz- und Bankwesen in Saudisch-Arabien. In: Orient (Opladen) 8: 5-7.   -21, 1oo, H 223-

1o9 Ahmad, Y.J., 1974: Oil revenues in the Gulf: a preliminary estimate of absorptive capacity. Paris: Development Centre of the Organisation for Economic Co-operation and Development. 156 p.   -H 3, H 223, Bm 25-

11o Ahmed, S., 1976: Saudi Arabia. The Chase World Information Series on Developing Business in the Middle East and North Africa; New York: Chase World Information Corporation. 375 p.   -21-

111 Aitchison, C.U. (ed.), 1933: A collection of treaties, engagements and sanads relating to India and neighbouring countries. Vol. 11: Containing the treaties, &c., relating to Aden and the southwestern coast of Arabia, the Arab principalities in the Persian Gulf, Muscat (Oman), Balu-

chistan and the North-West Frontier Province. Rev. ed. New Delhi: Manager of Publications (reprint: Nendeln: Kraus-Thomson, 1973). XXXI, 633, LXXXVI p.   -21-

112 Aiz, M. & Y. Abdulwahid, 1977: The final report on the second agricultural defense line in al-Hasa Sand Stabilization Project. Riyadh: Forestry Department, Ministry of Agriculture and Water.

113 Ajami, F., 1977/78: Stress in the Arab triangle. In: Foreign Policy (29): 9o-1o8.   -21, 24/213, 18o-

114 A jewel of a university on the Arabian sands. In: Fortune 92.1975(5): 126-133.   -24, 3o, 89-

115 Ajroush, H.A. al-, 1981: A historical development of the public secondary school curriculum in Saudi Arabia from 193o to the present. Ph.D., University of Oklahoma, Norman. 222 p.

116 Akhdar, F.M.H., 1974: Multinational firms and developing countries: a case study of the impact of the Arabian American Oil Company 'ARAMCO' on the development of the Saudi Arabian economy. Ph.D., University of California, Riverside. XIV, 357 p.   -2o6-

117 Akkad, A.A.-H. al-, 1974: Socio-economic impact of development schemes on a rural community, al-Umran al-Shamaliyah, al-Hasa oasis, Saudi Arabia. M.Sc., School of Agriculture, University College of North Wales, Bangor.

118 Akkad, H.A.-H. al-, 1967: The nomads problem and the implementation of a nomadic settlement scheme in Saudi Arabia. In: M.R. el-Ghonemy (ed.): Land policy in the Near East. Proceedings of the Development Center on Land Policy and Settlement for the Near East, held in Tripoli, Libya, from 16 to 28 Oct. 1965, organized by the Food and Agriculture Organization of the United Nations. Rome: FAO, 1967: 296-3o5.   -Gö 153-

119 Alabbadi, A.H., 1981: Nomadic settlements in Saudi Arabia: a socio-historical and analytical case study. Ph.D., Michigan State University, East Lansing. 269 p.

12o Alageel, K.M.N., 198o: The flexibility of Saudi Arabian oil production through 1983. Ph.D., Pennsylvania State University, University Park. 385 p.

121 Alaki, M.A., 1972: Industrial-vocational education in Saudi Arabia: 'problems and prospects'. Ph.D., University of Arizona, Tucson. XVIII, 331 p.

122 Alam, M.S., 1982: Basic macro-economics of oil economies. In: The Journal of Development Studies 18: 2o5-216.   -3o, 18o, 2o6-

123 Alami, J., 1977: Education in the Hijaz under Turkish and Sharifian rule. In: Arabic and Islamic garland. Historical, educational and literary papers presented to Abdul-Latif Tibawi by colleagues, friends and students. London: The Islamic Cultural Centre, 1977: 48-53.   -21-

124 Alan, R., 1955: Saudi Arabia: oil, sand, and royalties. In: The Reporter (New York) 13(Dec. 1, 1955): 18-2o.   -18o-

125 Albaharna, H.M., 1968: A note on the Kuwait-Saudi Arabia Neutral Zone agreement of July 7, 1965, relating to the partition of the zone. In: The International and Comparative Law Quarterly IV, 17: 73o-735.   -21, 24, 18o-

126 ——, 1975: The Arabian Gulf states: their legal and political status and their international problems. 2nd, rev.ed. Beirut: Librairie du Liban (1st ed.: The legal status of the Arabian Gulf states: a study of their treaty relations and their international problems. Manchester: Manchester University Press; Dobbs Ferry: Oceana Publications, 1968). LXII, 428 p.   -21-

127 Albers, H.H., 1981: Beduinen und Manager: Saudi-Arabiens Weg in die Zukunft. Düsseldorf: Erb. 214 S.   -12, 21, 212-

128 Albraikan, S.M., 198o: OPEC foreign investment: the case of Saudi Arabia. Ph.D., University of Colorado, Boulder. 197

129 Algawad, M.A., 1977: Co-operative principles and practices in Saudi Arabia. In: Review of International Co-operation 7o: 1o7-112.   -2o6, H 3-

13o Algemene Bank Nederland, 1976: Country report for Saudi Arabia. Amsterdam. 4o p.

131 Alghamdi, A.A.S.B., 1982: Selected factors associated with intermediate and high school dropouts in rural southwestern Saudi Arabia. Ph.D., George Peabody College for Teachers at Vanderbilt University, Nashville. 284 p.

132 Alghofaily, I.F., 1980: Saudi youth attitudes towards work and vocational education: a constraint on economic development. Ph.D., Florida State University, Tallahassee. 163 p.

133 Alhumaid, A.I., 1981: An empirical study of the characteristics of the governmental budgetary process in rich and uncertain environments: the case of Saudi Arabia. Ph.D., Louisiana State University and Agricultural and Mechanical College, Baton Rouge. 182 p.

134 Ali, A.M., 1968: The systems of financial administration in Saudi Arabia and New York State: a study of contrasts, with special reference to pre audit. D.P.A., State University of New York, Albany. 359 p.

135 Ali, M.I.A. & A.W. Abu Sulaiman, 1969: Recent judicial developments in Saudi Arabia. In: Journal of Islamic and Comparative Law 3: 11-20.   -1a-

136 ——, 1970: Social responsibilities of the individual and the State in Saudi Arabian law. Ph.D., Department of Law, School of Oriental and African Studies, University of London.

137 Ali, S.R., 1976: Saudi Arabia and oil diplomacy. Praeger Special Studies in International Politics and Government; New York, Washington, London: Praeger (zugleich: The use of oil as a weapon of diplomacy: a case study of Saudi Arabia. Ph.D., School of International Service, American University, Washington, 1975). XVIII, 197 p.   -212, Tü 17-

138 Aliboni, R., 1980: Saudi economic development: the case for regional integration. In: Istituto Affari Internazionali (ed.): Red Sea conflicts and cooperation. Regional balance and strategic implications. Rome. 33 p.   -H 223-

139 Alireza, M., 1971: At the drop of a veil. Boston: Houghton Mifflin (London: Hale, 1972). 240 p.   -21-

140 Alkazaz, A., 1977: Arabische Entwicklungshilfe-Institutionen. Organisationsform und Leistungen. Hamburg: Deutsches Orient-Institut. 172 S.   -212, H 223-

141 —— & T. Koszinowski, 1978: Das System der technischen und beruflichen Ausbildung in den arabischen Golfstaaten und Saudi-Arabien. Bisherige Entwicklung, Aufbauplanung und

Grundprobleme. In: Orient (Opladen) 19(4): 92-122. -21, 1oo, H 223-

142 Alkazaz, A., 1981: Sultan Ben Abdel Aziz Al Saud (Sulṭân ibn 'Abd al-'Azîz Âl Sa'ûd). In: Orient (Opladen) 22: 5-8. -1a, 4, 21-

143 Alkhowaiter, H., 1967: Saudi Arabia: educational developments in 1966-1967. In: International Yearbook of Education 29: 358-362. -1a, 12, 212-

144 ——, 1968: Saudi Arabia: educational developments in 1967-1968. In: International Yearbook of Education 3o: 416-422. -1a, 12, 18-

145 Allan, M., 1972: Palgrave of Arabia. The life of William Gifford Palgrave, 1826-88. London, Basingstoke: Macmillan. 318 p. -21-

146 Allen, M.J.S. & G.R. Smith, 1975: Some notes on hunting techniques and practices in the Arabian Peninsula. In: Arabian Studies 2: 1o8-147. -12, 21, 212-

147 Allen, R., 1981: Getting to grips with joint ventures. In: J. Whelan (ed.): Japan and the Middle East. A MEED Special Report; London: Middle East Economic Digest, 1981: 46, 48. -H 223-

148 Allen, T.E., 1951: Arabia and Aramco. In: Harvard School of Public Health: Industry and Tropical Health 1951: 25-29.

149 Almana, M., 198o: Arabia unified: a portrait of Ibn Saud. London: Hutchinson Benham. 328 p. -21-

15o Alnassar, S.N., 1981: Professional job knowledge and skills needed by extension personnel in the central region of Saudi Arabia. Ph.D., Ohio State University, Columbus. 214 p.

151 Alnimir, S.M., 1981: Present and future bureaucrats in Saudi Arabia: a survey research. Ph.D., Florida State University, Tallahassee. 197 p.

152 Alohaly, M.N., 1977: The spatial impact of government funding in Saudi Arabia: a study in rapid economic growth, with special reference to the Myrdal development model. Ph.D., University of Oklahoma, Norman. VIII, 285 p. -H 223-

153 Alraegi, A.H., 1981: A study of the predictive validity of twelfth grade transcript data on freshman college GPA for science majors, colleges of education, Saudi Arabia. Ed.D., University of Northern Colorado, Greeley. 63 p.

154 Alrashid, S.A., 1980: Darb Zubaydah: the pilgrim road from Kufa to Mecca. Riyadh.

155 Alstyne, R.W. van, 1947: Arabian-American entente. In: Current History 13(73): 135-139.   -61, 180, 352-

156 Alter, H.W., 1965a: Al-Dammām. In: EI^2 2: 108-109.   -21, 24, 212-

157 ——, 1965b: Al-Djubayl. In: EI^2 2: 568-569.   -21, 24, 212-

158 Altorki, S., 1973: Religion and social organization of elite families in urban Saudi Arabia. Ph.D., Department of Anthropology, University of California, Berkeley.

159 ——, 1977: Family organization and women's power in urban Saudi Arabian society. In: Journal of Anthropological Research 33: 277-287.   -1a, 12, 31-

160 ——, 1980: Milk-kinship in Arab society: an unexplored problem in the ethnography of marriage. In: Ethnology 19: 233-244.   -7, 12, 30-

161 Alyahya, K.A.M., 1981: Constructing a comprehensive orientation program for Saudi Arabian students in the United States. Ph.D., University of Pittsburgh. 284 p.

162 Alyami, A.H., 1977a: The impact of modernization on the stability of the Saudi monarchy. Ph.D., Graduate Faculty of Government, Claremont Graduate School. XII, 206 p.

163 ——, 1977b: The coming instability in Saudi Arabia. In: New Outlook (Tel Aviv) 20(6): 19-26.   -21, 30-

164 Alzamel, I.A., 1974: An analysis of the role and scope of the adult basic education program in Saudi Arabia. Ed.D., University of Arkansas, Fayetteville. 126 p.

165 Alzamil, A., 1978: Future plans for iron and steel production in Saudi Arabia. In: International Iron and Steel Institute: Report of proceedings 1977. Eleventh annual conference. Rome, Italy, October 10-12, 1977. Brussels, 1978: 114-116.   -H 3-

166 A matter of foresight. In: Aramco World 13.1962(4): 3-6.   -H 223-

167 Ambah, S., 1969: The role of the College of Petroleum and Minerals in the industrialization of Saudi Arabia. In:

C. Nader & A.B. Zahlan, with the assistance of S. Antonius (eds.): Science and technology in developing countries. Proceedings of a conference held at the American University of Beirut, Lebanon, 27 November - 2 December 1967. Cambridge/England: Cambridge University Press, 1969: 249-271.   -21, Gö 153-

168 Amelunxen, C., 1973: Rechtsleben in Saudi-Arabien. Juristischer Reisebericht. In: Deutsche Richterzeitung 51: 155-159.   -21, 24-

169 Amer, M., 1921: The oases of central Arabia, with special reference to trade and pilgrim routes. M.A., Department of Geography, University of Liverpool. 119 p.

17o ——, 1932: An Egyptian explorer in Arabia in the 19th century. In: Bulletin de la Société Royale de Géographie d'Égypte 18: 29-45.   -18o, 385-

171 American Friends of the Middle East, 1966: Basic facts on education in the Middle East - North Africa. Washington. 94 p.

172 Americans in Arabia. In: Standard (of California) Oil Bulletin 1946(Autumn): 8-9.

173 Americans in Arabia. In: The Texaco Star 35.1948(1): 18-19.

174 Amin, G.A., 1972: Arab economic growth and imbalances, 1945-197o. In: L'Égypte Contemporaine 63(35o): 5-43.   -21, 2o6-

175 ——, 1974: The modernization of poverty: a study in the political economy of growth in nine Arab countries, 1945-197o. Social, Economic and Political Studies of the Middle East / Études Sociales, Économiques et Politiques du Moyen-Orient 13; Leiden: Brill. XV, 124 p.   -21, 212-

176 Amin, H., 1973: Notes, reports and comments on nomadic settlements in some Arab countries; II. In: Geographical Review of Afghanistan 12(2): 18-3o.   -Tü 17-

177 Amin, M., 1976: The Hajj - the most sacred journey. In: The Middle East (London) (18): 18-21.   -12, 3o-

178 ——, 1978: Pilgrimage to Mecca. London: Macdonald & Jane's. 256 p.   -21-

179 Amr, S.M. al-, 1978: The Hijaz under Ottoman rule 1869-1914: Ottoman Vali, the Sharif of Mecca, and the growth of British influence. Riyadh: Riyad University Press (zugleich: Ph.D., University of Leeds, 1974). 255 p.

180 Amry, M.-A.Y., 1976: Program budgeting model for Saudi Arabian elementary education: an emphasis on program costs for decisions. Ph.D., University of Arizona, Tucson. 3o7 p.

181 Anani, F.M., 1972: Desalination: its potentials and limitations in the economic development process of Saudi Arabia. Ph.D., Syracuse University. VIII, 327 p.   -2o6, H 223-

182 Anastos, D., A. Bêdos & B.W. Seaman, 198o: The development of modern management practices in Saudi Arabia. In: Columbia Journal of World Business 15(2): 81-92.   -2o6, H 3-

183 Andere Länder - andere Sitten: Saudi-Arabien. In: Export-Dienst 1955(7): 48-5o.   -212-

184 Anderer, K., R. Dörflinger, J. Iwanowitsch, H. Nölker & E. Schwarz, 198o: Neukonzeptionierung der Ausbildung von Lehr- und Führungskräften im Higher Technical Institute (HTI) in Er-Riyadh, Saudi-Arabien. Gutachten im Auftrag der GTZ. Eschborn. 378 S.   -Esb 2ooo-

185 Anderson, I.H., 1979: Lend-lease for Saudi Arabia: a comment on alternative conceptualizations. In: Diplomatic History 3: 413-423.   -7, 12, 3o-

186 ——, 1981: Aramco, the United States, and Saudi Arabia: a study of the dynamics of foreign oil policy, 1933-195o. Princeton: Princeton University Press. XIII, 259 p.   -21-

187 Anderson, J.N.D., 1959: Islamic law in the modern world. New York: New York University Press. XX, 1o6 p.   -21-

188 André, M., 19o7: La question de l'eau en Arabie. In: Revue du Monde Musulman 3: 294-298.   -21, 25-

189 Ankary, K.M. al-, 1977: Geographical evolution of the urban structure of Mecca, Saudi Arabia. M.A., Department of Geography, University of Oregon, Eugene. XI, 179 p.   -Tü 17-

*** Annesley, G.: s. Valentia, G.

19o Ansari, 'A.-Q. al-, 1964: Medina, second city of Islam. In: Aramco World 15(4): 3o-33.   -H 223-

191 Ansary, A.O.T. el-, 1974: The demand and supply for managerial skills by Saudi Arabia under agreements of ownership participation in oil companies concessioned in its territory. Ph.D., American University, Washington. 234 p.
192 Anthony, J.D., 1979: Foreign policy: the view from Riyadh. In: The Wilson Quarterly 3(1): 73-82.   -7, 12, 3o-
193 ———, 198o: The US-Saudi Arabian Joint Commission on Economic Co-operation. In: W.A. Beling (ed.): King Faisal and the modernisation of Saudi Arabia. London: Croom Helm; Boulder: Westview Press, 198o: 1o2-1o9.   -21, H 223-
194 ———, 1982: Aspects of Saudi Arabia's relations with other Gulf states. In: T. Niblock (ed.): State, society and economy in Saudi Arabia. London: Croom Helm; Exeter: Centre for Arab Gulf Studies, 1982: 148-17o.   -21-
195 Anti-strike legislation in Saudi Arabia. In: The Muslim World 46.1956: 366.   -21, 291, 352-
196 Antonius, G., 1946: The Arab awakening. The story of the Arab national movement. New York: Putnam ([1]New York: Lippincott, 1939). 471 p.   -21-
197 A partnership in oil and progress. In: Standard (of California) Oil Bulletin 1946(Autumn): 3.
198 Apgar, M., 1977: Succeeding in Saudi Arabia. In: Harvard Business Review 55: 14-33, 166-168.   -2o6-
199 Appelman, H., 197o: A fattening trial with yearling rams of the Saudi Arabian Najdi and Arabi breeds. In: Netherlands Journal of Agricultural Science 18: 84-88.   -1a, 89, 1oo-
2oo Arab fleet growth. In: Lloyd's Shipping Economist 4.1982(6): 12-13.   -H 3-
2o1 Arab Information Center, 1966: Education in the Arab states. Information Paper 25 (I-XIII); New York. 3o5 p.   -H 223-
2o2 'Arabian middle class growing rapidly. In: The Oil Forum 3.1949: 2o7.   -9o-
2o3 Arabien. Frankfurt/M.: Umschau, 1977 (Originalausgabe: An Arabian portfolio. Zug: First Azimuth, 1976). 223 S. -21, 212-
*** Arabie Séoudite: s. auch Royaume d'Arabie Séoudite.
2o4 Arabie Séoudite. Agriculture and water resources. In: Syrie et Monde Arabe 28.198o(32o): 24-31.   -2o6, H 3, H 223-

2o5 Arabie Séoudite à l'heure de la guerre. In: L'Économie des Pays Arabes 16.1973(189): 6o-62.   -2o6-

2o6 Arabie Séoudite. Conjoncture économique et financière en 1967 et perspectives 1968. In: Étude Mensuelle sur l'Économie et les Finances des Pays Arabes 11.1968(5/131): 65-97.   -2o6-

2o7 Arabie Séoudite. Développement des secteurs économique et social. In: Le Commerce du Levant 1973(15o): 45-46.
-H 3, H 223-

2o8 Arabie Séoudite. Enquête sur les établissements industriels dans la Province Occidentale. In: L'Économie des Pays Arabes 14.1971(163): 22-31.   -2o6-

2o9 Arabie Séoudite. Foreign trade and balance of payments. In: Syrie et Monde Arabe 27.198o(319): 7-16.   -2o6, H 3, H 223-

21o Arabie Séoudite. Horizons 1985 et budget-record pour 198o. In: L'Économiste Arabe 23.198o(259): 19-22.   -2o6, H 223-

211 Arabie Séoudite. Importations accrues de 35,5 %, 21,4 milliards au lieu de 16,7 %. In: Syrie et Monde Arabe 29.1982 (346): 37-39.   -2o6, H 3, H 223-

212 Arabie Séoudite. Inauguration d'une seconde étape de développement économique. In: Étude Mensuelle sur l'Économie et les Finances des Pays Arabes 12.1969(17/143): 7-12. -H 223-

213 Arabie Séoudite. L'activité pétrolière en 1967. In: Étude Mensuelle sur l'Économie et les Finances des Pays Arabes 11.1968(3/129): 7-12.   -2o6-

214 Arabie Séoudite. L'arbitrage en matière commerciale. In: L'Économiste Arabe 25.1982(281): 18-2o.   -2o6, H 223-

215 Arabie Séoudite. La situation de l'énergie. In: Syrie et Monde Arabe 29.1982(346): 5-8.   -2o6, H 3, H 223-

216 Arabie Séoudite. Le budget 1982-1983. In: Syrie et Monde Arabe 29.1982(341): 16-23.   -2o6, H 3, H 223-

217 Arabie Séoudite. L'économie séoudienne en plein boom. In: Syrie et Monde Arabe 29.1982(343): 59-6o.   -2o6, H 3-

218 Arabie Séoudite. Le début d'application d'une 'réforme agraire'. In: Étude Mensuelle sur l'Économie et les Finances des Pays Arabes 12.1969(7/133): 25-33.   -2o6-

219 Arabie Séoudite. Le pétrole et l'économie séoudienne. In: Syrie et Monde Arabe 29.1982(337): 2o-3o. -2o6, H 3, H 223-

22o Arabie Séoudite. Les chances du nouveau plan quinquennal. In: L'Économiste Arabe 23.198o(258): 25. -2o6, H 223-

221 Arabie Séoudite. Les chances du Second Plan 1975-8o. In: L'Économie des Pays Arabes 18.1975(211): 16-17. -2o6, H 223-

222 Arabie Séoudite. L'industrie séoudienne en 198o. In: Syrie et Monde Arabe 29.1982(335): 9-13. -2o6, H 3, H 223-

223 Arabie Séoudite. Money and banking. In: Syrie et Monde Arabe 27.198o(318): 9-25. -2o6, H 3-

224 Arabie Séoudite. Nouveau projet pour le chemin de fer du Hedjaz. In: L'Économiste Arabe 24.1981(267): 22-24. -2o6, H 223-

225 Arabie Séoudite. Nouvelles voies pour l'industrie. In: L'Économie des Pays Arabes 17.1974(198): 2o-23. -2o6-

226 Arabie Séoudite. Oil concession areas at end 1974. In: Syrie et Monde Arabe 23.1977(275): Annexe 5. -H 3-

227 Arabie Séoudite. PETROMIN: Bilan de cinq ans et perspectives d'avenir. In: Étude Mensuelle sur l'Économie et les Finances des Pays Arabes 11.1968(1/127): 5-1o. -2o6-

228 Arabie Séoudite. Priorité absolue à l'agriculture. In: L'Économiste Arabe 25.1982(28o): 27-28. -2o6, H 223-

229 Arabie Séoudite. Rapport économique. In: L'Économie des Pays Arabes 18.1975(212): 3o-32. -2o6, H 223-

23o Arabie Séoudite. Réglementation des activités des compagnies commerciales dans le domaine des hydrocarbures. In: Étude Mensuelle sur l'Économie et les Finances des Pays Arabes 12.1969(138): 111-113. -H 223-

231 Arabie Séoudite. Regulations for the encouragement of national industries. In: Étude Mensuelle sur l'Économie et les Finances des Pays Arabes 11.1968(5/131): 128-132. -2o6-

232 Arabie Séoudite. Ressources et perspectives. In: Industries et Travaux d'Outremer 21.1973: 518-519. -2o6-

233 Arabie Séoudite. Situation économique. In: Bulletin de la Société Générale de Banque 14.1975(148): 17-19. -H 3-

234 Arabie Séoudite. Situation et perspectives. In: Industries et Travaux d'Outremer 25.1977: 524-535. -1a, 89, 2o6-

235 Arabie Séoudite. Situation et perspectives de l'économie. In: Bulletin de la Société Générale de Banque 16.1977 (163): 9-12. -H 3-

236 Arabie Séoudite. Steady growth in economy. In: Syrie et Monde Arabe 27.198o(321): 15-23. -2o6, H 3, H 223-

237 Arabie Séoudite. Un budget établi en fonction des besoins du développement. In: Le Commerce du Levant 1973(158): 42-44. -H 3-

238 Arabie Séoudite. Un budget record consacrant la reprise. In: Étude Mensuelle sur l'Économie et les Finances des Pays Arabes 11.1968(5/131): 22-27. -2o6-

239 Arabie Séoudite. Vers la répartition équitable des richesses. In: Le Commerce du Levant 1973(153): 47-5o. -H 3-

24o ARAMCO, 1948a: Arabian oil and its relation to world oil needs. N.p. 48 p.

241 ——, 1948b: Summary of Middle East oil developments. 2nd ed. ([1]1947?). N.p. (Dhahran?). II, 3o p. -2o6-

242 ——, 1948c: 15 years; a story of achievement, 1933-1948. New York. 12 p.

243 ——, 1951: Arabian oil and its relation to world shortages. New York ([1]1949). 47 p.

244 ——, 1952-1961: Report of operations to the Saudi Arab Government by the Arabian American Oil Company. 1951-196o; Dammam: Al-Mutawa Press. -H 223-

245 ——, Industrial Relations Department, Planning Division, 1954: Census of Aramco's Saudi Arab employees 1954. 2 vols. N.p. (Dhahran).

246 ——, Arabian Research Division, 1955a: Al Murrah. Dhahran. 24 p.

247 ——, 1955b: Memorial of Arabian American Oil Company; arbitration between the Government of Saudi Arabia and Arabian American Oil Company. Cairo. VII, 614, XLVIII p.

248 ——, 1956a: Aramco camera; a photographic survey of life and work with Aramco in Saudi Arabia. Dhahran. 44 p.

249 ARAMCO, Local Government Relations Department, Arabian Research Division, 1956b: Municipalities in Saudi Arabia. Preliminary report. N.p. (Dhahran).

250 ——, 1957: Policies and programs for the training of Saudi employees: a general guide. Dhahran.

251 ——, n.d.: Natural gas for sale in Saudi Arabia. New York. 15 p.

252 —— (ed.), 1960a: Natural gas in Saudi Arabia: economics, sales, utilization. Dhahran.   -H 223-

253 ——, Arab Development Department / The Associate Consulting Engineers, 1960b: A study of the most profitable uses of the date crop of the Eastern Province of Saudi Arabia. Report S 5810; Dhahran.

254 ——, 1962-: Aramco 1961-; a review of operations by the Arabian American Oil Company. Dammam: Al-Mutawa Press. -H 223-

255 ——, 1963: Saudi Arabian Government; ministries, principal directorates and other important agencies; April 1, 1963. N.p. (Dhahran?). Var.pag.

256 ——, Local Government Relations Department, Arabian Affairs Division, [6]1964: Directory of the Saudi Royal Family and officials of the Saudi Arabian Government. ([1]1953). Dhahran. Var.pag.   -21-

257 ——, 1965: Aramco's role in the development of the Eastern Province. Dammam. 12 p.   -Documentation Section, Centre for Middle Eastern and Islamic Studies, University of Durham-

258 ——, Arab Industrial Development Department, 1966a: Report of activities, 1965. Dhahran.

259 ——, General Office, Employee Relations Department, 1966b: Industrial relations survey of Saudi Arab employees 1965. N.p. (Dhahran).

260 ——, 1968: Aramco handbook: oil and the Middle East. Rev.ed. ([1]1960). Dhahran. 279 p.

261 ——, Medical Department, 1970: Morbidity and mortality, 1970. Dhahran. 22 p.

262 ——, n.d. (1973?): Transport of oil. Dammam: Al-Mutawa Press. 20 p.

263 ARAMCO, 1974: Ras Tanura Refinery. Dammam: Al-Mutawa Press. 14 p.

264 ——, n.d. (1974): Facts and figures 1973. Dammam: Al-Mutawa Press. 22 p.

265 Aramco in Eastern Region development. In: Emergent Nations 2.1966(2): 64-65. -H 223-

266 Aramco '72. In: Aramco World Magazine 24.1973(3): 8-11. -24-

267 Aramco TV on the air. In: Aramco World 14.1963(5): 2-7. -H 223-

268 Arbose, J., 1977a: Running Saudi Arabia's biggest trading group. In: International Management 32(5): 1o-13. -3o, 18o, H 3-

269 ——, 1977b: How Jiddah port cleared up its congestion. In: International Management 32(6): 26-28. -3o, 18o, H 3-

27o ——, 1977c: Petromin's sink or swim style. In: International Management 32(8): 54-56. -3o, 18o, H 3-

271 Arendonk, C. van, 1934: S̄harīf. In: EI[1] 4: 349-354. -21, 24-

272 —— & K.N. Chaudhuri, 1978: Ḳahwa. In: EI[2] 4: 449-455. -21, 24, 212-

273 A review of Saudi Arabia's third development plan. In: International Currency Review 13.1981(1): 43-5o. -38, 18o, H 3-

274 Arfaj, N.A., 198o: Saudi Arabia's maritime policy, 1948-1978: a study in the law of the sea as applied by developing countries. Ph.D., Claremont Graduate School. 217 p.

275 Arkadakshi, A.F., 1978: Oil doesn't flow by itself. In: World Trade Union Movement (6): 16-18.

276 Arle, M. d' (Pseud., d.i. Bochskandl, M.), 1958: Ich war in Mekka. Unter Fremdenlegionären, Haschisch-Schmugglern und Pilgern. Wien, Stuttgart, Zürich: Europa (deutsche Ausgabe: Ich war in Mekka. Eine Frau entdeckt die geheimnisvolle Welt des Orients. Berlin: Universitäts-Verlag, 1958). 3o4 S. -24-

277 ——, 1964: El Harem. Roman aus Saudi-Arabien. Berlin: Universitas. 2oo S.

278 Armstrong, H.C., 1936: Ibn Saud, König im Morgenland. Leipzig: List (Originalausgabe: Lord of Arabia, Ibn Saud. An intimate study of a king. London: Barker, 1934). 339 S. -21, 24/213, Tü 17-

279 Arndt, R., 1977: Partners in growth: Saudi Arabia. In: Aramco World Magazine 28(1): 12-25. -21, 24-

280 Arnold, J., 1964: Golden swords and pots and pans. London: Gollancz. 239 p.

281 Arnon, Y. & P. Weissman, 1971: Saudi Arabia. In: Middle East Record 3: 449-460. -12, 21, 24-

282 ——, H. Shaked & M. Efrat, 1973: Saudi Arabia. In: Middle East Record 4: 683-690. -12, 21, 24-

283 Asaad, M.M.A., 1981: Saudi Arabia's national security: a perspective derived from political, economic, and defense policies. Ph.D., Claremont Graduate School. 494 p.

284 As'ad, I. A., 1982: Government budgeting and its control in Saudi Arabia. Ph.D., Claremont Graduate School. 150 p.

285 Asad, Mohammed A., 1981: The possibility of change in Bedouin society: a study of current developments in Saudi Arabia. Ph.D., Claremont Graduate School. VIII, 142 p.

286 Asad, Muhammad (früherer Name: Weiss, L.), 1955: Der Weg nach Mekka. Berlin, Frankfurt/M.: S. Fischer (Originalausgabe: The road to Mecca. New York: Simon & Schuster, 1954. Französische Ausgabe: Le chemin de La Mecque. La Bibliothèque des Voyageurs; Paris: Fayard, 1979). 436 S. -24/213-

287 Asadallah, M.M., 1971: Land and maritime boundaries of north and north-eastern Saudi Arabia. Ph.D., University of Newcastle upon Tyne.

288 Asadullah, M., 1931: Medina, die Stadt des Propheten. In: Atlantis 3: 385-389. -21, 24, 180-

289 Asfour, E.Y., with the collaboration of G.S. Medawar, H.M. al-Kaylani & L. Takieddine, 1965: Saudi Arabia. Long-term projections of supply of and demand for agricultural products. Beirut: Economic Research Institute, American University of Beirut. 180 p. -Gö 153-

290　Asfour, E.Y., 1971: Saudi Arabia. In: M. Adams (ed.): The Middle East: a handbook. Handbooks to the Modern World; New York, Washington: Praeger, 1971: 271-281.　-93, 1oo-

291　——, 1972: Prospects and problems of economic development of Saudi Arabia, Kuwait, and the Gulf Principalities. In: C.A. Cooper & S.S. Alexander (eds.): Economic development and population growth in the Middle East. The Middle East Economic and Political Problems and Prospects; New York: Elsevier, 1972: 367-398.　-21, Gö 153-

292　Ashiry, H. el-, 198o: The rehabilitation of ad-Dar'iyya. In: M. Meinecke (ed.): Islamic Cairo: architectural conservation and urban development of the historic centre. Proceedings of a seminar organised by the Goethe-Institute, Cairo (October 1-5, 1978). German Institute of Archaeology, Art and Archaeology Research Papers 18; London, 198o: 81-83.　-21-

293　Ashkenazi, T., 1946/49: La tribu arabe: ses éléments. In: Anthropos (Freiburg/Schweiz) 41-44: 657-672.　-16, 21, 291-

294　——, 1948: The 'Anazah tribes. In: Southwestern Journal of Anthropology 4: 222-239.　-1a, 3o, 385-

295　——, 1965: Social and historical problems of the 'Anazeh tribes. In: Journal of the Economic and Social History of the Orient 8: 93-1oo.　-21-

296　Ashoor, M.-S.J., 1978: A survey of user's attitudes toward the resources and services of three university libraries in Saudi Arabia. Ph.D., University of Pittsburgh. X, 233 p.　-H 223-

297　'Asīr. In: EI[1] 1.1913: 5o5-5o6.　-21, 24-

298　Assa, A., 1969: Miracle dans les sables, Arabie saoudite. Paris: Maisonneuve (englische Ausgabe: Assah, A.: Miracle of the Desert Kingdom. London: Johnson, 1969). 291 p.　-21, 24/213, Tü 17-

299　Assaf, I.A., 1982: The economic impact of guestworkers in Saudi Arabia. Ph.D., Colorado State University, Fort Collins. 151 p.

300 Assaf, S.H., 1982: Factors influencing secondary male teachers in Saudi Arabia to leave teaching. Ph.D., Michigan State University, East Lansing. 231 p.

301 Assaneea, A.A., 1975: An analysis of the contemporary geographic education in the public schools of Saudi Arabia. Ph.D., University of Oklahoma, Norman. 174 p.

302 A steel plant in Su'ûdí Arabia. In: The Islamic Review and Arab Affairs 57.1969(6): 19-22.   -1a, 21-

303 Ateque, H.I., 1973a: The College of Law and Islamic Studies in Mecca. In: University News 1973(Jan.): 1-3.

304 ——, 1973b: The College of Education in Mecca. In: University News 1973(Feb.): 1-3.

305 Atiyah, A.M., 1980: The perception of noise: a survey study around Jeddah International Airport, Saudi Arabia. M.A., Michigan State University, East Lansing. 131 p.

306 Attar, M.S. el-, 1972: Réflexions sur la situation en Arabie. In: Politique Étrangère 37: 333-350.   -21, 24, 25-

307 At war with trachoma. In: Aramco World 11.1960(8): 3-6.   -H 223-

308 Aulas, M.-C., 1977: Pétrodollars et stabilisation du monde arabe. La diplomatie saoudienne à l'épreuve. In: Le Monde Diplomatique 24(277): 2-3.   -30, 352, 385-

309 Auler, K., 1906, 1908: Die Hedschasbahn. 2 Bde., Ergänzungshefte 154 und 161 zu 'Petermanns Mitteilungen'; Gotha: Perthes. IV, 80 bzw. 65 S.   -21-

310 Aurada, F., 1962: Entwicklung und Bedeutung der Erdölfelder Ostarabiens. In: Mitteilungen der Österreichischen Geographischen Gesellschaft 104: 230-239, 379-397.   -1a, 5, Tü 17-

311 ——, 1964: Jordanien-Saudi-Arabien: Wiederaufbau der Hedschasbahn. In: Mitteilungen der Österreichischen Geographischen Gesellschaft 106: 267-268.   -1a, 5, Tü 17-

312 ——, 1965: Kuwait-Saudi-Arabien: Teilung der Neutralen Zone. In: Mitteilungen der Österreichischen Geographischen Gesellschaft 107: 115-116.   -1a, 5, Tü 17-

313 Australian Department of Overseas Trade, 1973: Arabian Peninsula: an export market. Report of the Australian Trade Mission to the Arabian Peninsula, March 1973. Overseas Trading 25(22, Suppl.); Canberra. 16 p.   -H 3-

314 Avril, A. d', 1868: L'Arabie contemporaine avec la description du pèlerinage de la Mecque. Paris: Maillet, Challamel. 313 p.   -16-

315 ———, 1901: Quelques notes sur l'Arabie. In: Questions Diplomatiques et Coloniales 12: 281-296.   -3o-

316 Awad, M., 1959: Settlement of nomadic and semi-nomadic tribal groups in the Middle East. In: International Labour Review 79: 25-56.   -24, 188, 2o6-

317 ———, 1966: Report on slavery. New York: UNO. XIX, 314 p.   -1a, 12, 46-

318 Awaji, I.M. al-, 1971: Bureaucracy and society in Saudi Arabia. Ph.D., University of Virginia, Charlottesville. XI, 292 p.   -21, H 223-

319 Ayoob, M., 1981: Oil, Arabism and Islam: the Persian Gulf in world politics. In: M. Ayoob (ed.): The Middle East in world politics. London: Croom Helm; Canberra: Australian Institute of International Affairs, 1981: 118-135.   -21-

32o Ayouti, Y. el- & J. Flint, 1982: Move over. In: Forbes 129 (8): 81-9o.   -38, 18o, H 3-

321 Ayubi, N.N.M., 1982/83: The politics of militant Islamic movements in the Middle East. In: Journal of International Affairs 36: 271-283.   -24, 25, 352-

322 Azouz, A.-A.H., 1977: An attitudinal study of mathematics teacher effectiveness in the intermediate schools of the Kingdom of Saudi Arabia. Ed.D., University of Northern Colorado, Greeley. 239 p.

323 Azzi, R., 1978: The Saudis go for broke. In: Fortune 98(2): 11o-113.   -24, 3o, 89-

324 ———, 198o: Saudi Arabia: the kingdom and its power. In: National Geographic 158: 286-333.   -21, 24, 93-

B

325 Babtain, A.-A.A.-W. al-, 1982: An investigation of the relationship between teacher effectiveness and indirect and direct teaching of intermediate social science teachers in Riyadh, Saudi Arabia. Ed.D., University of Northern Colorado, Greeley. 155 p.

326 Babtain, I.A. al-, 1980: A survey of mathematics curriculum of junior colleges for training of the elementary teacher in Saudi Arabia. Ph.D., University of Wyoming, Laramie. 149 p.

327 Backer, A.S., 1982: Analysis and recommendations for restructuring the administrative configuration of King Abdul Aziz University, Saudi Arabia. Ed.D., Oklahoma State University, Stillwater. 104 p.

*** Badia y Leblich, D.: s. Abbassi, A. el-.

328 Badr, F.I., 1968: Developmental planning in Saudi Arabia: a multidimensional study. Ph.D., University of Southern California, Los Angeles. XIII, 344 p.

329 ——, 1979/80: Operational and technical development of Saudi Arabia's ports. In: Jahrbuch der Hafenbautechnischen Gesellschaft 37: 99-112.   -206-

330 Badr, H.A. al-, 1972: Public relations activities at two Saudi Arabian universities. Ph.D., Michigan State University, East Lansing. 227 p.

331 Badre, A.Y. & S.G. Siksek, 1960: Manpower and oil in Arab countries. Beirut: Economic Research Institute, American University of Beirut (reprint: Westport: Hyperion Press, 1981). VIII, 270 p.   -21-

332 Bagader, A.A., 1978: Literacy and social change: the case of Saudi Arabia. Ph.D., University of Wisconsin, Madison. 153 p.

333 Bagais, M.O., 1979: Public junior colleges for the Kingdom of Saudi Arabia. Ed.D., Indiana University, Bloomington. 301 p.

*** Baharna, H.M. al-: s. Albaharna, H.M.

334 Bahry, L., 1982: The new Saudi woman: modernizing in an Islamic framework. In: The Middle East Journal 36: 5o2-515. -12, 21, 2o6-

335 Baker, A.J. el-, 1952: Date cultivation in Saudi Arabia. FAO Expanded Technical Assistance Program 31; Rome: FAO. 25 p.

336 Baker, P.R., 1979: King Husain and the Kingdom of Hejaz. Arabia Past and Present 1o; Cambridge/England, New York: The Oleander Press. XIV, 243 p. -21, 24/213-

337 Baldry, J., 1975: The powers and mineral concessions in the Idrīsī Imāmate of 'Asīr 191o-1929. In: Arabian Studies 2: 76-1o7. -12, 21, 212-

338 ——, 1976a: The Turkish-Italian War in the Yemen, 1911-12. In: Arabian Studies 3: 51-65. -12, 21, 212-

339 ——, 1976b: Al-Yaman and the Turkish occupation 1849-1914. In: Arabica 23: 156-196. -1a, 18, 21-

34o ——, 1976/77: Anglo-Italian rivalry in Yaman and 'Asīr, 193o-34. In: Die Welt des Islams N.S. 17: 155-193. -16, 21, 2o6-

341 ——, 1982: Al-Ḥudaydah and the powers during the Sa'ūdī-Yemeni War of 1934. In: Arabian Studies 6: 7-34. -21, Tü 17-

342 Baleela, M.M., 1975: Design for livability: the housing requirements of middle income families in Saudi Arabia. Ph.D., University of Pennsylvania, Philadelphia. 151 p.

343 Ballantine, J., 1966: At the turn of a tap. In: Aramco World 17(2): 14-21. -24-

344 Ballool, M.M., 1981: Economic analysis of the long-term planning investment strategies for the oil surplus funds in Saudi Arabia: an optimal control approach. Ph.D., University of Houston. 384 p.

345 Balta, P., 1973: L'Arabie Saoudite: de la tribu à l'État. In: Défense Nationale 3o(Mai): 77-88. -12, 18, 21-

346 Bancal, J.-C., 1977: L'imposition des bénéfices des sociétés étrangères en Arabie Saoudite. In: Droit et Pratique du Commerce International 3: 6o3-614. -1a, 12, 2o6-

347 Bankwesen in Saudi-Arabien und Offshore-Zentrum Bahrain. In: Aussenhandels-Blätter, Commerzbank 34.1982(3): 3-5. -H 3-

348 Banyan, A.S. al-, 1980: Saudi students in the United States: a study of cross cultural education and attitude change. London: Ithaca Press. VI, 91 p.

349 Barakat, H., 1960: Relationship between degree of education and dissatisfaction with traditional family values among Saudi employees of Aramco. M.A., Department of Sociology and Anthropology, American University of Beirut. 92 p. -Jafet Library, American University of Beirut-

350 Baram, P.J., 1978: The Department of State in the Middle East, 1919-1945. N.p. (Philadelphia): University of Pennsylvania Press. XXIV, 343 p.   -21-

351 Barcata, L., 1968: Arabien nach der Stunde Null. Ein Augenzeugenbericht. Wien, Frankfurt/M., Zürich: Molden. 383 S. -12, 21, 24/213-

352 Barham, P.R., 1970: Agricultural development and the changing landuse pattern in the Ha'il oasis, Jebel Shammar, Saudi Arabia. M.A., University of London.   -Documentation Section, Centre for Middle Eastern and Islamic Studies, University of Durham-

353 Barker, P., 1982: Saudi Arabia: the development dilemma. Special Report 116; London: Economist Intelligence Unit. 92 p.   -206-

354 Barois, J., 1939: Mekka, die verbotene Stadt des Islam. Bern, Stuttgart: Hallwag (Originalausgabe: La Mecque, ville interdite. Paris: Corrêa, 1938). 144 S.   -180, Tü 17-

355 Baroody, G.M., 1961: Crime and punishment under Hanbali law. Dhahran: ARAMCO.

356 ——, 1966: Shari'ah law of Islam. In: Aramco World 17(6): 26-35.   -24-

357 ——, 1980: The practice of law in Saudi Arabia. In: W.A. Beling (ed.): King Faisal and the modernisation of Saudi Arabia. London: Croom Helm; Boulder: Westview Press, 1980: 113-124.   -12, 21, H 223-

358 Baroudi, E., 1979: L'Arabie Séoudite, une source fondamentale pour l'octroi des prêts aux pays industriels à la fin de ce siècle. In: Syrie et Monde Arabe 26(301): 19-27. -206, H 3, H 223-

359 Barrak, I.A. al-, 1981: Application of Islamic law to taxation in Saudi Arabia. Ph.D., University of Exeter.

360 Barré, P., 1903: L'Arabie. In: Revue de Géographie (Paris) 52: 123-134, 250-263. -180-

361 Barreveld, W.H., 1963: Report to the Government of the Kingdom of Saudi Arabia on a survey of the Saudi Arabian date industry. Rome: FAO.

362 Barrows, G.H., 1959: Middle East: basic oil laws and concession contracts; original text. 2 vols. New York: Petroleum Legislation.

363 Barry, Z.A.A., 1980: Inflation in Saudi Arabia: 1964 to 1978. Ph.D., University of Colorado, Boulder. 167 p.

364 Barth, H.K., 1976a: Landesnatur. In: H. Blume (Hg.): Saudi-Arabien. Natur, Geschichte, Mensch und Wirtschaft. Ländermonographien 7; Tübingen, Basel: Erdmann, 1976: 23-88. -12, 21, 24-

365 ——, 1976b: Probleme der Wasserversorgung in Saudi-Arabien. Erdkundliches Wissen 45 (zugleich: Geographische Zeitschrift, Beih. 45); Wiesbaden: Steiner. 33 S. -21, 24, 100-

366 ——, 1980: Saudi-Arabien. Auf dem Weg vom Wüstenstaat zur Industrienation. In: Übersee-Rundschau 32(3): 9-17. -24, 212, Tü 17-

367 ——, 1983: Die physisch-geographischen Grundlagen modernen Kulturlandschaftswandels. In: T. Koszinowski (Hg.): Saudi-Arabien: Ölmacht und Entwicklungsland. Beiträge zur Geschichte, Politik, Wirtschaft und Gesellschaft. Mitteilungen des Deutschen Orient-Instituts 20; Hamburg: Deutsches Orient-Institut, 1983: 7-59. -21, H 223-

368 Bashir, F.S. al-, 1977: A structural econometric model of the Saudi Arabian economy, 1960-1970. A Wiley-Interscience Publication; New York: Wiley (zugleich: Ph.D., University of Arizona, Tucson, 1973). IX, 134 p.

369 Bassam, I.A. al-, 1973: A study of selected factors contributing to students' failure at the freshman level at Riyadh University. Ph.D., Michigan State University, East Lansing. 131 p.

370 Bassam, N.A. al-, 1968: The development of elementary education of women in the Eastern Province of Saudi Arabia - problems and trends. M.A., Education Department, American University of Beirut. 134 p.   -Jafet Library, American University of Beirut-

371 Bassi, U., 1932: L'Italia e l'Arabia Centrale. I trattati col Higiaz-Negd. Modena: Bassi & Nipoti. 51 S.

372 Bates, B.S., 1967: Search in the sand mountains. In: Aramco World Magazine 18(5): 1-5.   -24-

373 ——, 1968: The long look forward. In: Aramco World Magazine 19(6): 2o-27.   -24-

374 ——, 1969: To share the burden. In: Aramco World Magazine 2o(6): 22-23.   -24-

375 ——, 197oa: New GOSP for Safaniya. In: Aramco World Magazine 21(4): 28-29.   -24-

376 ——, 197ob: 'Through the Hawse Pipe' - a story of Ras Tanura. In: Aramco World Magazine 21(5): 22-32.   -24-

377 ——, 1971: Shutdown at Plant 11. In: Aramco World Magazine 22(1): 12-13.   -24-

378 ——, 1972a: Road show. In: Aramco World Magazine 23(1): 32-36.   -24-

379 ——, 1972b: Positively earthshaking. In: Aramco World Magazine 23(6): 8-11.   -24-

380 ——, 1972c: They are off! in Riyadh. Saudi Arabia's King's Cup is on the inside and coming up fast. In: Aramco World Magazine 23(Special): 26-27.   -24-

381 ——, 1973: Oscar for an oil field. In: Aramco World Magazine 24(6): 14-15.   -24-

382 Batrik, A.H.M. el-, 1947: Turkish and Egyptian rule in Arabia, 181o-1841. Ph.D., School of Oriental and African Studies, University of London.

383 Baumer, M., 1964: Scientific research aimed at developing the arid zone of Saudi Arabia. In: UNESCO Newsletter about Arid Zone 24: 4-12.

384 Baumhauer, O. (Hg.), 1965: Arabien. Dokumente zur Entdeckungsgeschichte 1; Stuttgart: Goverts. 345 S.   -24, 93-

385 Bawazeer, S.A., 1979: Curriculum renewal: guidelines for the Saudi Arabian secondary schools. Ed.D., Columbia University Teachers College, New York. 175 p.

386 Bawden, E., 1945/46: An Arabian journey. In: The Geographical Magazine 18: 16-22.   -7, 18o, 464-

387 Baz, F., 1981a: L'évolution du système bancaire séoudien. In: L'Économiste Arabe 24(267): 3o-32.   -2o6, H 223-

388 ——, 1981b: Le sponsorship et le cadre juridique des investissements. In: L'Économiste Arabe 24(268): 38-4o. -2o6, H 223-

389 Beaumont, A. de, 1854: Le pèlerinage de la Mekke. In: Revue de l'Orient et de l'Algérie 16: 13-31.   -1a-

39o Beaumont, P., 1977: Water and development in Saudi Arabia. In: The Geographical Journal 143: 42-6o.   -21, 291, 352-

391 Becher, R., 1976: Pre-feasibility study on a six year afforestation programme for the south-western region of the Kingdom of Saudi Arabia, Abha District. Submitted by GTZ. Eschborn. 21, 5 p.

392 Bechtel Corporation, 1947: Transportation report, Dammam to Riyadh, Kingdom of Saudi Arabia. A study for the development of a transportation system between the Persian Gulf and the city of Riyadh. San Francisco. XII, 66 p.

393 ——, 1966: Desalination in the Eastern Province of Saudi Arabia. Prepared for the Government of Saudi Arabia by the Bechtel Corporation appointed under the United Nations Programme of Technical Assistance. Report TAO/SAU/6; New York: Department of Economic and Social Affairs, Commissioner for Technical Co-operation, UNO. VIII, 134 p. -2o6-

394 Bechtel Power Corporation, 198o: Assignment: Saudi Arabia. Rev.ed. San Francisco.

395 Becker, H. & G. Mittelmann, 1973: Herstellung von Bewässerungskanälen in Saudi Arabien unter Verwendung von Beton-Einbauzügen. In: Strassen- und Tiefbau 27: 612-616. -21, 93, 2o6-

396 Becker, K., 1963: Die Hedschasbahn. In: Orient (Opladen) 4: 193-195.   -21, 1oo, H 223-

397 Beckingham, C.F., 1965: Farasān (Farsān). In: EI^2 2: 787-788. -21, 24, 212-

398 Bedore, J. & L. Turner, 1977: Die Industrialisierung der ölproduzierenden Länder im Mittleren Osten. In: Europa-Archiv, Beiträge und Berichte 32: 538-546 (zugleich: The industrialization of the Middle Eastern oil producers. In: The World Today 1977(33): 326-334). -21, 24, 25-

399 ——, 1978a: Saudi Arabia: greatness thrust upon them. In: Middle East International (79): 14-16. -3o-

4oo ——, 1978b: Saudi Arabia in a changing world. In: National Westminster Bank Quarterly Review 1978(1): 13-23. -2o6-

4o1 Behrens, G., 197o: Zur Eheschliessung eines Saudi-Arabers mit einer Christin. In: Das Standesamt (Frankfurt/M.) 23: 81-82. -1a, 3o, 291-

4o2 Békri, C., S. Ismail & M.A. el-Ghannam, 1974: Rapport provisoire du Bureau Régional pour l'Éducation dans les Pays Arabes sur les besoins éducatifs en Arabie Séoudite. Beyrouth: Bureau Régional pour l'Éducation dans les Pays Arabes, UNESCO.

4o3 Belgrave, C.D., 196o: Personal column. London, Melbourne, Sydney, Auckland, Bombay, Toronto, Johannesburg, New York: Hutchinson. 247 p. -18o-

4o4 Beling, W.A. (ed.), 198o: King Faisal and the modernisation of Saudi Arabia. London: Croom Helm; Boulder: Westview Press. 253 p. -12, 21, H 223-

4o5 Bell, F.E.E. (ed.), 1927: The letters of Gertrude Bell. Vol. 1. New York: Boni & Liveright (englische Ausgabe: London: Benn, 1947). 328 p. -21, 24/213, 352-

4o6 Bell, G.L., 1914: A journey in northern Arabia. In: The Geographical Journal 44: 76-77. -21, 291, 352-

4o7 Bell, M. & R. Schamberger, 1979: The experience of one inland transport company in Saudi Arabia. In: Arab ports in the 7o's: a survey of ports, shipping and maritime trade in the Middle East, based on the proceedings of the Arab Ports Conference, London. London: International Communications, 1979: 62-66.

408 Bell, S.D., Jr., D.E. McComb, E.S. Murray, R.S.-M. Chang & J.C. Snyder, 1959: Adenoviruses isolated from Saudi Arabia. I. Epidemiologic features. In: The American Journal of Tropical Medicine and Hygiene 8: 492-500. -16, 25, 100-

409 Bellotti, F., 1960: La cittadella di Allah. Viaggio nell' Arabia Saudita. Milano: Cino del Duca. 235 S. -12-

410 Beltran, A.A., 1956: Books in the desert. In: The Library Journal 81: 585-589. -21, 24, 352-

411 Ben Cheneb, M., 1927: Ibn Taimīya. In: EI[1] 2: 447-450. -21, 24-

412 Ben Chérif, 1919: Aux villes saintes de l'Islam. Paris: Hachette. II, 252 p. -1a-

413 Benderly, B.L., 1982: Education and the arts. - Religious life. - Social systems. In: R.F. Nyrop et al.: Saudi Arabia: a country study. Area Handbook Series DA Pam. 550-51; Washington: US Government Printing Office, 1982: 91-155. -H 223-

414 Ben Gabr, A.A., 1972: Die Bedeutung der Staatstätigkeit für die wirtschaftliche Entwicklung Saudi-Arabiens. Diss., Rechts-, wirtschafts- und sozialwissenschaftliche Fakultät, Universität Freiburg/Schweiz. Reinheim. XV, 184 S. -21, 212, 291-

415 Benoist-Méchin, J., 1956: Ibn Sa'ud und die arabische Welt. Düsseldorf, Köln: Diederichs (Originalausgabe: Ibn-Séoud ou la naissance d'un royaume. Paris: Michel, 1955. Englische Ausgabe: Arabian destiny. London: Elek Books, 1957). 389 S. -24, 93-

416 ——, 1960: Le roi Saud ou l'Orient à l'heure des relèves. Paris: Michel. 575 p. -21, 24-

417 ——, 1961: Arabie, carrefour des siècles. Paris: Michel. 138 p. -212-

418 ——, 1975: Fayçal, roi d'Arabie. L'homme, le souverain, sa place dans le monde (1906-1975). Paris: Michel. 302 p. -21, 24/213, H 223-

419 Benton, G., 1979: Saudi oasis grows into model farm. In: The Middle East (54): 94, 96. -1a, 12, 30-

420 Berger, J., 1982: Das Konzept der Neuen Internationalen Militärordnung der 'International Peace Research Association' und seine Relevanz für die arabischen Golfstaaten. Diplomarbeit, Fachbereich Politische Wissenschaft, Freie Universität Berlin.

421 Bergmann, W., 1977: Gutachten für die Errichtung einer Ausbildungswerkstatt für Industrie-Elektronik an der SVS in Er Riyadh, Saudi-Arabien. Im Auftrag der GTZ. Hamburg. 29 S.   -Esb 2ooo-

422 Bergsten, C.F., 1981: The world economy in the 1980s: selected papers of C.F. Bergsten, 1980. Lexington Books; Lexington, Toronto: Heath. X, 179 p.   -H 3-

423 Bernasconi, P., 1932: I Trattati fra l'Italia e l'Heggiaz. In: L'Oltremare 6: 95-98.   -1a, 3o-

424 Bernleithner, E., 1968: Gedenkfeier für das Ehrenmitglied Univ.-Prof. Dr. Alois Musil. In: Mitteilungen der Österreichischen Geographischen Gesellschaft 11o: 277-279.   -1a, 5, Tü 17-

425 Berreby, J.-J., 1958: La péninsule Arabique. Terre Sainte de l'Islam, patrie de l'arabisme et empire du pétrole. Bibliothèque Historique; Paris: Payot. 27o p.   -21-

426 Besson, Y., 1979: Hussein ou Ibn Sa'ûd, une fausse alternative. La politique de l'émir du Najd durant la première guerre mondiale. In: Relations Internationales 19: 241-261.   -25, 18o, 291-

427 ——, 1980: Ibn Sa'ûd, roi bédouin. La naissance du Royaume d'Arabie Saoudite (zugleich: La fondation du Royaume d'Arabie Saoudite. Essai sur la stratégie d'Abdul 'Aziz Ibn Abdul Rahman Al Sa'ûd. Thèse présentée à l'Université de Genève pour l'obtention du grade de Docteur ès sciences politiques). Lausanne: Trois Continents. 284 p.   -21, 24, H 223-

428 Bethmann, E.W., 1967: Basic facts on Saudi Arabia. Washington: Department of Research, American Friends of the Middle East. 27 p.

429 Bianchini, M., 1951: Il petrolio nell'Arabia Saudita. In: Bollettino della Società Geografica Italiana 85 (zugleich: VIII, 4): 144-155.   -21, 188, 352-

430 Bidwell, R. (ed.), 1971a: The affairs of Kuwait, 1896-1905. Foreign Office Confidential Print: Correspondence respecting affairs at Koweit, 1896-1905. 2 vols. London: Cass. XXXVII, VII, 94, VI, 68, XIV, 169 resp. V, XVI, 168, VI, 72, X, 111, 11 p.   -21-

431 —— (ed.), 1971b: The affairs of Arabia, 1905-1906. Foreign Office Confidential Print: Correspondence respecting the affairs of Arabia, 1905-1906. 2 vols. London: Cass. LIII, IX, 154, XI, 116, VI, 82 resp. V, VII, 78, VI, 81, V, 93, VI, 79, IV, 40 p.   -21-

432 —— (ed.), 1973: Bidwell's guide to government ministers. Vol. 2: The Arab world, 1900-1972. London: Cass. XI, 124 p.   -21, 24-

433 ——, 1976: Travellers in Arabia. London, New York, Sydney, Toronto: The Hamlyn Publishing Group. 224 p.   -12, 21, 24-

434 Biggest-ever hospital management contract. In: Achievement 45.1978(10): 17.   -1a-

435 Bilainkin, G., 1947: Guest of King Ibn Saud. In: The Contemporary Review 172: 16-20.   -25, 180, 188-

436 ——, 1950: Cairo to Riyadh diary. London: Williams & Norgate. X, 233 p.   -21-

437 Bilimatsis, J.S., 1980: Small states as major powers: a case study of Saudi Arabia. Ph.D., George Washington University, Washington. 432 p.

438 Bill, J.A. & C. Leiden, [2]1975: The Middle East: politics and power. Boston, London, Sydney: Allyn & Bacon ([1]1974). XII, 287 p.   -12, 21-

439 Binsaleh, A.M., 1982: The civil service and its regulation in the Kingdom of Saudi Arabia. Ph.D., Claremont Graduate School. 195 p.

440 Binzagr, S., 1979: Arabie Séoudite. Peintures récentes d'un temps révolu. Lausanne: Trois Continents; Genève: Arabian Resource Management (Originalausgabe: Saudi Arabia: an artist's view of the past. Lausanne: Three Continents, 1979). 139 p.   -21-

441 Biographical section. In: Who's who in the Arab world. 6th, thoroughly rev. & completed ed. Beirut: Publitec Publications, 1981/82: 651-1367.   -93-

442 Bird, K., 1980: Co-opting the Third World elites. Trilateralism and Saudi Arabia. In: Trilateralism. The Trilateral Commission and elite planning for world management; Boston, 1980: 341-351.   -206-

443 Birken, A., 1976: Saudi-Arabien in der Neuzeit. - Regierungsform und Verfassung. In: H. Blume (Hg.): Saudi-Arabien. Natur, Geschichte, Mensch und Wirtschaft. Ländermonographien 7; Tübingen, Basel: Erdmann, 1976: 123-150 bzw. 152-155.   -12, 21, 24-

444 Birket-Smith, K., 1940: Raunkiaer, Anders Christian Barclay. In: Dansk Biografisk Leksikon 19; Kopenhagen: Schultz, 1940: 264-265.   -7, 21-

445 Birks, J.S., 1975: Overland pilgrimage in the savanna lands of Africa. In: L.A. Kosiński & R.M. Prothero (eds.): People on the move: studies on internal migration. London: Methuen, 1975: 297-307.   -21-

446 ——, 1977a: The Mecca Pilgrimage by West African pastoral nomads. In: The Journal of Modern African Studies 15: 47-58.   -21, 212, 291-

447 ——, 1977b: Overland pilgrimage from West Africa to Mecca: anachronism or fashion? In: Geography 62: 215-217.   -25, 180, Tü 17-

448 ——, 1978: Across the savannas to Mecca: the overland pilgrimage route from West Africa. London: Hurst (zugleich: Aspects of overland pilgrimage from West Africa to Mecca, with special reference to Darfur Province, The Republic of the Sudan. Ph.D., Department of Geography, University of Liverpool, 1975). XIII, 161 p.   -12, 21-

449 —— & C.A. Sinclair, 1978: The Kingdom of Saudi Arabia. International Migration Project, Country Case Studies; Centre for Middle Eastern and Islamic Studies, University of Durham.

450 —— & ——, 1980a: Arab manpower: the crisis of development. London: Croom Helm; New York: St. Martin's Press. 391 p.   -12, 21-

451 Birks, J.S. & C.A. Sinclair, 1980b: International migration and development in the Arab region. A WEP Study; Geneva: ILO. XII, 175 p.   -12, 21-

452 —— & ——, 1982: The domestic political economy of development in Saudi Arabia. In: T. Niblock (ed.): State, society and economy in Saudi Arabia. London: Croom Helm; Exeter: Centre for Arab Gulf Studies, 1982: 198-213.   -21-

453 Bishara, G., 1978: The Middle East arms package. A survey of the Congressional debates. In: Journal of Palestine Studies 7(4): 67-78.   -21, 25, 2o6-

454 Bitsch, J., 1961: Hinter Arabiens Schleier. Land der Wüste zwischen gestern und morgen. Berlin, Frankfurt/M., Wien: Ullstein (Originalausgabe: Bag Arabiens Slør. Kopenhagen: Gyldendal, 1961. Amerikanische Ausgabe: Behind the veil of Arabia. New York: Dutton, 1962). 2o3 S.   -21-

455 Björkman, W., 1944: Die Haschimiten. In: W. Björkman, R. Hüber, E. Klingmüller, D. von Mikusch & H.H. Schaeder: Arabische Führergestalten. Arabische Welt 5; Heidelberg, Berlin, Magdeburg: Vowinckel, 1944: 88-114.   -Tü 17-

456 Blackwood, P., 1935: The pilgrimage in 1934. In: The Moslem World 25: 287-292.   -21, 291, 352-

457 Blake, G.H. & R. King, 1972: The Hijaz railway and the pilgrimage to Mecca. In: Asian Affairs (London) 59/N.S. 3: 317-325.   -5, 21, 9o-

458 ——, 1976: Rising tide in Mecca. In: Middle East International 1976(April): 16-17.   -3o-

459 Blanckenhorn, M., 19o7: Die Hedschāz-Bahn auf Grund eigener Reisestudien. In: Zeitschrift der Gesellschaft für Erdkunde zu Berlin 19o7: 218-245, 288-32o.   -7, 16, 21-

46o ——, 1912: Die Hedschāzbahn. In: Geographische Zeitschrift 18: 15-29.   -16, 21, 25-

461 Blandford, L., 1977: Oil sheikhs. London: Weidenfeld & Nicolson. XI, 286 p.   -12, 21-

462 Bligh, A. & S.E. Plaut, 1982: Saudi moderation in oil and foreign policies in the post-AWACS-sale period. In: Middle East Review 14(3/4): 24-32.   -21, H 223-

463 Blume, H., 1976: Lage, Grösse und Grenzen des Staatsgebietes. - Bildungs-, Gesundheits- und Sozialwesen. - Moderne Wirtschaftsentwicklung und Weltverflechtung. In: H. Blume (Hg.): Saudi-Arabien. Natur, Geschichte, Mensch und Wirtschaft. Ländermonographien 7; Tübingen, Basel: Erdmann, 1976: 151-152, 161-166 bzw. 253-333. -12, 21, 24-

464 ——, 1980: Saudi-Arabien - Der Wüstenstaat im Wandel. In: Geoökodynamik 1: 65-79. -7-

465 Blunt, A., 1881: A pilgrimage to Nejd, the cradle of the Arab race. A visit to the court of the Arab Emir, and 'Our Persian Campaign'. 2 vols. London: Murray. XXXI, 273 resp. IX, 283 p. -21, 24, Tü 17-

*** Bochskandl, M.: s. Arle, M. d'

466 Bökemeier, R., 1980: Saudi-Arabien: Die Beduinen werden bequem. In: Geo 1980(7): 8o-1oo. -24, 1o7, 18o-

467 Boesch, H., 1949: Erdöl im Mittleren Osten. In: Erdkunde 3: 68-82. -16, 21, 24-

468 ——, 1963: Neutrales Land - in Nordarabien. In: Geographica Helvetica 18: 2o9-211. -21, 24, Tü 17-

469 Bokhari, A.Y., 1978: Jeddah: a study in urban formation. Ph.D., University of Pennsylvania, Philadelphia. 537 p.

47o Bonin, C.-E., 19o9: Le chemin de fer du Hedjaz. In: Annales de Géographie 18: 416-432. -21, 24, 291-

471 Bonnenfant, P., 1977a: L'évolution de la vie bédouine en Arabie centrale. Notes sociologiques. In: Revue de l'Occident Musulman et de la Méditerranée (23): 111-178. -21, H 223-

472 ——, 1977b: Mouvements migratoires en Arabie centrale. Notes sociologiques. In: Revue de l'Occident Musulman et de la Méditerranée (23): 179-223. -21, H 223-

473 ——, 1982: La capitale saoudienne: Riyadh. In: Centre d'Études et des Recherches sur l'Orient Arabe Contemporain (éd.): La péninsule Arabique d'aujourd'hui. Sous la direction de P. Bonnenfant. T. 2: Études par pays. Paris: Centre National de la Recherche Scientifique, 1982: 655-7o5. -21-

474 Bono, S., 1966: Giovanni Finati, militare e archeologo in Levante ai tempi di Mohammed Ali. In: Levante 13(2): 3-2o.-21-

475 Borel, F., 1904: Choléra et peste dans le pèlerinage musulman (1860-1903). Étude d'hygiène internationale. Paris: Masson. III, 197 p.   -1a-

476 Boswinkle, E., 1971: Investigations of crop responses to fertilizer in Saudi Arabia. Rome: FAO.

477 Botschaft der Bundesrepublik Deutschland, ca. 1975: German companies operating in the Kingdom of Saudi Arabia and their projects. Dschidda. 8 S.

478 Bouaoula, A., 1977: Politiques et plans d'éducation dans les états Arabes dans les années 7o: résumés et synthèse. Reports, Studies C.55; Geneva: Division of Educational Policy and Planning, UNESCO. 48 p.

479 Boucheman, A. de, 1935: Matériel de la vie bédouine. Recueilli dans le désert de Syrie (tribu des Arabes Sba'a). Documents d'Études Orientales de l'Institut Français de Damas 3; Damas. 139 p.   -21-

480 Boucher, B.P. & H. Singh, 1975: Plow-back: the use of Arab money. In: Aramco World Magazine 26(5): 22-25.   -21, 24-

481 Boulicaut, A. le, 1913: Au pays des mystères. Pèlerinage d'un chrétien à la Mecque et à Médine. $2^{\text{ème}}$ éd. Paris: Plon-Nourrit. XXVII, 289 p.   -12-

482 Bouteiller, G. de, 1978a: Mes entretiens avec le roi Fayçal d'Arabie. In: Revue des Deux Mondes 1978(Juillet): 62-7o.   -16, 21, 24-

483 ——, 1978b: L'Arabie Saoudite, aujourd'hui et demain. In: Défense Nationale 34(Nov.): 91-1o7, (Déc.): 41-6o.   -18, 21, 291-

484 ——, 1979: L'Arabie Saoudite: quel avenir économique? In: Défense Nationale 35(Août/Sept.): 29-4o.   -18, 21, H 223-

485 ——, 1981: L'Arabie Saoudite. Cité de Dieu, cité des affaires, puissance internationale. Politique d'aujourd'hui; Paris: Presses Universitaires de France. 218 p.   -12, 25-

486 Bouvat, L., 1908: Le premier journal publié à La Mecque. In: Revue du Monde Musulman 6: 561-562.   -21, 25-

487 ——, 1917/18: 'Al-Kibla', journal arabe de La Mecque. In: Revue du Monde Musulman 34: 32o-328.   -21, 25-

488 Bouygues builds on its Saudi foundations. In: J. Whelan (ed.): France and the Middle East. A MEED Special Report; London: Middle East Economic Digest, 1982: 23-24.   -H 223-

489 Bowen, R.L., Jr., 1949: Arab dhows of eastern Arabia. In: The American Neptune 9: 87-132.   -18-

490 ——, 1951a: The dhow sailor. In: The American Neptune 11: 161-2o2.   -18-

491 ——, 1951b: The pearl fisheries of the Persian Gulf. In: The Middle East Journal 5: 161-18o.   -12, 21, 2o6-

492 ——, 1951c: Marine industries of eastern Arabia. In: The Geographical Review 41: 384-4oo.   -21, 93, 2o6-

493 Bowen-Jones, H., 198o: Agriculture and the use of water resources in the Eastern Province of Saudi Arabia. In: M. Ziwar-Daftari (ed.): Issues and development: the Arab Gulf States. London: MD Research & Services,198o: 118-137.   -21-

494 Bowler, R.M., 198o: Expatriates in Saudi Arabia: stress, social support, modernity and coping. Ph.D., Graduate School of Psychosocial Development and Education, The Wright Institute, San Francisco State University. XVIII, 16o p.   -291-

495 Boxhall, P., 1976: Pilgrims across Africa. In: Middle East International 1976(Apr.): 18-19.   -3o-

496 Boyd, D.A., 197o/71: Saudi Arabian television. In: The Journal of Broadcasting 15(1): 73-78.   -3o, 352-

497 ——, 1972: An historical and descriptive analysis of the evolution and development of Saudi Arabian television: 1963-1972. Ph.D., University of Minnesota, Minneapolis. VIII, 369 p.   -21-

498 ——, 198o: Saudi Arabian broadcasting: radio and television in a wealthy Islamic state. In: Middle East Review 12(4)/13(1): 2o-27.   -21-

499 Boylan, F.T., 1967: Taif: city of color. In: Aramco World 18(4): 34-36.   -24-

5oo Bräunlich, E., 1928: Philby's Reisen in Arabien. In: Orientalistische Literaturzeitung 31: Sp. 342-348.   -1a, 21, 24-

5o1 Bräunlich, E., 1933: Beiträge zur Gesellschaftsordnung der arabischen Beduinenstämme. In: Islamica 6: 68-111, 182-229. -16, 21, 25-

5o2 Braibanti, R. & F.A.-S. al-Farsy, 1977: Saudi Arabia: a developmental perspective. In: Journal of South Asian and Middle Eastern Studies 1(1): 3-43. -1a, H 223-

5o3 ——, 198o: Saudi Arabia in the context of political development theory. In: W.A. Beling (ed.): King Faisal and the modernisation of Saudi Arabia. London: Croom Helm; Boulder: Westview Press, 198o: 35-57. -21, H 223-

5o4 Branscheid, V., 1973: Grossraum-Bewässerungsanlagen in Saudi-Arabien. In: Deutsche Gesellschaft für Agrar- und Ernährungshilfe in Entwicklungsländern e.V.: Agrarhilfe 1972. 1. Teil. O.O.: 16-18.

5o5 Braun, U., 1973: Wachsende Polarisierung in der Region des Arabischen Golfs. In: Europa-Archiv, Beiträge und Berichte 28: 6o3-612. -21, 24, 25-

5o6 ——, 1976: Der Entscheidungsprozess innerhalb der OPEC und die Stabilität der Organisation. In: Europa-Archiv, Beiträge und Berichte 31: 225-234. -21, 24, 25-

5o7 ——, 1979: Die innenpolitische Entwicklung in Saudi-Arabien und die Frage der internen Stabilität. Stiftung Wissenschaft und Politik, Forschungsinstitut für Internationale Politik und Sicherheit SWP-AP 22o4; Ebenhausen. 36 S. -H 223-

5o8 ——, 198oa: Saudi-Arabien im Spannungsfeld zwischen Nahost, Golf und Rotem Meer unter besonderer Berücksichtigung des saudi-arabisch-amerikanischen Verhältnisses. Stiftung Wissenschaft und Politik, Forschungsinstitut für Internationale Politik und Sicherheit SWP-AZ 2248; Ebenhausen. 69 S. -H 223-

5o9 ——, 198ob: Saudi-Arabiens veränderter Standort: Auswirkungen für den Westen. In: Europa-Archiv, Beiträge und Berichte 35: 539-546. -21, 24, 25-

51o ——, 1981: Die Aussen- und Sicherheitspolitik Saudi-Arabiens. In: Orient (Opladen) 22: 219-24o. -21, 2o6, H 223-

511 Braun, U., 1983: Die Aussen- und Sicherheitspolitik. In: T. Koszinowski (Hg.): Saudi-Arabien: Ölmacht und Entwicklungsland. Beiträge zur Geschichte, Politik, Wirtschaft und Gesellschaft. Mitteilungen des Deutschen Orient-Instituts 2o; Hamburg: Deutsches Orient-Institut, 1983: 121-15o.   -21, H 223-

512 Bray, N.N.E., 1934: Shifting sands. London: Unicorn Press (reprint: New York: AMS Press, 1974). XII, 312 p.   -21, 24/213-

513 ——, 1936: A paladin of Arabia. The biography of Brevet Lieut.-Colonel G.E. Leachman, C.I.E., D.S.O., of the Royal Sussex Regiment. London: Heritage. XVI, 429 p.   -1a, 294-

514 Bredi, D., 1978: Incidenza politica delle relazioni Arabia Saudiana-Pakistan negli ultimi anni. In: Oriente Moderno 58: 475-483.   -5, 7, 21-

515 Brémond, É., 1931: Le Hedjaz dans la Guerre Mondiale. Collection de Mémoires, Études et Documents pour servir à l' Histoire de la Guerre Mondiale 32; Paris: Payot. 351 p.   -12, 24/213-

516 ——, 1935: Marins à chameau. Les Allemands en Arabie en 1915-1916. Paris, Limoges, Nancy: Charles-Lavauzelle. 12o p.   -12-

517 ——, 1937: Yémen et Saoudia. L'Arabie actuelle. Paris, Limoges, Nancy: Charles-Lavauzelle. IX, 141 p.   -H 3-

518 Brent, P., 1977: Far Arabia: explorers of the myth. London: Weidenfeld & Nicolson. XI, 239 p.   -21-

519 Bretholz, W., 196o: Aufstand der Araber. Neue Sammlung Desch 7; Wien, München, Basel: Desch. 559 S.   -12, 21, 24/213-

52o Bricault, G.C., M. Donovan with contributions from E. Cotran, J.F.K. Lee, M. Wahby & Sarabex Ltd., 1976: The Arab business yearbook. London: Graham & Trotman. XVII, 579 p.   -212-

521 Bridge, J.N., 1974: Financial structure and development in Saudi Arabia. In: Proceedings of the Seminar for Arabian Studies 4: 1o-18.   -1a, 21-

522 Briner, E.K., 1975: Steuersysteme der wichtigsten OPEC-Staaten: Saudi-Arabien. In: Steuer-Revue 31: 194-2oo.   -38, 18o, 2o6-

523 Brinton, J.Y., 1967: The Saudi-Kuwaiti Neutral Zone (with: Comment from Mr. El Ghoneimy). In: The International and Comparative Law Quarterly IV, 16: 820-823. -21, 24, 291-

524 British Council, 1979: Education profile: Saudi Arabia. London. 25 p. -H 108-

525 British National Export Council, Committee for Middle East Trade, c. 1970: The market for household goods in Saudi Arabia. A report prepared by Industrial Export Surveys for the British National Export Council, Committee for Middle East Trade. London. Var.pag. -H 3-

526 British Overseas Trade Board, 1974: Saudi Arabia. London. 76 p.

527 ——, Export Services and Promotion Division, 1977: Hints to businessmen visiting Saudi Arabia. London. 48 p.

528 British scientists fighting locust plague. Worst invasion for years. In: Great Britain and the East 68.1952(1830): 43, 45. -21-

529 Brittain, M.Z., 1947: Doughty's mirror of Arabia. In: The Moslem World 37: 42-48. -21, 291, 352-

530 Brizard, S., 1974: Les Grands Ulémas d'Arabie Saoudite en Europe. In: L'Afrique et l'Asie Modernes (103): 45-47. -1a, 21, 30-

531 Broadfoot, W., 1895: Pelly, Sir Lewis. In: Dictionary of National Biography 44; London: Smith, Elder, 1895: 275-277. -7, 21-

532 Brodie, D.M., 1958: Crane, Charles Richard. In: Dictionary of American Biography 22, Suppl. 2; London: Oxford University Press, 1958: 128-130. -21-

533 Broucke, J., 1929: L'empire arabe d'Ibn Séoud. Bruxelles: Falk. 89 p. -1a-

534 Brown, E.H., 1963: The Saudi Arabia-Kuwait Neutral Zone. Middle East Oil Monographs 4; Beirut: The Middle East Research & Publishing Center. XIV, 150 p. -206-

535 Brown, Glen F., 1949: The geology and ground water of Al Kharj district, Nejd, Saudi Arabia. Ph.D., Department of Geology, Northwestern University, Evanston. V, 125 p. -21-

536 Brown, Grover F., 1964: Date cultivation in the Eastern Province. Qatif: Qatif Experimental Station. 7 p.

537 Brown, W., J. Farnworth & S.H. Clarke, 1976a: A feasibility study of three farming types in Saudi Arabia. 1. Forage production for sale. University College of North Wales, Bangor and Ministry of Agriculture and Water, Saudi Arabia: Joint Agricultural Research and Development Project, Publication 71; n.p. (Bangor?). 9 p.

538 ———, ——— & ———, 1976b: A feasibility study of three farming types in Saudi Arabia. 2. 1oo cow dairy farm. University College of North Wales, Bangor and Ministry of Agriculture and Water, Saudi Arabia: Joint Agricultural Research and Development Project, Publication 72; n.p. (Bangor?). 14 p.

539 ———, ——— & ———, 1976c: A feasibility study of three farming types in Saudi Arabia. 3. 1,ooo ewe sheep farm. University College of North Wales, Bangor and Ministry of Agriculture and Water, Saudi Arabia: Joint Agricultural Research and Development Project, Publication 73; n.p. (Bangor?). 13 p.

54o Brownell, G.A., 1947: American aviation in the Middle East. In: The Middle East Journal 1: 4o1-416. -12, 21, 38-

541 Bruce, J., 179o: Reisen zur Entdeckung der Quellen des Nils in den Jahren 1768, 1769, 177o, 1771, 1772 und 1773. Bd. 1. Leipzig: Weidmann (Originalausgabe: Travels to discover the source of the Nile in the years 1768, 1769, 177o, 1771, 1772 and 1773. Vol. 1. Edinburgh: Ruthven; London: Robinson, 179o. Französische Ausgabe: Voyage en Nubie et en Abyssinie entrepris pour découvrir les sources du Nil, pendant les années 1768, 1769, 177o, 1771, 1772 & 1773. T. 1. Paris, 179o). XXVIII, 579 S. -24-

542 Bruyn, P. de, 1982: Saoedi-Arabië en de Wahhābīya. In: Inforiënt-Dossier 2(2): 17-21, (3): 12-16.

543 ———, 1983: De Islam in de buitenlandse politiek von Saoedi-Arabië. In: Inforiënt-Dossier 2(6): 12-15.

544 Bruzonsky, M.A., 1982: The peace plans compared. In: Middle East International (169): 11-13. -12, 3o-

545 Brydges, H.J., 1834: An account of the transactions of His Majesty's mission to the court of Persia, in the years

1807-11, to which is appended a brief history of the Wahauby. Vol. 2. London: Bohn. V, 238 p.   -21-

546 Bryson, T.A., 1981: Seeds of Mideast crisis: the United States diplomatic role in the Middle East during World War II. Jefferson: McFarland. VIII, 216 p.   -21-

547 Buchan, J., 1980a: Saudi promise and performance. In: Middle East International (133): 6-7.   -12, 30-

548 ——, 1980b: Saudi Arabia. In: Middle East Annual Review 6: 307, 309, 311-312, 317, 319, 321, 323, 325, 327, 329, 331-332.   -21, 24/213, 206-

549 ——, 1982: Secular and religious opposition in Saudi Arabia. In: T. Niblock (ed.): State, society and economy in Saudi Arabia. London: Croom Helm; Exeter: Centre for Arab Gulf Studies, 1982: 106-124.   -21-

550 Buckingham, J.S., 1855: Autobiography of James Silk Buckingham; including his voyages, travels, adventures, speculations, successes and failures, faithfully and frankly narrated; interspersed with characteristic sketches of public men with whom he has had intercourse, during a period of more than fifty years. Vol. 2. London: Longman, Brown, Green & Longmans. XII, 424 p.   -1a-

551 Buckner, R.G., 1978: Investment prospects in Saudi Arabia. In: North Carolina Journal of International Law and Commercial Regulation 3: 71-103.   -B 208, B 212-

552 Buddenberg, J., 1980: Saudi-Arabien, Markt für deutsche Konsumgüter. In: Markenartikel 42: 367-369.   -H 3-

553 Büren, R., 1973: Regierung, Parteien und Demokratie in modernen islamischen Staaten. In: Orient (Opladen) 14: 5-6.   -21, 100, H 223-

554 Bürkner, F.C., 1977: Saudi-Arabien, Emirate, Jemen. Polyglott-Reiseführer 875; München: Polyglott. 63 S.   -12, 212-

555 Buez, E.A., 1873: Une mission au Hedjaz, Arabie. Contributions à l'histoire du choléra, etc. Paris: Masson (extrait de: Gazette Hebdomadaire de Médecine et de Chirurgie II, 10.1873: 265-272, 281-288, 537-547, 601-611, 633-644, 649-656, 665-677, 681-689, 697-705, 729-741, 745-754, 793-805). II, 126 p.

556 Buhl, F., 1936: Al-Madīna. In: EI[1] 3: 88-98.   -21, 24-

557 Building Management and Marketing Consultants, Ltd., 1976: Guide to the construction industry in Saudi Arabia. 6 vols. London.

558 Bullard, R. (ed.), [3]1958: The Middle East: a political and economic survey. London, New York, Toronto: Oxford University Press ([1]1950). XVIII, 569 p.   -21, Gö 153-

559 ——, 1961: The camels must go: an autobiography. London: Faber & Faber. 3oo p.   -21-

56o Bundesstelle für Aussenhandelsinformation, 1957, 196o: Kurzmerkblatt Saudisch-Arabien. Mitteilungen; Köln. 4 bzw. 4 S.

561 ——, 1961: Regionale Märkte in Saudisch-Arabien I, II. Mitteilungen 45-46; Köln. 4 bzw. 4 S.

562 ——, 1961, 1963, 1966, 1969, 1975, 198o: Kurzmerkblatt Saudi-Arabien. Mitteilungen; Köln. 4-12 S.

563 ——, 1969-: Weltwirtschaft am Jahreswechsel (1968/69-) - Saudi-Arabien. Mitteilungen; Köln. 4-8 S.   -212-

564 ——, 197o-: Weltwirtschaft zur Jahresmitte (197o-) - Saudi-Arabien. Mitteilungen; Köln. 2-6 S.   -212

565 ——, 1974, 1976, 1982: Zollvorschriften (Kurzfassung). Zollinformation; Köln.

566 ——, 1975a: Auskunfts- und Kontaktstellen in Entwicklungsländern - Saudi-Arabien. Mitteilungen 11o; Köln. 2 S.

567 ——, 1975b: Industrialisierung: Saudi-Arabien und 9 weitere Länder Ostarabiens. Marktinformationen A/273; Köln.

568 ——, 1975c: Saudi-Arabien, Kuwait, Vereinigte Arabische Emirate: Anregungen zur Markterschliessung. Marktinformationen A/282; Köln. II, 41 S.   -212-

569 ——, 1975d: Wirtschaftsdaten: Saudi-Arabien und 9 andere Länder Asiens. Marktinformationen A/284; Köln.

57o ——, 1975e: Finanzierungs- und Investment-Institutionen Saudi-Arabiens. Dokumente 1o2; Köln. 1o S.

571 ——, 1975f: Liste saudi-arabischer Firmen, die neue Industriebetriebe gründen wollen. Dokumente 124; Köln. 11 S.

572 Bundesstelle für Aussenhandelsinformation, 1975g: Liste der Feasibiliy Studies für Industrieprojekte in Saudi-Arabien. Dokumente 168; Köln. 8 S.

573 ——, 1975h: Die Agrarpolitik Saudi-Arabiens. Dokumente 222; Köln. 2 S.

574 ——, 1975i: Saudi-Arabien: Markterschliessung und industrielle Kooperation. Dokumente 281; Köln. 22 S.

575 ——, 1975j: Saudi-Arabien: Entwicklung der Landwirtschaft. Dokumente 347; Köln. 4 S.

576 ——, 1975k: Saudi-Arabien: Angaben zum 2. Fünfjahresplan (Juli 1975 bis Mai 198o). Dokumente 351; Köln. 12 S.

577 ——, 1975l: List of pending applications by Saudi investors. Dokumente 422; Köln. 8 p.

578 ——, 1975m: Investitionen in Saudi-Arabien. Rechtsinformationen 66; Köln. 34 S. -H 3-

579 ——, 1976-: Saudi-Arabien: wirtschaftliche Entwicklung 1975-. Marktinformationen; Köln.

58o ——, 1976a: Wirtschaftslage in Saudi-Arabien. Mitteilungen 138; Köln. 8 S.

581 ——, 1976b: Hinweise für das Saudi-Arabien-Geschäft (Einreise-, Unterkunfts- und Verkehrsfragen). Mitteilungen 237; Köln. 2 S. -212-

582 ——, 1976c: Der Markt für Baumaterialien in Saudi-Arabien. Mitteilungen 247; Köln. 2 S. -212-

583 ——, 1976d: Arabische Staaten Asiens: Entwicklungstendenzen der Nahrungsmittelindustrie. Marktinformationen B/844; Köln. 34 S. -H 3-

584 ——, 1976e: Handelsvertreterrecht in Irak, Kuwait, Saudi-Arabien, Bahrain, Qatar, Jordanien und der AR Jemen. Rechtsinformationen 89; Köln. II, 87 S. -H 3-

585 ——, 1976f: Saudi-Arabien: Entwicklungsprojekte im Rahmen des zweiten saudiarabischen Fünfjahresplans 1975/76-1979/8o. Köln. IV, 22 S. -Bm 25-

586 ——, 1977-: Saudi-Arabien: Wirtschaftsdaten und Wirtschaftsdokumentation 1977-. Marktinformationen; Köln.

587 Bundesstelle für Aussenhandelsinformation, 1977a: Hinweise für das Saudi-Arabien-Geschäft (Fragen der Wirtschaftsmentalität). Mitteilungen 119; Köln. 4 S. -212-

588 ——, 1977b: Der Markt für Baumaschinen in Saudi-Arabien. Mitteilungen 247; Köln. 2 S.

589 ——, 1977c: Investitionen in Saudi-Arabien und Kuwait. Wirtschafts- und Steuerrecht 53; Köln. 384 S. -2o6, H 3, H 223-

59o ——, 1978a: Der Markt für Uhren in Saudi-Arabien. Marktinformationen 184; Köln. 2 S.

591 ——, 1978b: Der Markt für Glaserzeugnisse in Saudi-Arabien. Marktinformationen 299; Köln. 2 S.

592 ——, 1979-: Saudi-Arabien: Energiewirtschaft 1977-. Marktinformationen; Köln.

593 ——, 1979a: Saudi-Arabien: Investitions- und Handelsvertreterrecht. Rechtsinformationen 112; Köln. I, 19 S. -H 3-

594 ——, 1979b: Saudi-Arabien: Durchführungsbestimmungen zu dem Investitionsförderungsgesetz 1979. Rechtsinformationen 118; Köln. I, 17 S. -H 3-

595 ——, 1979c: Der Markt für Hebezeuge und Fördermittel in Saudi-Arabien. Mitteilungen 165; Köln. 2 S.

596 ——, 1979d: Der Markt für Pumpen in Saudi-Arabien. Mitteilungen 168; Köln. 2 S.

597 ——, 1979e: Hinweise für das Saudi-Arabien-Geschäft (Einreise-, Unterkunfts- und Verkehrsfragen). Mitteilungen 19o; Köln. 4 S. -212-

598 ——, 1979f: Der Saudi-Markt für Industrie und Bauwesen. Broschüre 55.oo1.79.472; Köln. 22o S.

599 ——, 198oa: Der Markt für Kühlschränke und Kühlanlagen in Saudi-Arabien. Mitteilungen 267; Köln. 2 S.

6oo ——, 198ob: Der Markt für Papier und Pappe in Saudi-Arabien. Mitteilungen 285; Köln. 2 S.

6o1 ——, 1981a: Handelsvertreterrecht in Saudi-Arabien. Rechtsinformationen 15o; Köln. II, 35 S. -H 3-

6o2 ——, 1981b: Der Markt für Zweiräder und Zweiradteile in Saudi-Arabien. Mitteilungen 117; Köln. 2 S.

603 Bundesstelle für Aussenhandelsinformation, 1981c: Der Markt für Werkzeuge in Saudi-Arabien. Mitteilungen 228; Köln. 2 S.

604 ——, 1981d: Der Markt für elektrische Haushaltsgeräte in Saudi-Arabien. Mitteilungen 229; Köln. 2 S.

605 ——, 1982a: Kurzberichte zum Investitionsrecht der arabischen Staaten. Rechtsinformationen 153; Köln. 53 S.

606 ——, 1982b: Der Markt für Spielzeug und Sportartikel in Saudi-Arabien. Mitteilungen 163; Köln. 2 S.

607 ——, 1982c: Der Markt für Möbel in Saudi-Arabien. Mitteilungen 164; Köln. 2 S.

608 ——, 1983a: Der Markt für Oberbekleidung in Saudi-Arabien. Mitteilungen 47; Köln. 2 S.

609 ——, 1983b: Der Markt für Tonaufnahme- und -wiedergabegeräte sowie Tonträger in Saudi-Arabien. Mitteilungen 118; Köln. 2 S.

610 ——, 1983c: Hinweise für das Saudi-Arabien-Geschäft (Einreise-, Unterkunfts- und Verkehrsfragen). Mitteilungen 131; Köln. 4 S.

611 ——, 1983d: Der Markt für Heizungs-, Lüftungs- und Klimageräte sowie Anlagen in Saudi-Arabien. Mitteilungen 137; Köln. 2 S.

612 ——, 1983e: Der Markt für Uhren in Saudi-Arabien. Mitteilungen 162; Köln. 2 S.

613 Bundesverband der Deutschen Industrie e.V., 1977: Industrielle Zusammenarbeit mit Saudi-Arabien. Informationsgespräch mit dem saudischen Industrieminister al-Ghosaibi. Köln. 3o S. -F 2oo3-

614 —— (Hg.), ca. 1981: Saudi-Arabien-Seminar. Möglichkeiten und Grenzen der wirtschaftlichen Zusammenarbeit unter besonderer Berücksichtigung mittelständischer Unternehmen. Veranstaltet vom Bundesverband der Deutschen Industrie gemeinsam mit der Deutsch-Arabischen Gesellschaft am 26. Januar 1981 im Haus der Deutschen Industrie, Köln. Köln. 72 S.

615 Bunyan, J., 1982: Insurance developments in Saudi Arabia. In: The Review (London) (April 16, 1982): 6-7. -H 3-

616 Burchardt, H., 1906: Ost-Arabien von Basra bis Maskat auf Grund eigener Reisen. In: Zeitschrift der Gesellschaft für Erdkunde zu Berlin 1906: 305-322. -7, 16, 21-

617 Burckhardt, J.L., 1830: Reisen in Arabien, enthaltend eine Beschreibung derjenigen Gebiete in Hedjaz, welche die Mohammedaner für heilig achten. Von der Londoner Gesellschaft zur Beförderung der Entdeckung des Innern von Africa hg. Neue Bibliothek der wichtigsten Reisebeschreibungen zur Erweiterung der Erd- und Völkerkunde 54; Weimar: Landes-Industrie-Comptoir (Nachdruck: Stuttgart: Brockhaus, 1963. Originalausgabe: Travels in Arabia, comprehending an account of those territories in Hedjaz, which the Mohammedans regard as sacred, by the late John Lewis Burckhardt. Ed. by W. Ouseley. 2 vols. London: Colburn, 1829; London: Murray, ^{2}1830; reprint: New York: Barnes & Noble, 1968. Französische Ausgabe: Voyages en Arabie, contenant la description des parties du Hedjaz, regardées comme sacrées par les musulmans, suivis de notes sur les Bédouins et d'un essai sur l'histoire des Wahhabites. 3 vols. Paris: Bertrand, 1835. Italienische Ausgabe: Viaggi in Arabia di J.L. Burckhardt. Prato: Giachetti, 1844). XIV, 706 S. -24, 212, Tü 17-

618 ——, 1831: Bemerkungen über die Beduinen und Wahaby, gesammelt während seinen Reisen im Morgenlande von dem verstorbenen Johann Ludwig Burckhardt. Von der Gesellschaft zur Beförderung der Entdeckung des innern Africa hg. Neue Bibliothek der wichtigsten Reisebeschreibungen zur Erweiterung der Erd- und Völkerkunde 57; Weimar: Landes-Industrie-Comptoir (Originalausgabe: Notes on the Bedouins and Wahábys, collected during his travels in the East, by the late John Lewis Burckhardt. 2 vols. London: Colburn & Bentley, 1830). XII, 608 S. -21, 24-

619 Bureau des Affaires Économiques et Sociales de l'ONU à Beyrouth, 1969: Étude de certains problèmes que pose le développement dans divers pays du Moyen-Orient, 1968. New York: UNO.

620 Burki, S.J., 1980: What migration to the Middle East may mean for Pakistan. In: Journal of South Asian and Middle Eastern Studies 3(3): 47-66. -1a, 12, 21-

621 Burningham, C.W.M., 1952: Report to the Government of Saudi Arabia on land and water resources. Report 4o; Rome: FAO.

622 Burrell, R.M., 1976: Saudi Arabia. In: R.M. Burrell, S. Hoyle, K.S. McLachlan & C. Parker: The developing agriculture of the Middle East - opportunities and prospects. London: Graham & Trotman, 1976: 55-63.   -21, 1oo-

623 Burton, I., 1893: The life of Captain Sir Richard Burton, K.C.M.G., F.R.G.S. 2 vols. London: Chapman & Hall.   -156-

624 Burton, R.F., 1855-56: Personal narrative of a pilgrimage to El-Medinah and Meccah. 3 vols. London: Longman, Brown, Green & Longmans (amerikanische Ausgabe: New York: Dover Publications, 1964. Gekürzte deutsche Ausgabe: Meine Wallfahrt nach Medina und Mekka. Wege zum Wissen 13; Berlin: Ullstein, 1924). XIV, 388; IV, 426 resp. X, 448 p.   -21, 24-

625 ——, 1878: The gold-mines of Midian and the ruined Midianite cities. A fortnight's tour in north-western Arabia. London: Kegan Paul (reprint: Arabia Past and Present 9; Cambridge/England, New York: The Oleander Press, Naples: The Falcon Press, 1979). XVI, 398 p.   -21, 24-

626 ——, 1879: The land of Midian (revisited). 2 vols. London: Kegan Paul (reprint: Arabia Past and Present 11; Cambridge/England, New York: The Oleander Press; Naples: The Falcon Press, 1979). XXVIII, 338 resp. VII, 319 p.   -24-

627 Busch, B.C., 1967: Britain and the Persian Gulf, 1894-1914. Berkeley, Los Angeles: University of California Press; London: Cambridge University Press. XIV, 432 p.   -21, 24/213-

628 ——, 1971: Britain, India, and the Arabs, 1914-1921. Berkeley, Los Angeles, London: University of California Press. IX, 522 p.   -24/213-

629 Buschow, R., 1979: The Prince & I. London: Futura Publications. 268 p.   -21-

630 Bushnak, M.'U., 1972: The influence of environment on the development of the Saudi administration system. M.A., Political Studies and Public Administration Department, American University of Beirut.   -Jafet Library, American University of Beirut-

631 Bushra, S. el-, 1980: The distribution of population and medical facilities in Saudi Arabia. In: Erdkunde 34: 215-218. -7, 212, Tü 17-

632 Business International S.A., 1981: Saudi Arabia: issues for growth. An inside view of an economic power in the making. Geneva, New York, Hongkong, Tokyo.

633 Business laws of Saudi Arabia. Middle East Business Law Series; London: Graham & Trotman, 1977. Var.pag. -2o6, H 223-

634 Bustani, E., 1958: Doubts and dynamite. The Middle East today. London: Wingate. 159 p. -352-

635 Butler, G.C., 1960: Kings and camels: an American in Saudi Arabia. New York: Devin-Adair. X, 2o6 p. -12, 21-

636 ——, 1964: Beyond Arabian sands: the people, places, and politics of the Arab world. New York: Devin-Adair. 223 p. -21-

637 Butler, S.S., 1909: Baghdad to Damascus viâ El Jauf, northern Arabia. In: The Geographical Journal 33: 517-535. -21, 24, 291-

638 Buxton, J., 1981: Saudi Arabia. In: Middle East Review 7: 275-276, 278, 28o, 282-284, 287, 29o. -1a, 21, 24/213-

639 Buxton, P.A., 1944: Rough notes: Anopheles mosquitoes and malaria in Arabia. In: Transactions of the Royal Society of Tropical Medicine and Hygiene 38: 2o5-214. -1a, 38-

C

640 Cain, L.F., et al., 1975: Expansion of higher education: an implementation strategy for Saudi Arabia. Washington: US Department of Health, Education, and Welfare.

641 Calverley, Edwin E., 1921: The doctrines of the Arabian 'Brethren'. In: The Moslem World 11: 364-376. -21, 291, 352-

642 ——, 1952: Samuel Marinus Zwemer. In: The Muslim World 42: 157-159. -21, 291, 352-

643 Calverley, Eleanor T., 1963: A tribute: Paul Wilberforce Harrison, M.D., F.A.C.S., January 12, 1883 - November 3o, 1962. In: The Muslim World 53: 263-264. -21, 291, 352-

644 Campaign against the shifting sands. In: Aramco World 13.1962 (2): 12-14.   -H 223-

645 Campbell, C.P., 198o: Trade training in Saudi Arabia: a strategy for the eighties. In: Islam and the Modern Age 11: 149-165.   -21-

646 Campbell, J.C., 1977: Oil power in the Middle East. In: Foreign Affairs (New York) 56: 89-11o.   -21, 24, 25-

647 ——, 1979: The Middle East: the burden of empire. In: Foreign Affairs (New York) 57: 613-632.   -21, 24, 2o6-

648 Candler, E., 1925: Lawrence and the Hejaz. In: Blackwood's Magazine 218: 733-761.   -1a, 18o, 385-

649 Capece Galeota Zuccoli, V., 1954: Arabia Saudiana 1954. Milano: Ceschina. 187 S.   -12-

65o Carami, M.S.G., 1978: Development and proposed reorganization of educational administration in Saudi Arabia. Ed.D., University of Cincinnati. XVII, 359 p.   -H 223-

651 Carmichael, J., 1942: Prince of Arabs. In: Foreign Affairs (New York) 2o: 719-731.   -24, 25, 18o-

652 Carmichael, K., 1967: To pull the cork. In: Aramco World 18 (1): 8-17.   -24-

653 Caroe, O., 1951: Wells of power: the oilfields of south-western Asia. A regional and global study. London: Macmillan. XX, 24o p.   -21, 24/213-

654 Carpet of the desert. A new road links Dhahran and Abqaiq. In: Aramco World 5.1954(12): 2-4.

655 Carré, O., 1982: Idéologie et pouvoir en Arabie Saoudite et dans son entourage. In: Centre d'Études et de Recherches sur l'Orient Arabe Contemporain (éd.): La péninsule Arabique d'aujourd'hui. Sous la direction de P. Bonnenfant. T. 1. Paris: Centre National de la Recherche Scientifique, 1982: 219-244.   -21-

656 Carruthers, D., 191o: A journey in north-western Arabia. In: The Geographical Journal 35: 225-248.   -21, 24, 291-

657 ——, 1922: Captain Shakespear's last journey. In: The Geographical Journal 59: 321-334, 4o1-418.   -21, 24, 291-

658 Carruthers, D., 1935: Arabian adventure: to the Great Nafud in quest of the Oryx. London: Witherby. XII, 2o8 p. -12, 21-

659 Carter, J.R.L., 1979: Leading merchant families of Saudi Arabia. London: Scorpion Publications in association with the Llewellyn Group. 19o p. -21, H 223-

66o ——, 1981: Investors in Saudi Arabia: a reference to private investment. London: Scorpion Publications. 452 p.

661 Carter, L.N., 1982: Geography and population. - Living conditions. In: R.F. Nyrop et al.: Saudi Arabia: a country study. Area Handbook Series DA Pam. 55o-51; Washington: US Government Printing Office, 1982: 45-89. -H 223-

662 Carter, W., 1966: The pilgrim railway. In: The Geographical Magazine 39: 422-433. -93, 464-

663 Caskel, W., 1929: Altes und neues Wahhabitentum. In: Ephemerides Orientales (38): 1-1o (Nachdruck in: Bustan 196o(2): 6-9). -4, 5, 21-

664 ——, 1933: Das Farraschen-Amt in Medina. In: Aus fünf Jahrtausenden morgenländischer Kultur. Festschrift Max Freiherrn von Oppenheim zum 7o. Geburtstage gewidmet von Freunden und Mitarbeitern. Archiv für Orientforschung, Beibd. 1; Berlin: Weidner, 1933: 138-147. -21-

665 ——, 1953: Zur Beduinisierung Arabiens. In: Zeitschrift der Deutschen Morgenländischen Gesellschaft 1o3/N.F. 28: *28*-*36*. -21, 291, 352-

666 Cattan, H., 1967: The evolution of oil concessions in the Middle East and North Africa. Dobbs Ferry: Oceana Publications. XVI, 173 p. -12-

667 Cattin, J., 197o: Les incidences de la production pétrolière de l'Arabie Séoudite sur son économie. Thèse, Faculté de Droit et des Sciences Économiques, Université de Neuchâtel. 187 p. -2o6-

668 Cayre, G., 1982: La solidarité des pays producteurs de pétrole et des pays en voie de développement: les fonds d'aide multilatérale. In: Centre d'Études et de Recherches sur l'Orient Arabe Contemporain (éd.): La péninsule Arabi-

que d'aujourd'hui. Sous la direction de P. Bonnenfant. T. 1. Paris: Centre National de la Recherche Scientifique, 1982: 277-3o6. -21-

669 Cement shortage and construction bottlenecks. In: The Arab Economist 7.1975(83): 36. -2o6, H 3-

67o Central Office of Information, Reference Division, 196o: Saudi Arabia. Reference Paper R.4869; London. 36 p. -212-

671 Centre d'Études et de Recherches sur l'Orient Arabe Contemporain (éd.), 1982: La péninsule Arabique d'aujourd'hui. Sous la direction de P. Bonnenfant. 2 vols. Paris: Centre National de la Recherche Scientifique. XV, 379 resp. XVII, 724 p. -21-

672 Centre for Economic, Financial and Social Research and Documentation S.A.L., 1974: Saudi Arabia. The Arab Economist 65, Suppl.; Beirut. 7o p. -H 3-

673 Champenois, L. & G.J.-L. Soulié, 1982: Le royaume d'Arabie Saoudite. - Chronologie du royaume d'Arabie Saoudite. In: Centre d'Études et de Recherches sur l'Orient Arabe Contemporain (éd.): La péninsule Arabique d'aujourd'hui. Sous la direction de P. Bonnenfant. T. 2: Études par pays. Paris: Centre National de la Recherche Scientifique, 1982: 565-6o2. -21-

674 Chapman, R.A., 1974: Administrative reform in Saudi Arabia. In: Journal of Administration Overseas 13: 332-347. -2o6-

675 Chatelus, M., 1975: Pétrole et perspectives de développement. Analyse de quelques états du Moyen-Orient. In: Mondes en Développement (1o): 221-241. -2o6-

676 ——, 1982: Brève histoire du pétrole dans la péninsule Arabique. - De la rente pétrolière au développement économique: perspectives et contradictions de l'évolution économique dans la péninsule. In: Centre d'Études et de Recherches sur l'Orient Arabe Contemporain (éd.): La péninsule Arabique d'aujourd'hui. Sous la direction de P. Bonnenfant. T. 1. Paris: Centre National de la Recherche Scientifique, 1982: 59-154. -21-

677 Cheesman, R.E., 1923: From Oqair to the ruins of Salwa. In: The Geographical Journal 62: 321-335. -21, 291, 352-

678 Cheesman, R.E., 1925: The deserts of Jafura and Jabrin. In:
The Geographical Journal 65: 112-141.   -21, 291, 352-

679 ——, 1926: In unknown Arabia. London: Macmillan. XX, 447 p.
-7, 21-

680 Chelhod, J., 1971: Le droit dans la société bédouine. Recherches ethnologiques sur le 'orf ou droit coutumier des Bédouins. Petite Bibliothèque Sociologique Internationale A,
12; Paris: Rivière. XIII, 461 p.   -21, Tü 17-

681 Cheney, M.S., 1958: Big oilman from Arabia. London, Melbourne,
Toronto: Heinemann; New York: Ballantine Books. 32o p.
-18o-

682 Chérif, A., 193o: Le Pèlerinage de la Mecque. Essai d'histoire, de psychologie et d'hygiène sur le voyage sacré de l'
Islam. Beyrouth: Angeliil. VI, 71 p.   -21-

683 Childs, J.R., 195o: Saudi Arabia's libraries still largely
unknown. In: The Library Journal 75: 1276-1277, 1382-1383.
-21, 24, 352-

684 ——, 1969: Foreign service farewell: my years in the Near
East. Charlottesville: The University Press of Virginia.
192 p.   -21-

685 Chisholm, A.H.T., 1975: The first Kuwait oil concession agreement: a record of the negotiations 1911-1934. London: Cass.
XV, 254 p.   -212, Tü 17-

686 Christian, V., 1924: Der Pariastamm der Ṣlêb. In: Mitteilungen
der Anthropologischen Gesellschaft in Wien 54: (27)-(29).
-16, 21, 24-

687 Chu, C.K., S.K. Djazzar & M.H. Adham, 1963: Report on a
health survey of Saudi Arabia from 1o November 1962 to
31 January 1963. EM/PHA/11o; Alexandria: WHO Regional Office.

688 Chubin, S., 1982: Security in the Persian Gulf 4: the role of
outside powers. Aldershot: Gower. V, 18o p.   -21-

689 Cipriani, L., 1939: Arabi dello Yemen e dell'Higiāz. Firenze:
Centro di Studi Coloniale. 58 S.

69o Clarke, J., 1972: Labor law and practice in the Kingdom of
Saudi Arabia. US Department of Labor, Bureau of Labor

Statistics Report 4o7; Washington: US Government Printing Office. VII, 1o3 p.   -2o6, B 212, B 851-

691 Clawson, M., H.H. Landsberg & L.T. Alexander, 1971: The agricultural potential of the Middle East. The Middle East: Economic and Political Problems and Prospects; New York: American Elsevier. XX, 312 p.   -21, 1oo-

692 Clayton, G.F., 1929: Arabia and the Arabs. In: Journal of the Royal Institute of International Affairs 8: 8-2o. -18o, 291, 352-

693 ——, 1969: An Arabian diary. Introduced and ed. by R.O. Collins. Berkeley, Los Angeles: University of California Press. XIV, 379 p.   -12, 21-

694 Clements, F., 1976: Training for development. In: Middle East International (76): 19-21.   -3o-

695 Clemow, F.G., 191o: Étude sur la défense sanitaire du chemin de fer du Hedjaz. In: Revue d'Hygiène et de Police Sanitaire 32: 213-244, 341-361.   -F 1-

696 ——, 1913: A visit to the rock-tombs of Medain-i-Salih, and the southern section of the Hejaz railway. In: The Geographical Journal 42: 534-54o.   -21, 291, 352-

697 Cleron, J.P., 1978: Saudi Arabia 2ooo: a strategy for growth. London: Croom Helm. 168 p.   -21, 1oo, 2o6-

698 Clifford, M.L., 1977: The land and people of the Arabian Peninsula. Portrait of the Nations Series; Philadelphia, New York: Lippincott. 191 p.   -21-

699 Clifford, R.L., 1976: The Arabian American Oil Company. In: Asian Affairs (London) 63/N.S. 7: 178-182.   -5, 21, 9o-

7oo Cobbold, E., 1934: Pilgrimage to Mecca. London: Murray. XIX, 259 p.   -1a-

7o1 ——, 1935: Pilgrim to Mecca. In: The Geographical Magazine 1: 1o7-116.   -7, 18o, 464-

7o2 Colby, C.B., 1948a: Pipeline air force. In: Flying 42(Jan.): 15.   -89-

7o3 ——, 1948b: Aramco takes to the air. In: World Oil 127 (Apr.): 227-228.   -89, 9o-

704 Cole, D.P., 1970: Sociological factors affecting the success of the Faisal Settlement Project. Presented to WAKUTI GmbH. Zug: WAKUTI GmbH, Consulting Engineers. 23 p.

705 ——, 1973a: The enmeshment of nomads in Sa'udi Arabian society: the case of Āl Murrah. In: C. Nelson (ed.): The desert and the sown; nomads in the wider society. Research Series 21; Berkeley: Institute of International Studies, University of California, 1973: 113-128.   -7, Tü 17-

706 ——, 1973b: Bedouins of the oil fields. In: Natural History 82(5): 94-103.   -89, 291, F 1-

707 ——, 1974: Bedouins of the oil field. In: Ekistics 37: 268-270.   -1a, 89, 206-

708 ——, 1975: Nomads of the nomads: the Āl Murrah Bedouin of the Empty Quarter. Worlds of Man, Studies in Cultural Ecology; Chicago: Aldine (completely rev. version of: The social and economic structure of the Āl Murrah: a Saudi Arabian Bedouin tribe. Ph.D., Department of Anthropology, University of California, Berkeley). 179 p.   -21-

709 ——, 1979: The design and management of pastoral development. Pastoral nomads in a rapidly changing economy: the case of Saudi Arabia. Pastoral Network Paper 7e; London: Agricultural Administration Unit, Overseas Development Institute. 23 p.

710 ——, 1980: Pastoral nomads in a rapidly changing economy: the case of Saudi Arabia. In: T. Niblock (ed.): Social and economic development in the Arab Gulf. London: Croom Helm; Exeter: Centre for Arab Gulf Studies, 1980: 106-121.   -21, Bo 149, H 3-

711 ——, 1981: Bedouin and social change in Saudi Arabia. In: Journal of Asian and African Studies 16: 128-149.   -1a, 21, 206-

712 Collins, M., 1981: Riyadh: the Saud balance. In: The Washington Quarterly 4(1): 200-208.   -12, 24/213, 206-

713 Colloques de Ryad sur le dogme musulman et les droits de l'homme en Islam. En annexe, le memorandum du Gouvernement du Royaume d'Arabie Saoudite relatif au dogme des droits de l'homme en Islam et son application dans le Royaume,

adressé aux organisations internationales intéressées. Beyrouth: Dar al-Kitab Allubnani, 1973. 258 p.   -2o6-

714 Committee for Middle East Trade, 1979: Development of the industrial cities of Jubail and Yanbu in the Kingdom of Saudi Arabia. Opportunities for British consultants, contractors and suppliers. COMET Report; London. XI, 1oo p. -21, H 3-

715 Communications in the Middle East. In: Middle Eastern Affairs 7.1956: 248-256.   -21, 25, 18o-

716 Coneybear, J.F., D. Skilbeck & H.E. Myers, 1943: The organisation of transport and food in the Kingdom of Saudi Arabia. British Middle East Office MS. 43/13/1; Beirut.

717 ——, —— & ——, 1944: Survey of the organisation of agriculture, food supplies and transport in Saudi Arabia. British Middle East Office MS. 1o/13/2; Beirut.

718 Confederation of British Industry, 1975: Foreign investment in Saudi Arabia: legislation, procedure and related conditions. London. 35 p.

719 ——, 1978: Investing and working in Saudi Arabia: the proceedings of a CBI conference held in the CBI Council Chamber, Tothill Street, London, on 5 July 1977. London.

72o Container facilities expanded at Saudi Arabian ports. In: Middle East Construction 4.1979(1): 11.   -89-

721 Cooke, H.V., 1952: Challenge and response in the Middle East. The quest for prosperity 1919-1951. New York: Harper. XIII, 366 p.   -21-

722 Cooksey, J.J., 1939: The new day in Arabia. In: World Dominion 17: 67-72.   -21-

723 Cooley, J., 1979: Riyadh-Washington axis in jeopardy. In: Middle East International (94): 2-3.   -12, 3o-

724 Coon, C.S., 1955a: The nomads. In: S.N. Fisher (ed.): Social forces in the Middle East. Ithaca: Cornell University Press, 1955: 23-42.   -24, Tü 17-

725 —— (supplemented by P.G. Franck), 1955b: Operation Bultiste: promoting industrial development in Saudi Arabia. In:

H.M. Teaf, Jr. & P.G. Franck (eds.): Hands across frontiers: case studies in technical cooperation. Publications of the Netherlands Universities Foundation for International Cooperation; Leiden: Sijthoff, 1955: 3o7-361.   -38-

726 Coon, C.S., H. von Wissmann, F. Kussmaul & W.M. Watt, 196o: Badw. In: EI^2 1: 872-892.   -21, 24, 212-

727 Cooper, W.W. & F.M. al-Tamimi, 1982: An integrated multi-objective family-housing model: the case for Saudi Arabia. In: Institute of Management Sciences (ed.): Planning processes in developing countries. Techniques and achievements. TIMS Studies in Management Sciences 17; Amsterdam, New York, 1982: 199-224.   -2o6-

728 Coppock, J.D., 1966: Foreign trade of the Middle East: instability and growth, 1946-1962. Beirut: Economic Research Institute, American University of Beirut. XIV, 255 p.   -21-

729 Corancez, L.A.O. de, 181o: Histoire des Wahabis, depuis leur origine jusqu'à la fin de 18o9. Paris: Crapart. VIII, 2oo p.

73o Corcoran, K.R. (ed.), 1981: Saudi Arabia: keys to business success. London (etc.): McGraw-Hill. XIII, 225 p.   -21, 2o6, H 3-

731 Cornwall, P.B., 1942: Arabia awaits the outcome. In: Asia (New York) 42: 275-278.   -1a-

732 Cornwallis, K., 1976: Asir before World War I: a handbook. Arabia Past and Present 3; Cambridge/England, New York: The Oleander Press; Naples: The Falcon Press (Originalausgabe: Handbook to Asir. Cairo: Arab Bureau, 1916). 155 p.   -21, Tü 17-

733 Coulon, C., 1973: La tournée africaine du roi Fayçal. In: Revue Française d'Études Politiques Africaines 8(85): 18-21.   -18, 21, 2o6-

734 Council of British Manufacturers of Petroleum Equipment (CBMPE), 1967: Report of mission to Saudi Arabia and Kuwait. London. 27 p.   -H 3-

735 'Country store' in Saudi Arabia. In: Aramco World 12.1961 (7): 3-6.   -H 223-

736 Crane, C.R., 1928: Visit to the Red Sea littoral and the Yaman. In: Journal of the Royal Central Asian Society 15: 48-67.   -21, 9o, 291-

737 Crane, R.D., 1975: Saudi Arabia's plans for national development: a systematic analysis for U.S. negotiators. Washington: Native American Economic Development Corporation. 392 p.

738 ——, 1978a: Planning the future of Saudi Arabia: a model for achieving national priorities. Praeger Special Studies in International Economics and Development; New York, London, Sydney, Toronto: Praeger. XII, 241 p.   -21, 2o6, H 223-

739 ——, 1978b: Planning, Islamic style: the Saudis are learning financial discipline, while they try to protect their social system from disruptions. In: Fortune 98(2): 114-116.   -24, 3o, 89-

74o Crary, D.D., 1951: Recent agricultural developments in Saudi Arabia. In: The Geographical Review 41: 366-383.   -24, 93, 2o6-

741 ——, 1955: The villager. In: S.N. Fisher (ed.): Social forces in the Middle East. Ithaca: Cornell University Press, 1955: 43-59.   -24, Tü 17-

742 Craufurd, C.E.V., 1921: Hodeida before and after the war. In: The Geographical Journal 58: 464-465.   -21, 291, 352-

743 ——, 1933: Yemen and Assir: El Dorado? In: Journal of the Royal Central Asian Society 2o: 568-576.   -21, 9o, 291-

744 Crawford, M.J., 1982: Civil war, foreign intervention, and the question of political legitimacy: a nineteenth-century Sa'ūdī Qadī's dilemma. In: International Journal of Middle East Studies 14: 227-248.   -1a, 12, 21-

745 Création du Conseil de Coopération du Golfe. In: L'Économiste Arabe 24.1981(271): 15-16.   -2o6, H 223-

746 Cressey, G.B., 196o: Crossroads: land and life in southwest Asia. The Lippincott Geography Series; Chicago, New York, Philadelphia: Lippincott. XIV, 593 p.   -24-

747 Crichton, A., ^{2}1852: History of Arabia and its people, containing an account of the country and its inhabitants,

the life and religion of Mohammed, the conquests, arts, and literature of the Saracens, the Caliphs of Damascus, Bagdad, Africa, and Spain, the modern Arabs, the Wahabees, the Bedouins &c. &c. Edinburgh Cabinet Library; London, Edinburgh: Nelson (^{1}1833) (amerikanische Ausgabe: New York: Harper, 1845). XVI, 652 p.   -21-

748 Crowe, P., 1964: A trip to Madain Salih. In: Journal of the Royal Central Asian Society 51: 291-3oo.   -21, 9o, 291-

749 Crowe, T. (ed.), 1967: Gathering moss: a memoir of Owen Tweedy. London: Sidgwick & Jackson. XV, 345 p.

75o Cruz, D. da, 1964a: A drop of rain. In: Aramco World 15(4): 24-29.   -H 223-

751 ——, 1964b: The long steel shortcut. In: Aramco World 15(5): 16-25.   -H 223-

752 ——, 1965: Pilgrim's road. In: Aramco World 16(5): 24-33.   -24-

753 ——, 1967a: TV in the Middle East. In: Aramco World 18(5): 18-25.   -24-

754 ——, 1967b: Plague across the land. In: Aramco World 18(6): 2o-25.   -24-

755 ——, 1969: Silver threads among the old. In: Aramco World Magazine 2o(4): 28-36.   -24-

756 Cummings, J.T., H.G. Askari & M. Skinner, 198o: Military expenditures and manpower requirements in the Arabian peninsula. In: Arab Studies Quarterly 2: 38-49.   -21-

D

757 D., 193o: Ein Funknetz in Hedschaz und Nedschd. In: Mitteilungen der Geographischen Gesellschaft in Wien 73: 3oo-3o1.   -16, 21, 291-

758 Dabbagh, T.H. al-, 198o: Analysis of managerial training and development within Saudi Arabian Airlines. Ph.D., North Texas State University, Denton. 344 p.

759 Dabbagh, Z.M.A. al-, 198o: The establishment of average body measurements for Saudi Arabian women and the development

of basic garment patterns. Ph.D., Texas Woman's University, Denton. 233 p.

760 Daggy, R.H., 1957: Oasis malaria. In: Harvard School of Public Health: Industry and Tropical Health 3: 42-55.

761 ——, 1959: Malaria in oases of eastern Saudi Arabia. In: The American Journal of Tropical Medicine and Hygiene 8: 223-291.   -16, 25, 1oo-

762 Daghistani, Abdal-Majeed I., 1976: At-Taif city and amirate, al-Hijaz, Saudi Arabia: a geographical study. Ph.D., University of Durham.

763 ——, c. 1981: At-Taif, a city in transition. N.p.: Ministry of Education, Kingdom of Saudi Arabia. 1o6 p. -212-

764 Daghistani, Abdulaziz I., 1979: Economic development in Saudi Arabia: problems and prospects. Ph.D., Department of Economics, University of Houston. XVII, 2o2 p.   -21, Tü 17-

765 Daguillon, L., 1866: Le Hedjaz (pèlerinage à la Mecque). Étude sur sa géographie, ses caravanes, ses épidemies (typhus, choléra, pestes, etc.). Paris: Challamel.

766 Daham, A.A., 1981: The growth of the Hajj: a decade of increasing pilgrims to Mecca, 197o-1979. Ph.D., Claremont Graduate School. 231 p.

767 Daihan, M.A.-R. al-, 1982: Perceptions of Saudi male academic public secondary school administrators and teachers in Riyadh concerning shortages of Saudi male teachers in secondary public schools in Saudi Arabia. Ed.D., University of Northern Colorado, Greeley. 262 p.

768 Dalenberg, C., 1946: Changes in Arabia. In: The Moslem World 36: 65-68.   -21, 291, 352-

769 Dall, R.F., 198o: A SAGE analysis of overseas employee contract failure in Saudi Arabia. Ed.D., Brigham Young University, Provo. 24o p.

77o Dame, L.P., 1924: Four months in Nejd. In: The Moslem World 14: 353-362.   -21, 291, 352-

771 ——, 193o: Objectives in Arabia. In: The Moslem World 2o: 179-184.   -21, 291, 352-

772 Dame, L.P., 1933: From Bahrain to Taif: a missionary journey across Arabia. In: The Moslem World 23: 164-178. -21, 291, 352-

773 Dame, Mrs. L.P., 1934: A woman's trip to central Arabia. In: The Missionary Review of the World 57: 517-526. -1a, 4, 21-

774 Dammann, T., 1970: Saudi Arabia's dilemma. An interview with King Faisal. In: Interplay 1970(Sept.): 16-19. -206, 281-

775 Dana, L.P., 1923: Arab Asia; a geography of Syria, Palestine, Irak, and Arabia. Beirut: American Press. XI, 170 p.

776 Darrat, A.F., 1982: A monetarist approach to inflation for Saudi Arabia. Ph.D., Indiana University, Bloomington. 221 p.

777 Das Königreich Saudi-Arabien (mit einer Reihe von Untertiteln). In: Auslandskurier - Diplomatischer Kurier 21.1980 (11): 15-42. -21, 206, 212-

778 Das Rote Meer verschmutzt. In: Naturwissenschaftliche Rundschau 34.1981: 76-77. -16, 21, 100-

779 Davidson, B., 1981: Zivilluftfahrt in Saudi-Arabien. Ein integraler Bestandteil der Entwicklungspläne. In: Interavia 36: 916-917. -206, H 3-

780 Davies, P., 1981: Prospects for agricultural and project business in the Kingdom of Saudi Arabia. A Report on Saudi Arabia for Agricultural Exporters; London: British Agricultural Export Council. IV, 51 p. -H 3-

781 Dawisha, A.I., 1979a: Internal values and external threats: the making of Saudi foreign policy. In: Orbis (Philadelphia) 23: 129-143. -21, 24, 206-

782 ——, 1979b: Saudi Arabia in the limelight. In: Middle East International (103): 9-10. -12, 30-

783 ——, 1979c: Saudi Arabia's search for security. Adelphi Papers 158; London: The International Institute for Strategic Studies (deutsche Ausgabe: Saudi-Arabien und seine Sicherheitspolitik. Hg. vom Internationalen Institut für Strategische Studien, London. Bernard & Graefe aktuell 24; München: Bernard & Graefe, 1981). 36 p. -12, 21, H 223-

784 Dawn, C.E., 1957: 'Abdullāh ibn al-Ḥusein, Lord Kitchener e l'idea della rivolta araba. In: Oriente Moderno 37: 1-12. -7, 18, 21-

785 ——, 1959: Ideological influences in the Arab Revolt. In: J. Kritzeck & R.B. Winder (eds.): The world of Islam. Studies in honour of Philip K. Hitti. London: Macmillan; New York: St. Martin's Press, 1959: 233-248. -12, 21-

786 ——, 1960: The Amir of Mecca al-Ḥusayn ibn-'Ali and the origin of the Arab Revolt. In: Proceedings of the American Philosophical Society 1o4(1): 11-34. -25, 93, 18o-

787 ——, 1971: Hāsh̲imids. In: EI^2 3: 263-265. -21, 24, 212-

788 Dayil, A.S.M. al-, 1979: A study to identify ways of increasing the enrollment of Saudi male elementary school teachers in Teacher Training Institutes and improving the quality of instruction in those institutes. Ph.D., University of Idaho, Moscow. 236 p.

789 Deane, C., 1978: Saudi Arabia – business opportunities II. London: Metra Consulting Group. 223 p.

790 Dearden, A. (ed.), 1977: Arab women. Minority Rights Group Report 27; London: Minority Rights Group.

791 Deaver, S., 1980: The contemporary Saudi woman. In: A world of women. Anthropological studies of women in the societies of the world. Praeger Special Studies, Praeger Scientific; New York: Praeger, 1980: 19-42. -2o6-

792 Decken, H. von der, 1973: Saudi-Arabien: Wirtschaftsstruktur. Mitteilungen A/197; Köln: Bundesstelle für Aussenhandelsinformation. 127 S. -2o6-

793 Deeik, K.G., 1972: Business education curriculum practices in Saudi Arabia public secondary schools and certain recommendations for change. Ph.D., University of Southern California, Los Angeles. 224 p.

794 DeGolyer, E., 1946: Oil exploration in the Middle East. In: The Mines Magazine 36: 493-495, 575. -89-

795 De invoer van enkele landbouwprodukten in Saoedi-Arabië in 1974 en 1978. In: Landbouwwereldnieuws 35.1980: 2o4-2o7. -H 3-

796 Delegation der Liga der Arabischen Staaten (Hg.), 1961: Das Königreich Saudi-Arabien. Veröffentlichungen der Liga der Arabischen Staaten 4(1o); Bonn. 59 S. -212, H 3-

797 Demeuse, P., 1978: L'Arabie des Saoud. Le royaume des sables et du vent... Bruxelles: Musin. 217 p. -12, 21-

*** Denis de Rivoyre, B.L.: s. Rivoyre, D. de.

798 Den økonomiske udvikling i Saudi Arabien 1967/68. In: Udenrigsministeriets Tidsskrift 5o.1969: 627-63o. -2o6-

799 DeNovo, J.A., 1963: American interests and policies in the Middle East 19oo-1939. Minneapolis: The University of Minnesota Press. XII, 447 p. -21-

8oo Dequin, H., 1962: Erfahrungen mit deutschem Torf im tropischen Teil der arabischen Halbinsel. In: Der Deutsche Tropenlandwirt 63: 51-52. -7, 26, 1oo-

8o1 ——, 1962/63: Die Landwirtschaft Saudisch-Arabiens und ihre Entwicklungsmöglichkeiten. Zeitschrift für Ausländische Landwirtschaft Sh. 1; Frankfurt/M.: DLG (zugleich: Diss., Fakultät für Landbau, Technische Universität Berlin, 1962). XII, 259 S. -1oo, 212, H 223-

8o2 ——, 1972: Prüfung von möglichen landwirtschaftlichen Anschlussmassnahmen im Zusammenhang mit der Förderung der Oase Al Hasa, Saudisch-Arabien. Riad. IV, 8o S.

8o3 ——, 1973a: A basis of operation for the nomads of the desert. Development aspects of settling nomads in Saudi Arabia and Ethiopia. In: Orient (Opladen) 14: 177-178. -21, 1oo, H 223-

8o4 ——, 1973b: Saudi Arabia. In: World Atlas of Agriculture, under the aegis of the International Association of Agricultural Economists. Monographs ed. by the Committee for the World Atlas of Agriculture. Vol. 2: Asia and Oceania. Novara: Istituto Geografico De Agostini, 1973: 459-468. -1oo-

8o5 ——, 1976: The challenge of Saudi Arabia. The regional setting and economic development as a result of the conquest of the Arabian Peninsula by King 'Abdul 'Aziz Al Sa'ud. Hofuf. VII, 147 p. -12, 93, 1oo-

806 Destrées, M.(?), 1874: Note sur l'arrondissement d'El Haça. In: Bulletin de la Société de Géographie (Paris) VI, 8: 314-315.   -1a, 4, 291-

807 Desvergers, N., 1847: Arabie. L'Univers, ou Histoire et Description de tous les Peuples, de leurs Religions, Moeurs, Coutumes, etc.; Paris: Didot. 522 p.   -Tü 17-

808 Deutsche Bank, 1974: Investieren in Saudi-Arabien. Frankfurt/ M. II, 31 S.   -H 223-

809 ——, 1979: Saudi-Arabien. Frankfurt/M.: Frankfurter Societäts-Druckerei (Originalausgabe: Saudi Arabia. Frankfurt/M.: Frankfurter Societäts-Druckerei, 1978). 94 S.   -21, 93, 2o6-

810 Deutsche Gesellschaft für Technische Zusammenarbeit (GTZ) GmbH, 1979: GTZ-Mitarbeiter in Saudi-Arabien. Informationen für deutsche Agrar-Fachkräfte, die sich für einen Einsatz im Projekt 'Landwirtschaftsschule Buraydah' interessieren (Aussentitel: GTZ-Mitarbeiter in Saudi-Arabien. Informationen für interessierte deutsche Agrar-Fachkräfte). Eschborn. 23 S.

811 Deutsches Hydrographisches Institut, ^{5}1976: Handbuch des Persischen Golfs. Seehandbuch 2o35; Hamburg (^{4}1964). 281 S.   -21-

812 ——, ^{5}1978: Handbuch für das Rote Meer und den Golf von Aden. Seehandbuch 2o34; Hamburg (^{4}1963). 388 S.   -H 2-

813 Deux nouvelles institutions arabes. In: L'Économie des Pays Arabes 17.1974(2o2): 17.   -2o6-

814 Development of the port of Jeddah. In: The Dock and Harbour Authority 51.1971: 418-42o.   -1a, 18, 89-

815 Développement économique et social en Arabie Séoudite. In: Syrie et Monde Arabe 2o.1973(229): 79-84.   -2o6, H 3-

816 Dhaher, A.J., 1981: Culture and politics in the Arab Gulf states. In: Journal of South Asian and Middle Eastern Studies 4(4): 21-36.   -1a, 12, 21-

817 Dhahran International Airport. In: Aramco World 14.1963(3): 17-21.   -H 223-

818 Dhanani, G., 198o: Political institutions in Saudi Arabia. In: International Studies (New Delhi) 19: 59-69.   -1a, 7, 18-

819 Dhanani, G., 1981: Saudi Arabia and non-alignment. In: International Studies (New Delhi) 2o: 361-369. -1a, 7, 2o6-

82o Diab, M.A., 1964: Inter-Arab economic cooperation, 1951-196o. Beirut: Economic Research Institute, American University of Beirut. VIII, 319 p. -21-

821 Dickson, H.R.P., ^{2}1951: The Arab of the desert: a glimpse into Badawin life in Kuwait and Sa'udi Arabia. London: Allen & Unwin (^{1}1949). 648 p. -21, 24, Tü 17-

822 ——, 1956: Kuwait and her neighbours. London: Allen & Unwin. 627 p. -21, 24-

823 Dickson, V., 1949: Artistic house-decoration in Riyadh. In: Man (London) 49: 76-77. -26, 3o, 291-

824 ——, 1971: Forty years in Kuwait. London: Allen & Unwin. 335 p. -21-

825 Didier, C.E., 1857: Séjour chez le Grand-Chérif de la Mekke. Bibliothèque des Chemins de Fer; Paris: Hachette (deutsche Ausgabe: Ein Aufenthalt bei dem Gross-Scherif von Mekka. Leipzig: Schlicke, 1862). VII, 31o p. -1a-

826 Dieckmann, P., 1922: Die Hedjasbahn und die syrischen Privatbahnen im Weltkriege und ihre gegenwärtige Lage. In: Zwischen Kaukasus und Sinai, Jahrbuch des Bundes der Asienkämpfer 2: 47-69. -24-

827 Die Hedschasbahn. In: Archiv für Eisenbahnwesen 39.1916: 289-315. -25, 93, 18o-

828 Dietvorst, A.G.J., 1973: De Moslempelgrimage naar Mekka. In: Geografisch Tijdschrift N.S. 7: 5o-53. -21, 291, 352-

829 Diffelen, R.W. van, 1927: De leer der Wahhabieten. Leiden: Brill (zugleich: Diss., Juristische Fakultät, Universität Leiden, 1927). 83, 4 S. -12-

83o Dilger, K., 1978: Grundbegriffe der Eigentumsordnung zwischen Wandel und Tradition auf der Arabischen Halbinsel. In: Zeitschrift für vergleichende Rechtswissenschaft einschliesslich des Rechts der Entwicklungsländer und der ethnologischen Rechtsforschung 77: 21-78. -21, 24-

831 Dimock, W.C., 1961: Preliminary program for aeolian sand control, area of al-Hasa, Eastern Province, Saudi Arabia. Dhahran: Exploration Department, ARAMCO. 49 p.

832 Dinet, A.É. & S.I. Baâmer, ²1962: Le pèlerinage à la Maison Sacrée d'Allah. Paris: Adrien-Maisonneuve (Paris: Hachette, ¹1930). 218 p.   -Kn 28-

833 Dingelstedt, V., 1916: Arabia and the Arabs. In: Scottish Geographical Magazine 32: 321-330, 366-373.

834 Dingemans, H.H., 1973: Bij Allah's buren. Rotterdam: Donker. 306 S.   -21-

835 Dixon, H., 1964: Cool, dense and ship. In: Aramco World 15 (4): 9-14.   -H 223-

836 Djemal, A., 1922: Erinnerungen eines türkischen Staatsmannes. Bücherei für Politik und Geschichte; München: Drei Masken. 390 S.   -24/213-

837 Dodson, R.H.T., 1978: Icebergs to Arabia? In: Focus (New York) 29(1): 10-16.   -89, 385, Tü 17-

838 Dohaish, A.A. & M.J.L. Young, 1969/73: Modes of address and epistolary forms in Saudi Arabia. In: The Annual of Leeds University Oriental Society 7: 110-117.   -21-

839 ——, 1973: A critical and comparative study of the history of education in the Hijaz during the periods of Ottoman and Sharifian rule between 1869-1925. Ph.D., University of Leeds.

840 —— & M.J.L. Young, 1975: An unpublished educational document from the Ḥijāz (A.H. 1299). In: Annali dell'Istituto Orientale di Napoli 35/N.S. 25: 133-137.   -21-

841 Donini, P.G., 1979: Saudi Arabia's hegemonic policy and economic development in the Yemen Arab Republic. In: Arab Studies Quarterly 1: 299-308.   -21-

842 ——, 1982: Trasformazione economica della Penisola araba ed egemonia saudiana. In: S. Boni & A. Tramontana (Hgg.): Italia e Paesi arabi nell'economia internazionale. Milano: Angeli, 1982: 83-97.

843 Donkan, R. (Pseud., d.i. Zischka, A.), 1935: Die Auferstehung Arabiens. Ibn Sauds Weg und Ziel. Bern, Leipzig, Wien: Goldmann (²1942) (Originalausgabe: Zischka, A.: Ibn Séoud, roi de l'Arabie. Paris: Payot, 1934). 260 S.   -24, Tü 17-

844 Dorozynski, A., 1976: Science, technology and education on the Arabian Peninsula. In: Impact of Science on Society 26: 193-197. -24, 93-

845 Dostal, W., 1956: Die Ṣulubba und ihre Bedeutung für die Kulturgeschichte Arabiens. Monographische Zusammenfassung der Ergebnisse meines Studienaufenthaltes in Kuwēt (1956). In: Archiv für Völkerkunde 11: 15-42. -7, 21, 352-

846 ——, 1958a: Arabia. In: Bulletin of the International Committee on Urgent Anthropological and Ethnological Research (1): 42-45. -1a, 3o, 385-

847 ——, 1958b: Zum Problem der Mädchenbeschneidung in Arabien. In: Wiener Völkerkundliche Mitteilungen 6/N.F. 1: 83-89.

848 ——, 1959: The evolution of Bedouin life. In: F. Gabrieli (Hg.): L'antica società beduina. Studi Semitici 2; Roma: Centro di Studi Semitici, 1959: 11-34. -12-

849 ——, 1964: Paria-Gruppen in Vorderasien. In: Zeitschrift für Ethnologie 89: 19o-2o3. -16, 21, 3o-

85o ——, with contributions by A. Gingrich & H. Riedl, 1983: Ethnographic atlas of 'Asīr. Preliminary report. Österreichische Akademie der Wissenschaften, Philosophisch-historische Klasse, Sitzungsberichte 4o6; Wien: Österreichische Akademie der Wissenschaften. 144 p. -Tü 17-

851 Doughty, C.M., 1884: Travels in north-western Arabia and Nejd. In: Proceedings of the Royal Geographical Society and Monthly Record of Geography N.S. 6: 382-399. -1a, 21, 24-

852 ——, 1888: Travels in Arabia Deserta. 2 vols. Cambridge/England: Cambridge University Press (gekürzte deutsche Ausgaben: Die Offenbarung Arabiens (Arabia Deserta). Leipzig: List, 1937. Reisen in Arabia Deserta; Wanderungen in der Arabischen Wüste 1876-1878. Hg. von H.-T. Gosciniak. Reiseberichte in der Reihe DuMont Dokumente; Köln: DuMont, 1979). XX, 623 resp. XIV, 69o p. -21, 24-

853 ——, 19o8: Wanderings in Arabia. Being an abridgment of 'Travels in Arabia Deserta'. Arranged with introduction by E. Garnett. 2 vols. London: Duckworth. XX, 3o9 resp. X, 296 p. -212-

854 Doughty, C.M., 1931: Passages from Arabia Deserta. Selected by E. Garnett. Life and Letters Series 21; London, Toronto: Cape (reprinted 1935, 1949). 32o p. -24-

855 Dowson, V.H.W., 1949: The date and the Arab. In: Journal of the Royal Central Asian Society 36: 34-41. -21, 9o, 291-

856 ——, 1952: To Arabia in search of date-palm offshoots. In: Journal of the Royal Central Asian Society 39: 45-56. -21, 9o, 291-

857 Doxiadis Associates, 1968: Riyadh - existing conditions. A report prepared for the Ministry of Interior for Municipalities, Kingdom of Saudi Arabia. DOX-SAU-A 2; n.p.

858 ——, 1973: Central Region. Regional physical plan and development program. Existing conditions. Vol. 1. A report prepared for the Ministry of Interior for Municipal Affairs, General Directorate, City and Regional Planning, Kingdom of Saudi Arabia. Report 1, DOX-SAU-RD 4; Riyadh.

859 ——, 1974: Central Region. Regional physical plan and development program. Final report. Vol. 1. A report prepared for the Ministry of Interior for Municipal Affairs, General Directorate, City and Regional Planning, Kingdom of Saudi Arabia. Report 25, DOX-SAU-RD 53; Riyadh. 3o9 p.

86o ——, 1977: Formulating a housing program for Saudi Arabia. In: Ekistics 44: 1o5-1o8. -1a, 89, 2o6-

861 Drees, I.A. al-, 1982: Leadership as a variable in institution building: toward qualified leadership for junior colleges in Saudi Arabia. Ph.D., Indiana University, Bloomington. 251 p.

862 Drouhin, G., R. Ambroggi & A. Kroon, 1963: Report to the Government of Saudi Arabia on future prospects for hydro-agricultural development and technical organization. Suggestions for an administrative and technical organization. Report 1638; Rome: FAO. Var.pag. -Gö 153-

863 Drucker, J., 1975: Divers of Arabia. In: Aramco World Magazine 26(3): 4-11. -21, 24-

864 Du développement de l'économie séoudienne. In: Cahiers de l'Orient Contemporain 26.1969(76): 12-2o. -21-

865 Duguet, F., 1932: Le pèlerinage de la Mecque au point de vue religieux, social et sanitaire. Paris: Rieder. XII, 337 p. -12, 21-

866 Duguid, S., 1970: A biographical approach to the study of social change in the Middle East. Abdullah Tariki as a new man. In: International Journal of Middle East Studies 1: 195-220.   -21, 180, H 223-

867 Duheash, O.A., 1976: Analysis of some Saudi Arabian soils and associated crops. M.Sc., School of Agriculture, University College of North Wales, Bangor.

868 Dunbar, G.S., 1977: West African pilgrims to Mecca. In: The Geographical Review 67: 483-484.   -21, 24, 206-

869 Duncan, P.D. & R.H. Reeser, 1969: Bedouin policy. 2 vols. Riyadh: Stanford Research Institute for Central Planning Organization.

870 Dyer, G., 1979: Saudi Arabia. In: J. Keegan (ed.): World armies. London, Basingstoke: Macmillan, 1979: 607-620.   -21-

E

871 Earle, M.W., 1980: Conservation of neighborhoods in traditional cities. In: Ekistics 47: 36.   -1a, 89, 206-

872 Ebert, C.H.V., 1965: Water resources and land use in the Qaṭīf oasis of Saudi Arabia. In: The Geographical Review 55: 496-509.   -21, 93, 206-

873 Ebrahim, M.H.S., 1981: Problems of nomad settlement in the Middle East with special reference to Saudi Arabia and the Haradh Project. Ph.D., Cornell University, Ithaca. 266 p.

874 Economic Commission for Western Asia (ECWA), 1979: The population situation in the ECWA region: Saudi Arabia. Beirut. 27 p.   -H 223-

875 Economic relations between Saudi Arabia and Japan. In: The Oriental Economist 45.1977(802): 10-13.   -1a, 21, 206-

876 Eddy, W.A., 1954: F.D.R. meets Ibn Saud. Kohinur Series 1; New York: American Friends of the Middle East. 45 p. -Library of Congress, Washington-

877 ——, 1963: King Ibn Sa'ūd: 'Our faith and your iron'. In: The Middle East Journal 17: 257-263.   -21, 206, 352-

878 Edens, D.G. & W.P. Snavely, 1970: Planning for economic development in Saudi Arabia. In: The Middle East Journal 24: 17-30. -21, 38, H 223-

879 ——, 1974: The anatomy of the Saudi revolution. In: International Journal of Middle East Studies 5: 50-64. -21, 212, H 223-

880 Educational administration in Saudi Arabia. In: Educational Trends 1970(13): 8-10.

881 Education of women in Saudi Arabia. In: The Muslim World 46.1956: 366-367. -21, 291, 352-

882 Edwards, F.M., 1957: George Forster Sadleir (1789-1859) of the 47th Regt., the first European to cross Arabia. In: Journal of the Royal Central Asian Society 44: 38-49 (reprinted in: G.F. Sadleir: Diary of a journey across Arabia (1819). Arabia Past and Present 5; Cambridge/England, New York: The Oleander Press; Naples: The Falcon Press, 1977: 3-18). -21, 90, 291-

883 Egbert, R., et al., 1975: Education in Saudi Arabia: findings, recommendations, and proposed projects. Washington: US Department of Health, Education, and Welfare.

884 Eglin, D.R., 1982: Character and structure of the economy. - The petroleum industry and foreign trade. - Agriculture, industry and domestic trade. In: R.F. Nyrop et al.: Saudi Arabia: a country study. Area Handbook Series DA Pam. 550-51; Washington: US Government Printing Office, 1982: 219-309. -H 223-

885 Ehlers, E., 1978: Die Erdölförderländer des Mittleren Ostens 1960-1976. Zum Wert- und Bedeutungswandel einer Wirtschaftsregion. In: Die Erde 109: 457-491. -16, 21, 24-

886 Ehrenberg, C.G., 1827: Vorläufige Bemerkungen über eine, noch unbekannte, grössere Insel im rothen Meere, von den Insulanern Farsan genannt. In: Hertha 9: 312-319. -21, 30, 90-

887 —— (Hg.), 1828: Naturgeschichtliche Reisen durch Nord-Afrika und West-Asien in den Jahren 1820 bis 1825 von Dr. W.F. Hemprich und Dr. C.G. Ehrenberg. Historischer Theil. Berlin, Posen, Bromberg: Mittler.

888 Ehrenberg, E. & W. Mallmann, 1978: Rüstung und Wirtschaft am Golf. Iran und seine Nachbarn (1965-1978). Mitteilungen des Deutschen Orient-Instituts 11; Hamburg: Deutsches Orient-Institut. 224 S.   -21, 24, H 223-

889 Eigeland, T., 1970: The twice-used water. In: Aramco World Magazine 21(6): 22-29.   -24-

890 ——, 1975: Scenic Arabia - a personal view (The changing coast. The not-so-empty quarter. The historic heartland. The busy gateway. The green highlands). In: Aramco World Magazine 26(1): 2-37.   -21, 24, 212-

891 ——, with P. Lunde, 1980: Back to the highlands. In: Aramco World Magazine 31(5): 12-21.   -21, 24, 212-

892 Eilts, H.F., 1971: Social revolution in Saudi Arabia. In: Parameters 1(1): 4-18, (2): 22-33.   -24/213-

893 ——, 1980: Security considerations in the Persian Gulf. In: International Security 5(2): 79-113.   -12, 21, 24/213-

894 ——, 1981: A rejoinder to J.B. Kelly. In: International Security 5(4): 195-203.   -12, 21, 24/213-

895 Eisenberger, J., 1928: Indië en de bedevaart naar Mekka. Leiden: Dubbeldeman. 3, 220 S.

896 Elefteriadès, E., 1944: Les chemins de fer en Syrie et au Liban. Étude historique, financière et économique. Beyrouth: Imprimerie Catholique. XVI, 420 p.   -12-

897 Elektrizitäts-Actien-Gesellschaft, vorm. W. Lahmeyer & Co., Consulting Engineers, 1961: General survey report on the development of industries in Saudi Arabia. Prepared for the Supreme Planning Board of The Government of the Kingdom of Saudi Arabia. Frankfurt/M. 279 p.   -H 223-

898 Elham, M., 1982: Fahd Ben Abdel Aziz Al Saud (Fahd ibn 'Abd al-'Azîz Âl Sa'ûd). In: Orient (Opladen) 23: 170-172.   -1a, 4, 21-

899 Eliseit, H., 1955: Halbmond um Israel. Das neue Gesicht Arabiens. Berlin: Safari. 578 S.   -24-

900 Ellwood, W., N. Cumming-Bruce & D. Shirreff, 1978: Milking the energy cow. In: New Internationalist (61): 20-24.   -30-

901 Elmgren, S.G., 1864-66: Georg August Wallin's Reseanteckningar från Orienten åren 1843-1849. 4 Bde. Helsingfors: Frenckell.   -1a-

9o2 Elwan, O., 1983: Das Rechtswesen. In: T. Koszinowski (Hg.):
Saudi-Arabien: Ölmacht und Entwicklungsland. Beiträge zur
Geschichte, Politik, Wirtschaft und Gesellschaft. Mitteilungen des Deutschen Orient-Instituts 2o; Hamburg: Deutsches Orient-Institut, 1983: 177-217.   -21, H 223-

9o3 Emilia, A.D., 1952: Intorno al Codice di Commercio dell'Arabia Saudiana. In: Oriente Moderno 32: 316-325.   -7, 18, 21-

9o4 Employment of expatriate managers in Saudi Arabia. In: Middle
East Construction 1.1976(6). 5 p.   -89-

9o5 Ende, W., 1981-82: Religion, Politik und Literatur in Saudi-Arabien. Der geistesgeschichtliche Hintergrund der heutigen religiösen und kulturpolitischen Situation. In: Orient
(Opladen) 22: 377-39o, 23: 21-35, 378-393, 524-539.   -21,
1oo, H 223-

9o6 Entelis, J.P., 1976: Oil wealth and the prospects for democratization in the Arabian Peninsula: the case of Saudi
Arabia. In: N.A. Sherbiny & M.A. Tessler (eds.): Arab oil:
impact on the Arab countries and global implications.
Praeger Special Studies in International Business, Finance, and Trade; New York, Washington, London: Praeger,
1976: 77-111.   -12, 21, 2o6-

9o7 Erb, R.D., 198o: Saudi Arabia: economic developments. In:
A.E.I. Foreign Policy and Defense Review 2(4): 21-3o.
-2o6-

9o8 Erdman, D.S., 195o: Fishing in Arabia. In: The Scientific
Monthly 7o: 58-65.   -1a, 7, 1oo-

9o9 Erris, T.S. el-, 1965: Saudi Arabia: a study in nation building. Ph.D., Faculty of the Graduate School, American University, Washington. VI, 23o p.   -21-

91o Esin, E., 1964: Mekka und Medina. Frankfurt/M.: Umschau (Originalausgabe: Mecca the Blessed, Madinah the Radiant. London: Elek, 1963). 223 S.   -12, 212-

911 España, Ministerio de Asuntos Exteriores, Dirección de Cooperación Técnica Internacional, o.J.: An agroindustrial development strategy for the Kingdom of Saudi Arabia. Study
offered to the Kingdom of Saudi Arabia. O.O. V, 121 S.
-H 223-

912 Ess, J. van, 1943: Meet the Arab. New York: Day. VII, 229 p. -21-

913 Establishment of a Higher Institute of Technology in Saudi Arabia. In: Unesco Chronicle 8.1962: 3o4.   -24, 25, 291-

914 Euting, J., 1886: Ueber seine Reise in Inner-Arabien 1883/84. In: Verhandlungen der Gesellschaft für Erdkunde zu Berlin 13: 262-284.   -21, 24-

915 ——, 1896, 1914: Tagebuch einer Reise in Inner-Arabien. 2 Bde Leiden: Brill. VIII, 251 bzw. XIII, 3o4 S.   -7, 24, Tü 17-

916 Exchange Control Law in Saudi Arabia. In: Middle East Law Review 1.1958: 3o9-317.   -291-

917 Expansion of activities of the Saudi Development Fund. In: OAPEC News Bulletin 4.1978(4): 19-23.   -3o, 38, 2o6-

F

918 Fabietti, U., 1982a: Transformations économiques et leurs effets sur l'organisation sociale des groupes nomades d'Arabie Saoudite. In: F. Scholz & J. Janzen (Hgg.): Nomadismus - ein Entwicklungsproblem? Abhandlungen des Geographischen Instituts, Anthropogeographie 33; Berlin: Reimer, 1982: 159-166.   -1a-

919 ——, 1982b: Sedentarisation as a means of detribalisation: some policies of the Saudi Arabian government towards the nomads. In: T. Niblock (ed.): State, society and economy in Saudi Arabia. London: Croom Helm; Exeter: Centre for Arab Gulf Studies, 1982: 186-197.   -21-

92o Faheem, M.E., 1982: Higher education and nation building: a case study of King Abdulaziz University. Ph.D., University of Illinois at Urbana-Champaign. 231 p.

921 Faisal, K. al-, 1969: Die geschichtliche und wirtschaftliche Entwicklung Saudi-Arabiens. Diss., Universität Graz. 127 S

922 Faisal, M. al-, 1967: Desalination program for Saudi Arabia. In: International Conference on <u>Water for Peace</u>. Vol. 2: Water supply technology. Washington: US Government Printing Office, 1967: 37-42.

923 Faisal, M. al-, 1977: New water resources for desert development from icebergs. In: UNITER/State of California Conference: Alternative strategies for desert development and management. Sacramento, 1977.

924 Faisal: monarch, statesman and patriarch: 19o5-1975. In: Aramco World Magazine 26.1975(4): 18-23.   -21, 24-

925 Fakieh, O.A., 1979: Audit profile: Saudi Arabia. In: International Journal of Government Auditing 6: 14.   -5, 12, 18o-

926 Faleh, N.A. al-, 1981: Effects of lecture-demonstration and small group experimentation teaching methods on Saudi Arabian students' chemistry achievement and attitudes toward science learning. Ph.D., Indiana University, Bloomington. 122 p.

927 Falk, A., 1958: Visa pour l'Arabie. L'Air du Temps; Paris: Gallimard. 255 p.

928 Fallatah, I.M., 198o: Toward a conceptual model for planning the elementary school curriculum in Saudi Arabia. Ph.D., University of Wisconsin, Madison. 246 p.

929 Fallon, N., 1976: Winning business in Saudi Arabia. London: Graham & Trotman. 63 p.   -21-

93o —— & G.C. Bricault, 1976: Development projects and street maps of the Arab world and Iran. London: Graham & Trotman. 31 p.   -21-

931 Falls, C.B. & A.F. Becke, 193o: Military operations, Egypt & Palestine, from June 1917 to the end of the war. Part II. History of the Great War based on Official Documents 7; London: HMSO. XXIII, 394, XIV p.   -24/213-

932 Fanning, L.M., 1947: American oil operations abroad. New York, London: McGraw-Hill. VII, 27o p.   -24-

933 FAO, 1953a: Agriculture in the Near East: development and outlook. Rome. IV, 33, IV, 78 p.   -21-

934 ——, 1953b: Rapport sur les activités de la FAO au titre du Programme élargi d'assistance technique pendant 1952/53. Rome. 96 p.   -5-

935 FAO, 1955: Problems of food and agricultural expansion in the Near East. Rome. 53 p.

936 ——, 1966: Land and water surveys in the Wadi Jizan, Kingdom of Saudi Arabia. Final report. FAO/UNDP/SF: 23/SAA; Rome. 1oo p.   -12-

937 Farag, W., 197o: Saudi Arabia. In: C.E. Beck (ed.): Perspectives on world education. Dubuque: Brown, 197o: 262-269.

938 Faraj, A.H., 1981: Saudi Arabian educators' perceptions of educational goals for secondary schools. Ph.D., Indiana University, Bloomington. 233 p.

939 Fargues, P., 198o: Réserves de main-d'oeuvre et rente pétrolière. Étude démographique des migrations de travail vers les pays arabes du Golfe. Beyrouth: Centre d'Études et de Recherches sur le Moyen-Orient Contemporain; Lyon: Presses Universitaires de Lyon. 143 p.   -2o6, H 223-

94o ——, 1982: Présentation démographique des pays de la péninsule Arabique. In: Centre d'Études et de Recherches sur l'Orient Arabe Contemporain (éd.): La péninsule Arabique d'aujourd'hui. Sous la direction de P. Bonnenfant. T. 1. Paris: Centre National de la Recherche Scientifique, 1982: 155-189.   -21-

941 Farid, M.A., 1956: Implications of the Mecca pilgrimage for a regional malaria eradication programme. In: Bulletin of the World Health Organization / Bulletin de l'Organisation Mondiale de la Santé 15: 828-833.   -1a, 24, 26-

942 Faris, A. el-, 1972: The role of the Extension and Agriculture Services Department in the field of agricultural development. Paper presented at the Hofuf Agricultural Research Seminar, 24-3o March 1972. 13 p.

943 Faris, B., 1969: Descriptive paper on Bedouins. Riyadh: Central Planning Organization.

944 Farmer, L., 1971: The Arab woman - a traditional view. In: Aramco World Magazine 22(2): 2-4, 7-9, 11.   -24-

945 Farmer, R.N., 1959: Local entrepreneurship in Saudi Arabia. In: Business History Review 33: 73-86.   -38-

946 Farmer, R.N., 1962: Inland freight transportation pricing in eastern Saudi Arabia. In: The Journal of Industrial Economics 1o: 174-187.   -18o, 291, 352-

947 Farnworth, J., A.F. Halfpenny & W.I. Robinson, 1973: A survey of Hofuf Oasis farmers. University College of North Wales, Bangor and Ministry of Agriculture and Water, Saudi Arabia: Joint Agricultural Research and Development Project, Publication 24; n.p. (Bangor?). 5 p.

948 ——, 1974: Saudi-Wales co-operation in arid zone agriculture. In: Overseas Development Institute Review 1974(July): 3.   -2o6-

949 ——, 1976: Irrigated forage production under extreme arid zone conditions in Saudi Arabia. In: Experimental Agriculture 12: 177-187.   -1a, 7, 1oo-

95o —— & P. Bartholomew, 198o: A sprinkle of hope for Saudi farmers. In: Farmers Weekly (Sutton) 92(11): XX-XXI, XXIII.   -98-

951 Farra, T.O.M. el-, 1973: The effects of detribalizing the Bedouins on the internal cohesion of an emerging state: the Kingdom of Saudi Arabia. Ph.D., University of Pittsburgh. VIII, 243 p.   -21, H 223, Tü 17-

952 Farsy, F.A.-S. al-, 1978: Saudi Arabia: a case study in development. London: Stacey International (^{2}198o) (zugleich: Ph.D., Duke University, 1976). 22o p.   -12, 21, 93-

953 ——, 198o: King Faisal and the first five year development plan. In: W.A. Beling (ed.): King Faisal and the modernisation of Saudi Arabia. London: Croom Helm; Boulder: Westview Press, 198o: 58-71.   -12, 21, H 223-

954 Faruki, M.T. el-, 1966: Eastern Province. Governmental report, Ministry of Agriculture and Water. Riyadh.

955 Faruqi, Z.-H., 1973: A note on the Wahhabiyah. In: Islam and the Modern Age 4(1): 38-5o.   -21-

956 Fathy, H., 1966a: Main lines for implementation of first stage of rural reconstruction project in Saudi Arabia. New York: UNO.

957 Fathy, H., 1966b: Model houses for El Daréeya, Saudi Arabia. In: Ekistics 21: 214-219.   -1a, 385-

958 ——, 1966c: Model of rural housing for Saudi Arabia. In: Ekistics 22: 2o3-2o4.   -1a, 385-

959 Felemban, A.A.H. (auch: A.H.S.), 1976: Regional physical planning in Saudi Arabia: an evaluation of the Western Region plan and a proposal for a methodology for the Kingdom. Ph.D., University of East Anglia.

960 ——, 198o: Regional development planning as an essential part of regional development: a case study in Saudi Arabia. In: Ekistics 47: 36o-368.   -1a, 89, 2o6-

961 Fellmann, W., 1964: Saudi-Arabien im Umbruch. In: Deutsche Aussenpolitik 9: 411-419.   -24, 18o, 2o6-

962 Feoktistov, A., 1977: Saudi Arabia and the Arab world. In: International Affairs (Moscow) 1977(7): 1o1-1o7.   -7, 2o6, 212-

963 Ferguson, N., 1978: Consolidation: the new phase in the Middle East business world - I. In: Bankers' Magazine (London) 222(6): 15-18.   -18o, 2o6-

964 Ferret, P.V.A. & J.G. Galinier, 1847: Voyage en Abyssinie dans les provinces du Tigré, du Samen et de l'Amhara. T. 1. Paris: Paulin. 526 p.   -16, 21-

965 Ferris, H.J., 1953: Report to the Government of Saudi Arabia on reconnaissance soil and land classification of the South Asir Tihama. Report 69; Rome: FAO.

966 Fiander, W., 1979: Saudi Arabia and the Gulf states. Rev.ed. Sources of Statistics and Market Information 7; London: Statistics and Market Intelligence Library. 48 p.

967 Fiar, M.H. al-, 1977: The Faisal Settlement Project at Haradh, Saudi Arabia: a study in nomad attitudes toward sedentarization. Ph.D., Michigan State University, East Lansing. VII, 24o p.   -21, H 223-

968 Fiedler, V., 1982: Saudi-Arabien: Strukturelle Probleme durch Ölmilliarden. In: Übersee-Rundschau 34(3): 9-13.   -24, 3o, 18o-

969   Field, H., 1931: Among the Beduins of North Arabia. In: The Open Court 45: 577-595.   -1a-

97o   —, 1932: The ancient and modern inhabitants of Arabia. In: The Open Court 46: 847-872.   -1a-

971   —, 1952: Camel brands and graffiti from Iraq, Syria, Jordan, Iran, and Arabia. Journal of the American Oriental Society, Suppl. 15. 41 p.   -21-

972   —, 1971: Contributions to the anthropology of Saudi Arabia. Coconut Grove: Field Research Projects. VIII, 62 p. -Tü 17-

973   Field, M., 1975: A hundred million dollars a day. London: Sidgwick & Jackson (französische Ausgabe: Cent millions de dollars par jour. Paris: Fayard, 1975). 24o p.   -21-

974   —, 1978a: Saudi Arabia - the Eastern Province before oil. In: Middle East International (86): 27-28.   -12, 3o-

975   —, 1978b: Saudi Arabia. In: Middle East Annual Review 4: 311, 315, 317, 319, 321, 323, 325, 327, 329-33o, 335, 337, 339-341.   -21, 24/213, 2o6-

976   —, 1982: Society, the royal family and the military in Saudi Arabia. In: Vierteljahresberichte, Probleme der Entwicklungsländer (89): 217-226.   -2o6, H 3-

977   Field, P., 1978: Joint Trade Commissions - little help to the exporter. In: Middle East Annual Review 4: 69, 71-73. -21, 24/213, 2o6-

978   Fields, J.H., 1975: Pawn of empires: a study of United States-Middle East policy, 1945-1953. Ph.D., Rutgers University, The State University of New Jersey, New Brunswick. VIII, 52o p.   -21-

979   Filali, M., 1966: Structures sociales en Arabie Séoudite. In: Revue Tunisienne de Sciences Sociales 3(6): 69-84.   -38, 2o6-

98o   Finati, G., 183o: Narrative of the life and adventures of Giovanni Finati, native of Ferrara; who, under the assumed name of Mahomet, made the campaigns against the Wahabees for the recovery of Mecca and Medina; and since acted as interpreter to European travellers in some of the parts least visited of Asia and Africa. Translated from the Italian, as dictated by himself, and ed. by W.J. Bankes. 2 vols. London: Murray. XXVII, 296 resp. VIII, 43o p.   -12-

981 Finch, E., 1938: Wilfrid Scawen Blunt, 184o-1922. London: Cape. 415 p.   -77-

982 Fingar, P., 1977: EDP training perspectives from Saudi Arabia. In: Journal of Educational Data Processing 14(1): 1-7. -1a, 89-

983 Finielz, C.I., 1955: Summary report to the Government of Saudi Arabia on a soils study - classification and utilization with respect to irrigation in the Wadi Jizan area. Report 13: Rome: FAO.

984 Finnie, D.H., 1958a: Recruitment and training of labor: the Middle East oil industry. In: The Middle East Journal 12: 127-143.   -12, 21, 38-

985 ——, 1958b: Desert enterprise: the Middle East oil industry in its local environment. Harvard Middle Eastern Studies; Cambridge/Mass.: Harvard University Press. X, 224 p. -12, 21-

986 Finnie, R., 1958: Bechtel in Arab lands. A fifteenth year review of engineering and construction projects. San Francisco: Bechtel Corporation. 15o p.

987 Fire at Dammam No. 12. In: Aramco World 13.1962(7): 15-17. -H 223-

988 First instalment of Labour and Workers Act. N.p., 1969. N.pag.   -H 223-

989 First National City Bank, 1974: Saudi Arabia: a new economic survey. New York. 42, 7 p.   -H 3-

99o ——, 1977: Investment guide to the Kingdom of Saudi Arabia. New York. 42 p.

991 Fisher, S.N., ^{3}1979: The Middle East: a history. New York: Knopf (^{1}1959). XIX, 811 p.   -21-

992 Fisher, W.B., 1978: The Middle East: a physical, social and regional geography. 7th ed., completely rev. & reset. London: Methuen (^{1}195o). XIV, 615 p.   -21-

993 Fisher, W.B., 1979: The good life in modern Saudi Arabia. In: The Geographical Magazine 51: 762-768.   -18o, 464-

994 Fitchett, J., 1973: From khans to khiltons. In: Aramco World Magazine 24(6): 16-29.   -24-

995 Fleischer, H.L., 1857: Briefwechsel zwischen den Anführern der Wahhabiten und dem Paśa von Damaskus. In: Zeitschrift der Deutschen Morgenländischen Gesellschaft 11: 427-443 (Nachdruck in: H.L. Fleischer: Kleinere Schriften. Gesammelt, durchgesehen und vermehrt. Bd. 3. Leipzig: Hirzel, 1888: 341-36o).   -21, 291, 352-

996 FMC's Al Hasa vegetable farm will supply thirty crops. In: World Crops (Horne) 31.1979(3): VI.   -98-

997 Fogel, M.M., 1964: Irrigation plan for the Qatif Oasis, Eastern Province, Saudi Arabia. Dhahran: Arab Industrial Development Department, ARAMCO. IV, 44 p.

998 Foley, J.A., 1979: Problems of understanding science and technology textbooks in English for first year students at the University of Petroleum and Minerals in Saudi Arabia. Ph.D., Institute of Education, University of London.

999 Forbes(-McGrath), R., 1923: A visit to the Idrisi territory in 'Asir and Yemen. In: The Geographical Journal 62: 271-278.   -21, 24, 291-

1ooo ——, 1944: Gypsy in the sun. London, Toronto, Melbourne, Sydney: Cassell. 381 p.

1oo1 Forde, C.D., 1933: The habitat and economy of the North Arabian Badawîn. In: Geography 18: 2o5-219.   -3o, 385, 464-

1oo2 Forder, A., 19o2: With the Arabs in tent and town. An account of missionary work, life and experiences in Moab and Edom and the first missionary journey into Arabia from the North. London: Marshall. X, 243 p.   -12-

1oo3 ——, 19o9a: Ventures among the Arabs in desert, tent, and town. Thirteen years of pioneer missionary life with the Ishmaelites of Moab, Edom, and Arabia. New York: Gospel, X, 292 p.   -21-

1oo4 ——, 19o9b: Arabia, the desert of the sea. In: The National Geographic Magazine 2o: 1o39-1o62.   -21, 93-

1oo5 Forschungsinstitut der Friedrich-Ebert-Stiftung (Hg.), 1982: Saudi-Arabien in den 8oer Jahren. Expertengespräch in Bonn, 26. und 27. April 1982. Internationale Politik; Bonn. 34 S.   -21, H 223-

1oo6 Forsythe, D.W., 1978: The adjustment problems of adolescent expatriates returning to the United States from Saudi Arabia to attend boarding school. Ed.D., University of Montana, Missoula. 136 p.

1oo7 Foss, M., 1969: Dangerous guides: English writers and the desert. In: The New Middle East (9): 38-42.   -1a-

1oo8 Fouad, M.H., 1978: Petrodollars and economic development in the Middle East. In: The Middle East Journal 32: 3o7-321.   -21, 38, 2o6-

1oo9 Frade, F., 1974: Faisal Ibn Saud, cumbre de una familia esforzada. In: Revista de Política Internacional (133): 171-2o2.   -16, 18, 2o6-

1o1o ——, 1976: La herencia de Faisal. In: Revista de Política Internacional (144): 111-136.   -16, 18, 2o6-

1o11 ——, 1977: La andadura del rey Jaled de Arabia Saudita. In: Revista Política Internacional (15o): 77-96.   -16, 18, 2o6-

1o12 France and the Arab world. In: The Arab Economist 6.1974 (67): 19-26.   -H 3-

1o13 France, Ministère de la Guerre, État-Major de l'Armée - Service Historique, 1935: Les Armées françaises dans la Grande Guerre. T. 9: Les fronts secondaires. 1er vol.: Théâtre d'opérations du Levant (Égypte-Palestine-Syrie-Hedjaz). La propagande allemande au Maroc. Avec: Vol. d'annexes. Paris: Imprimerie Nationale. XI, 239 resp. 1o36 p.   -21-

1o14 ——, Ministère des Affaires Étrangères, Direction des Consulats et Affaires Commerciales, 1875: Histoire de la fondation, en 1824, de la ville de Riad, capitale actuelle du Nedjd, et description géographique de ce pays. In: Bulletin de la Société de Géographie (Paris) VI, 1o: 71-77.   -1a, 291-

1o15   France, Service Hydrographique, 1933: Golfe d'Oman, Golfe Persique et côte ouest de l'Inde. Instructions Nautiques 367; Paris: Imprimerie Nationale. 6o3 p.

1o16   ——, ——, 1942: Mer Rouge et Golfe d'Aden. Instructions Nautiques 412; Paris: Imprimerie Nationale. XVI, 471 p.

1o17   Frankel, G.S., 1975: Arabia saudita. Un difficile post-Feisal. In: Affari Esteri 7: 3oo-316.   -1a, 18, 188-

1o18   Franzmathes, F., 1981: Saudi-Arabien im Zeichen des Technologie-Transfers. Ein entscheidungstheoretischer Versuch. In: Orient (Opladen) 22: 241-256.   -21, 1oo, H 223-

1o19   (——, 1982:) Königreich Saudi-Arabien. In: Nah- und Mittelost-Verein e.V.: Jahresbericht 1981/82. Hamburg, 1982: 39-4o.   -H 223-

1o2o   Freeth, Z., 1956: Kuwait was my home. London: Allen & Unwin. 164 p.   -21-

1o21   —— & H.V.F. Winstone, 1978: Explorers of Arabia from the Renaissance to the end of the Victorian Era. London, Boston, Sydney: Allen & Unwin. 3o8 p.   -21, 24, Tü 17-

1o22   —— & ——, 198o: A journey to Hail. In: Aramco World Magazine 31(3): 8-13.   -21, 24-

1o23   ——, 198o: Rashid of Saudi Arabia. How they live now 2; Guildford: Lutterworth. 3o p.

1o24   Fresnel, F., 1839: L'Arabie. In: Revue des Deux Mondes 17: 241-257.   -24, 18o, 291-

1o25   ——, 1871: L'Arabie vue en 1837-1838. In: Journal Asiatique VI, 17: 5-164.   -24, 25, 291-

1o26   Fridolin, 1854: Scènes de la vie religieuse en Orient. Damas.-Jérusalem.-Le désert.-La caravane de La Mecque. In: Revue des Deux Mondes 24 (zugleich: II, 6): 73-99.   -24, 18o, 188-

1o27   Friedemann, J., 1974: Die Scheiche kommen: Arabien, Zentrum neuer Macht. Bergisch Gladbach: Lübbe. 263 S.   -24/213, 212-

1o28   Friedman, I., 197oa: The McMahon-Hussein Correspondence and the question of Palestine. In: Journal of Contemporary History 5(2): 83-122.   -21, 18o, 385-

1029 Friedman, I., 1970b: Isaiah Friedman replies (a reply to Professor Arnold Toynbee's comments). In: Journal of Contemporary History 5(4): 193-201.   -21, 180, 385-

1030 Frith, D.E., Kirk & W. Spinks, 1976: The construction industry and market in Saudi Arabia and the Gulf States. London: Graham & Trotman.

1031 ——, 1978: New priorities and developments in the construction market in Saudi Arabia and Iraq. London: Graham & Trotman. 125 p.   -21-

1032 From a correspondent in Jeddah... Restrictions against Western women. In: Women's International Network News 4.1978 (3): 41.

1033 Frood, A.M., 1937: A geographical study of Saudi Arabia. M.A., University of Liverpool.

1034 ——, 1939: Recent economic and social developments in Sa'udi Arabia. In: Geography 24: 162-170.   -7, 30, 464-

1035 Fürstenmühl, R. von, 1980: Saudi Arabien: Kultur- und Reiseführer. SYRO-Individualreiseführer 18; Göttingen: SYRO. IV, 167 S.   -21, 93-

G

**** G.: s. Gobée, E.

1036 Gabriel, E., 1966/67: Zur Erdölwirtschaft am Persischen Golf. Eine wirtschaftsgeographische Betrachtung. In: Mitteilungen der Fränkischen Geographischen Gesellschaft 13/14: 111-123.   -1a, 21, 24-

1037 ——, 1975: Strukturwandel der Wirtschaftslandschaften am Persergolf. In: E. Otremba et al.: 25 Jahre Forschung und Lehre im Wirtschafts- und Sozialgeographischen Institut der Universität zu Köln. Kölner Forschungen zur Wirtschafts- und Sozialgeographie 21: 283-295.   -5, 83, 89-

1038 Gälli, A., 1979: Die sozio-ökonomische Entwicklung der OPEC-Staaten. Auswirkungen und Perspektiven des Devisenreichtums. Ifo-Studien zur Entwicklungsforschung 4; München, London: Weltforum. XIX, 396 S.   -100, 212-

1039   Gahtani, T.M.S. al-, 1981: Sponsoring Saudi male graduates in the United States and their academic commitment: King Abdulaziz University case. Ed.D., University of Virginia, Charlottesville. 169 p.

1040   (Galinier, J.G. & P.V.A. Ferret, 1843:) Géographie de l'Arabie. Notice rédigée d'après M. Chédufau, par MM. Galinier et Ferret. In: Bulletin de la Société de Géographie (Paris) II, 19: 1o6-118.   -291-

1041   Gamie, M.N., 1982: The need for balanced development of rural and urban sectors in Saudi Arabia. In: Addarah 7(4): 1-17.

1042   Ganoubi, A.I., 1976: Irqah: a village community in Najd. Ph.D., University of Hull.

1043   Garantieprobleme mit arabischen Ländern. In: Die Bank 1978: 4o2-4o3.   -H 3-

1044   Garbe, C.W., 1977: Coast development in the Middle East in general, Saudi Arabia in particular. In: World Dredging and Marine Construction 13(5): 18-2o.

1045   Gaspard, J., 1969: Feisal's Arab alternative. In: The New Middle East (6): 15-19.   -1a-

1046   Gaudefroy-Demombynes, M., 1923: Le pèlerinage à la Mekke. Étude d'histoire religieuse. Annales du Musée Guimet, Bibliothèque d'Études 33; Paris: Geuthner. VIII, 332 p. -21, 25-

1047   Gaulis, B.-G., 193o: La Question arabe. De l'Arabie du Roi Ibn Sa'oud à l'Indépendance syrienne. Paris: Berger-Levrault. 3o8 p.   -21-

1048   Gaury, G. de, 1943: A Saudi Arabian notebook. Cairo: Misr. 5o p.

1049   ——, 1944: Arabia and the future. In: Journal of the Royal Central Asian Society 31: 4o-47.   -21, 9o, 291-

1050   ——, 1946a: Arabia Phoenix; an account of a visit to Ibn Saud, chieftain of the austere Wahhabis and powerful Arabian king. London: Harrap. 169 p.   -21-

1051   ——, 1946b: The end of Arabian isolation. In: Foreign Affairs (New York) 25: 82-89.   -21, 24, 18o-

1o52 Gaury, G. de, 195o: Arabian journey and other desert travels. London, Sydney, Toronto, Bombay: Harrap. 19o p.

1o53 ——, 1951: Rulers of Mecca. London, Sydney, Toronto, Bombay: Harrap. 317 p. -21-

1o54 ——, 1966: Faisal, king of Saudi Arabia. London: Barker (New York: Praeger, 1967). XIV, 191 p. -24/213, Frei 119-

1o55 ——, 1975: Memories and impressions of the Arabia of Ibn Saud. In: Arabian Studies 2: 19-32. -12, 21, 212-

1o56 Gautier, É.-F., 1918: Les villes saintes de l'Arabie. In: Annales de Géographie 27: 115-131. -21, 24, 291-

1o57 ——, 1931: Moeurs et coutumes des Musulmans. Collection d'Études, de Documents et de Témoignages pour servir à l'Histoire de notre Temps; Paris: Payot. 3o5 p. -1a-

1o58 Gawawi, 1936: La Mecque 1935. In: Renseignements Coloniaux et Documents, Suppl. à l'Afrique Française 46.1936: 17-26, 39-48, 51-64, 79-8o. -1a, 3o-

1o59 Gelpi, A.P., 1965: Glucose-6-phosphate dehydrogenase deficiency in Saudi Arabia: a survey. In: Blood 25: 486-493. -1a, 21, 1oo-

1o6o ——, 1966: Q fever in Saudi Arabia. In: The American Journal of Tropical Medicine and Hygiene 15: 785-798. -12, 25, 1oo-

1o61 ——, 1967: Glucose-6-phosphate dehydrogenase deficiency, the sickling trait and malaria in Saudi Arabia. In: The Journal of Pediatrics 71: 138-146. -21, 24, 289-

1o62 —— & M.C. King, 1967: Screening for abnormal haemoglobins in the Middle East: new data on haemoglobin S and the presence of haemoglobin C in Saudi Arabia. In: Acta Haematologica 36: 334-337. -21, 25, 352-

1o63 —— & A. Mustafa, 1967: Seasonal pneumonitis with eosinophilia: a study of larval Ascariasis in Saudi Arabs. In: The American Journal of Tropical Medicine and Hygiene 16: 646-657. -1a, 16, 1oo-

1o64 ——, 197oa: Sickle cell disease in Saudi Arabs. In: Acta Haematologica 43: 89-99. -21, 25, 352-

1o65   Gelpi, A.P., 197ob: Malignant lymphona in the Saudi Arab. In: Cancer 25: 892-895.   -1a, 7, 21-

1o66   General instructions for pilgrims to the Hedjaz. Calcutta: Government Printers, 1922. 69 p.

1o67   George, A.R., 1972: Processes of nomadic sedentarization in the Middle East. M.A., Faculty of Social Sciences, University of Durham.

1o68   ——, 1975: Bedouin settlement in Saudi Arabia. In: Middle East International (51): 27-3o.   -3o-

1o69   Gerakis, A.S., 1976: Bindung an SZR. Die Erfahrungen Irans, Jordaniens, Qatars und Saudi-Arabiens. In: Finanzierung und Entwicklung 13(1): 35-38.   -21, 24, H 3-

1o7o   Gerken, Lipsmeier & Puttendörfer, 1977: Bauprogrammierung von Berufsschulzentren in Saudi-Arabien. Unterlagen zu einem Bauwettbewerb für drei Bauschulen und eine Landwirtschaftsschule. Gutachten im Auftrag der GTZ. Hannover. Getr.Pag.   -Esb 2ooo-

1o71   Germanus, J., 1938: Allah Akbar. Im Banne des Islams. Berlin: Holle (Originalausgabe: Allah Akbar. 2 Bde. Budapest, 1936. Italienische Ausgabe: Sulle orme di Maometto. Vita e pensiero dei Musulmani. 2 Bde. Milano: Treves, 1938). 718 S.   -21-

1o72   ——, 1952: Modern Sa'udi Arabian literature. In: The Islamic Review 4o(11): 31-33.   -1a, 21-

1o73   Gervais-Courtellemont, J.-C., 1896: Mon voyage à la Mecque. Paris: Hachette. 236 p.   -21-

1o74   Ghaith, A., 1967: The marching caravan: the story of modern Saudi Arabia. Jeddah: Almadina Almonawarra. 2o9 p. -H 223-

1o75   Ghamdi, Abdullah A. al-, 1982: Action research and the dynamics of organizational environment in the Kingdom of Saudi Arabia. Ph.D., University of Southern California, Los Angeles.

1o76   Ghamdi, Abdulrahim M. al-, 1982: The professional development of in-service teachers in Saudi Arabia: a study of the practice and needs. Ph.D., Michigan State University, East Lansing. 225 p.

1o77 Ghamdi, Mohammed A.H., 1977: A study of selected factors related to student dropouts in the secondary schools of Saudi Arabia. Ph.D., Michigan State University, East Lansing. 5, XI, 2o1 p. -H 223-

1o78 Ghanayem, M.A., 1981: The theory of foreign direct investment in capital-rich, labor-short economies: the case for Saudi Arabia, Kuwait and United Arab Emirates. Ph.D., Southern Methodist University, Dallas. 1o4 p.

1o79 Ghandoura, A.H., 1982: Achievement effects of teacher comments on homework in mathematics classes in Saudi Arabia. Ph.D., Oregon State University, Corvallis. 139 p.

1o8o Ghoneimy, M.T. el-, 1966: The legal status of the Saudi-Kuwaiti Neutral Zone. In: The International and Comparative Law Quarterly IV, 15: 69o-717. -7, 21, 2o6-

1o81 Ghorban, N., 1978: A study of the emergence, development and role of the national oil companies of Iran, Kuwait and Saudi Arabia. Ph.D., School of Oriental and African Studies, University of London.

1o82 Giannini, A., 1921: La questione Orientale alle Conferenza della Pace. Cap. V. - La questione Araba (Higiaz e Luoghi Santi dell'Islam, Mesopotamia, Siria. In: Oriente Moderno 1: 193-21o. -1a, 5, 21-

1o83 ——, 1932: Gli Accordi di Gedda fra il Regno d'Italia ed il Regno del Ḥigiāz e Neǵd. In: Oriente Moderno 12: 225-227. -1a, 5, 21-

1o84 ——, 1935: Il Trattato di eṭ-Ṭā'if e l'equilibrio dell'Arabia. In: Oriente Moderno 15: 489-498. -1a, 5, 21-

1o85 ——, 1942: Saggi di storia diplomatica. Milano: ISPI. 259 S. -Biblioteca Nazionale Centrale, Roma-

1o86 Giant on the seas. In: Aramco World 13.1962(1o): 3-6. -H 223-

1o87 Gil Benumeya, R., 1974: Actualidad y continuidad en la Arabia del rey Faisal. In: Revista de Política Internacional (132): 141-15o. -18, 2o6, 352-

1o88 Gilbert, A.L., 198o: Health policy for Saudi Arabia's third plan. M.A., University of Oregon, Eugene. 1o2 p.

1089   Girvin, E., 1968: Al-Hajj '...they will come from every remote path...'. In: The Arab World 14(3/4): 7-9.   -2o6-

1o9o   Gleisbau in Saudi-Arabien. In: Bauen macht Freude, Heitkamp-Mitteilungen 1978(1): 22.

1o91   Glick, L.A., 198o: Trading with Saudi Arabia. A guide to the shipping, trade, investment, and tax laws of Saudi Arabia. Montclair: Allanhead, Osmun; London: Croom Helm. XV, 595 p.   -2o6, B 851, H 3-

1o92   Glubb, F., 1975: Saudi Arabia: the same but different. In: Middle East International (49): 15-17.   -3o-

1o93   Glubb, J.B., 1935: The bedouins of Northern Arabia. In: Journal of the Royal Central Asian Society 22: 13-31.   -21, 9o, 291-

1o94   ——, 1943: The Sulubba and other ignoble tribes of southwestern Asia. In: H. Field & J.B. Glubb: The Yezidis, Sulubba, and other tribes of Iraq and adjacent regions. General Series in Anthropology 1o; Menasha: Banta, 1943: 14-16.

1o95   ——, 1948: The story of the Arab Legion. London: Hodder & Stoughton. 371 p.   -24/213-

1o96   ——, 196o: War in the desert: an R.A.F. frontier campaign. London: Hodder & Stoughton. 352 p.   -24/213-

1o97   ——, 1978: Arabian adventures: ten years of joyful service. London: Cassell. 224 p.   -21, 24/213-

1o98   Gobée, E., 1921: Indrukken over het schoolwezen in de Hidjāz. In: Tijdschrift voor Indische Taal-, Land- en Volkenkunde 6o: 187-2o6.   -1a, 5-

1o99   ——, 1921, 1922, 1924: Textes historiques sur le réveil arabe au Hedjaz. In: Revue du Monde Musulman 46: 1-22, 47: 1-27, 5o: 74-1oo, 57: 158-167.   -21, 25-

11oo   Gobineau, J.A. de, 19o5: Trois ans en Asie (de 1855 à 1858). Nouvelle éd. (^{1}1859). Paris: Leroux. VI, 5oo p.   -1a, 21, 24-

11o1   Goedde, H.W., H.G. Benkmann, D.P. Agarwal, L. Hirth, U. Bienzle, M. Dietrich, H.H. Hoppe, J. Orlowski, E. Kohne & E. Kleinhauer, 1979: Genetic studies in Saudi Arabia: red cell enzyme, haemoglobin and serum protein polymorphisms. In: American Journal of Physical Anthropology 5o: 271-277.   -26, 289, 352-

1102 Goeje, M.J. de, F. Hommel, B. Moritz, A. Schaade, G. Kampffmeyer & C. Brockelmann, 1913: Arabien. In: EI[1] 1: 384-432.   -21, 24-

1103 Goellner, W.A., 1974: A guide to Hofuf and the Jebal Qara area of the al-Hasa Oasis, Saudi Arabia. Dhahran. 15 p.

1104 ——, 1975: Hofuf and the Jebel al-Qara area of the al-Hasa Oasis, Saudi Arabia. Dhahran. 19 p.

1105 Götz, W., 1977: Gutachten zur Errichtung einer Ausbildungsabteilung für Giessereifachkräfte an der SVS Er Riyadh, Saudi-Arabien. Im Auftrag der GTZ. O.O. 16 S.   -Esb 2ooo-

1106 Goglia, L., 1978: Un documento inedito della Regia Marina a proposito di una poco nota Missione Italiana presso il Re Husein 'Ali del Ḥigiāz nel luglio 192o. In: Oriente Moderno 58: 527-532.   -7, 18, 21-

1107 Gohaidan, M.S.S., 1981: Organizational innovations in developing countries: the case of Saudi Arabia. Ph.D., University of California, Santa Barbara. 272 p.

1108 Gold, F.R. & M.A. Conant, 1977: Access to oil - the United States relationships with Saudi Arabia and Iran. Publication 95-7o; Washington: US Government Printing Office. XIII, 113 p.   -12-

1109 Goldberg, D., 1978: The foreign policy of the third Saudi state, 19o2-1918. Ph.D., Harvard University.

1110 Goldberg, J. & Y. Gal, 1981: The Saudi Arabian Kingdom. In: Middle East Contemporary Survey 4: 681-721.   -12, 21, 1oc

1111 —— & I.E. Hoffman, 198o: The Saudi Arabian Kingdom. In: Middle East Contemporary Survey 3: 736-769.   -12, 21, 1o

1112 ——, 1982: The 1913 Saudi occupation of Hasa reconsidered. In: Middle Eastern Studies 18: 21-29.   -1a, 21, 2o6-

1113 Goldrup, L.P., 1971: Saudi Arabia: 19o2-1932: the development of a Wahhabi society. Ph.D., University of Southern California, Los Angeles. X, 476 p.   -21, H 223-

1114 Goldziher, I., 1925: Vorlesungen über den Islam. 2., von F. Babinger umgearb. Aufl. Religionswissenschaftliche Bibliothek 1; Heidelberg: Winter (Nachdruck: 1963). XIII, 4o6 S. -21-

1115 Gómez Aparicio, P., 1962: Dos Arabias Saudíes. In: Revista de Política Internacional (62/63): 4o3-432. -16, 18, 2o6-

1116 Gondrecourt, A. de, 1858: Scénes de la vie arabe. Médine. 2 vols. Paris: Cadot. -Library of Congress, Washington-

1117 Gordon, E., 1973: Saudi Arabia in pictures. Visual Geography Series; New York: Sterling; London, Sydney: Oak Tree Press. 64 p. -21-

1118 Gordon, N., 1953: The strangest railroad on earth. In: The Saturday Evening Post 226(2o): 28-29, 114-116. -24, 18o-

1119 Gormly, J.L., 198o: Keeping the door open in Saudi Arabia: the United States and the Dhahran airfield, 1945-46. In: Diplomatic History 4: 189-2o5. -7, 12, 18o-

112o Gosaibi, G.A. al-, 197o: The 1962 revolution in Yemen and its impact on the foreign policies of the U.A.R. and Saudi Arabia. Ph.D., University College, University of London.

1121 Gottheil, F.M., 1978: The manufacture of Saudi Arabian economic power. In: Middle East Review 11(1): 18-23. -1a, 21, 1oo-

1122 Gouilly, A., 1964: Pèlerinage à la Mecque. In: Revue Juridique et Politique 18: 33-1o6. -1a, 12, 291-

1123 Gouldrup, L., 1982: The Ikhwān movement of Central Arabia. In: Arabian Studies 6: 161-169. -21, Tü 17-

1124 Goy, R., 1957: L'affaire de l'oasis de Buraïmi. In: Annuaire Français de Droit International 3: 188-2o5. -24, 2o6, 291-

1125 Gräf, E., 1952: Das Rechtswesen der heutigen Beduinen. Beiträge zur Sprach- und Kulturgeschichte des Orients 5; Walldorf/Hessen: Verlag für Orientkunde (zugleich: Das Gerichtswesen der heutigen Beduinen. Diss., Philosophische Fakultät, Universität Bonn, 1948). 198 S. -1a, 5, Bo 139-

1126 Gräf, E., 1960: 'Anaza. In: EI^2 1: 482-483.   -21, 24, 212-

1127 Gräfen, R., 1980: Saudi-Arabien. In: Energiewirtschaftliche Tagesfragen 30: 678-687.   -H 3-

1128 Grafftey-Smith, L., 1970: Bright Levant. London: Murray. XII, 295 p.   -7-

1129 Grandguillaume, G., 1975: L'institution coopérative en Arabie Séoudite. In: Archives Internationales de Sociologie de la Coopération et du Développement (37): 42-63.   -206-

1130 ——, 1982: Valorisation et dévalorisation liées aux contacts de cultures en Arabie Saoudite. In: Centre d'Études et de Recherches sur l'Orient Arabe Contemporain (éd.): La péninsule Arabique d'aujourd'hui. Sous la direction de P. Bonnenfant. T. 2: Études par pays. Paris: Centre National de la Recherche Scientifique, 1982: 623-654.   -21-

1131 Grant, C.P., 1937: The Syrian Desert: caravans, travel and exploration. London: Black. XV, 410 p.   -21-

1132 Graves, P.P. (ed.), 1950: Memoirs of King Abdullah of Transjordan. London: Cape. 278 p.   -7-

1133 Great Britain, 1955: Memorial of the Government of the United Kingdom and Northern Ireland in arbitration concerning Buraimi and the common frontier between Abu Dhabi and Saudi Arabia. 2 vols. London.

1134 ——, Admiralty, Hydrographic Department, ^{8}1932: Red Sea and Gulf of Aden pilot. Comprising the Suez Canal, the Gulfs of Suez and Aqaba, the Red Sea, the Gulf of Aden, the south-east coast of Arabia from Ras Baghashwa to Ras al Hadd, the coast of Africa from Ras Asir to Ras Hafun, Socotra and its adjacent islands. London: HMSO. XXXIV, 540 p.   -24/213-

1135 ——, ——, Naval Intelligence Division, 1916-17: A handbook of Arabia. Prepared on behalf of the Admiralty and the War Office, Admiralty War Staff, Intelligence Division. 2 vols. London. 708 resp. 519 p.   -21-

1136 ——, ——, ——, 1944: Iraq and the Persian Gulf. Geographical Handbook Series B.R. 524; London: HMSO. XVIII, 682 p.   -24, 26-

1137 Great Britain, Admiralty, Naval Intelligence Division, 1946: Western Arabia and the Red Sea. Geographical Handbook Series B.R. 527; London: HMSO. XIX, 659 p.   -24, 212-

1138 ——, Board of Trade, 1954: Markets in the Middle East: report of the United Kingdom Trade Mission to Iraq, Kuwait, the Lebanon, Syria and Saudi Arabia, November - December 1953. London: HMSO.

1139 ——, Foreign Office, 1920: Arabia. Handbooks prepared under the direction of the Historical Section of the Foreign Office 61; London: HMSO. 122 p.   -1a-

1140 ——, Parliament, 1954: Arbitration agreement between the Government of the United Kingdom (acting on behalf of Abu Dhabi and His Highness the Sultan Sa'id bin Taimur) and the Government of Saudi Arabia, with exchange of notes, Jedda, July 30, 1954. In: Accounts and Papers 33; London: HMSO, 1954: 961-970.

1141 Greenidge, C.W.W., 1956: Slavery in the Middle East. Report by the Secretary of the Anti-Slavery Society (London). In: Middle Eastern Affairs 7: 435-440.   -21, 25, 180-

1142 Greenip, W.E., 1966: Jordan-Saudi border demarcation. In: Viewpoints (Washington) 6(1): 24.   -1a-

1143 Griessbauer, L., 1907: Die internationalen Verkehrs- und Machtfragen an den Küsten Arabiens. Schriften der Deutsch-Asiatischen Gesellschaft 4; Berlin: Paetel. 25 S.   -1a-

1144 Griffith, W.E., 1977: Der Nahe Osten, die Energiefrage und die Grossmächte. In: Europa-Archiv, Beiträge und Berichte 32: 181-192.   -21, 24, 25-

1145 Griffith-Jones, S., 1981: The impact of the massive 1981 Saudi Arabian loan to the International Monetary Fund. Discussion Paper 170; Brighton: Institute of Development Studies, University of Sussex. 23 p.   -Bo 149-

1146 ——, 1982: The Saudi loan to the IMF: a new route for recycling. In: Third World Quarterly 4: 304-311.   -206, H 3, H 223-

1147 Griggs, L., 1972: The best of both. In: Aramco World Magazine 23(Special): 30-35.   -24-

1148 Grimaldi, F., 1982: All roads lead to Rome. In: The Middle East (97): 43-48. -Frei 119, H 3, H 223-

1149 Grindley, W., et al., 1977: The construction industry in Saudi Arabia. Menlo Park: SRI International. 55 p.

1150 Grobba, F., 1967: Männer und Mächte im Orient. 25 Jahre diplomatischer Tätigkeit im Orient. Göttingen, Zürich, Berlin, Frankfurt/M.: Musterschmidt. 339 S. -21-

1151 Grohmann, A., 1927: Al-Ḳaṭif. In: EI[1] 2: 879-880. -21, 24-

1152 ——, 1934a: Tihāma. In: EI[1] 4: 827-829. -21, 24-

1153 ——, 1934b: Yanbu'. In: EI[1] 4: 1253-1254. -21, 24-

1154 ——, 1936: Nedjd. In: EI[1] 3: 964-967. -21, 24-

1155 ——, 1938: Al-Riyāḍ. In: EI[1] Ergänzungsbd.: 199-201. -21, 24-

1156 Grosse Sportanlagen in Saudi-Arabien. Stuttgarter Consultingfirma plant Sportstätte. In: Übersee-Rundschau 31. 1979(2): 20-21. -24, 30, 180-

1157 Grunwald, K. & J.O. Ronall, 1960: Industrialization in the Middle East. New York: Council for Middle Eastern Affairs Press. XX, 394 p. -24, 38-

1158 Guaiz, S.A. al-, 1974: Hygienische Verhältnisse in Saudi-Arabien. Diss., Medizinische Fakultät, Universität des Saarlandes, Saarbrücken. 131 S. -291-

1159 Guarmani, C., 1971: Northern Najd. Ed. by D. Carruthers. Argonaut Press 16; Amsterdam: Israel; New York: Da Capo Press (1st ed.: Northern Najd; a journey from Jerusalem to Anaiza in Qasim. London: Argonaut Press, 1938. Originalausgabe: Il Neged settentrionale. Itinerario da Gerusalemme a Aneizeh nel Cassim. Gerusalemme, 1866). XLIV, 134 p. -21-

1160 Güldner, W., 1963: Industrielle Strukturplanung in Nah- und Mittelost. In: Deutsche Stiftung für Entwicklungsländer, Programm-Abteilung: Zweite wissenschaftliche Regionaltagung über Nah- und Mittelost, Bericht über eine Tagung in Hamburg vom 28. bis 30. November 1962. Dok. 93/62/DT 4; Berlin, Bonn, 1963: 112-115. -H 3-

1161 Guellouz, E., 1978: Pilgrimage to Mecca. Tunis: Sud Éditions (deutschsprachige Ausgabe: Pilgerfahrt nach Mekka. Zürich: Atlantis, 1978. Französische Ausgabe: Pèlerinage à La Mecque. Lausanne, Paris: La Bibliothèque des Arts, 1980). 2o8 p.   -24, 93-

1162 Günthardt, W., 1978: Arabiens Sprung in die Zukunft. In: Nationale Schweizerische UNESCO-Kommission (Hg.): Seminar über Saudi-Arabien für Sekundar-, Gymnasial- und Seminarlehrer der deutschen Schweiz, Gwatt bei Thun, 16. bis 18. Juni 1977: Schlussbericht. Bern, 1978: 1-35.   -H 223-

1163 Gulf States. Importance of cement industries. In: The Arab Economist 12.198o(131): 24-25.   -2o6, H 3, H 223-

1164 Gulf States. Oil producers face deficit prospects. In: The Arab Economist 14.1982(153): 2o-23.   -2o6, H 3-

1165 Gurashi, H.D. al-, 1982: Proposed goals for adult basic education programs in the Western Province of Saudi Arabia as perceived by teachers and administrators. Ed.D., University of Northern Colorado, Greeley. 16o p.

1166 Guthe, H., 1917: Die Hedschasbahn von Damaskus nach Medina, ihr Bau und ihre Bedeutung. Länder und Völker der Türkei N.F. 7; Leipzig: Gaeblers Geographisches Institut. 37 S.   -1a-

H

1167 H., A.S., 1968: Saudi Arabia's PETROMIN. In: Orient (Opladen) 9: 89-92.   -21, 1oo, H 223-

1168 Habeeb, M.M.S. al-, 1981: Evaluation of instructional supervision of English programs at the intermediate and secondary schools in Saudi Arabia. Ph.D., University of Wisconsin, Madison. 199 p.

1169 Habib, J.S., 1978: Ibn Sa'ud's warriors of Islam. The Ikhwan of Najd and their role in the creation of the Sa'udi Kingdom, 191o-193o. Social, Economic and Political Studies of the Middle East 27; Leiden: Brill. XVI, 196 p.   -21, H 223-

117o Hablützel, H., 1963: Final report to the Government of Saudi Arabia on farm mechanization problems and services. Report 1611; Rome: FAO. II, 27 p.   -Gö 153-

1171 Hablützel, R., 1981: Probleme einer wirtschaftlichen Diversifizierung der Ölländer. In: Finanzierung und Entwicklung 18(2): 1o-13. -24, 93, 18o-

1172 Habshush, H., 1941: Travels in Yemen. An account of Joseph Halévy's journey to Najran in the year 187o written in San'ani Arabic by his guide Hayyim Habshush. Ed. with a detailed glossary in English and a glossary of vernacular words by S.D. Goitein. Jerusalem: Hebrew University Press. VI, 1o2 p. -1a-

1173 Haddad, G.A., 1969: Industrialisation et archaïsme économique: l'example de l'Arabie Séoudite. Thèse pour le Doctorat ès Sciences Économiques, American University of Beirut. -Jafet Library, American University of Beirut-

1174 Haddad, H.S. & B.K. Nijim (eds.), 1978: The Arab world: a handbook. AAUG Monograph Series 9; Wilmette: Medina Press. VI, 249 p. -21-

1175 Haddad, M.S. al-, 1981: The effect of detribalization and sedentarization on the socio-economic structure of the tribes of the Arabian Peninsula: Ajman tribe as a case study. Ph.D., Department of Anthropology, University of Kansas, Lawrence. X, 265 p. -Tü 17-

1176 Haddadeen, M.S., 1982: Investment and Company Law in Saudi Arabia. University of Durham, Centre for Middle Eastern and Islamic Studies, Economic Research Papers on the Middle East 1o; Durham.

1177 Hadie, M.A., 1959: Saudi Arabia: educational progress in 1958-1959. In: International Yearbook of Education 21: 364-367. -1a, 24, 18o-

1178 ——, 196o: Saudi Arabia: educational developments in 1959-196o. In: International Yearbook of Education 22: 338-342. -1a, 24-

1179 ——, 1961: Saudi Arabia: educational developments in 196o-1961. In: International Yearbook of Education 23: 326-33o. -1a, 24-

118o ——, 1962: Saudi Arabia: educational developments in 1961-1962. In: International Yearbook of Education 24: 313-317. -1a, 7, 18-

1181   Hadj, E. el- & R. Sarkis, 1981: The Saudi banking system and the role of the riyal. In: P. Field & A. Moore (eds.): Arab financial markets. London: Euromoney Publications, 1981, 1o1-1o8.   -H 3-

1182   Hafidh, N. al-, 1973: Saudi Arabia educational profile. Beirut: UNESCO Regional Office for Education in the Arab Countries. 1o2 p.

1183   Hafiz, F.A., 198o: Changes in Saudi Arabian foreign policy behavior 1964-1975: a study of the underlying factors and determinants. Ph.D., University of Nebraska, Lincoln. 273 p.

1184   Hafiz, O.Z., 1981: A foreign trade model for Saudi Arabia: an econometric approach. Ph.D., Indiana University, Bloomington. 2o4 p.

1185   Hafiz, T.K., 1976: The potential role of educational/instructional television in higher education and human resources development for the Kingdom of Saudi Arabia. Ph.D., University of Colorado, Boulder. 413 p.

1186   Hague, B.C., 197o: New fertilizer plants in Saudi Arabia and Kuwait to reduce gas wastes. In: World Petroleum 41(7): 28-32.   -2o6-

1187   Haidar, M., 1945: Arabesque. London, New York, Melbourne, Sydney: Hutchinson. 244 p.   -12-

1188   Haj, F.M. al-, n.d.: Proposed agricultural extension system in Al-Hassa Irrigation and Drainage Project (Draft). N.p. (Riyadh?): Ministry of Agriculture and Water, Kingdom of Saudi Arabia. 45, 5 p.

1189   Hajrah, H.H., 1982: Public land distribution in Saudi Arabia. London, New York: Longman (zugleich: Ph.D., Department of Geography, University of Durham, 1974). XXI, 28o p.   -1a, 21, Tü 17-

119o   Hakim, M.H., 1974: The preparation of educational administrators in Saudi Arabia. Ph.D., University of Arizona, Tucson. 176 p.

****   Hakima, A.M.: s. Abu-Hakima, A.M.

1191   Hakken, B.D., 1933: Sunni-Shia discord in eastern Arabia.
       In: The Moslem World 23: 3o2-3o5.   -21, 291, 352-

1192   Halawani, A.W. el- & C.S. Leithead, 1964: Heat illness. I.
       Heat illness during the Mecca Pilgrimage. II. Heat illness and some related problems. In: WHO Chronicle 18:
       283-3o4.   -1a, 26, 2o6-

1193   Halévy, J., 1873, 1877: Voyage au Nedjran. In: Bulletin de
       la Société de Géographie (Paris) VI, 6: 5-31, 249-273,
       581-6o6, 13: 466-479.   -291-

1194   Hallam, H.M., 198o: Saudi Arabia and the economic and political control of the world. Albuquerque: Institute for
       Economic and Political World Strategic Studies. 38 p.

1195   Hallawani, E.A.-R., 1982: Working women in Saudi Arabia:
       problems and solutions. Ph.D., Claremont Graduate School.
       2o5 p.

1196   Halliday, F., 1973: Saudi Arabia: bonanza and repression.
       In: New Left Review (8o): 3-26.   -21, 2o6, 352-

1197   ——, 1975: Arabia without Sultans. Penguin Books; Manchester:
       Nicholls, The Philips Park Press. 528 p.   -24, 212-

1198   ——, 198o: The shifting sands beneath the house of Saud. In:
       The Progressive 44(3): 39-4o.   -3o-

1199   ——, 1982: A curious and close liaison: Saudi Arabia's relations with the United States. In: T. Niblock (ed.): State,
       society and economy in Saudi Arabia. London: Croom Helm;
       Exeter: Centre for Arab Gulf Studies, 1982: 125-147.   -21-

12oo   Halm, H., 1976: Der Islam als staatstragende Religion. In:
       H. Blume (Hg.): Saudi-Arabien. Natur, Geschichte, Mensch
       und Wirtschaft. Ländermonographien 7; Tübingen, Basel:
       Erdmann, 1976: 155-16o.   -12, 21, 24-

12o1   Halmos, E.E., Jr., 1977: Doing business in Saudi Arabia. In:
       Civil Engineering (London) 1977(6): 76-79.   -17, 89, 93-

12o2   Hamad, H.S. al-, 1973: The legislative process and the development of Saudi Arabia. Ph.D., University of Southern
       California, Los Angeles. III, 2oo p.   -21-

1203 Hamadi, A.M., 1981: Saudi Arabia's territorial limits: a study in law and politics. Ph.D., Indiana University, Bloomington. 145 p.

1204 Hambleton, H.G., 1982: The Saudi Arabian petrochemical industry: its rationale and effectiveness. In: T. Niblock (ed.): State, society and economy in Saudi Arabia. London: Croom Helm; Exeter: Centre for Arab Gulf Studies, 1982: 235-277. -21-

1205 Hamdan, A.I., 1980: Competencies needed for teachers of the mentally retarded in Saudi Arabia: a need assessment study. Ph.D., Michigan State University, East Lansing. 199 p.

1206 Hamdan, M.Z., 1977: A comparative analysis of Saudi male elementary teacher preparation programs. Ph.D., Kent State University. 214 p.

1207 Hamdan, Y.A. al-, 1971: Development of local government in Saudi Arabia. M.P.A., University of Pittsburgh.

1208 Hamid Daoud, S.A. el-, 1979: Demographic developments in Saudi Arabia during the present century. Ph.D., University of City.

1209 Hamidi, A.S., 1975: Motivational factors toward literacy in Riyadh, Saudi Arabia. Ph.D., Arizona State University, Tempe. 193 p.

1210 Hamídulláh, M., 1969: The pilgrimage to Mecca: its history, rites and philosophy. In: The Islamic Review and Arab Affairs 57(9): 17-34. -1a, 21-

1211 Hamilton, C.W., 1962: Americans and oil in the Middle East. Houston: Gulf. XI, 307 p. -21-

1212 Hamilton, J.A., 1895: Palgrave, William Gifford. In: Dictionary of National Biography 43; London: Smith, Elder, 1895: 109-110. -7, 21-

1213 Hamilton, P., 1971: Seas of sand. Aldus Encyclopedia of Discovery and Exploration; London: Aldus. 191 p. -21-

1214 Hammad, M.A., 1973: The educational system and planning for manpower development in Saudi Arabia. Ph.D., Department of Political Science, Indiana University, Bloomington. 371 p. -21-

1215 Hammoudeh, S.M., 1979: The future oil price behaviour of OPEC and Saudi Arabia: a survey of optimization models. In: Energy Economics 1(3): 156-166. -H 3-

1216 —, 1980: Optimal oil pricing policy for Saudi Arabia. Ph.D., University of Kansas, Lawrence. 305 p.

1217 Hamza, F., 1935: Najran. In: Journal of the Royal Central Asian Society 22: 631-640. -21, 90, 291-

1218 Hans, J., 1938: Aus der Finanzwelt des Islams. Wien: Selbstverlag. 80 S. -31-

1219 —, 1939: Mekkapilger. Auswirkungen auf die Wirtschaft von Sa'udi-Arabien. In: Orient-Nachrichten 5: 54-55. -12, 206-

1220 —, ^{2}1961: Maria-Theresien-Taler. Zwei Jahrhunderte 1751-1951. Epilog 1951-1960. Leiden: Brill (Klagenfurt: Hans, ^{1}1950). 91 S. -24-

1221 Hansen, H., 1949: Arabien, das reichste Land der Welt. In: Passat 1(1): 14-18. -212-

1222 Hansen, T., 1965: Reise nach Arabien. Die Geschichte der königlich dänischen Jemen-Expedition 1761-1767. Hamburg: Hoffmann & Campe (Originalausgabe: Det lykkelige Arabien. Kopenhagen: Gyldendal, 1962. Englische Ausgabe: Arabia Felix: the Danish expedition of 1761-1767. London: Collins, 1964; amerikanische Ausgabe: New York: Harper & Row, 1964). 423 S. -21, 24-

1223 Hanson, W.G., 1933: A Nasrany in Arabia. In: London Quarterly Review VI, 158(2): 65-75. -Cambridge University Library-

1224 Harby, M.K., M.H. Afifi & M.A. el-Ghannam, 1965a: Technical education in the Arab states. Paris: UNESCO. 57 p.

1225 —, — & —, 1965b: Technical education in the changing Arab world. Cairo: Anglo-Egyptian Bookshop. 196 p.

1226 Hare, R.A., 1972: The great divide: World War II. In: The Annals of The American Academy of Political and Social Science 401: 23-30. -12, 180-

1227 Hariri, H.B., 1982: School climate, competency and training of principals in intermediate schools in Saudi Arabia. Ed.D., University of Northern Colorado, Greeley. 174 p.

1228 Harrell, G.T., 1976: Medical education in Saudi Arabia. In: Annals of Internal Medicine 85: 677-678. -21, 291-

1229 Harrington, C.W., 1958: The Saudi Arabian Council of Ministers. In: The Middle East Journal 12: 1-19. -12, 21, 2o6-

123o Harrison, P.W., 1918: Al Riadh, the capital of Nejd. In: The Moslem World 8: 412-419. -21, 291, 352-

1231 ——, 1919: Neglected Arabia. In: The Moslem World 9: 191. -21, 291, 352-

1232 ——, 192o: The situation in Arabia. In: The Atlantic Monthly 126: 849-855. -21, 25, 385-

1233 ——, 1921: What the Arab thinks of the missionary. In: The Missionary Review of the World 34: 759-76o. -1a, 4-

1234 ——, 1924a: Economic and social conditions in east Arabia. In: The Moslem World 14: 163-171. -21, 291, 352-

1235 ——, 1924b: The Arab at home. New York: Crowell. XII, 345 p. -7-

1236 ——, 1924c: The charm of the Arab. In: The International Review of Missions 11: 436-441. -21, 24, 25-

1237 ——, 1926: Facing difficulties in Arabia. In: World Dominion 4: 63-7o. -21-

1238 Harsaghy, F.J., Jr., 1965: The administration of American cultural projects abroad: a developmental study with case histories of community relations in administering educational and informational projects in Japan and Saudi Arabia. Ph.D., New York University. 82o p.

1239 ——, 1972: Design for future service in a developing country. In: Special Libraries 63: 4oo-4o3. -21, 9o, 18o-

124o Harsham, P., 1977: Partners in growth: the United States 1. 2. In: Aramco World Magazine 28(1): 26-48. -21, 24-

1241 Hart, P.T., 1953: Application of Hanbalite and decree law to foreigners in Saudi Arabia. In: The George Washington Law Review 22: 165-175. -1a-

1242 Hartmann, M., 19o8: Die Mekkabahn. In: Orientalistische Literatur-Zeitung 11(1): Sp. 1-7. -16, 21, 291-

1243 Hartmann, M., 1912: Die Mekkabahn, ihre Aussichten und ihre Bedeutung für den Islam. In: Asien (Berlin) 11: 148-150, 163-167.   -1a, 90-

1244 ——, 1914: Irak und Arabien. In: Die Welt des Islams 2: 24-63, 295-332.   -16, 21, 206-

1245 Hartmann, R., 1913: Djidda. In: EI^1 1: 1086-1087.   -21, 24-

1246 ——, 1917: Arabien im Weltkrieg. In: A. Petermann's Mitteilungen aus Justus Perthes' Geographischer Anstalt 63: 54-57, 84-87.   -16, 21, 24-

1247 ——, 1918: Die arabische Frage und das Türkische Reich. In: Beiträge zur Kenntnis des Orients, Jahrbücher der Deutschen Vorderasiengesellschaft 15: 1-31.   -1a, 21, 25-

1248 ——, 1921: Die nationalen Bestrebungen der Araber, ihre Grundlagen, Ziele und Aussichten. In: Deutsche Revue 46, Bd. 3: 247-256, Bd. 4: 53-62.   -25, 212, 291-

1249 ——, 1924/25: Die Wahhābiten. In: Zeitschrift der Deutschen Morgenländischen Gesellschaft 78/N.F. 3: 176-213.   -21, 291, 352-

1250 ——, 1938: Zur heutigen Lage des Beduinentums. In: Die Welt des Islams 20: 51-73.   -16, 21, 206-

1251 —— & P.A. Marr, 1965: Djudda. In: EI^2 2: 571-573.   -21, 24, 212-

1252 Hartzell, M.E., 1957: A special library in an Arab culture. In: Special Libraries 48: 56-63.   -21, 30, 180-

1253 Harvest without end. Modern methods and hard work have Hofuf and Qatif farmlands blooming again. In: Aramco World 11. 1960(9): 3-5.

1254 Hashagen, J., 1937: Doughty's klassisches Arabienwerk. In: Zeitschrift der Deutschen Morgenländischen Gesellschaft 91/N.F. 16: 605-615.   -19, 21, 291-

1255 Hashe, A.M., 1965: Analysis of community development programs in Saudi Arabia, 1963-65. New York: UNO.

1256 Hassan, H.M., 1966: UN cooperates in Saudi development plans In: Emergent Nations 2(2): 59-60.   -H 223-

1257   Hassanain, M.A., 1971: An economic review of the Saudi Arabian planning framework. Ph.D., Graduate Faculty of Arts and Sciences, University of Pittsburgh. VI, 241 p.   -21-

1258   Hasson, R.C., 1955: Explorations in Saudi Arabia. In: The Mines Magazine 45(1o): 119-128.   -89, 1o4-

1259   Hatem, M.A.-K., 1977: Land of the Arabs. London, New York: Longman. 323 p.   -21-

1260   Haupert, J.S., 1966: Saudi Arabia. In: Focus (New York) 16 (9): 1-6.   -Tü 17-

1261   ——, 1972: Saudi Arabia. In: A. Taylor (ed.): The Middle East. Newton Abbot: David & Charles, 1972: 99-111.   -93-

1262   Hayes, S.D., 1977: Joint Economic Commissions as instruments of US foreign policy in the Middle East. In: The Middle East Journal 31: 16-3o.   -12, 21, 2o6-

1263   ——, 198o: Riyadh on the move. In: Aramco World Magazine 31 (4): 26-32.   -21, 24-

1264   Hayil. In: Aramco World 14.1963(7): 13-17.   -H 223-

1265   Hayit, B., 1961: Mekka. In: Bustan 1961(2): 9-1o.   -12, 212-

1266   Hazard, H.W., 1956a: Saudi Arabia. Subcontractor's Monograph, HRAF-5o; New Haven: HRAF. VIII, 369 p.   -Library of Congress, Washington-

1267   ——, 1956b: Eastern Arabia. Subcontractor's Monograph, HRAF-51; New Haven: HRAF. IX, 365 p.

1268   ——, 1959: Islamic philately as an ancillary discipline. In: J. Kritzeck & R.B. Winder (eds.): The world of Islam. Studies in honour of Philip K. Hitti. London: Macmillan; New York: St. Martin's Press, 1959: 199-232.   -12, 21-

1269   ——, 1964: The Arabian Peninsula. Around the World Program; Garden City: Doubleday. 61 p.

127o   Hazzam Dawsari, F.S. al-, 1975: Descriptive study of the development of the College of Petroleum and Minerals, Dhahran, Saudi Arabia. Ph.D., Arizona State University, Tempe. 373 p.

1271   Headley, R.L., 196o: Burayda. In: EI^2 1: 1312-1313.   -21, 24, 212-

1272   Headley, R.L., W.E. Mulligan & G. Rentz, 1960: 'Asīr. In: EI^2 1: 707-710.   -21, 24, 212-

1273   ———, 1964: People of the camel. In: Aramco World 15(5): 10-15.   -H 223-

1274   Heady, H.F., 1963: Report to the Government of Saudi Arabia on grazing resources and problems. Report 1614; Rome: FAO. 30 p.   -Gö 153-

1275   Health guardians. Malaria drops 95% in eastern Saudi Arabia. In: Aramco World 1.1949/50(8): 1, 5-6.

1276   Hebshi, H.B. el-, 1974: Preparation of high school mathematics teachers in Saudi Arabia. Ed.D., Arizona State University, Tempe. 69 p.

1277   Hecker, M., 1914: Die Eisenbahnen der asiatischen Türkei. In: Archiv für Eisenbahnwesen 37: 744-800, 1057-1087, 1283-1321, 1539-1584.   -25, 93, 180-

1278   Hecklau, H., 1978: Saudi-Arabien. In: D. Nohlen & F. Nuscheler (Hgg.): Handbuch der Dritten Welt 4, II. Hamburg: Hoffmann & Campe, 1978: 575-588.   -21, 24-

1279   Hejailan, S., 1979: Saudi Arabia. In: Yearbook Commercial Arbitration 4: 162-173.   -1a, 24, 25-

1280   Helaissi, A.S., 1959: The Bedouins and tribal life in Saudi Arabia. In: International Social Science Journal 11: 532-538.   -21, 24, 180-

1281   Helmensdorfer, E., 1972: Hartöstlich von Suez. Die feudale Halbinsel. München, Percha: Schulz. 448 S.   -12, 212-

1282   Helms, C.M., 1981: The cohesion of Saudi Arabia: evolution of political identity. London: Croom Helm (zugleich: Evolution of political identity in Saudi Arabia: delineation of a nation-state, 1901-1932. Ph.D., University of Oxford, 1979). 313 p.   -12, 21, 24-

1283   Henin, C., 1980: Financial infrastructure of Saudi Arabia. Discussion Papers 804; Ottawa: Institut de Coopération Internationale, Université d'Ottawa. 39 p.   -Bo 149, H 3

1284   Hennig, R., 1915: Die deutschen Bahnbauten in der Türkei, ihr politischer, militärischer und wirtschaftlicher Wert. In: H. Grothe (Hg.): Länder und Völker der Türkei. Schrif tensammlung des Deutschen Vorderasienkomitees; Leipzig: Veit, 1915: 333-364.   -21-

1285 Henninger, J., 1939a: Pariastämme in Arabien. In: Festschrift zum 5ojährigen Bestandsjubiläum des Missionshauses St. Gabriel, Wien-Mödling. Sankt Gabrieler Studien 8; Wien-Mödling: Missionsdruckerei St. Gabriel, 1939: 5o1-539.   -Anthropos-Institut, St. Augustin-

1286 ——, 1939b: Zur Verbreitung des Brautpreises bei den arabischen Beduinen. In: Anthropos (Freiburg/Schweiz) 34: 38o-388.   -16, 21, 291-

1287 ——, 1941: Fell- und Lederkleidung in Arabien. In: Internationales Archiv für Ethnographie 4o: 41-5o.   -12, 16, 3o-

1288 ——, 1943: Die Familie bei den heutigen Beduinen Arabiens und seiner Randgebiete. Ein Beitrag zur Frage der ursprünglichen Familienform der Semiten. Internationales Archiv für Ethnographie 42, Leiden: Brill (zugleich: Neubearb. Fassung der Diss., Philosophische Fakultät, Universität Wien, 1937). VIII, 188 S.   -21-

1289 ——, 1945: La famille chez les Bédouins d'Arabie. In: Nova et Vetera (Fribourg) 2o: 275-3o1.

129o ——, 1951a: Werbungsformen und Liebespoesie bei den Arabern. In: Anthropos (Freiburg/Schweiz) 46: 998-1oo5.   -16, 21, 291-

1291 ——, 1951b: Tribus et classes de Parias en Arabie et en Égypte. In: Actes du 14e Congrès International de Sociologie, Rome, 3o août - 3 septembre 195o, Vol. 4: 266-278.

1292 ——, 1959: Das Eigentumsrecht bei den heutigen Beduinen Arabiens. In: Zeitschrift für vergleichende Rechtswissenschaft einschliesslich der ethnologischen Rechtsforschung 61: 6-56.   -21, 24-

1293 ——, 1968: Zum Erstgeborenenrecht bei den Semiten. In: E. Gräf (Hg.): Festschrift Werner Caskel. Zum 7o. Geburtstag, 5. März 1966, gewidmet von Freunden und Schülern. Leiden: Brill, 1968: 162-183.   -21, 291-

1294 ——, 1981: Arabica Sacra. Aufsätze zur Religionsgeschichte Arabiens und seiner Randgebiete. Contributions à l'histoire religieuse de l'Arabie et de ses régions limitrophes. Orbis Biblicus et Orientalis 4o; Freiburg/Schweiz: Universitätsverlag; Göttingen: Vandenhoeck & Ruprecht. 347 S. -21, 24-

1295 Henry, J.C., 1948a: Ibn Saud rubs a magic lamp. In: Nation's Business 36: 36-38, 83-84. -1a, 18o, H 3-

1296 ——, 1948b: Industrial revolution in the desert. In: Popular Mechanics 89: 89-93, 264, 268. -2o1, 21o-

1297 Hentig, W.O. von, 1955a: Saudisch Arabien - Ein Staat tritt in die Geschichte ein. In: Zeitschrift für Geopolitik 26: 3o8-31o. -21, 93, Tü 17-

1298 ——, 1955b: Bericht über Saudien. In: Zeitschrift für Geopolitik 26: 434-438. -21, 93, Tü 17-

1299 ——, 1955c: Die Gesellschaftsstruktur Saudi-Arabiens. In: Frankfurter Hefte 1o: 436-439. -16, 21, 24-

13oo Herzog, R., 1963: Sesshaftwerden von Nomaden. Geschichte, gegenwärtiger Stand eines wirtschaftlichen wie sozialen Prozesses und Möglichkeiten der sinnvollen technischen Unterstützung. Forschungsberichte des Landes Nordrhein-Westfalen 1238; Köln, Opladen: Westdeutscher Verlag. 2o7 S. -12, 24-

13o1 Hess, J.J., 19o2: Bemerkungen zu Doughty's Travels in Arabia. In: Wiener Zeitschrift für die Kunde des Morgenlandes 16: 45-62. -5, 21, 25-

13o2 ——, 1938: Von den Beduinen des innern Arabiens. Erzählungen / Lieder / Sitten und Gebräuche. Zürich, Leipzig: Niehans. 177 S. -21-

13o3 Heykal, M.H., 1977: The Saudi era. In: Journal of Palestine Studies 6(4): 158-164. -12, 21, 25-

13o4 Hibshy, M.A., 1967: The development of higher education in Saudi Arabia, 1945-1965. M.Phil., Institute of Education, University of London.

13o5 ——, 1975: The development of teacher education in Saudi Arabia, 1928-1972. Ph.D., Institute of Education, University of London.

13o6 Hidjāz, al-. In: EI[1] 2.1927: 319. -21, 24-

1307   Hidore, J.J. & Y. Albokhair (wohl identisch mit dem Verfasser
       von Nr. 79), 1982: Sand encroachment in al-Hasa oasis,
       Saudi Arabia. In: The Geographical Review 72: 350-356.
       -21, 24, 206-

1308   Higher education in Saudi Arabia. In: Orient (Opladen) 7.
       1966: 43-44.   -21, 100, H 223-

1309   Hilaisi, N. el-, 1965: Wirtschaftliche und soziale Entwick-
       lung Saudi-Arabiens nach dem Zweiten Weltkrieg. Diss.,
       Rechts- und Staatswissenschaftliche Fakultät, Universität
       Graz. 168, 5 S.   -Universitätsbibliothek Graz-

1310   Hilālī, T.D. al-, 1940: Die Kasten in Arabien. In: Die Welt
       des Islams 22: 102-110.   -16, 21, 206-

1311   Hill, A.G., 1981: Population, migration and development in
       the Gulf states. In: S. Chubin (ed.): Security in the
       Persian Gulf 1: Domestic political forces. Westmead,
       Farnborough: Gower, 1981: 58-83.   -21-

1312   Hirashima, H.Y., 1972: The road to Holy Mecca. This Beauti-
       ful World 31; Tokyo, Palo Alto, New York: Kodansha Inter-
       national. 129 p.

1313   Hirst, D., 1966: Oil and public opinion in the Middle East.
       London: Faber & Faber. 127 p.   -212-

1314   Hirszowicz, Ł., 1966: The Third Reich and the Arab East.
       Studies in Political History; London: Routledge & Kegan
       Paul; Toronto: University of Toronto Press (Originalaus-
       gabe: III Rzesza i arabski wschód. Warszawa, 1963). XI,
       403 p.   -24/213-

1315   Hitti, S.H. & G.T. Abed, 1974: The economy and finances of
       Saudi Arabia. In: International Monetary Fund Staff Pa-
       pers 21: 247-306.   -206, H 3-

1316   Hoagland, J. & J.P. Smith, 1978: Saudi Arabia and the United
       States: security and interdependence. In: Survival 20:
       80-83.   -24/213, 180-

1317   Hobday, P., ²1979a: Saudi Arabia today: an introduction in
       the richest oil power. London, Basingstoke: Macmillan
       (¹1978). X, 133 p.   -12, 21, Tü 17-

1318   ——, 1979b: Saudi Arabia spends her oil wealth. In: National
       Westminster Bank Quarterly Review 1979(Feb.): 43-50.
       -38, 206-

1319 Hobson, R., with J. Lawton, R. Fraga, A. Clark & M. Love, 1982: Science in the modern age. In: Aramco World Magazine 33(3): 34-48.   -21, 24-

1320 Hoel, L., 1959: 'Probably my greatest experience'. In: The Arab World 5(6): 8-11.   -206-

1321 Hörder, M.-H. & T. Hanf, 1963: Krankheit und Sozialstruktur in Saudi-Arabien. Eine Studie zur Frage der Bedeutung sozialer Faktoren für die Medizin in Entwicklungsländern. Freiburger Universitätsblätter 3; Freiburg i.Br.: Rombach. 12 S.   -21-

1322 Hogarth, D.G., 1904: The penetration of Arabia. A record of the development of Western knowledge concerning the Arabian Peninsula. London: Lawrence & Bullen; New York: Stokes (reprint: Khayats Oriental Reprint 22; Beirut: Khayats, 1966). XIII, 359 p.   -7, 21, Tü 17-

1323 ——, 1905: The Nearer East. The Regions of the World 2; London, Edinburgh, Glasgow: Frowde (Originalausgabe: New York: Appleton, 1902, ^{2}1915). XVI, 296 p.   -21-

1324 ——, 1908: Problems in exploration. I. Western Asia. In: The Geographical Journal 32: 549-570.   -21, 24, 291-

1325 ——, 1920a: War and discovery in Arabia. In: The Geographical Journal 55: 422-439.   -21, 291, 352-

1326 ——, 1920b: Obituary: Lieut.-Colonel G.E. Leachman, C.I.E., D.S.O. In: The Geographical Journal 56: 325-326.   -21, 291, 352-

1327 ——, 1921: Some recent Arabian explorations. In: The Geographical Review 11: 321-337.   -21, 24, 206-

1328 ——, 1922: Arabia (Aussentitel: A history of Arabia). Oxford: Clarendon Press. 139 p.   -12-

1329 ——, 1925: Wahabism and British interests. In: Journal of the British Institute of International Affairs 4: 70-81.   -25, 180, 291-

1330 ——, 1927: Gertrude Bell's journey to Hayil. In: The Geographical Journal 70: 1-21.   -21, 24, 291-

1331 ——, 1928: The life of Charles M. Doughty. London: Oxford University Press. VIII, 216 p.   -21-

1332   Hogarth, D.G., 1978: Hejaz before World War I: a handbook.
       Arabia Past and Present 7; Cambridge/England, New York:
       The Oleander Press; Naples: The Falcon Press (Original-
       ausgabe: Handbook to Hejaz. 2nd, impr. & enl. ed. Cairo:
       Arab Bureau, 1917). 155 p.   -21, Tü 17-

1333   Holden, D., 1966: Farewell to Arabia. London: Faber & Faber.
       268 p.   -21-

1334   —— & R. Johns, 1981: The house of Saud. London: Sidgwick &
       Jackson (deutsche Ausgabe: Die Dynastie der Sauds. Wüsten-
       krieger und Weltfinanziers. Düsseldorf: Econ, 1983).
       XIV, 569 p.   -21-

1335   Holm, H.M., 1955: The agricultural resources of the Arabian
       Peninsula. Washington: Foreign Agricultural Service, US
       Department of Agriculture. 14 p.

1336   Holman, J.K., 1967: Sacred line to Madina: the history of
       the Hijaz Railway. B.A., Princeton University.

1337   Holt, A.L., 1923: The future of the north Arabian desert.
       In: The Geographical Journal 62: 259-271.   -21, 291, 352-

1338   Holzhausen, R., 1936: Die Mission Stotzingen und der Beginn
       des Arabischen Aufstandes (1916). Eine Kriegserinnerung
       aus der alten Türkei. In: Süddeutsche Monatshefte 33:
       56o-568.   -7, 18o, 2o6-

1339   Hoog, P.H. van der, 1935: Pelgrims naar Mekka. Den Haag:
       Leopold. VIII, 252 S.   -21-

134o   ——, 1939: De bedevaart naar Mekka. In: Oostersch Genoot-
       schap in Nederland, Verslag van het 9^e congres. Leiden:
       Brill, 1939: 4-6.

1341   Hope, W.E.S., 1951: Arabian adventurer: the story of Haji
       Williamson. London: Hale. 335 p.   -21-

1342   Hopper, H., 1956: Jiddah, Saudi Arabia's 'west coast boom
       town'. In: Lands East 1956(Feb.): 4-6, 16-17.   -Library
       of Congress, Washington-

1343   Hopwood, D., 1982: The ideological basis: Ibn Abd al-Wah-
       hab's Muslim revivalism. In: T. Niblock (ed.): State,
       society and economy in Saudi Arabia. London: Croom Helm;
       Exeter: Centre for Arab Gulf Studies, 1982: 23-35.   -21-

1344 Hoskins, H.L., 1947: Background of the British position in Arabia. In: The Middle East Journal 1: 137-147. -12, 21, 38-

1345 ——, 1950a: Point Four with reference to the Middle East. In: The Annals of The American Academy of Political and Social Science 268: 85-95. -12, 21, 25-

1346 ——, 1950b: Middle East oil in United States foreign policy. Public Affairs Bulletin 89; Washington: Legislative Reference Service, Library of Congress.

1347 Hosni, S.M., 1966: The partition of the Neutral Zone. In: The American Journal of International Law 6o: 735-749. -1a, 21, 352-

1348 Hottinger, A., 1958: Saudiarabien. In: Atlantis 3o: 245-247. -21, 24, 212-

1349 ——, 1966: Appell an den Islam. König Feisals Gegenoffensive in der arabischen Staatenwelt. In: Wort und Wahrheit 21: 296-3o4. -21, 188, 352-

1350 ——, 197o: 1omal Nahost. Panoramen der modernen Welt 4; München: Piper. 452 S. -12, 21, 24/213-

1351 ——, 1974: Im Hauptquartier der Erdölproduzenten. In: Merkur 28: 266-278. -21, 1oo-

1352 ——, 1979a: Saudi Arabia: on the brink? In: Swiss Review of World Affairs 29(2): 8-12. -21-

1353 ——, 1979b: Does Saudi Arabia face revolution? In: The New York Review of Books 26(June): 14-17. -1a, 25, 18o-

1354 ——, 198oa: Behind the Grand Mosque incident. In: Swiss Review of World Affairs 29(1o): 9-1o. -21-

1355 ——, 198ob: Internal problems in the wealthy Arab states. In: Swiss Review of World Affairs 29(12): 1o-12. -21-

1356 ——, 198oc: Notes from Saudi Arabia. In: Swiss Review of World Affairs 3o(5): 1o-15, 18-2o. -21-

1357 ——, 198od: Camels, Toyotas and tank trucks. In: Swiss Review of World Affairs 3o(9): 8-13. -21-

1358 ——, 1981a: Political institutions in Saudi Arabia, Kuwait and Bahrain. In: S. Chubin (ed.): Security in the Persian

Gulf 1: Domestic political factors. Westmead, Farnborough: Gower, 1981: 1-18. -21, 24/213, H 223-

1359 Hottinger, A., 1981b: Wie stabil ist das Regime in Saudi-Arabien? Die inneren Probleme der Staaten auf der Arabischen Halbinsel. In: Der Bürger im Staat 31(1): 5o-55 (zugleich in: Brennpunkt Mittel-Ost. Mit Beiträgen von F. Scholz et al. Kohlhammer Taschenbücher 1o55; Stuttgart, Berlin, Köln, Mainz: Kohlhammer, 1981: 142-159). -12, 18, 21-

136o Hourani, A., 1962: The life and ideas of Wilfrid Scawen Blunt. In: Middle East Forum 38(8): 21-27. -212-

1361 Howard, B., Jr., 1958: Buraimi: a study diplomacy by default. In: The Reporter (New York) 18(Jan. 23, 1958): 13-16. -18o-

1362 Howarth, D.A., 1964: The desert king; a life of Ibn Saud. London: Collins (amerikanische Ausgabe: The desert king: Ibn Saud and his Arabia. New York: McGraw-Hill, 1964). 251 p. -21-

1363 ——, 1965: To light a flame. In: Aramco World 16(1): 1-7. -24-

1364 Hoye, P.F., 1968: The great Badanah flood. In: Aramco World Magazine 19(2): 32-39. -24-

1365 ——, 1969: Tom Barger: myth or man? In: Aramco World Magazine 2o(5): 4-13. -24-

1366 Hoyt, C.K., 1976: Caudill Rowlett Scott's ongoing work for a Saudi Arabian university. In: Architectural Record 159 (4): 95-1oo. -89, 93-

1367 Hoyt, M.P., 1978: Saudi Arabia - a legal update - a U.S. lawyer's perspective. In: Corporate law and practice. Course Handbook 266; New York, 1978: 97-115.

1368 H.St.J.B. Philby. In: The Middle East Journal 14.196o: 433. -12, 21, 2o6-

1369 Huber, C., 1884-85: Voyage dans l'Arabie centrale, 1878-1882 (Hamâd, Šammar, Qaçîm, Hedjâz). In: Bulletin de la Société de Géographie (Paris) VII, 5: 3o4-363, 468-53o, 6: 92-148. -291-

1370 Huber, C., 1891: Journal d'un voyage en Arabie (1883-1884), publié par la Société asiatique et la Société de géographie sous les auspices du Ministère de l'instruction publique. Paris: Imprimerie Nationale. XII, 778 p.   -16, 21, Tü 17-

1371 Hüber, R., 1940: Deutsche Pioniertaten im nahöstlichen Bahnbau. In: Der Nahe Osten 1(1/2): 8-10.   -21, 206-

1372 ——, 1943: Arabisches Wirtschaftsleben. Arabische Welt 4; Heidelberg, Berlin, Magdeburg: Vowinckel. VIII, 140 S. -93, Tü 17-

1373 ——, 1952: Ein Königstraum wird Wirklichkeit: Die Transarabische Eisenbahn entsteht. In: Übersee-Rundschau 4: 316-318, 364-365.   -24, 30, 180-

1374 Hübner, G., 1970: Saudi-Arabien. In: Autorenkollektiv: Die arabischen Länder. Hg. von G. Nötzold. Gotha, Leipzig: VEB Haack, 1970: 228-236.   -21, 93, 100-

1375 Hughes, H., 1981: Middle East railways. Harrow: The Continental Railway Circle. 128 p.   -21-

1376 Hulais, H.Y., 1982: The interactional variation of motivational factors, abilities and support in relation to perceived intensity of stress conditions among medical students. Ph.D., Kansas State University, Manhattan. 209 p.

1377 Hull, C., 1948: The memoirs of Cordell Hull. Vol. 2. New York: Macmillan. VI, 919-1804 p.   -16, 24/213-

1378 Humaidan, S.H., 1980: Policies and management guidelines for optimum resource utilization at Al Hasa Irrigation and Drainage Project, Saudi Arabia. Ph.D., Oklahoma State University, Stillwater. 217 p.

1379 Humphreys, R.S., 1979: Islām and political values in Saudi Arabia, Egypt and Syria. In: The Middle East Journal 33: 1-19.   -21, 38, 206-

1380 **Hurewitz, J.C.**, 1969: Middle East politics: the military dimension. New York, Washington, London: Praeger. XVIII, 550 p.   -24/213-

1381 ——, 1979: The Middle East and North Africa in world politics: a documentary record. Vol. 2: British-French supremacy, 1914-1945. 2nd ed., rev. & enl. New Haven, London: Yale University Press (Princeton: Van Nostrand, ¹1956). XXII, 861 p.   -12-

1382 Hurren, B.J., 1965: Pilgrim railway into Arabia. In: The Railway Magazine 111: 346-351, 4o8-412.   -89-

1383 Hussain, M.K., 197o: Copper Taweelah of Ḥaṣa. In: The Indian Numismatic Chronicle 8: 41-44.   -21-

1384 Hussain, Z., 1978a: Land and water use in Saudi Arabia. In: World Crops (Horne) 3o: 58-61.   -98-

1385 ——, 1978b: Alfalfa cultivation in Saudi Arabia. In: World Crops (Horne) 3o: 26o-261.   -98-

1386 Hussaini, A., 1972: Football Arabian style: on the Saudi sports scene, soccer has no rival. In: Aramco World Magazine 23(Special): 12-13.   -24-

1387 Hussayen, M.A. al-, 198o: Building problems in Saudi Arabia: the need for building research, and the development of building research approaches. Arch.D., University of Michigan, Ann Arbor. 227 p.

1388 Hussein, F., 1975: Date culture in Saudi Arabia. Ministry of Agriculture and Water, Department of Research and Development; Jeddah: Dar al-Asfahani. II, 33 p.   -H 223-

1389 Hussein, T., 1932: Literary life in the Arabian Peninsula. In: The Open Court 46: 828-846.   -1a-

I

139o IBA. Internationales Biographisches Archiv, ca. 197o-: Abdullah Ben Abdulasis.-Fahd.-Faisal.-Khaled.-Saud Ben Faisal (etc.). Ravensburg: Munzinger Archiv. Getr.Pag.   -21-

1391 Ibrahim, A.A. al-, 1982: Regional and urban development in Saudi Arabia. Ph.D., University of Colorado, Boulder. 424 p.

1392 Ibrahim, S.E. & D.P. Cole, 1978: Saudi Arabian bedouin: an assessment of their needs. Cairo Papers in Social Science 1, Monograph 5; Cairo: American University of Cairo. II, 115 p.   -H 223-

1393   Ibrahim, S.E., 1981: Superpowers in the Arab world. In: The
        Washington Quarterly 4(3): 81-96.   -12, 24/213, 2o6-

1394   ——, 1982: The new Arab social order: a study of the social
        impact of oil wealth. Westview's Special Studies on the
        Middle East; Boulder: Westview Press; London: Croom Helm.
        XIV, 2o8 p.   -21-

1395   Idries Shah, H., 1957: Destination Mecca. London: Rider.
        192 p.   -294-

1396   Ilam, H.M.D.M., 198o: Aspects of the urban geography of Makkah and Al-Madinah, Saudi Arabia. Ph.D., University of Durham.

1397   Ileri, M., 1979: Hamburg und der Nahe Osten. Struktur und Perspektiven der Wirtschaftsbeziehungen. Mitteilungen des Deutschen Orient-Instituts 12; Hamburg: Deutsches Orient-Institut. 148 S.   -21, H 223-

1398   ILO, Petroleum Committee, 195o: Social conditions in the petroleum industry. IC/P/3/II; Geneva.

1399   ILO/UNDP, 1968: Report to the Government of the Kingdom of Saudi Arabia on the organisation of a manpower assessment and training programmes. Geneva.

14oo   Il Trattato di amicizia e buon vicinato tra il Regno del Ḥigiāz, Neǵd e dipendenze ed il Yemen, del 15 dicembre 1931. In: Oriente Moderno 12.1932: 13o-131.   -1a, 5, 21-

14o1   Il Trattato di eṭ-Ṭā'if del 2o maggio 1934 fra il Regno Arabo Sa'ūdiano e il Yemen. In: Oriente Moderno 14.1934: 315-323.   -1a, 5, 21-

14o2   Il Trattato di fratellanza araba e d'alleanza fra l''Iraq e il Regno Arabo Sa'ūdiano. In: Oriente Moderno 16.1936: 257-259.   -1a, 5, 21-

14o3   Il Trattato 7 maggio 1936 di amicizia tra l'Egitto e il Regno Arabo Sa'ūdiano. In: Oriente Moderno 16.1936: 3o9-31o.   -1a, 5, 21-

14o4   In a country where education is as valuable as oil, Aramco tries to share the burden. In: Aramco World Magazine 2o.1969(6): 22-23.   -24-

1405 Inayatullah, S., 1942: Geographical factors in Arabian life and history. Lahore: Ashraf (zugleich (?): The influence of physical environment upon Arabian life and institutions. Ph.D., University of London, 1931/32). 16o p.

1406 Inshaullah, M., 19o8: The history of the Hamidia Hedjaz Railway project, in Urdu, Arabic and English. Lahore: Central Printing Works. -38-

1407 Institute of Practitioners in Advertising, 1979: Advertising conditions in the Gulf States and Saudi Arabia. London. 28 p.

1408 Intercambio comercial México-Arabia Saudita. In: Comercio Exterior 27.1977: 139o-1399. -H 3-

1409 International Bank for Reconstruction and Development (IBRD), 196o: Approach to the economic development of Saudi Arabia. Mission Report AS 82a; Washington. 91 p.

1410 (International) Institute for Strategic Studies (IISS), 1977: Saudi Arabia's emerging role as financier and arbiter. In: The Middle East (32): 78-8o. -12, 3o, 188-

1411 International Land Development Consultants B.V. (ILACO), 1975a: Socio-economic development plan 1395/96-1399/14oo AH (1975/76-1979/8o AD) for the Eastern Region of Saudi Arabia. Annex 3: Agriculture and water. Ed. by the Central Planning Organization, Kingdom of Saudi Arabia. Arnhem.

1412 ——, 1975b: Socio-economic development plan 1395/96-1399/14oo AH (1975/76-1979/8o AD) for the Eastern Region of Saudi Arabia. Main report. Ed. by the Central Planning Organization, Kingdom of Saudi Arabia. Arnhem.

1413 International university chronicle: Saudi Arabia. In: International Association of Universities Bulletin 26.1978 (Feb.): 29-3o.

1414 Ioffe, A.Y., 1966: Early Soviet contacts with Arab and African countries. In: Mizan 8(2): 87-91. -3o, 2o6, H 3-

1415 Iqbal, M., 1977: Emergence of Saudi Arabia: a political study of King Abd al-Aziz ibn Saud, 19o1-1953. Srinagar: Saudiyah. XIX, 276, XXXIII p. -21-

1416 Iron deficiency in children in Saudi Arabia. In: Nutrition Reviews 27.1969: 6-7. -1oo, 289, 352-

1417 Irving, T.B., 1977: Mecca in the spring. In: Modern Age 21: 418-420. -30, 281, 464-

1418 Irwin, E., 1780: A series of adventures in the course of a voyage up the Red-Sea, on the coasts of Arabia and Egypt; and of a route through the desarts of Thebais, hitherto unknown to the European traveller, in the year M.DCC.LXXVII. In letters to a lady. London: Dodsley (deutsche Ausgabe: Begebenheiten einer Reise auf dem rothen Meere, der arabischen und ägyptischen Küste und durch die thebaische Wüste. Leipzig: Weidmann, 1781. Französische Ausgabe: Voyage à la mer Rouge, sur les côtes de l'Arabie, en Égypte, et dans les déserts de la Thébaïde; suivi d'un autre, de Venise à Bassorah par Latiqué, Alep, les déserts, etc., dans les années 1780 et 1781. 2 vols. Paris: Briand, 1792). XV, 400 p. -21-

1419 Isa, A.S., 1982: Proposed standards for university libraries in Saudi Arabia. Ph.D., University of Pittsburgh. 210 p.

1420 Isaac, E., 1973: The pilgrimage to Mecca. In: The Geographical Review 63: 405-409. -21, 24, 206-

1421 Iseman, P.A., 1978: The Arabian ethos. I. In: Harpers 256 (1533): 37-56. -180-

1422 Iskandar, M.M., 1978: Le pétrole séoudien et son rôle local et international. In: Syrie et Monde Arabe 25(297): 10-20. -206, H 3, H 223-

1423 Ismaeel, A.U. al-, 1981: Selected social studies teaching strategies in Saudi Arabian secondary schools. Ph.D., University of Kansas, Lawrence. 132 p.

1424 Issa-Fullata, M.M., 1982: An experimental study for modernizing instruction through educational technology: the case of Saudi Arabia. Ph.D., State University of New York at Buffalo. 263 p.

1425 Issawi, C.P. & M. Yeganeh, 1962: The economics of Middle Eastern oil. New York: Praeger. XIV, 230 p. -24/213-

1426 —— (ed.), 1966: The economic history of the Middle East, 1800-1914. A book of readings. Chicago, London: The University of Chicago Press. XV, 543 p. -21, 24-

1427 Issawi, C.P., 1973: Economic development in the Middle East. In: International Journal 28: 729-747. -12, 18, 352-

1428 Issawi, H.F. el-, 1979: Report on a consultancy mission to Wadi Gizan agricultural development project, Hakua Agricultural Station, Saudi Arabia (September 16 - October 25, 1979). - Future development of poultry industry in Gizan Region. Rome: FAO. 35 p.

1429 ITALCONSULT, 1966: Water and agricultural development studies for Area IV. Program of work and budget report. Rome.

1430 ——, 1969: Water and agricultural development studies for Areas II and III. Final report. Rome.

1431 Ives, G.O., 1948: Tapline work in second phase. In: World Oil 128: 2o5-2o8. -89, 9o-

1432 Izzard, M., 1979: The Gulf: Arabia's western approaches. London: Murray. XIII, 314 p. -21-

1433 Izzedine, C., 19o9: Le choléra et l'hygiène à la Mecque. Paris: Maloine. 119 p.

1434 ——, 1918: Les épidemies de choléra au Hedjaz. Constantinople: Amiré. IV, 235 p.

J

1435 Jabbur, J., 1959: Abu-al-Duhūr, the Ruwalah 'uṭfah. In: J. Kritzeck & R.B. Winder (eds.): The world of Islam. Studies in honour of Philip K. Hitti. London: Macmillan; New York: St. Martin's Press, 1959: 195-198. -12, 21-

1436 Jabr, S.M., 1981: Individualizing geography instruction in Saudi Arabian secondary schools. Ph.D., University of Kansas, Lawrence. 223 p.

1437 Jacob, H.F., 1923: Kings of Arabia. The rise and set of the Turkish sovranty in the Arabian Peninsula. London: Mills & Boon. 294 p. -21-

1438 Jacobs, A., 1963: Entwicklungshilfe und Entwicklungspläne im Nahen und Mittleren Osten. In: Deutsche Stiftung für Entwicklungsländer, Programm-Abteilung: Zweite wissenschaftliche Regionaltagung über Nah- und Mittelost. Bericht über eine Tagung in Hamburg vom 28. bis 3o. November 1962. Dok. 93/62/DT 4; Berlin, Bonn, 1963: 1o5-112.   -H 3-

1439 Jadallah, S.M., 1978: Performance auditing and program evaluation in government - Saudi Arabia as a field of application. D.B.A., Texas Tech University, Lubbock. 283 p.

144o Jäschke, G., 1967: Ein scherifisches Bündnisangebot an Mustafa Kemal. Ein Beitrag zur Geschichte der arabisch-türkischen Beziehungen. In: W. Hoenerbach (Hg.): Der Orient in der Forschung, Festschrift für Otto Spies zum 5. April 1967. Wiesbaden: Harrassowitz, 1967: 371-394. -21-

1441 Jafary, A.A. al-, 1979: Management systems and organizational effectiveness in selected multinational organizations in the Arabian Gulf region. Ph.D., University of Oklahoma, Norman. 162 p.

1442 Jakubiak, H.E. & M.T. Dajani, 1976: Öleinnahmen und Finanzpolitik im Iran und Saudi-Arabien. In: Finanzierung und Entwicklung 13(4): 12-15, 42.   -24, 25, 18o-

1443 Jalabert, L., 1937: La maître de l'heure en Arabie: Ibn Séoud. In: Études 233: 289-3o7, 465-484.   -1a, 21, 3o-

1444 Jallal, A.A. al-, 1973: Evaluation of the vocational schools in Saudi Arabia in social and economical context. Ph.D., School of Education, University of Colorado, Boulder. 152 p.   -21-

1445 Jamjoom, M.A., 197o: International trade and balance of payments in a mono-product economy: a case study of the Saudi Arabian Kingdom. Ph.D., Faculty of the Graduate School, University of Southern California, Los Angeles. 3o4 p. -2o6-

1446 Jamjoum, A.S., 1964: L'avenir des investissements dans le Royaume d'Arabie Saoudite. In: Revue de la Société d'Études et d'Expansion 63: 55o-554.   -212-

1447 Jammaz, S.I., 1973: Riyadh University: historical foundations, current status, critical problems, and suggested solutions. Ph.D., University of Southern California, Los Angeles. 2o5 p.

1448 Jarrar, G., 1983: Das Erziehungs- und Ausbildungswesen. In: T. Koszinowski (Hg.): Saudi-Arabien: Ölmacht und Entwicklungsland. Beiträge zur Geschichte, Politik, Wirtschaft und Gesellschaft. Mitteilungen des Deutschen Orient-Instituts 2o; Hamburg: Deutsches Orient-Institut, 1983: 219-249. -21, H 223-

1449 Jarvis, C.S., 1942: Arab command. The biography of Lieutenant-Colonel F.W. Peake Pasha, C.M.G., C.B.E. London, New York, Melbourne: Hutchinson. 158 p. -12-

145o Jasir, A.S. al-, 1982: An analytical and descriptive comparison of international communication systems in the United States and the Soviet Union as adapted for use in Saudi Arabia. Ph.D., University of Oklahoma, Norman. 24o p.

1451 Jaussen, J.A. & R. Savignac, 19o9, 1914, 192o: Mission archéologique en Arabie. 2 vols., Suppl. Publications de la Société des Fouilles Archéologiques; Paris: Leroux / Geuthner. XIV, 5o7; XV, 689 resp. 98 p. -21-

1452 Jazairi, M.Z. al-, 197o: Saudi Arabia: a diplomatic history, 1924-1964. Ph.D., Department of History, University of Utah, Salt Lake City. 412 p. -25-

1453 Jeandet, N., 1957: En Arabie séoudite (choses vues). In: Orient (Paris) 1(2): 92-99. -7, 18, 21-

1454 Jeddah 68/69; the first and only definitive introduction to Jeddah, Saudi Arabia's most modern and varied city. Nairobi: University Press of Africa & Arabia, 1968. 174 p.

1455 Jeffery, A., 1929: Christians at Mecca. In: The Moslem World 19: 221-235. -21, 291, 352-

1456 Jerash, M.A. al-, 1968: Soils and agricultural development in the region of al-Qassim, Saudi Arabia. M.A., Department of Geography, University of Durham. XX, 232 p. -Tü 17-

1457 ——, 1972: Irrigation and agricultural development in the region of Jizan, Sa'udi Arabia. Ph.D., University of Manchester. XX, 312 p. -Tü 17-

1458 Jiabajee, N.A., 1957: Saudi Arabia - water supply of important towns. In: Pakistan Journal of Science 9: 189-2o1. -2o, F 1-

1459   Jiddah: the port of Mecca. In: The Islamic Review 44.1956 (7): 32.   -21-

1460   Johany, A.D., 1980: The myth of the OPEC cartel: the role of Saudi Arabia. Dhahran: University of Petroleum and Minerals; Chichester, New York, Brisbane, Toronto: Wiley. XI, 107 p.   -21, H 3-

1461   Johns, R., 1977: Saudi Arabia. In: Middle East Annual Review 3: 287-288, 290-292, 297-299, 301-303, 305-308.   -21, 24/213, 206-

1462   Joint statement on Saudi Arabian-United States cooperation, Washington, D.C., June 8, 1974. In: The Middle East Journal 28.1974: 305-307.   -21, 206, H 223-

1463   Jomard, E.F., 1823: Notice géographique sur le pays de Nedjd ou Arabie centrale, accompagnée d'une carte, suivie de notes sur l'histoire de l'Égypte sous Mohammed-Aly. Paris: Rignoud. 65 p.

1464   ——, 1839: Études géographiques et historiques sur l'Arabie, accompagnées d'une carte de l'Acyr et d'une carte générale de l'Arabie; suivies de la relation du voyage de Mohammed-Aly dans le Fazoql, avec des observations sur l'état des affaires en Arabie et en Égypte. Paris: Didot.

1465   Jomier, J., 1953: Le maḥmal et la caravane égyptienne des pèlerins de La Mecque (XIIIe-XXe siècles). Recherches d'Archéologie, de Philologie et d'Histoire 20; Le Caire: Institut Français d'Archéologie Orientale. XVI, 241 p.   -12, 21-

1466   Jones, K.W., 1951: The miracle of Aramco. In: The National and English Review 137: 344-348.   -1a, 21-

1467   ——, 1952: The romance of Saudi Arabia's oilfields. Changes in Arab way of life. In: Great Britain and the East N.S. 68(1826): 21-23.   -12, 206-

1468   Jong, F. de, 1980: The proclamations of al-Ḥusayn b. 'Alī and 'Alī Ḥaydar. Some observations and notes relative to MS Leiden Or. 12.482. In: Der Islam 57: 281-287.   -16, 21, 25-

1469   Jong, G.E. de, 1934: Slavery in Arabia. In: The Moslem World 24: 126-144.   -21, 291, 352-

1470 Jordan, A.A., Jr., 1979: Saudi Arabia: the next Iran? In: Parameters 9(1): 2-8. -12, 24/213-

1471 Jouin, Y., 1967: Hedjaz 1916-1918. Les compagnons français de Lawrence. In: Revue Historique de l'Armée 23(4): 1o7-121. -1a, 12, 24/213-

1472 Jovelet, L., 1933: L'évolution sociale et politique des 'Pays arabes' (193o-1933). In: Revue des Études Islamiques 7: 425-644. -16, 21, 352-

1473 Jubail and Yanbu: planning and implementation of two macroprojects. In: Middle East Construction 3.1978(9): 7o-77. -89-

1474 Jubail industrial complex: bigger slices, smaller cake. In: The Middle East 1979(58): 78-8o. -1a, 12, 3o-

1475 Jum'ah, 'A.S., 1979: Al-Khubar. In: EI^2 5, Fasc. 79-8o: 4o. -21, 24, 212-

1476 Jung, E., 19o6: Les puissances devant la révolte arabe. La crise mondiale de demain. Paris: Hachette. 232 p.

1477 ——, 1924-25: La Révolte arabe. 2 vols. Paris: Colbert, Bohrer. 199 resp. 221 p. -12-

1478 Jung, H., 1958: Arabien im Aufbruch. Eine Reise in die neue Zeit. München: Bruckmann. 143 S. -24, 93, 212-

1479 Juynboll, G.H.A., 198o: Westerlingen in Mekka en Medina. In: De Gids 143: 648-658. -12, 294-

148o Jwaideh, Z.E., 1976: Legal, tax and regulatory aspects of doing business in Saudi Arabia. Law Library Studies 76-3LL; Washington: Law Library, Library of Congress. 17 p.

K

1481 Kabbaa, A.S., 1979: Saudi Arabia and the United Nations. Ph.D., Southern Illinois University, Carbondale. 228 p.

1482 Kaddoura, M., 197o: Wirtschaftsprojekte in Saudi-Arabien. Diss., Universität Graz. 133, IV S.

1483 Kadi, M.A. al-, 198o: The cultural, social and environmental needs for Saudi Arabian educated urban planners - an undergraduate and graduate curriculum design for the University of Petroleum and Minerals, Dhahran, Saudi Arabia. Ph.D., Rensselaer Polytechnic Institute, Troy. 416 p.

1484 Käselau, A., 1927: Die freien Beduinen Nord- und Zentral-Arabiens. Diss., Mathematisch-Naturwissenschaftliche Fakultät, Universität Hamburg. 139 S.   -21, Tü 17-

1485 Kahlenberg, C., 1885: Djeddah und Hodeidah. In: Österreichische Monatsschrift für den Orient 11: 127-13o.   -1a, 4, 3o-

1486 Kaïdi, H., avec la collaboration de N.O.-D. Bammate & H. Tidjani, 198o: La Mecque et Médine aujourd'hui. Paris: Les Éditions. 224 p.   -21-

1487 Kaikati, J.G., 1976: The marketing environment in Saudi Arabia. In: Akron Business and Economic Review 7(2): 5-13.   -2o6-

1488 ——, 1979: Marketing practices in Iran vis-à-vis Saudi Arabia. In: Management International Review 19(4): 31-37.   -H 3-

1489 Kamal, Ahmad, 1954: Ich war in der verbotenen Stadt. In: Das Beste aus Reader's Digest 7(Feb.): 21-27.   -12, 26, 212-

149o ——, 1964: The Sacred Journey, being Pilgrimage to Makkah. London: Allen & Unwin. XX, 1o8 p.   -21-

1491 Kamal, A.H., 1971: Al-Hidjar. In: EI^2 3: 361-362.   -21, 24, 212-

1492 Kamookh, A.A. al-, 1981: A survey of the English language teachers' perceptions of the English language teaching methods in the intermediate and secondary schools of the Eastern Province in Saudi Arabia. Ph.D., University of Kansas, Lawrence. 135 p.

1493 Kanoo, A.L., 1971: A study of the need for housing and the development of a housing system for Saudi Arabia and the Arabian Gulf. Ph.D., University of Texas at Austin. 333 p

1494 Kanovsky, E., 1977: Saudi Arabia's moderation in oil pricing - political or economic? Tel Aviv: Shiloah Center of Middle Eastern and African Studies.

1495   Kanovsky, E., 1979: Deficits in Saudi Arabia: their meaning and possible implications. In: Middle East Contemporary Survey 2: 318-359.   -21, 1oo-

1496   ——, 198o: Saudi Arabia in the red. In: The Jerusalem Quarterly 16: 137-144.   -21, 24/213, 3o-

1497   ——, 1981: Middle East oil: an economic analysis. - Oil statistics. In: Middle East Contemporary Survey 4: 295-321.   -21, 1oo-

1498   Kapoor, A., with the assistance of A. Haast, 1975: Foreign investments and the new Middle East: a survey of prospects, problems and planning strategies (Aussentitel: Foreign investments and the new Middle East: Egypt, Iran, Saudi Arabia). Princeton: The Darwin Press. 118 p.   -21, H 3-

1499   Karout, Z.I., 1979: Anti-wahhabitische Polemik im XIX. Jahrhundert. Diss., Philosophische Fakultät, Universität Bonn. 91 S.   -5, 21-

15oo   Kashmeeri, B.O., 1971: The role of Saudi Arabia in the Arab League. M.A., University of Virginia, Charlottesville.

15o1   Kashmeeri, M.O., 1977: A study of college and university goals in Saudi Arabia. Ph.D., University of Oklahoma, Norman. 129 p.

15o2   Katakura, M., 1973: Some social aspects of Bedouin settlements in Wādi Fātima, Saudi Arabia. In: Orient (Tokyo) 9: 67-1o8.   -1a, 21, 3o-

15o3   ——, 1974: Socioeconomic structure of a Bedouin settlement - a case study of Bushūr, Saudi Arabia. In: Bulletin of the Department of Geography, University of Tokyo (6): 41-91.   -Tü 17-

15o4   ——, 1977: Bedouin village: a study of a Saudi Arabian people in transition. The Modern Middle East Series 8; Tokyo: University of Tokyo Press. XX, 189 p.   -21, 1oo, Tü 17-

15o5   Katanani, A.K., 1971: Policies and models for planning the economic development of the non-oil sector in Saudi Arabia. Ph.D., Iowa State University, Ames. VI, 187 p.   -H 223-

1506 Kay, E., 1979: Legal aspects of business in Saudi Arabia. London: Graham & Trotman. 162 p.

1507 Kay, S., 1978: The Bedouin. This Changing World; New York: Crane, Russak; Newton Abbot, London, Vancouver: David & Charles. 159 p.   -21-

1508 ——, 1979a: Something old, something new. In: The Middle East (54): 74-75.   -1a, 12, 3o-

1509 ——, 1979b: Saudi Arabia (Aussentitel: Saudi Arabia: past and present). London: Namara Publications. 149 p.   -21, 212-

151o ——, 1982: Social change in modern Saudi Arabia. In: T. Niblock (ed.): State, society and economy in Saudi Arabia. London: Croom Helm; Exeter: Centre for Arab Gulf Studies, 1982: 171-185.   -21-

1511 Kaylani, H.M. al-, 1964a: Estimate of the rate of increase of the population of Saudi Arabia. Beirut: Economic Research Institute, American University of Beirut.

1512 ——, 1964b: Sample surveys of household food consumption in Riyadh, Saudi Arabia. Beirut: Economic Research Institute, American University of Beirut.

1513 Kayser, M., 1936: Die Verkehrsstellung des Persischen Golfes. Bochum-Langendreer: Pöppinghaus (zugleich: Diss., Philosophische und Naturwissenschaftliche Fakultät, Universität Münster, 1937). VI, 114 S.   -6, 21, H 2-

1514 Kazem Zadeh, H., 1912: Relation d'un pèlerinage à la Mecque. In: Revue du Monde Musulman 19: 144-227.   -21, 25-

1515 Kazmi, Z.A. al-, 1981: Student perceptions of parental influence in choice of college and academic field of study at King Abdulaziz University in Saudi Arabia. Ph.D., Michigan State University, East Lansing. 198 p.

1516 Keane, J.F., 1881a: Six months in Meccah: an account of the Mohammedan pilgrimage to Meccah. Recently accomplished by an Englishman professing Mohammedanism. London: Tinsley. 212 p.   -21-

1517 ——, 1881b: My journey to Medinah: describing a pilgrimage to Medinah, performed by the author disguised as a Mohammedan. London: Tinsley. VIII, 212 p.   -21-

1518   Keane, J.F., 1887: Six months in the Hejaz: an account of the Mohammedan pilgrimages to Meccah and Medinah. Accomplished by an Englishman professing Mohammedanism. London: Ward & Downey. XI, 307 p. (Reprint of nos. 1516 and 1517).   -7-

1519   Kedourie, E., 1976: In the Anglo-Arab labyrinth: the McMahon-Husayn correspondence and its interpretations, 1914-1939. Cambridge/England, London, New York, Melbourne: Cambridge University Press. XII, 330 p.   -21-

1520   ———, 1977: The surrender of Medina, January 1919. In: Middle Eastern Studies 13: 124-143.   -1a, 21, H 223-

1521   ———, 1978: England and the Middle East: the destruction of the Ottoman Empire, 1914-1921. Hassocks: The Harvester Press (^{1}1956). X, 236 p.   -21-

1522   ———, 1980: Islam in the modern world, and other studies. London: Mansell. 332 p.   -21-

1523   Keiser, H., 1971: Arabien. Zürich: Silva. 121 S.   -21-

1524   ———, 1977: Abenteuer Schwarzes Gold. Erlebnisse und Begegnungen in Saudi Arabien. Luzern, München: Rex. 199 S.   -12, 21-

1525   Kelidar, A.R., 1972: The Arabian Peninsula in Arab and power politics. In: D. Hopwood (ed.): The Arabian Peninsula: society and politics. Studies on Modern Asia and Africa 8; London: Allen & Unwin, 1972: 145-159.   -21, 24-

1526   ———, 1978: The problem of succession in Saudi Arabia. In: Asian Affairs 65/N.S. 9: 23-30.   -5, 21, 90-

1527   Kelly, J.B., 1956: The Buraimi Oasis dispute. In: International Affairs (London) 32: 318-326.   -21, 180, 291-

1528   ———, 1958: Sovereignty and jurisdiction in Eastern Arabia. In: International Affairs (London) 34: 16-24.   -21, 180, 291-

1529   ———, 1964: Eastern Arabian frontiers. London: Faber & Faber. 319 p.   -7, 21, 212-

1530   ———, 1965: Mehemet 'Ali's expedition to the Persian Gulf 1837-1840. I, II. In: Middle Eastern Studies 1: 350-381, 2: 31-65.   -1a, 21, 30-

1531   ———, 1966: The future in Arabia. In: International Affairs (London) 42: 619-640.   -21, 180, 291-

1532 Kelly, J.B., 1968: Britain and the Persian Gulf, 1795-1880. Oxford: Clarendon Press. XIV, 911 p. -21-

1533 ——, 1976: Saudi Arabia and the Gulf States. In: A.L. Udovitch (ed.): The Middle East: oil, conflict & hope. Critical Choices for Americans 1o, Lexington Books; Lexington, Toronto: Heath, 1976: 427-462. -H 3-

1534 ——, 1980: Arabia, the Gulf and the West. London: Weidenfeld & Nicolson (deutsche Ausgabe: Brennpunkt Golf: die Ölstaaten und der Westen. Berlin, Frankfurt/M., Wien: Ullstein, 1981). IX, 530 p. -21-

1535 ——, 1981: A response to Hermann Eilts' 'Security considerations in the Persian Gulf'. In: International Security 5(4): 186-195. -21, 24/213, 3o-

1536 Kelly, K. & R.T. Schnadelbach, 1976: Landscaping the Saudi Arabian desert. Philadelphia: The Delancey Press. 182 p. -21-

1537 Kennedy, E.M., 1975: The Persian Gulf: arms race or arms control. In: Foreign Affairs (New York) 54: 14-35. -21, 24, 18o-

1538 Kennedy, W.J. (ed.), 1979: Secret history of the oil companies in the Middle East. 2 vols. Salisbury/N.C.: Documentary Publications. XI, 232 resp. II, 233-466 p. -21, 2o6-

1539 Kershaw, R.M., 1973: Attitudes toward religion of Saudi Arabian students in the United States. Ph.D., University of Southern California, Los Angeles. 258 p.

1540 Keun de Hoogerwoerd, R.C., 1889: Die Häfen und Handelsverhältnisse des Persischen Golfs und des Golfs von Oman. In: Annalen der Hydrographie und maritimen Meteorologie 17: 189-2o7. -7, 21, 24-

1541 Khadduri, M., 1973: Arab contemporaries: the role of personalities in politics. Baltimore, London: The Johns Hopkins University Press. X, 255 p. -21-

1542 ——, 1981: Arab personalities in politics. Washington: The Middle East Institute. 353 p. -21-

1543 Khadra, O.A., 1963: Saudi Arabia: educational developments in 1962-1963. In: International Yearbook of Education 25: 322-325. -1a, 24, 18o-

1544   Khalaf, A.A., 1975: The changing political, economic and military status of the littoral Arabian/Persian Gulf states: its impact on world powers. Ph.D., University of Idaho, Moscow. 344 p.

1545   Khalil, M., 1962: The Arab states and the Arab League: a documentary record. 2 vols. Beirut: Khayats. XXXVII, 7o5 resp. XXXVI, 1o19 p.   -21-

1546   Khammas, M., 1953: Die Besonderheiten des Exportes nach dem Orient. Vertriebswirtschaftliche Schriftenreihe; Berlin: Schiele & Schön. 278 S.   -12-

1547   Khan, G.A. & W. Sparroy, 19o5: With the pilgrims to Mecca: the great pilgrimage of A.H. 1319; A.D. 19o2. London, New York: Lane. 314 p.   -38-

1548   Khan, M.A.A., 1977: The assumption of caliphal office by Sharīf Ḥussain - its reaction in India. In: Journal of the Asiatic Society of Bangladesh 22: 211-232.   -21-

****   Khan, M.A. Saleem: s. Saleem Khan, M.A.

1549   Khan, Mohammad S., 1973: Development of educational statistics: Saudi Arabia (Mission November 197o - April 1973). Paris: UNESCO. 44 p.

155o   Khan, Mu'Inuddīn A., 1968: A diplomat's report on Wahhabism of Arabia. In: Islamic Studies 7(1): 33-46.   -1a, 21, 3o-

1551   Khashoggi, H.Y., 1979: Local administration in Saudi Arabia. Ph.D., Claremont Graduate School. 211 p.   -21-

1552   Khatib, Abdel Basset el-, 1974: Seven green spikes 1965-1972: water and agricultural development. Ed. by the Ministry of Agriculture and Water, Kingdom of Saudi Arabia. Beirut: Dar al-Qalam Press. XXVII, 226 p.   -H 3, H 223-

1553   Khatib, Abd al-Hamid al-, 1951: The harbinger of justice. Biography of His Majesty King Abdul Aziz Ibn Saud. 2 vols. Karachi: Al-Arab Printing Press.

1554   Khatib, Abdul Rahman, H.M. Munif & F.A. Ruwayha, 1963: Aramco's participation in Saudi Arabian development. In: Fourth Arab Petroleum Congress, organized by The Secretariat General of the League of Arab States, Beirut, November 5th-12th, 1963. Serial 53 (A-1); Beirut, 1963: 175-188.   -2o6-

1555 Khatib, M.F. el-, 1958: The Saudi Cabinett comes of age. In: The Arab World 4(8): 1o-11.   -2o6-

1556 Khattab, M.K., 1982: The importance of telecommunications media used in conjunction with non-formal education and rural community centers for rural development in Saudi Arabia. Ph.D., University of Wisconsin, Madison. 2o2 p.

1557 Khayat, A.A., 1981: A study of institutional environment at King Abdulaziz University as perceived by upper division students. Ph.D., Michigan State University, East Lansing. 2o3 p.

1558 Khedaire, K.S. al-, 1978: Cultural perception and attitudinal differences among Saudi Arabian male college students in the United States. Ph.D., University of Arizona, Tucson. 296 p.

1559 Kheirallah, G., 1952a: Arabia reborn. Albuquerque: The University of New Mexico Press. VIII, 3o7 p.   -21-

156o ——, 1952b: Saudi Arabia, the safest land on earth. In: The Islamic Review 4o(9): 17-19.   -21-

1561 Khudr, A.S., 1974: Industrial development of Saudi Arabia. In: The Arab Economist 6(63): 14-19.   -H 3-

1562 Khurshid, Z., 1979: Libraries and information centers in Saudi Arabia. In: International Library Review 11: 4o9-419.   -24, 18o, 291-

1563 Khuthaila, H.M., 1981: Developing a plan for Saudi Arabian women's higher education. Ph.D., Syracuse University. 321 p.

1564 Kiernan, R.H., 1937: The unveiling of Arabia. The story of Arabian travel and discovery. London, Bombay, Sydney: Harrap. 36o p.   -21-

1565 Kilner, P., J. Wallace & S. Milmo (eds.), 1977: The Gulf handbook 1978: a guide for businessmen and visitors. 2nd, rev.ed. (^{1}1976). Bath: Trade & Travel Publications; London: Middle East Economic Digest. XII, 574 p.   -21-

1566 Kimball, S.T., 1956: American culture in Saudi Arabia. In: Transactions of the New York Academy of Sciences II, 18: 469-484.   -16, 93, 352-

1567 Kimche, D., 1972: The opening of the Red Sea to European ships in the late eighteenth century. In: Middle Eastern Studies 8: 63-71.   -1a, 21, 3o-

1568 Kimche, J., 1977: The Saudi connection. In: Midstream 23 (1o): 3-8.   -3o, 18o-

1569 ——, 198o: What price Saudi stability? In: Midstream 26(2): 3-8.   -3o, 18o-

157o King, G.R.D., 1976: Some observations on the architecture of south-west Saudi Arabia. In: Architectural Association Quarterly 8(1): 2o-29.   -89, 289-

1571 ——, 1977: Traditional architecture in Najd, Saudi Arabia. In: Proceedings of the Seminar for Arabian Studies 7: 9o-1oo.   -1a, 21-

1572 ——, 1978a: Islamic architecture in eastern Arabia. In: Proceedings of the Seminar for Arabian Studies 8: 15-28. -1a, 21-

1573 ——, 1978b: Traditional Najdī mosques. In: Bulletin of the School of Oriental and African Studies 41: 464-498. -12, 21, 291-

1574 ——, 198o: Notes on some mosques in Eastern and Western Saudi Arabia. In: Bulletin of the School of Oriental and African Studies 43: 251-276.   -12, 21, 291-

1575 ——, 1982: Some examples of the secular architecture of Najd. In: Arabian Studies 6: 113-142.   -21, Tü 17-

1576 King, R., 1972: The pilgrimage to Mecca: some geographical and historical aspects. In: Erdkunde 26: 61-73.   -16, 24, 1oo-

1577 King Faisal of Saudi Arabia. In: The Middle East 1975(8): 8-11.   -12, 3o-

1578 King Faysal of Su'udi Arabia calls the Muslims to <u>Jihad</u> to save Palestine and the Holy Places on the occasion of the <u>Hajj</u> (1388 A.H.) - 28th February 1969. In: The Islamic Review and Arab Affairs 57.1969(2): 3-5.   -1a, 21-

1579 King Sa'ud I of Arabia on a state visit to the Republic of India, 27th November - 13th December, 1955. A short description of King Sa'ud's visit to India. In: The Islamic Review 44.1956(3): 2o-23.   -1a, 21-

1580 Kingdom of Saudi Arabia, 1955: Memorial of the Government of Saudi Arabia; arbitration for the settlement of the territorial dispute between Muscat and Abu Dhabi on one side and Saudi Arabia on the other. 3 vols. N.p. (Cairo). 539; 343 resp. 624 p. -21-

1581 ——, 1956a: Final memorial of the Royal Government of Saudi Arabia; arbitration between the Royal Government of Saudi Arabia and Arabian American Oil Company. Geneva. 79, 15 p.

1582 ——, 1956b: Arbitration between Government of Saudi Arabia and Arabian American Oil Company. 12 vols. Geneva.

1583 ——, 1958a: Arbitral award given on 23 August 1958, in the arbitration between the State of Saudi Arabia and the Arabian American Oil Company. Geneva. VII, 169 p.

1584 ——, 1958b: Arbitration between the State of Saudi Arabia and the Arabian American Oil Company. New York(?). Var.pag.

1585 ——, 1970: Labor and Workmen Law. This law has been approved by the Council of Ministers on his decision No. 745 dated 23/24/8/1389 and issued under Royal Decree No. M/21 dated 6/9/1389 and published on Um-Alqura news-paper No. 2299 dated 19/9/1389. Mecca: Government Press. 62 p. -H 223-

1586 ——, 1980: Saudi Arabia's policy on foreign investment in the Kingdom. N.p. 39 p. -212-

1587 ——, Agency for Technical Cooperation Administration, 1966: The impact of technical cooperation between Saudi Arabia and the United Nations and its specialized agencies. Riyadh. 68 p. -206-

1588 ——, Central Planning Office, 1963: Population and housing census 1962/63. Riyadh.

1589 ——, Central Planning Organization, 1970: Development plan, 1390 A.H. Tayif. 277 p. -H 3-

1590 ——, ——, n.d. (1974): Report of the Central Planning Organization. N.p. (Riyadh). 187 p.

1591 ——, General Directorate for Girls' Education, Statistics Section, 1977: Girls' education during the seventeen years A.H. 1380-97. N.p. (Riyadh?).

1592 Kingdom of Saudi Arabia, General Directorate for Girls' Education, Statistics Section, 1977/78: General statistical guide. N.p. (Riyadh?).

1593 ——, General Directorate of Girls' Schools, 1967(?): Statistical directory of Saudi girls' education during seven years, 1380/81-1386/87 (1960/61-1966/67). Riyadh.

1594 ——, General Directorate for the National Guard, 1979: Report on the Department of Culture of the National Guard. N.p. (Riyadh?).

1595 ——, General Directorate for Youth Sponsorship, 1979: Report on the work of the directorate in the field of culture and administrative affairs. N.p. (Riyadh?).

1596 ——, General Organisation for Social Insurance, 1975: Social Insurance Law. Regulations implementing the law. Decisions concerning application of the law. Social Insurance at Your Service 9; Riyadh: Hanifa Printing Press. 232 p. -2o6, H 223-

1597 ——, General Petroleum and Mineral Organization (PETROMIN), c. 1970: Annual report 1969. N.p. 56 p.

1598 ——, ——, c. 1971: Progress report 1970-1971. Dammam: Al-Mutawa Press. 48 p.

1599 ——, Imam Mohammad Ibn Saud Islamic University, Department of Studies and Information, c. 1977: Statistical yearbook, 1395/96 A.H., 1975/76 A.D. Riyadh: National Offset Printing Press. 20 p.   -H 223-

1600 ——, Industrial Studies and Development Centre, n.d.: The Industrial Studies and Development Centre in the service of industry. Jeddah: Matabi Dar al-Asfahani. 16 p. -H 223-

1601 ——, ——, c. 1969: Survey of industrial establishments in Saudi Arabia. 1. Riyadh. 2. Western Province; Jeddah, Mecca, Medina and Taif. 3. Eastern Province; Dammam, Al-Khobar and Al-Hassa. 3 vols. Riyadh.   -H 223-

1602 ——, ——, 1970: UNIDO-UNESOB Mission on agricultural machinery and metal working industries, Kingdom of Saudi Arabia. Riyadh, April 7 to April 19, 1970. Riyadh. 8, III, III p. -H 223-

1603 Kingdom of Saudi Arabia, Industrial Studies and Development Centre, 1971-: Guide to industrial investment in Saudi Arabia. Riyadh.   -2o6, H 3, H 223-

1604 ——, ——, 1971: Industrial opportunity study. Industries based on date palm cultivations. Riyadh. 16o p.   -H 223-

1605 ——, ——, 1972a: Specifications and conditions for hand-made carpets for the Holy Mosque of Mecca in accordance with international specifications. Riyadh. 22 p.

1606 ——, ——, 1972b: The possibility of manufacturing refrigerators in the kingdom. Riyadh. 15 p.

1607 ——, ——, 1972c: Consideration about the future development of the Portland cement industry. Riyadh. 122 p.

1608 ——, ——, 1972d: Standard specifications and test manual for concrete bricks. Riyadh. 22 p.

1609 ——, ——, 1972e: Standard specifications and test manual for hollow concrete masonry. Riyadh. 22 p.

161o ——, ——, 1972f: Standard specifications and test manual for structural clay blocks. Riyadh. 26 p.

1611 ——, ——, Industrial Research Department, 1972g: Pre-investment study, vegetable processing and canning. Riyadh. 4o p.   -H 223-

1612 ——, ——, 1973a: Pre-investment study on manufacturing of bicycles in Saudi Arabia. Riyadh. 31 p.

1613 ——, ——, 1973b: Industrial finance in the private sector in Saudi Arabia. Riyadh. 61 p.

1614 ——, ——, 1973c: Techno-economic industrial structure and growth prospects in Saudi Arabia. 2 vols. Riyadh. 227 resp. 163 p.

1615 ——, ——, 1974a: Prefeasibility study on manufacturing domestic washing machines. Riyadh 27 p.

1616 ——, ——, 1974b: The role of PVC in the market for pipes, fittings, fixtures in Saudi Arabia: a study of the market and the competitive situation. Riyadh. 246 p.

1617 ——, ——, 1974c: Projections for supply and demand of cement in Saudi Arabia. Riyadh. 13 p.

1618 Kingdom of Saudi Arabia, Industrial Studies and Development Centre, 1975a: Pre-investment study on jam processing and canning in Saudi Arabia. Riyadh. 4o p.

1619 ——, ——, 1975b: Pre-investment study on vegetable processing and canning in Saudi Arabia. Riyadh. 4o p.

1620 ——, ——, 1975c: A market study on gas cookers and stoves in Saudi Arabia. Riyadh. 25 p.

1621 ——, ——, 1975d: Study of the market for blankets in Saudi Arabia. Riyadh. 26 p.

1622 ——, ——, 1975e: Technical study on a motorcycle assembly plant. Riyadh. 15o p.

1623 ——, ——, 1975f: Pre-investment study on the manufacture of wire mesh and netting in Saudi Arabia. Riyadh. 43 p.

1624 ——, ——, 1975g: Projections for the consumption of marble in Saudi Arabia. Riyadh. 18 p.

1625 ——, ——, 1975h: Preliminary study on the manufacturing of polyester/cotton fabrics in Saudi Arabia. Riyadh. 56 p.

1626 ——, ——, 1975-76: List of industrial projects existing and under implementation in Saudi Arabia up to the end of the year 1394 (December 1974). Suppl.: List of industrial projects licensed during 1395 A.H. (1975 A.D.). 2 vols. Riyadh. 53 resp. IV, 54 p.   -H 223-

1627 ——, ——, n.d.: The Second Industrial Development Plan, A.H. 1395-14oo (1975-198o). N.p. N.pag.   -B 19, H 223-

1628 ——, ——, 1976a: The quality of concrete masonry units in Saudi Arabia: the Riyadh area; the Jeddah-Mecca area; the Eastern Province. Concrete bricks. Riyadh.

1629 ——, ——, 1976b: The aluminium door and window industry in Saudi Arabia. Riyadh. 31 p.

1630 ——, ——, 1976c: Saudi Arabian standard specifications, sampling and test procedures for asbestos cement sewer and drainage pipes. Riyadh. 42 p.

1631 ——, ——, 1976d: The market for cast iron pipes and fittings in Saudi Arabia. Riyadh. 87 p.

1632　Kingdom of Saudi Arabia, Industrial Studies and Development Centre, 1976e: The market for galvanized and block iron pipes in Saudi Arabia. Riyadh. 78 p.

1633　——, ——, 1976f: Market study of paints in Saudi Arabia. Riyadh. 63 p.

1634　——, ——, 1976g: The furniture market in Saudi Arabia. Riyadh. 36 p.

1635　——, ——, 1976h: Comments and proposals for a local furniture industry. Riyadh. 6o p.

1636　——, ——, 1976i: Study of the market for tyres in Saudi Arabia. Riyadh. 86 p.

1637　——, ——, 1976j: Investment opportunity for manufacturing tyres in Saudi Arabia. Riyadh. 55 p.

1638　——, ——, 1976k: Opportunity study on industries based on date palm cultivation in Saudi Arabia. Riyadh. 16o p.

1639　——, ——, 1976l: Industrial opportunities in the Southern region. Riyadh. XVII, 179 p.　-H 223-

1640　——, ——, Information and Documentation Department, 1976m: Government development financing agencies in Saudi Arabia. Riyadh. 1o p.　-H 223-

1641　——, ——, 1977a: Aspects of industrial technology transfer and development in Saudi Arabia. Riyadh. 15 p.

1642　——, ——, 1977b: Industrial structure and development in Saudi Arabia. Riyadh. IX, 78 p.

1643　——, ——, 1978: Transfer of technology to Saudi industries. In: N.A. Shilling (ed.): Arab markets - 1978. New York: Inter-Crescent Publishing & Information Corporation, 1978: 133-157.

1644　——, Institute of Public Administration, n.d.: Institute of Public Administration. Riyadh. 36 p.　-H 223-

1645　——, King Abdulaziz University, 1982: Annual statistical book, 198o-1981. Jeddah.

1646　——, Ministry of Agriculture, 1963: Annual report 1382-1383. N.p.　-H 223-

1647 Kingdom of Saudi Arabia, Ministry of Agriculture, Division of Statistics and Agricultural Economics, 1963: Ministry of Agriculture during one year. Budget, projects data and statistics about the ministry during the fiscal year 1382-1383. 78 p.   -H 223-

1648 ——, ——, ——, 1964: Ministry of Agriculture during one year. Budget, projects data and statistics about the ministry during the fiscal year 1383-1384 (1964). N.p. 51 p.

1649 ——, Ministry of Agriculture and Water, Department of Agricultural Research and Development, Division of Statistics and Agricultural Economics, 1966: Ministry of Agriculture and Water during one year, 1385/1386 (1966). N.p. 79 p. -H 223-

1650 ——, ——, ——, ——, 1968: Basic agricultural statistics. Summary of the skeletal statistical data of the agricultural survey undertaken in various districts in the period from 1380-1385 H., and of the survey undertaken in the Southern districts (Qunfudah, Jizan, Tihama) in the year 1385-86 H. N.p. (Riyadh). N.pag.

1651 ——, ——, Italconsult, 1968: Al Hasa Oasis irrigated agriculture expansion. Water and Agricultural Development Studies for Area IV; Rome.

1652 ——, ——, ——, 1969: Water and Agricultural Development Studies for Area IV. Final report. Rome.

1653 ——, ——, Planning Unit, 1969: Basic information about agriculture and water with a statement about major projects under execution. Riyadh.

1654 ——, ——, Minister of Agriculture and Water / WAKUTI GmbH, Consulting Engineers, 1969: Study on management organization for Al Hassa Irrigation and Drainage Scheme. Zug.

1655 ——, ——, —— / ——, 1970: Market study, Faisal Settlement Project Haradh. Zug.

1656 ——, ——, —— / ——, 1971: Final agricultural report and agricultural development program (pre-investment study phase II), Faisal Settlement Project Haradh. 2 vols. N.p.

1657 Kingdom of Saudi Arabia, Ministry of Agriculture and Water, c. 1971: Notes on some aspects of the work of the Ministry of Agriculture and Water. Riyadh. 2o, 22 p. -H 223-

1658 ——, ——, Minister of Agriculture and Water / WAKUTI GmbH, Consulting Engineers, 1972: Final completion report for supervision of the execution of the Al Hassa Irrigation and Drainage Project. Zug.

1659 ——, ——, Department of Agricultural Research and Development, Statistics Division, n.d.: Bulletin of agricultural current sample survey from 197o-71 to 1974-75. N.p. 57 p.

1660 ——, ——, 1975a: Proposed programs and projects to be included in the Second Five-Year Plan (agricultural sector) 1395/96-1399/14oo H. Riyadh. 7 p.

1661 ——, ——, 1975b: Proposed programs and projects to be included in the Second Five-Year Plan (water sector) 1395/96-1399/14oo H. Riyadh. 7 p.

1662 ——, Ministry of Agriculture and Water Extension, 1975: The agricultural policy of Saudi Arabia. N.p. (Riyadh). 18 p. -2o6-

1663 ——, Ministry of Commerce and Industry, 1964: Foreign capital investment regulation A.H. 1383/A.D. 1964. N.p. 7 p. -H 223-

1664 ——, Ministry of Communications / ETCO Consulting Engineers, 197oa: Engineering services for a feeder road network (Areas IV and V), Phase I: Reconnaissance - Masterplan - Feasibility study. General alphabetical index of all settlements in Areas IV and V. Tehran, Riyadh.

1665 ——, —— / ——, 197ob: Engineering services for a feeder road network (Areas IV and V), Phase I: Reconnaissance - Masterplan - Feasibility study. Masterplan - final report. Tehran, Riyadh.

1666 ——, ——, 1974: Roads and bridges in the Hajj areas. Preliminary report. Riyadh. 61 p.

1667 ——, ——, 1977: Roads and ports in Saudi Arabia. Riyadh.

1668 ——, ——, 1978: The great achievement - 19ooo kilometres. A report on the construction of Saudi Arabia's highway network. Riyadh.

1669 Kingdom of Saudi Arabia, Ministry of Defence and Aviation, 1979: The Department of Culture and Education and its role in the armed forces. N.p. (Riyadh?): Armed Forces Press.

1670 ——, Ministry of Education, 1957: Saudi Arabia: educational progress in 1956-1957. In: International Yearbook of Education 19: 336-339.   -1a, 7, 24-

1671 ——, ——, 1963: A brief report on the Ministry of Education for the year 1962-1963 submitted by the Saudi Arabian delegation to the XXVIth International Conference of Public Education, Geneva, 1963. Var.pag.   -H 223-

1672 ——, ——, 1964: Saudi Arabia: educational developments in 1963-1964. In: International Yearbook of Education 26: 283-288.   -1a, 7, 24-

1673 ——, ——, Centre for Statistical Information and Educational Documentation, 1967/68-: Educational statistics in the Kingdom of Saudi Arabia. 1.1967/68 A.D.-; Riyadh.

1674 ——, ——, 1969: Educational developments (national report): Saudi Arabia. In: International Yearbook of Education 31: 123.   -1a, 7, 18o-

1675 ——, ——, 197oa: Educational policy in the Kingdom of Saudi Arabia. Riyadh.

1676 ——, ——, 197ob: Report on special education for the education and rehabilitation of the handicapped. Riyadh. 52 p.

1677 ——, ——, 197oc: A brief report on adult education and the fight against illiteracy for the occasion of the International Education Year 197o. N.p. 25 p.   -H 223-

1678 ——, ——, 1972: Adult education and combatting of illiteracy. Riyadh. 29 p.

1679 ——, ——, 1973: Progress of education in Sa'udi Arabia, 138o-1392 A.H., 196o-1972 A.D. Riyadh.

1680 ——, ——, 1974: Towards an appropriate strategy for training skilled and semi-skilled workers. Riyadh.

1681 ——, ——, 1975: Report on the development of educational, cultural and scientific work in the Kingdom of Saudi Arabia in A.D. 1973-74. N.p.

1682   Kingdom of Saudi Arabia, Ministry of Education, n.d.: Objectives and educational work of the Ministry of Education during the second five-year plan, A.H. 1395-96 to A.H. 1399-14oo. N.p.

1683   ——, ——, 1975: The biannual report of the Ministry of Education, 1973 and 1974. Riyadh. 43 p.

1684   ——, ——, 1977: The bi-annual report of the Ministry of Education, 1975 and 1976. Riyadh. 25 p.

1685   ——, ——, 1978a: Twenty-five years of educational development at the Ministry of Education. In: Educational Documentation (Riyadh) (15).

1686   ——, ——, Centre for Statistical Information and Educational Documentation, Statistical Information Section, 1978b: Educational expenditure and pupil costs in the Ministry of Education, A.H. 1396-97. N.p.

1687   ——, ——, Department for Adult Education and Literacy, 1978c: The work of the kingdom in the field of literacy and adult education. N.p.

1688   ——, Ministry of Finance, Department of Economic Affairs, 1951/52: Economic bulletin 1. Mecca: Government Press.

1689   ——, Ministry of Finance and National Economy, 1957: Royal decree no. 17/2/28/576 dated 14 Rabi' I 1376 (19 October 1956) amending certain provisions of the income tax regulations. N.p. 18 p.

1690   ——, ——, Central Department of Statistics, 1963: Survey of population, buildings, and establishments in Saudi Arabia, 1963. N.p.

1691   ——, ——, ——, 1966-: Statistical yearbook. 1.1385 A.H., 1965 A.D.-; Dammam: Al-Mutawa Press.   -21, 212, H 223-

1692   ——, ——, ——, 1966-: (Quarterly) Digest of foreign trade statistics. Dammam: Al-Mutawa Press / Riyadh: Government Press.   -212, H 223-

1693   ——, ——, ——, 1969-: Foreign trade statistics. 1(?).1968-; Dammam: Al-Mutawa Press.   -21-

1694   ——, ——, ——, Census Unit, 1971: Guide to the towns, villages and hamlets of the Saudi provinces. Riyadh.

1695 Kingdom of Saudi Arabia, Ministry of Finance and National Economy, Central Department of Statistics, 1977-: The statistical indicator. N.p. (Riyadh).

1696 ——, ——, ——, 1977: Population census, 1974. 14 vols. Dammam.

1697 ——, ——, ——: Cost of living index. Quarterly; Riyadh.

1698 ——, ——, ——: Hajj statistics. Annual; Riyadh.

1699 ——, Ministry of Hajj: Pilgrim statistics. Annual; Riyadh.

1700 ——, Ministry of Higher Education, Directorate General for the Development of Higher Education, 1981: Statistics of Saudi students abroad, 1975/1976-1977/1978. Riyadh: Middle East Press. X, 218 p.   -21, 212-

1701 ——, ——, ——, c. 1982: Progress of higher education in the Kingdom of Saudi Arabia during ten years, 1970-1980. Riyadh: National Offset Printing Press. 141 p.   -21, 212-

1702 ——, Ministry of Industry, n.d.: Statement of the industrial policy of Saudi Arabia. Riyadh. 8 p.

1703 ——, Ministry of Industry and Electricity, Undersecretariat for Industry Affairs, 1975: Industrial firms licensed under regulations for the protection and encouragement of national industries and foreign capital investment regulations, up to the end of 1395 A.H. (1975 A.D.). Riyadh: Safir Offset Press. 54 p.   -H 3, H 223-

1704 ——, ——, ——, Statistics Section, 1977-: Industrial firms licensed under regulations for the protection and encouragement of national industries and foreign capital investment regulations. 1977-; Riyadh.   -H 3-

1705 ——, Ministry of Information, n.d.: Social welfare. N.p. N.pag.   -H 223-

1706 ——, ——, n.d.: The Kingdom of Saudi Arabia: progress & development. N.p. 66 p.   -H 3-

1707 ——, ——, Directorate General of Broadcasting, Press and Publications, 1958: Facts about Saudi Arabia, 1953-1958. Riyadh. 103 p.   -212-

1708 ——, ——, 1963a: King Faisal speaks. Riyadh. N.pag.   -21-

1709 Kingdom of Saudi Arabia, Ministry of Information, 1963b: Education for girls. Riyadh. N.pag.   -Bo 149-

1710 ——, ——, 1965a: Education in Saudi Arabia. Riyadh. 48 p.

1711 ——, ——, 1965b: Saudi Arabia as seen by the world press (französischsprachige Ausgabe: L'évolution de l'Arabie Séoudite, vue à travers la presse internationale). N.p. 147 p.   -21, 212-

1712 ——, ——, 1971a: Land distribution and settlement. The Kingdom of Saudi Arabia: Facts and Figures; n.p. 47 p.   -212-

1713 ——, ——, 1971b: The great water projects (französischsprachige Ausgabe: Les grands projets d'eau). The Kingdom of Saudi Arabia: Facts and Figures; n.p. 38 p.   -212-

1714 ——, ——, 1971c: Health for all (französischsprachige Ausgabe: La santé pour tous). Jeddah: Dar al-Asfahani. -H 223-

1715 ——, ——, 1972: The Kingdom of Saudi Arabia: facts and figures. Progress & Development 7; n.p. 47 p.   -H 3-

1716 ——, ——, 1975a: The Wadi Abha Dam and an outline about Assir. Al-Khobar: East Press, 82 p.   -H 223-

1717 ——, ——, General Directorate of Television, 1975b: Saudi Arabian television in its first decade. Riyadh. 41 p. -H 223-

1718 ——, ——, c. 1976: Saudi Arabia. Riyadh. N.pag.   -93-

1719 ——, ——, 1978: Higher education. On the Road to the Future; n.p.: The Falcon Press. 63 p.   -21-

1720 ——, ——, 1979a: Saudi Arabia and its place in the world. Lausanne: Three Continents. 196 p.   -Esb 2ooo-

1721 ——, ——, 1979b: Report on the work of the Ministry of Information. Riyadh.

1722 ——, Ministry of the Interior, Department of Municipal Affairs, 1976: Municipal services, 1385 to 1396 A.H. (1965 to 1976 A.D.). Vol. 2: 1395/96 A.H. (1975/76 A.D.). Riyadh: Safir Offset Press. 216 p.   -21, H 223-

1723 ——, Ministry of the Interior for Municipal Affairs, Regional and Town-Planning Office Dammam / G. Candilis Metra

International Consultants, 1975a: Eastern Region, Saudi Arabia. Al Hasa, existing conditions. Draft report. N.p. (Dammam).

1724 Kingdom of Saudi Arabia, Ministry of the Interior for Municipal Affairs, Regional and Town-Planning Office Dammam / G. Candilis Metra International Consultants, 1975b: Preliminary plan, Eastern Region, Saudi Arabia. N.p. (Dammam).

1725 ——, Ministry of Labour and Social Affairs, Deputy Ministry for Social Affairs, Planning and Social Studies Department, n.d.: Women's role in Saudi society. N.p. 11 p.
-H 223-

1726 ——, ——, Department of Social Affairs, 1968: Community development in Saudi Arabia. N.p.

1727 ——, ——, Labour Affairs Agency, Research and Statistics Department, 1973: Statistical publication about workers, occupations, wages and working hours in the private establishments during 1393 H./1973 G. 1. The northern region. 2. The southern region. 3. The central region. 4. The western region. 5. The eastern region. 5 vols. N.p.
-H 223-

1728 ——, ——, 1975: The Second Development Plan for the Vocational Training and Manpower Division of the Ministry of Labour and Social Affairs. N.p.

1729 ——, ——, 1978: Family and child welfare in Saudi Arabia. In: International Child Welfare Review (36): 51-54.

1730 ——, ——, Labour Affairs Agency: Statistical handbook. Annual; Riyadh.

1731 ——, Ministry of Municipal and Rural Affairs, 1978: Riyadh action master plans, technical report 5: Projects included in the Second Five Year Plan for the City of Riyadh. N.p.

1732 ——, Ministry of Petroleum and Mineral Resources, 1963: The impact of petroleum on the economy and social life of Saudi Arabia. Riyadh. 49 p.

1733   Kingdom of Saudi Arabia, Ministry of Petroleum and Mineral
        Resources, Directorate General of Mineral Resources,
        1965: Mineral resources of Saudi Arabia: a guide for in-
        vestment and development. Bulletin 1; Jeddah. 77 p.
        -H 223-

1734   ——, ——, ——, c. 1969: Mining Code of the Kingdom of Saudi
        Arabia. Jeddah. 22 p.   -H 223-

1735   ——, ——, ——, 1969-: Mineral resources research 1.1969-.
        Jeddah.   -H 223-

1736   ——, ——, Economic Department, 197o-: Petroleum statistical
        bulletin 1.197o-. Riyadh / Dammam: Al-Mutawa Press.

1737   ——, ——, Technical Affairs Department, 1971: Review of oil
        industry. Dammam. 127 p.

1738   ——, Ministry of Planning, 1975: Second development plan
        1395-14oo A.H./1975-198o A.D. Jeddah: Dar Okaz Printing
        and Publishing (Riyadh, 1976). XXII, 579 p.   -Bo 149,
        H 3, H 223-

1739   ——, ——, 1978: The strategies for the third development
        plan for 14oo-14o5 AH - 198o-1985 AD. Riyadh.

174o   ——, ——, 198o: Third development plan 14oo-14o5 A.H./198o-
        1985 A.D. Riyadh: Ministry of Planning Press. XXXIII, 5o3
        p.   -H 223-

1741   ——, Permanent Delegation to the UN, n.d. (1956?): The Burai
        mi dispute. New York.

1742   ——, Permanent Mission to the UN, 1975: The modern Saudi Ara
        bian women. Report of the Mission at the Conference of In
        ternational Women's Year, Mexico City.

1743   ——, Ports Authority, 1978-: Monthly statistics. Riyadh.

1744   ——, ——, 1979(?)-: Annual statistics. Riyadh: National Off
        set Printing Press.   -H 223-

1745   ——, ——, 198o: Ports information. N.p. N.pag.   -H 223-

1746   ——, ——, 1981: Kingdom of Saudi Arabia. Riyadh. N.pag.
        -H 3-

1747   ——, Riyad University, n.d.: University bulletin, Part I.
        Riyadh: Riyad University Press. XI, 67 p.

1748   Kingdom of Saudi Arabia, Royal Commission for Jubail and
        Yanbu, Directorate General for Jubail Project, 1979: Jubail Industrial City. Dammam: Al-Mutawa Press. 32 p.
        -212-

1749   ——, Saudi Arabian Agricultural Bank, 1965-: Annual report.
        1.1384-85 A.H./1964-65 A.D.-; n.p. (Riyadh).   -H 223-

1750   ——, Saudi Arabian Airlines, Public Relations Department,
        1973: Annual report. Jeddah. 22 p.   -H 223-

1751   ——, Saudi Arabian Monetary Agency, Research and Statistics
        Department, 1961-: Annual report. 1.1380 A.H./1960-61
        A.D.-; Jeddah / Riyadh: Saudi Arabian Printing Company
        Ammariyah.   -212, H 3-

1752   ——, ——, ——, 1964-: Statistical summary. 1.1963-; Jeddah.

1753   ——, ——, 1964: Foreign investment law for the Kingdom, approved by royal decree no. 35 dated 11-10-1383 (24-2-1964).
        Jeddah. 4 p.   -H 3-

1754   ——, Saudi Fund for Development, 1975-: Annual report. 1.
        1975-; Riyadh.   -H 223-

1755   ——, Saudi Industrial Development Fund, 1978: Annual report:
        fiscal year 1396-97. Riyadh.

1756   ——, ——, 1979: SIDF annual report 1398/99 and five years
        review: 1394/95-1398/99. Riyadh: Saudi Arabian Printing
        Company Ammariya. 40 p.   -212, H 223-

1757   ——, ——, Marketing Division, [4]1980: Saudi Arabian manufactured products. Riyadh: Saudi Arabian Printing Company
        Ammariyah. N.pag.

1758   Kirazian, H., 1981/82: Arabie saoudite. Deux préoccupations:
        le conflit israélo-arabe et le Golfe. In: Journal de
        l'Année 1981-82: 160-161.   -468-

1759   Kirchner, R.A., 1982: Government and politics. - Mass communications. - Foreign relations. - National defense and
        internal security. In: R.F. Nyrop et al.: Saudi Arabia:
        a country study. Area Handbook Series DA Pam. 550-51;
        Washington: US Government Printing Office, 1982: 157-217,
        311-346.   -H 223-

1760 Kirk, George E., 1952: Survey of international affairs 1939-1946: The Middle East in the War. London, New York, Toronto: Oxford University Press. XIII, 511 p.   -21, 24, 294-

1761 Kirk, Grayson L., 1934: Ibn Saud builds an empire. In: Current History (New York) 41: 291-297.   -21, 180-

1762 Klare, M.T., 1974: The political economy of arms sales: United States-Saudi Arabia. In: Society (New Brunswick) 11 (6): 41-49.   -206, 352-

1763 Klebnikoff, S. de, 1982: Les travailleurs immigrés de la péninsule. In: Centre d'Études et de Recherches sur l'Orient Arabe Contemporain (éd.): La péninsule Arabique d'aujourd'hui. Sous la direction de P. Bonnenfant. T. 1. Paris: Centre National de la Recherche Scientifique, 1982: 191-218.   -21-

1764 Kleist, H. von, 1906a: Die Hedjasbahn. In: Asien (Berlin) 5(6): 84-85.   -1a, 25, 90-

1765 ——, 1906b: England in Arabien. In: Geographische Zeitschrift 12: 425-439.   -21, 24, 93-

1766 Klemme, M., 1965: Report to the Government of Saudi Arabia on pasture development and range management. With respect to increasing livestock production. Report 1993; Rome: FAO. IV, 27 p.   -Gö 153-

1767 Klingmüller, E., 1943: Die Korrespondenz zwischen Sir Henry McMahon und dem Scherifen von Mekka, Huṣain. Ihre historischen Voraussetzungen und ihre juristisch-politische Bedeutung. Habil.-Schrift, Universität Berlin. III, 168 S.

1768 ——, 1981: Zu den Grundlagen von Schiedsgerichtsvereinbarungen in Saudisch-Arabien. In: K.-H. Böckstiegel (Hg.): Vertragspraxis und Streiterledigung im Wirtschaftsverkehr mit arabischen Staaten. Schriftenreihe des Deutschen Instituts für Schiedsgerichtswesen 2; Köln, Berlin, Bonn, München: Heymanns, 1981: 5-16.   -21-

1769 Klippel, E., 1940: Der weisse Beduine. Unter Karawanenleuten und Oasenmenschen. Braunschweig: Wenzel. 248 S.   -24-

1770 Klopp vom Hofe, P., 1934: Der arabische Krieg. In: Preussische Jahrbücher 237(Juli): 21-31.   -16, 93, 180-

1771 Knauerhase, R., 1973: A survey of the revenue structure of the Kingdom of Saudi Arabia, 1961/62 to 1971/72. In: Public Finance / Finances Publiques 28: 435-453. -1a, 7, H 3-

1772 ——, 1974: Saudi Arabia's economy at the beginning of the 1970s. In: The Middle East Journal 28: 126-140. -12, 21, 206-

1773 ——, 1975a: Saudi Arabia: a brief history. In: Current History (Philadelphia) 68: 74-79, 82-83, 88. -21, 180, 352-

1774 ——, 1975b: The Saudi Arabian economy. Praeger Special Studies in International Economics and Development; New York, Washington, London: Praeger. XXV, 359 p. -21, 100, H 3-

1775 ——, 1977: The economic development of Saudi Arabia: an overview. In: Current History (Philadelphia) 72: 6-10, 32-34. -18, 21, 206-

1776 ——, 1980: Saudi Arabia: our conservative Muslim ally. In: Current History (Philadelphia) 78: 17-21, 35-37. -18, 21, 206-

1777 ——, 1981: Saudi Arabia's foreign and domestic policy. In: Current History (Philadelphia) 80: 18-22, 37. -18, 21, 206-

1778 Knightly, P. & C. Simpson, 1969: Das Geheimleben des Lawrence von Arabien. Hamburg: Hoffmann & Campe (Originalausgabe: The secret lives of Lawrence of Arabia. London: Nelson, 1969. Amerikanische Ausgabe: New York, St. Louis, San Francisco, Mexico, Panama: McGraw-Hill, 1970). 322 S. -21, 24-

1779 Kobori, I., 1973: Some notes on diffusion of Qanat. In: Orient (Tokyo) 9: 43-66. -1a, 21, 30-

1780 Koch, K.-F., S. Altorki, A. Arno & L. Hickson, 1977: Ritual reconciliation and the obviation of grievances: a comparative study in the ethnography of law. In: Ethnology 16: 269-283. -7, 12, 30-

1781 Königreich Saudi-Arabien, Ministerium für Arbeit und Soziale Angelegenheiten, Generaldirektorium für Jugendbetreuung, 1972: Das Sportleben im Königreich Saudi-Arabien, Überblicke und Eindrücke / Glimpses of the sports movement in the Kingdom of Saudi Arabia / Aperçu sur la vie sportive dans le Royaume Arabe Séoudite. Beirut: Matabi Dar al-Kitab al-Lubnani. 7o S.   -H 223-

1782 Königreich Saudi-Arabien, Ministerium für Information, o.J.: Saudi Arabien, Land der Leistung (englischsprachige Ausgabe: The Kingdom of Saudi Arabia: land of achievement; französischsprachige Ausgabe: L'Arabie Séoudite, terre des réalisations; spanischsprachige Ausgabe: Reino de Arabia Saudi. Arabia Saudi - hace realizaciónes). Mehrere Auflagen. Dammam: Al-Mutawa Press.   -21, 212-

1783 ——, ——, ca. 1974: Königreich Saudi Arabien (auch in englisch-, französisch-, spanisch- und italienischsprachigen Ausgaben). Das ist unser Land; Dammam: Al-Mutawa Press. 111 S.

1784 ——, ——, ca. 1979: Industrie. Auf der Strasse der Zukunft; Neapel: The Falcon Press. 47 S.   -212-

1785 Koester, D. & S. al-Dossary, 1983: The new city of Jubail, Saudi Arabia. In: Public Management 65(Jan.): 2o-22.

1786 Kohn, H., 1929: Arabien 1924-1928. In: Zeitschrift für Politik 18: 171-183.   -16, 21, 2o6-

1787 ——, 1934a: The unification of Arabia. In: Foreign Affairs (New York) 13: 91-1o3.   -24, 25, 18o-

1788 ——, 1934b: The desert changes. In: Asia (New York) 34: 659-663.   -1a-

1789 Kolko, G., 1968: The politics of war: the world and United States foreign policy, 1943-1945. New York: Random House (englische Ausgabe: The politics of war: allied diplomacy and the world crisis of 1943-1945. London: Weidenfeld & Nicolson, 1969). X, 685 p.   -21, 25-

179o Koner, W., 1871: Adolph von Wrede. In: Zeitschrift der Gesellschaft für Erdkunde zu Berlin 6: 248-272.   -7, 16, 2

1791 Konzelmann, G., 1975: Die Reichen aus dem Morgenland. Wirtschaftsmacht Arabien. München: Desch. 197 S.   -21, 24/213, 212-

1792 ——, 1977: Mekka: Labbaika - Gott hier bin ich! In: Geo 1977 (8): 8-38.   -12, 3o, 18o-

1793   Korany, B., 1979: Pétro-puissance et système mondial. Le
        cas de l'Arabie saoudite. In: Études Internationales 1o:
        797-819.   -2o6-

1794   ——, 1982: Arabia Saudita. Da culla dell'Islam a petropo-
        tenza. In: Politica Internazionale 1982(1): 79-9o.
        -H 223-

1795   Kordi, K.A.K., 1979: The major characteristics of the ac-
        counting information systems in some selected Saudi in-
        dustrial businesses: a critical analysis. Ph.D., Univer-
        sity of Arizona, Tucson. 226 p.

1796   Kornrumpf, H.-J., 1978: Neuere Beschreibungen von al-Ḥasā
        in amtlichen osmanischen Veröffentlichungen. In: Der Is-
        lam 55: 74-92.   -16, 21, 25-

1797   ——, 198o: Die osmanische Herrschaft auf der Arabischen
        Halbinsel im 19. Jahrhundert. In: Saeculum 31: 399-4o8.
        -21, 24, 291-

1798   Koszinowski, T., 1973: Schwerpunkte der Aussenpolitik Saudi-
        Arabiens. In: Orient (Opladen) 14: 181-185.   -21, 1oo,
        H 223-

1799   ——, 1975: Die Bedeutung des Nahostkonflikts für die Aussen-
        politik Saudi-Arabiens. In: Orient (Opladen) 16(1): 87-
        98.   -21, 1oo, H 223-

18oo   ——, 1977: Arbeitsmarktprobleme in Saudi-Arabien. Die Bedeu-
        tung ausländischer Arbeitnehmer bei der Überwindung des
        Arbeitskräftemangels. In: Orient (Opladen) 18(1): 57-78.
        -21, 1oo, H 223-

18o1   ——, 1979: Saudi-Arabien. In: U. Steinbach, R. Hofmeier &
        M. Schönborn (Hgg.): Politisches Lexikon Nahost. Beck'-
        sche Schwarze Reihe 199; München: Beck, 1979 (2., neu-
        bearb. Aufl. 1981): 245-259.   -21, 93-

18o2   ——, 198oa: How stable is the political system of Saudi Ara-
        bia? In: Istituto Affari Internazionali (ed.): Red Sea
        conflicts and cooperation. Regional balance and strategic
        implications. Rome, 198o. 5 p.   -H 223-

18o3   ——, 198ob: Nach Mekka - droht Saudi-Arabien eine 'Islami-
        sche Revolution'? In: Geographische Rundschau 32: 543-
        547.   -21, 25, 212-

1804 Koszinowski, T., 1981: The Arabian peninsula in the 19th and 2oth centuries. In: The Muslim World, a historical survey. Part IV, Fasc. 1. Leiden: Brill, 1981: 199-235. -21-

1805 ——, 1983: Von der Gründung des ersten saudischen Staates bis zu König Abdalaziz - ein historischer Überblick. In: T. Koszinowski (Hg.): Saudi-Arabien: Ölmacht und Entwicklungsland. Beiträge zur Geschichte, Politik, Wirtschaft und Gesellschaft. Mitteilungen des Deutschen Orient-Instituts 2o; Hamburg: Deutsches Orient-Institut, 1983: 61-81. -21, H 223-

1806 —— (Hg.), 1983: Saudi-Arabien: Ölmacht und Entwicklungsland. Beiträge zur Geschichte, Politik, Wirtschaft und Gesellschaft. Mitteilungen des Deutschen Orient-Instituts 2o; Hamburg: Deutsches Orient-Institut. 365 S. -21, H 223-

1807 Kourouklis, S.D., 1977: Targeting sales to Saudi Arabia. In: International Trade Forum 13(2): 8-1o, 22. -1a, 2o6, H 3-

1808 Krajewski, L., 1926: Le Nedjd et les Wahabites. In: Revue Politique et Parlementaire 33: 1o5-122. -1a, 38, 2o6-

1809 Krause, W.W., 1953a: Im Reiche Ibn Sauds. Zwischen Rotem Meer und Persischen Golf. In: Übersee-Rundschau 5: 453-456. -3o, 18o, 212-

1810 ——, 1953b: Im Reiche Ibn Sauds. Bohrturm-Wälder am Persischen Golf. In: Übersee-Rundschau 5: 6o4-6o5. -3o, 18o, 212-

1811 Krenkow, F., 1937: Snouck Hurgronje. In: Islamic Culture 11: 142-143. -21, 3o-

1812 Krüger, H., 1979: Verbot von Rechtswahl-, Schieds- und Gerichtsstandsklauseln nach saudi-arabischem Recht. In: Recht der Internationalen Wirtschaft 25: 737-741. -12, 24, 18o-

1813 ——, 1981: Probleme des saudi-arabischen internationalen Vertrags- und Schiedsrechts. In: K.-H. Böckstiegel (Hg.): Vertragspraxis und Streiterledigung im Wirtschaftsverkehr mit arabischen Staaten. Schriftenreihe des Deutscher

Instituts für Schiedsgerichtswesen 2; Köln, Berlin, Bonn, München: Heymanns, 1981: 61-81.   -21-

1814   Krüger, K., 1951: Strassenbau in Arabien. In: Brücke und Strasse 3: 91.   -89, 9o, 93-

1815   Kruse, H., 1955: Saudi-arabisches Einkommensteuerrecht. In: Recht der Internationalen Wirtschaft 1: 95-96.   -24, 18o, 291-

1816   Kruyt, J.A., 1886: Consulaat der Nederlanden te Djeddah: Jaarlijksch verslag. In: Verzameling van Consulaire en andere Verslagen en Berichten over Nijverheid, Handel en Scheepvaart 1885: 61-63.   -Algemeen Rijksarchief 's-Gravenhage-

1817   Kuhaimi, S.A.A. al-, 1976: The treatment of young offenders in Saudi Arabia: a descriptive account set in a social and religious context. M.Phil., University of Southampton.

1818   Kumar, R., 196o: Abdul Aziz Al Saud and the genesis of Saudi Arabia (19o1-19o7). In: Bengal: Past and Present 79: 6o-66, 8o: 83-89.   -1a, 21-

1819   ——, 1965: India and the Persian Gulf region, 1858-19o7: a study in British imperial policy. London: Asia (zugleich: Ph.D., University of the Punjab, Lahore, 1961). 259 p.   -12, 21-

182o   Kuniholm, B.R., 1981: What the Saudis really want: a primer for the Reagan administration. In: Orbis (Philadelphia) 25: 1o7-121.   -21, 24/213, 2o6-

1821   Kurd, A.A. el-, 1963: The Hashemites' role in the Arab independence movement against the Turks. Ph.D., New York University. 261 p.

1822   Kurdi, M.A.M., 1982: Saudi Arabia: perspective on oil, foreign policy, and the Arab-Israeli conflict, 197o-198o. Ph.D., Claremont Graduate School. 21o p.

1823   Kurdi, T.M.K. & K.A. Hasan, 198o: Residential environment in the Hejaz. In: Ekistics 47: 38.   -1a, 89, 2o6-

1824   Kurian, G.T., 1978: Encyclopedia of the Third World. Vol. 2: Laos to Zambia. New York: Facts on File (London: Mansell, 1979). 817-1694 p.   -21-

1825 Kutschera, C., 1975: Arabie Séoudite / Saudi Arabia. Paris: Delroisse. 143 p. -21, H 223-

1826 Kuwait-Saudi Arabia agreement to partition the neutral zone (signed at Al-Hadda, Saudi Arabia, July 7, 1965). In: International Legal Materials 4.1965: 1134-1138. -1a, 3o, 291-

1827 Kuwaiz, A.I. el-, 1976: Comparison of the actual and optimal pricing policy for a public utility operating in a developing country: a case study of electricity in the city of Riyadh, Saudi Arabia. Ph.D., St. Louis University. 2oo p.

L

1828 L., 1927: Downing Street and Arab potentates. In: Foreign Affairs (New York) 5: 233-24o. -24, 25, 18o-

1829 Labaki, B., 1982: Croissance industrielle et pétrole en Arabie Saoudite. In: A. Bourgey et al.: Industrialisation et changements sociaux dans l'Orient arabe. Beyrouth: Centre d'Études et de Recherches sur le Moyen-Orient Contemporain; Lyon: Presses Universitaires de Lyon, 1982: 283-314.

183o Labban, S.A., 1974a: Agriculture in the main oases of the Eastern Province of Saudi Arabia. N.p.: Agricultural Assistance Division, Local Industrial Development Department, ARAMCO. 4o p.

1831 ——, 1974b: Vegetable production in the Eastern Province of Saudi Arabia. N.p.: Agricultural Assistance Division, Local Industrial Development Department, ARAMCO. 131 p.

1832 Labonne, M. & A. Hibon, 1978: Futur agricole et alimentaire de la Méditerranée Arabe. Paris: Institut National de la Recherche Agronomique. 145 p. -21-

1833 Lacey, R., 1981: The Kingdom: Arabia and the House of Saud. London: Hutchinson; New York: Harcourt Brace Jovanovich. XV, 63o p. -21-

1834 ——, 1982: Saudi Arabia: a more visible role in the Middle East. In: The World Today 38: 4-12. -12, 18, 352-

1835   Lacher, J., 1980: Die Joint Venture-Company in Saudi-Arabien. In: Recht der Internationalen Wirtschaft 26: 99-102.   -12, 24, 180-

1836   Lackner, H., 1978: A house built on sand: a political economy of Saudi Arabia. London: Ithaca Press. IV, 224 p.   -12, 24, H 223-

1837   L'activité pétrolière de l'Arabie Séoudite. In: Le Commerce du Levant 1970(117): 19-21.   -H 223-

1838   L'agriculture en Arabie saoudite. In: Maghreb-Machrek 1980 (89): 27-40.   -206, H 3, H 223-

1839   Lamare, P., 1924: Les explorations récentes de l'Arabie. In: La Géographie (Paris) 41: 162-166.   -206, 291, 294-

1840   Lambsdorff, O., 1981: Auf die Bedingungen des saudischen Marktes einstellen... In: Übersee-Rundschau 33(3): 8-11.   -24, 30, 180-

1841   Lamer, M., 1970: Milk and dairy products in Saudi Arabia. In: Dairy Industries 35: 429-432.   -100-

1842   Lammens, H., 1920: Le pèlerinage du dernier khédive d'Égypte. In: Revue du Monde Musulman 38: 58-84.   -21, 30-

1843   ——, 1934: Ṭā'if. In: EI^1 4: 672-673.   -21, 24-

1844   Lancaster, W.O., 1978: The Bedouin and 'progress'. In: Middle East International (79): 26-27.   -12, 30-

1845   ——, 1981: The Rwala Bedouin today. Changing Cultures; Cambridge/England, London, New York, New Rochelle, Melbourne, Sydney: Cambridge University Press. X, 179 p.   -21-

1846   ——, 1982: The development and function of the Shaykh in nomad/settler symbiosis. In: Arabian Studies 6: 195-204.   -21, Tü 17-

1847   Landau, J.M., 1971a: Ottoman propaganda in the Ḥijāz before the First World War - some new manuscript evidence. In: D. Sinor, with the assistance of T. Jacques, R. Larson & M.E. Meek (eds.): Proceedings of the twenty-seventh International Congress of Orientalists, Ann Arbor, Michigan, 13th-19th August 1967. Wiesbaden: Harrassowitz, 1971: 292.   -21-

1848 Landau, J.M., 1971b: The Hejaz Railway and the Muslim pilgrimage: a case of Ottoman political propaganda. Detroit: Wayne State University Press. 294 p.   -21-

1849 Landsberg, M., 1975: Black gold and the re-emergence of the dollar standard. In: The Review of Radical Political Economics 7(1): 69-78.   -2o6, 467, 468-

185o Lange, W., 1979: Saudi Arabien - Freundlicher Ölkrösus. In: Übersee-Rundschau 31(6): 4.   -24, 3o, 18o-

1851 ———, 1981: Saudi Arabien: Macht durch Erdöl. In: Übersee-Rundschau 33(2): 38-41.   -24, 3o, 18o-

1852 Langefeld-Wirth, K., 1982: Investieren in Saudi-Arabien. In: Recht der Internationalen Wirtschaft 28: 717-722.   -24, 2o6, H 3-

1853 Langella, V., 1964: Il petrolio nella geografia politica della penisola d'Arabia. In: Annali dell'Istituto Orientale di Napoli 24/N.S. 14: 2o3-232.   -21, 3o, 291-

1854 Langer, S., 1883: Reiseberichte aus Syrien und Arabien. Hg. von D.H. Müller. Wien (Leipzig: Kreysing, ²19o3). XXXVII, 1o3 S.   -21-

1855 Lanier, A.R., 198o: Update: Saudi Arabia. Chicago: Intercultural Press.

1856 Laoust, H., 1965: Les schismes dans l'Islam. Introduction à une étude de la religion musulmane. Bibliothèque Historique; Paris: Payot. XII, 466 p.   -21-

1857 L'Arabia Saudita. Un mercato favoloso per i nostri prodotti. In: Notiziario Ortofrutticolo e dei Prodotti Agricolo-Alimentari e Floricoli 31.1979(6): 27-34.   -H 3-

1858 L'Arabie Saoudite s'industrialise grâce au pétrole. In: Syrie et Monde Arabe 23.1976(272): 7-13.   -2o6, H 3-

1859 L'Arabie Séoudite à l'heure de la guerre. In: L'Économie des Pays Arabes 16.1973(189): 6o-62.   -2o6-

186o L'Arabie Séoudite entre le moyen âge et le XXe siècle. In: Problèmes Politiques et Sociaux, La Documentation Française 1974(23o). 48 p.   -2o6-

1861 Lasky, H., 1978: Saudi Arabia: a short history of an immoderate state. In: Middle East Review (New York) 11(1): 13-17. -21-

1862 Lateef, A., 1975: Feisal: from obscurity to international status. In: Pakistan Horizon 28(4): 116-13o. -21, 2o6-

1863 ——, 1978: Riyadh and Washington: a mutual reliance. In: Middle East International (85): 16-18. -12, 3o-

1864 Lateef, N.A., 1956a: Report to the Government of Saudi Arabia on agricultural extension. Report 518, Project SAU/Ag I; Rome: FAO. 13 p.

1865 ——, 1956b: Characteristics and problems of agriculture in Saudi Arabia. Background Country Studies 4; Rome: FAO.

1866 Laurent, F., 1958: L'Arabie séoudite à l'heure du choix. In: Orient (Paris) 2(2): 89-99. -7, 18, 21-

1867 Law, J., 1978: Arab aid: who gets it, for what, and how. New York: Chase World Information Corporation.

1868 Lawrence, T.E., 1927: Aufstand in der Wüste. Leipzig: List (Originalausgabe: Revolt in the desert. London: Cape, 1927). XII, 355 S. -21, 24-

1869 ——, 1936: Die sieben Säulen der Weisheit. Leipzig: List (Originalausgabe: Seven pillars of wisdom. A triumph. N.p. (London), 1926. Französische Ausgabe: Les sept piliers de la sagesse... Un triomphe. Paris: Payot, 1936). VIII, 848 S. -21, 1oo, Tü 17-

187o ——, 1939: Secret despatches from Arabia by T.E. Lawrence. Published by permission of the Foreign Office. London: The Golden Cockerel Press. 173 p. -12, 21-

1871 ——, 194o: Oriental assembly. Ed. by A.W. Lawrence. New York: Dutton. XII, 291 p. -21, 24-

1872 Lawton, J., 1978: Farming in the sand. In: Aramco World Magazine 29(3): 21-29. -21, 24-

1873 ——, 1981: A terminal in a tent. In: Aramco World Magazine 32(4): 8-17. -21, 24-

1874 ——, with A. Clark, 1982: Foundations - the introduction; the cornerstones; the pillars; the keystone; the underpinning; the new cities. In: Aramco World Magazine 33(6): 4-4o.   -21, 24-

1875 Leachman, G.E., 1911: A journey in North-Eastern Arabia. In: The Geographical Journal 37: 265-274.   -21, 24, 291-

1876 ———, 1914: A journey through Central Arabia. In: The Geographical Journal 43: 5oo-52o.   -21, 24, 291-

1877 Leaman, O., 1978: Power and women in Saudi Arabia: a note. In: Journal of Anthropological Research 34: 589-59o.   -1a, 12, 3o-

1878 Lean, O.B., 1965: FAO's contribution to the evolution of international control of the desert locust, 1951-1963. Rome: FAO. 142 p.

1879 Leatherdale, C.A., 1981: British policy towards Saudi Arabia, 1925-1939. Ph.D., University of Aberdeen.

188o Lebanon's choice. A Mideast Switzerland or Arab integration. In: The Arab Economist 7.1975(78): 74-75.   -H 3-

1881 Lebkicher, R., 1952: America's greatest Middle East oil venture. In: The Oil Forum 6: 389-396.   -9o-

1882 ———, G. Rentz & M. Steineke, 1952: The Arabia of Ibn Saud. New York: Moore. XIII, 192 p.   -12-

1883 ———, 1954: The training of Saudi Arab employees: Arabian American Oil Company. In: The Yearbook of Education 1954: 514-535.   -5, 21, 291-

1884 ———, G. Rentz, M. Steineke, with contributions by other Aramco employees, 196o: Aramco handbook. N.p.: ARAMCO. 343 p.   -21, 38-

1885 Lebling, B., 1979: Special relations headaches. In: The Middle East (55): 3o-31.   -1a, H 3, H 223-

1886 Le Chemin de fer du Hedjaz. In: Revue du Monde Musulman 6.19o8: 262-265.   -21, 25-

1887 L'économie séoudienne en 1969/197o. In: Étude Mensuelle sur l'Économie et les Finances des Pays Arabes 14.1971(32/158): 54-86.   -2o6-

1888 Lederer, A., 1971: Les ports de l'Arabie Séoudite à la côte du Golfe Persique. Leur rôle dans le développement du pays. Académie Royale des Sciences d'Outre-Mer, Classe des Sciences Techniques, Mémoire N.S. 17,1; Bruxelles: Classe des Sciences Techniques, Académie Royale des Sciences d'Outre-Mer. 55 p.   -21, Tü 17-

1889 Le développement agricole en Arabie Séoudite. In: Syrie et Monde Arabe 23.1977(275): 23-27.   -2o6, H 3-

189o Lee, E., ^{3}198o: The American in Saudi Arabia. Chicago: Intercultural Press (^{1}1977, ^{2}1978). VIII, 111 p.   -21-

1891 Lee, J.F.K., 1978: Tax aspects. In: R.M. Nelson (ed.): Corporate development in the Middle East. London: Oyez Publishing, 1978: 183-2o9.   -21-

1892 Lees, B., 198o: A handbook of the Al Sa'ud ruling family of Saudi Arabia. London: Royal Genealogies. 64 p.   -12, 21, H 223-

1893 Lefebvre, T., 1845, s.d.: Voyage en Abyssinie exécuté pendant les années 1839, 184o, 1841, 1842, 1843 par une commission scientifique composée de MM. T. Lefebvre, Lieutenant de vaisseau, Chevalier de la Légion d'honneur, A. Petit et Quartin-Dillon, Docteurs-Médecins, Naturalistes du Muséum, Vignaud, Dessinateur. 2 vols. Paris: Bertrand.   -7-

1894 Lehmann, H., G. Maranjian & A.E. Mourant, 1963: The distribution of sickle-cell haemoglobin in Saudi Arabia. In: Nature (London) 198: 492-493.   -16, 21, 93-

1895 Leichtweiss Institute, Technical University of Braunschweig, Department on Irrigation, Drainage and Water Resources, 197o-71: Technical report about the research work in the Al Hassa region (with appendices). 2 vols. Braunschweig.

1896 Leichtweiss-Institut für Wasserbau der Technischen Universität Braunschweig & Hofuf Agricultural Research Centre, 1968-79: Report on the work of the Leichtweiss Institute Research Team, Technical University Braunschweig. Publications 1-37; Hofuf.

1897 —— & Institut für Agrarsoziologie, landwirtschaftliche Beratung und angewandte Psychologie der Universität Hohenheim, 1975-83: Forschungsvorhaben 'Die Probleme bei der

Überführung des tradierten Oasenlandbaus in eine moderne Bewässerungswirtschaft', gefördert von der Stiftung Volkswagenwerk. 21 Forschungsberichte; Braunschweig.

1898 Leipold, L.E., 1974: Come along to Saudi Arabia. Minneapolis: Denison. 167 p.

1899 Le marché de l'Arabie Séoudite. In: La Revue du Commerce Extérieur 1974(97) (reproduction: L'économie de l'Arabie Séoudite. In: Problèmes Économiques 1974(1392): 14-22). -H 3-

1900 Lemaud, C., 1980: Arabie Saoudite. In: Revue des Deux Mondes 1980(Mars): 595-609. -16, 21, 24-

1901 Le miracle économique de l'Arabie Séoudite. In: L'Asie Nouvelle 19.1969(206): 20-23. -H 3-

1902 Lenczowski, G., 1960: Oil and state in the Middle East. Ithaca: Cornell University Press. XIX, 379 p. -24-

1903 ——, 1967: Tradition and reform in Saudi Arabia. In: Current History (Philadelphia) 52: 98-104, 115, 128. -21, 180, 352-

1904 ——, 1976: Middle East oil in the revolutionary age. National Energy Study 10; Washington: American Enterprise Institute for Public Policy Research. 36 p. -21-

1905 Le nouveau budget séoudien deux fois celui de 73. In: L'Économie des Pays Arabes 17.1974(199): 16-19. -206-

1906 Le pèlerinage à la Mekke. In: Maghreb-Machrek 1978(80): 74-78. -206, B 212, H 3-

1907 Les accords de Djeddah et la conférence de Harad dans la presse du Caire et de Damas. In: Orient (Paris) 9.1965 (4): 153-162. -7, 18-

1908 Lesch, W., 1931: Arabien. Eine landeskundliche Skizze. In: Mitteilungen der Geographischen Gesellschaft in München 24(1): 1-153. -107, 212-

1909 Les hotelleries françaises au Hedjaz. In: L'Afrique Française 27.1917: 137-138. -30, 291-

1910 Leveau, R. & T. Rifaï, 1974: L'arme du pétrole. In: Revue Française de Science Politique 24: 745-769. -21, 24, 206-

1911 Levi Della Vida, G., 1938: Carlo Alfonso Nallino (1872-
      1938). In: Oriente Moderno 18: 459-478.   -1a, 5, 21-

1912 Le voyage en Iran du souverain d'Arabie séoudite. In: Orient
      (Paris) 9.1965(4): 2o7-21o.   -7, 18-

1913 Levy, P., 1911: Die Betriebsmittel der Hedjazbahn. In: Organ
      für die Fortschritte des Eisenbahnwesens in technischer
      Beziehung N.F. 48: 82-86, 99-1o1.   -9o, 93, 465-

1914 Lewis, C.C., 1933: Ibn Sa'ūd and the future of Arabia. In:
      International Affairs (London) 12: 518-534. -18o, 291, 352-

1915 Lewis, W.H.W., 1979: U.S. debacle in the Horn. In: The Wash-
      ington Quarterly 2: 97-1o2.   -12, 24/213, 2o6-

1916 Liebesny, H.J., 195o: Legislation on the sea bed and terri-
      torial waters of the Persian Gulf (with: Royal pronounce-
      ment concerning the policy of the Kingdom of Saudi Arabia
      with respect to the subsoil and sea bed of areas in the
      Persian Gulf contiguous to the coasts of the Kingdom of
      Saudi Arabia, and: Decree regarding territorial waters
      of Saudi Arabia). In: The Middle East Journal 4: 94-98.
      -21, 38, 2o6-

1917 Light for the Holy City of Mecca. In: The Islamic Review
      41.1953(1): 2o-21.   -1a, 21-

1918 Linabury, G.O., 197o: British-Sa'udi Arab relations, 19o2-
      1927: a revisionist interpretation. Ph.D., Columbia Uni-
      versity, New York. 345 p.

1919 ——, 1978: The creation of Saudi Arabia and the erosion of
      Wahhabi conservatism. In: Middle East Review (New York)
      11(1): 5-12.   -21-

192o Linde, G., 1979: Saudi-Arabien auf neuem Kurs? Berichte des
      Bundesinstituts für ostwissenschaftliche und internatio-
      nale Studien 41-1979; Köln. 24 S.   -12, 212, H 223-

1921 L'insurrection du Hedjaz. Son caractère - ses résultats. In:
      L'Asie Française 1918(172): 34-38.   -1a, 291-

1922 Lippens, P., 1956: Expédition en Arabie centrale. Paris:
      Adrien-Maisonneuve. XI, 214 p.   -12, 21, Tü 17-

1923 Lipsky, G.A., in collaboration with M. Ani, M.C. Bigelow,
      F. Gillen, S.C. Gillen, T.J. Larson, A.T.J. Matthews &

C.H. Royce, 1959: Saudi Arabia. Its people, its society, its culture. Survey of World Cultures; New Haven: HRAF Press. 366 p.   -12, 21, 24-

1924 Lipsky, G.A., 1969: Public health and welfare in Saudi Arabia. In: A. Shiloh (ed.): Peoples and cultures of the Middle East. New York: Random House, 1969: 219-232 (reprinted from: G.A. Lipsky et al.: Saudi Arabia. Its people, its society, its culture. Survey of World Cultures; New Haven: HRAF Press, 1959: 262-276).   -1oo-

1925 Little, A.D., 1969: Report: agro-industry opportunities in Saudi Arabia. N.p.   -H 223-

1926 Littman, J., 1947: Hiring in Hofuf... In: The Standard Oiler 1947(Dec.): 1o-11, 21-22.

1927 Littmann, E., 1918: Ruinen in Ostarabien. In: Der Islam 8: 19-34.   -16, 21, 25-

1928 ——, 1936: Christiaan Snouck Hurgronje 1857-1936. Ehrenmitglied der DMG seit 1928. In: Zeitschrift der Deutschen Morgenländischen Gesellschaft 9o/N.F. 15: 445-458.   -21, 291, 352-

1929 Litwak, R., 1981: Security in the Persian Gulf 2: Sources of inter-state conflict. Aldershot: Gower. XII, 1o5 p.   -21-

193o Lloyd, E.M.H., 1956: Food and inflation in the Middle East, 194o-1945. Studies on Food, Agriculture, and World War II; Stanford: Stanford University Press. XIV, 375 p.   -21-

1931 Lloyds Bank, Overseas Department, International Trade Promotion Section, 1977: Economic report for Saudi Arabia. London. 24 p.

1932 Loans to developing countries. In: Rural Reconstruction 12. 1978(Jan.): 46-47.

1933 Lochner, R.K., 1979: Die Kaperfahrten des Kleinen Kreuzers 'Emden'. Tatsachenbericht. Heyne-Buch 554o; München: Heyne. 479 S.   -24/213-

1934 Loewenthal, N.P., 1979: Assignment Saudi Arabia: area handbook. San Francisco: Bechtel Corporation.

1935 Loir, R., 1955: Das ist Saudiarabien. In: Die Auslese (Berlin) 23: 146-148. -3o, 212-

1936 —, 1957: Escale à Djeddah. In: Géographia (72): 38-43. -1a, 7, 212-

1937 Lonchampt, J., 1982: La planification en Arabie Saoudite. In: Centre d'Études et de Recherches sur l'Orient Arabe Contemporain (éd.): La péninsule Arabique d'aujourd'hui. Sous la direction de P. Bonnenfant. T. 2: Études par pays. Paris: Centre National de la Recherche Scientifique, 1982: 6o3-622. -21-

1938 London Chamber of Commerce, 1963: Trade prospects in the Arabian Gulf. Report of the London Chamber of Commerce Trade Mission. London. VI, 66 p. -H 3-

1939 —, 1975: The construction industry in Saudi Arabia. The Construction Industry in the Middle East 3; London. 85 p.

1940 Londres, A., 1931: Pêcheurs de perles. Paris: Michel. 256 p. -12-

1941 Long, D.E., 1973: The Board of Grievances in Saudi Arabia. In: The Middle East Journal 27: 71-75. -21, 38, 2o6-

1942 —, 1976: Saudi Arabia. The Washington Papers 4(39); Beverly Hills, London: Sage Publications for The Center for Strategic and International Studies. 7o p. -12, 21-

1943 —, 1979a: The Hajj today: a survey of the contemporary Makkah pilgrimage. Albany: State University of New York Press; Washington: The Middle East Institute. XIII, 18o p. -21, Tü 17-

1944 —, 1979b: Saudi oil policy. In: The Wilson Quarterly 3 (1): 83-91. -7, 12, 3o-

1945 —, 198oa: Kingdom of Saudi Arabia. In: D.E. Long & B. Reich (eds.): The government and politics of the Middle East and North Africa. Boulder: Westview Press, 198o: 89-1o6. -12, 2o6, H 223-

1946 —, 198ob: King Faisal's world view. In: W.A. Beling (ed.): King Faisal and the modernisation of Saudi Arabia. London: Croom Helm; Boulder: Westview Press, 198o: 173-183. -12, 21, H 223-

1947 Longrigg, S.H., 1949: The liquid gold of Arabia. In: Journal of the Royal Central Asian Society 36: 2o-33. -21, 9o, 291-

1948 ——, ^{3}1968: Oil in the Middle East: its discovery and development. London, New York, Toronto: Oxford University Press (^{1}1954, ^{2}1961). XII, 519 p. -21-

1949 Looney, R.E., 1982a: Saudi Arabia's Islamic growth model. In: Journal of Economic Issues 16: 453-459. -12, 38, 2o6-

195o ——, 1982b: Saudi Arabia's development potential: application of an Islamic growth model. Lexington Books; Lexington, Toronto: Heath. XVII, 358 p. -21-

1951 Lorimer, J.G., 19o8, 1915: Gazetteer of the Persian Gulf, 'Omān, and Central Arabia. 2 vols. Calcutta: Superintendent Government Printing (reprint: Westmead, Farnborough: Gregg International; Shannon: Irish University Press, 197o). CXXX, 2741 resp. IV, 1952 p. -12, 21-

1952 Loustaunau, C.A., 1969: Foreign trade regulations of Saudi Arabia. Overseas Business Reports; Washington: Bureau of International Commerce, US Department of Commerce. 5 p. -2o6-

1953 Loutfi, Z.I., 19o6: La Politique sanitaire internationale. Thèse, Faculté de Droit, Université de Paris; Paris: Rousseau. XI, 255 p.

1954 Lucas, I.A.M. & M. el-Saadi, 1974: Saudi Arabia: towards increased livestock production. In: Span (London) 17(3): 1o6-1o8. -2o6-

1955 Lundbaek, T., 1962: A Sulubba tent. In: Folk 4: 91-94. -12, 16, 3o-

1956 Lupien, J.P., 1976: Infrastructure for national food control services. In: Food and Nutrition 2(1): 15-2o. -12, 1oo, 291-

1957 Lyautey, P., 1967: L'Arabie saoudite d'aujourd'hui. Paris: Julliard. 16o p. -12-

M

**** M.: s. Rousseau, J.B.L.J.

1958 Maaskola, P., 1979: Zu einigen sozialstrukturellen Entwicklungstendenzen in Saudi-Arabien bis zum Beginn der siebziger Jahre. Diplomarbeit, Fachbereich Gesellschaftswissenschaften, Universität Marburg. III, 148 S.   -H 223-

1959 ——, 198o: Volksbewegung gegen die Saudis. In: Antiimperialistisches Informationsbulletin 198o(7/8): 29-32.

196o MacDonald, C.G., 1978: The roles of Iran and Saudi Arabia in the development of the law of the sea. In: Journal of South Asian and Middle Eastern Studies 1(3): 3-1o.   -1a, H 223-

1961 ——, 198o: Iran, Saudi Arabia, and the law of the sea: political interaction and legal development in the Persian Gulf. Contributions in Political Science 48; Westport, London: Greenwood Press (zugleich: Iran and Saudi Arabia in the Persian Gulf: a study in the law of the sea. Ph.D., University of Virginia, Charlottesville, 1976). XV, 226 p.   -12-

1962 Macdonald, D.B., 1913: Bid'a. In: EI[1] 1: 742-743.   -21, 24-

1963 MacDonald, W.N., 1979: Arabia: the British connection. Part I: The awakening. In: The Contemporary Review 234: 193-199.   -25, 18o, 352-

1964 MacIntyre, R.R., 1981: Saudi Arabia. In: M. Ayoob (ed.): The politics of Islamic reassertion. London: Croom Helm, 1981: 9-29.   -21-

1965 Mackie, J.B., 1924: Hasa: an Arabian oasis. In: The Geographical Journal 63: 189-2o7.   -21, 24, 291-

1966 Maclean, H.W., 19o4: Report on the conditions and prospects of British trade in Oman, Bahrein, and Arab ports of the Persian Gulf. In: Parliamentary Papers, Cmd. 2281, Vol. 85; London: HMSO, 19o4: 733-74o.

1967 MacMunn, G.F. & C.B. Falls, 1928: Military operations, Egypt & Palestine, from the outbreak of war with Germany to June 1917. History of the Great War based on Official Documents 7; London: HMSO. XVIII, 445 p.   -24/213-

1968 Madani, A.R. al- & M. al-Fayez, 1976: The demographic situation in the Kingdom of Saudi Arabia. In: Population Bulletin of the United Nations Economic Commission for Western Asia, Special Issue (1o/11): 185-189. -2o6-

1969 Madani, G.O., 1972: An analysis of the capital expenditures of Saudi Arabian firms. Ph.D., University of Arizona, Tucson. 2o8 p.

197o Madani, M.O., 197o: The relationship between Saudi Arabian domestic law and international law: a study of the oil agreements with foreign companies. S.J.D., Faculty of the Law School, George Washington University, Washington. X, 254, 237A p. -21-

1971 Madani, N.O., 1977: The Islamic content of the foreign policy of Saudi Arabia. King Faisal's call for Islamic solidarity 1965-1975. Ph.D., American University, Washington. XIV, 216 p. -21, B 851, H 223-

1972 Madi, M.A.F., 1975: Developmental administration and the attitudes of middle management in Saudi Arabia. Ph.D., Southern Illinois University, Carbondale. 222 p. -21-

1973 Mahassni, H., 1981: Interpretation of contracts and settlement of disputes in Saudi Arabia. - Interpretation von Bauverträgen und Beilegung von Streitigkeiten in Saudi-Arabien. In: Der Streit bei internationalen Bauverträgen, Symposium Bad Ems 198o. Braunschweig, 1981: 79-123.

1974 Mahmoud, A., 1975: The oil world's disharmony: Yamani vs. OPEC. In: The Arab Economist 7(82): 22-28. -2o6-

1975 Mahnke, H.-J., 1979: Saudi-Arabien hat Geld, aber keine Arbeitskräfte. In: Übersee-Rundschau 31(2): 16. -24, 3o, 18o-

1976 Main outlines of the development plan of the Kingdom of Saudi Arabia. In: Economic Bulletin, National Bank of Egypt 25.1972(1/2): 45-6o. -H 3-

1977 Major companies of the Kingdom of Saudi Arabia. In: G.C. Bricault (ed.): Major companies of the Arab world, 1979/8o. London: Graham & Trotman, 1979: 297-356. -21-

1978 Makin, W.J., 1932: Red Sea nights. London: Jarrolds. 288 p. -291-

1979 Makki, M.S., 1982: Medina, Saudi Arabia: a geographic analysis of the region and city. N.p. (Amersham): Avebury. XIV, 231 p.   -21-

1980 Makky, G.A.W., 1978: Mecca, the pilgrimage city. A study of pilgrim accommodation. Hajj Research Centre Studies 1; London: Croom Helm for The Hajj Research Centre, King Abdul Aziz University, Jeddah. 95 p.   -21, Tü 17-

1981 ——, 1981: Characteristics of pilgrim accommodations in Mecca and recommendations for improvements. Ph.D., Michigan State University, East Lansing. 267 p.

1982 Makoshi, A.A.-R., 1978: An analysis of the mathematics curriculum in the public secondary schools in Saudi Arabia. Ed.D., University of Northern Colorado, Greeley. XIV, 172 p.   -H 223-

1983 Malaika, Y., 1977/78: Towards an optimal level of petroleum production in Saudi Arabia. Ph.D., University of Birmingham.

1984 Malak, J.A., A.K. Kurban, G.S. Bridi & J.D. Thaddeus, 1969: Systemic fungus infections in northern Saudi Arabia: a survey of the occurrence of positive skin tests in 1oo selected patients. In: Annals of Tropical Medicine and Parasitology 63: 143-146.   -61, 1oo, 289-

1985 Malik, A.A., 197o: The money supply process in Saudi Arabia. Ph.D., Indiana University, Bloomington. 3o6 p.   -2o6-

1986 Malik, S.A., 1968: A study of the social system of bedouins: its relation to the problems of settling in Saudi Arabia. M.A., Wayne State University, Detroit.

1987 ——, 1973: Rural migration and urban growth in Riyadh, Saudi Arabia. Ph.D., University of Michigan, Ann Arbor. X, 234 p.   -21-

1988 Mallakh, R. el-, 1966: Economic diversification is key to progress. In: Emergent Nations 2(2): 54-56.   -H 223-

1989 —— & M. Kadhim, 1976: Arab institutionalized development aid: an evaluation. In: The Middle East Journal 3o: 471-484.   -21, 2o6, 352-

1990 Mallakh, R. el-, 1982: Saudi Arabia, rush to development: profile of an energy economy and investment. London, Canberra: Croom Helm. 472 p.   -21-

1991 —— & D. el-Mallakh (eds.), 1982: Saudi Arabia, energy, developmental planning, and industrialization. Lexington Books; Lexington: Heath. XIII, 2o5 p.   -H 3-

1992 Malleess, S.M. al-, 198o: A comparative study of the collegiate success of graduates of comprehensive secondary schools and traditional secondary schools in selected regions of Saudi Arabia. Ph.D., Indiana University, Bloomington. 17o p.

1993 Malmignati, D., 1925: Through inner deserts to Medina. London: Allan. 187 p.   -1a-

1994 Malone, J.J., 1966: Islam in politics, a Muslim World Symposium; Saudi Arabia. In: The Muslim World 56: 29o-295.   -21, 291, 352-

1995 ——, 1973: The Arab lands of Western Asia. The Modern Nations in Historical Perspective Series; Englewood Cliffs: Prentice-Hall. X, 269 p.   -21, 3o-

1996 ——, 1976: America and the Arabian Peninsula: the first two hundred years. In: The Middle East Journal 3o: 4o6-424.   -12, 21, 2o6-

1997 ——, 198o: Involvement and change: the coming of the oil age to Saudi Arabia. In: T. Niblock (ed.): Social and economic development in the Arab Gulf. London: Croom Helm; Exeter: Centre for Arab Gulf Studies, 198o: 2o-48.   -21, H 3-

1998 Maltzan, H. von, 1865: Meine Wallfahrt nach Mekka. Reise in der Küstengegend und im Innern von Hedschas. 2 Bde. Leipzig: Dyk (gekürzte Ausgabe: Meine Wallfahrt nach Mekka. Hg. von G. Giertz. Alte abenteuerliche Reiseberichte; Tübingen: Erdmann, 1982). VI, 377 bzw. 373 S.   -25, 93, 212-

1999 ——, 1873: Reise nach Südarabien und geographische Forschungen im und über den südwestlichsten Theil Arabiens. Braunschweig: Vieweg. XVI, 4o7 S.   -24, Tü 17-

2ooo Mamméri, H., 1976: Le deuxième plan quinquennal saoudien et le budget de l'année 1975-1976. In: Maghreb (71): 75-79.   -2o6, H 3-

2oo1 Mandaville, James, 1965: Al-Djawf. In: EI^2 2: 492.   -21,
      24, 212-

2oo2 ——, 1971: Ḥayil. In: EI^2 3: 326-327.   -21, 24, 212-

2oo3 ——, 1972: Some experiments with solar ground stills in
      eastern Arabia. In: The Geographical Journal 138: 64-66.
      -21, 24, 291-

2oo4 Mandaville, Jon, 198o: The new historians. In: Aramco World
      Magazine 31(2): 2-7.   -21, 24-

2oo5 Mangat-Rai, C.R., 1964: Mission to Arabia. In: The Educa-
      tional Forum 29(Nov.): 5o-58.   -1a-

2oo6 Mani, M.A. al- & A.-R.S. as-Sbit, 1981: Cultural policy in
      the Kingdom of Saudi Arabia. Studies and Documents on
      Cultural Policies; Paris: The Unesco Press. 69 p.   -212,
      H 1o8-

2oo7 Manifa: oil field under the sea. In: Aramco World 11.196o
      (6): 16-19.   -H 223-

2oo8 Manifa: profile of a decision. In: Aramco World 14.1963(6):
      18-21.   -H 223-

2oo9 Mankour, N., 1958: Saudi Arabia: educational progress in
      1957-1958. In: International Yearbook of Education 2o:
      28o-283.   -1a, 24, 29-

2o1o Manners, I.R., 1978: The iceberg solution. In: The Geograph-
      ical Review 68: 224-226.   -21, 24, 2o6-

2o11 Mansfield, P., 1979: Islamic, yes... Republic, no. In: Middle
      East International (93): 6-7.   -12, 3o-

2o12 —— (ed.), ^{5}198o: The Middle East: a political and economic
      survey. Oxford, New York, Toronto, Melbourne: Oxford Uni-
      versity Press (^{4}1973). IX, 579 p.   -21, 93-

2o13 Mansour, H.O., 1973: The discovery of oil and its impact on
      the industrialization of Saudi Arabia: a historical anal-
      ysis. Ph.D., University of Arizona, Tucson. XI, 223 p.
      -21-

2o14 Mansur, A.K. (pseud.), 198o: The American threat to Saudi
      Arabia. In: Armed Forces Journal International 118(1):
      47-59 (reprinted in: Survival 23.1981(1): 36-41).   -89-

2o15 Manuie, M.A., 1976: A study of teacher-principal perceptions of the organizational climate in selected schools in Riyadh, Saudi Arabia. Ph.D., University of Oklahoma, Norman. 235 p.

2o16 Maranjian, G., E.W. Ikin, A.E. Mourant & H. Lehmann, 1966: The blood groups and haemoglobins of the Saudi Arabians. In: Human Biology 38: 394-42o. -1a, 12, 24-

2o17 Marble for Mecca. In: Aramco World 13.1962(9): 3-7. -H 223-

2o18 Marett, W.C., 1953: Some medical problems met in Saudi Arabia. In: United States Armed Forces Medical Journal 4(1): 31-38. -1a, 26, 38-

2o19 Margoliouth, D.S., 1934: Wahhābīya. In: EI[1] 4: 1175-118o. -21, 24-

2o2o Marimont, H. (Pseud.), 1969: Wüstensand und Exzellenzen. Die Frau eines Botschafters erzählt. München: List. 3o4 S. -H 223-

2o21 Marineleitung, [2]1926: Handbuch für das Rote Meer und den Golf von Aden. Berlin: Mittler. XII, 648 S. -H 2-

2o22 ——, [2]1929: Handbuch für den Persischen Golf. Berlin: Mittler. XIV, 275 S. -1a, H 2-

2o23 Marks, M.M., 198o: The American influence on the development of the universities in the Kingdom of Saudi Arabia. Ph.D., University of Oregon, Eugene. 163 p.

2o24 Marlowe, J., 1962: The Persian Gulf in the twentieth century. London: The Cresset Press. VIII, 28o p. -212-

2o25 Marotz, G., 198o: Die Oase Al Hassa in Saudi-Arabien. In: Wasserwirtschaft (Stuttgart) 7o: 353-354. -24, 9o, 1oo-

2o26 Marshall, D., 1979a: Work, but tarry not, in Saudi Arabia. In: Middle East International (95): 9-1o. -12, 3o-

2o27 ——, 1979b: Saudi women on the move. In: Middle East International (1o7): 11. -12, 3o-

2o28 ——, 198o: Saudi Arabia feels the draught. In: Middle East International (129): 8-9. -12, 3o-

2o29 Marsouqi, H.A. al-, 198o: A facet theory analysis of attitudes toward handicapped individuals in Saudi Arabia. Ph.D. Michigan State University, East Lansing. 221 p.

2o3o   Marston, T.E., 1961: Britain's imperial role in the Red Sea area, 18oo-1878. Foreign Area Studies 5; Hamden: The Shoe String Press. XIII, 55o p.   -18o-

2o31   Martan, S.S., 198o: Domestic development and the management of oil revenues in the economy of Saudi Arabia. Ph.D., University of Nebraska, Lincoln. 196 p.

2o32   Martin, L., 19o9: Die Hedschasbahn. In: Asien (Berlin) 8: 115-117.   -1a, 25, 9o-

2o33   Masia, M., 1937: L'Arabia Saudiana e la rinascita araba. Milano: Consociazione turistica italiana. 29 S.   -Biblioteca Nazionale Centrale, Roma-

2o34   Mason, A.D. & F.J. Barny, 1926: History of the Arabian Mission. New York: The Board of Foreign Missions, Reformed Church in America.   -Library of Congress, Washington-

2o35   Massignon, L., 1923: Nouveaux itinéraires en Arabie Centrale. In: La Géographie (Paris) 39: 2o8-21o.   -2o6, 291, 294-

2o36   ——, avec le concours de V. Monteil, ^{4}1955: Annuaire du monde musulman statistique, historique, social et économique. Éd. revisée et mise à jour. Paris: Presses Universitaires de France (^{1}1923, ^{2}1926, ^{3}1929). XVI, 431 p.   -21-

2o37   Massive investments on small base bring growth pains to Saudi Arabia. In: Commerce America 1.1976(1o): 25-26.   -1a, 2o6, H 3-

2o38   Mathieu, A.L. & S.T. Farouky, 1967: Technical education in agriculture in Saudi Arabia. Report of an agricultural education mission to the Kingdom of Saudi Arabia (4th-24th February, 1967). RU:MISC/67/11; Rome: FAO. III, 39 p.   -2o6-

2o39   Matossian, R.M., J.D. Thaddeus & G.A. Garabedian, 1963: Outbreak of epidemic typhus in the northern region of Saudi Arabia. In: The American Journal of Tropical Medicine and Hygiene 12: 82-9o.   -12, 25, 1oo-

2o4o   Matouk, A., 1976: 'Water rewrites the history of the desert': agriculture and water in Saudi Arabia. In: Irish Arab News 2(2): 1o-12.

2o41 Matthews, C.D., 1954: Research in Saudi Arabia. In: The Muslim World 44: 11o-125. -21, 291, 352-

2o42 ——, 196o: Bedouin life in contemporary Arabia. In: Rivista degli Studi Orientali 35: 31-61. -16, 21, 352-

2o43 Maull, H.W., 1979: Erdöl. Seine Bedeutung für die Zukunft der arabischen Welt. In: U. Steinbach (Hg.): Europäisch-arabische Zusammenarbeit. Arbeitspapiere zur Internationalen Politik 11; Bonn: Europa Union, 1979: 52-72. -2o6, H 3, H 223-

2o44 ——, 1983: Erdöl und Erdölpolitik. In: T. Koszinowski (Hg.): Saudi-Arabien: Ölmacht und Entwicklungsland. Beiträge zur Geschichte, Politik, Wirtschaft und Gesellschaft. Mitteilungen des Deutschen Orient-Instituts 2o; Hamburg: Deutsches Orient-Institut, 1983: 299-326. -21, H 233-

2o45 Maunsell, F.R., 19o8: The Hejaz Railway. In: The Geographical Journal 32: 57o-585. -21, 291, 352-

2o46 ——, 19o9: One thousand miles of railway built for pilgrims and not for dividends. In: The National Geographic Magazine 2o: 156-172. -21, 93-

2o47 Mazroa, S.A. al-, 198o: Public administration trends and prospects in the context of development in Saudi Arabia. Ph.D., Claremont Graduate School. 434 p.

2o48 Mazroe, H.M.H. al-, 1982: A comparison between Piagetian cognitive level and physics achievement for twelfth grade students in Saudi Arabia. Ed.D., University of Northern Colorado, Greeley. 133 p.

2o49 Mazyed, M.I. al-, 1975: Science education in public secondary schools in Saudi Arabia as perceived by science teachers and science students. Ph.D., University of Oregon, Eugene. 265 p.

2o5o Mazyed, S.M. al-, 1972: The structure and function of public personnel administration in Saudi Arabia. Ph.D., Graduate Faculty of Government, Claremont Graduate School. XI, 167 p. -21-

2o51 McConahay, M.-J., 1979: KFSH: showcase of the future. In: Aramco World Magazine 3o(4): 14-25, 28-29. -21, 24-

2o52 McDonald, J., with C. Burleson, 1981: Flight from Dhahran: the true experiences of an American businessman held hostage in Saudi Arabia. Englewood Cliffs: Prentice-Hall. IX, 26o p.   -21-

2o53 McDonald, W.N., 1977: Arabia Modernae. In: Contemporary Review 231(2): 87-92.   -25, 18o, 352-

2o54 McGregor, R., 1972: Saudi Arabia: population and the making of a modern state. In: J.I. Clarke & W.B. Fisher (eds.): Populations of the Middle East and North Africa. A geographical approach. London: University of London Press, 1972: 22o-241.   -7-

2o55 McHale, T.R., 198o: A prospect of Saudi Arabia. In: International Affairs (London) 56: 622-647.   -21, 2o6, H 223-

2o56 ———, 1982a: Whither Arabia in a changing world economy? Middle East Problem Paper 22; Washington: The Middle East Institute. 16 p.   -H 223-

2o57 ———, 1982b: The Saudi Arabian political system - its origins, evolution and current status. In: Vierteljahresberichte - Probleme der Entwicklungsländer 1982(9): 2o1-2o8.   -24, 2o6, H 3-

2o58 McKenna, J., 1979: Mission to Riyadh. In: Irish Arab News 5(1): 13-14.

2o59 McLoughlin, A., 1958: The Hejaz Railway. In: The Geographical Journal 124: 282-283.   -21, 24, 291-

2o6o McMaster, B., 198o: The definitive guide to living in Saudi Arabia. Tokyo: Dai Nippon Printing Company. 96 p.   -212-

2o61 McMillan, W.M., 1968: Report to the Government of the Kingdom of Saudi Arabia on requirements for marketing development in El Hassa and Haradh. Report FAO/TF 1o, Project: Saudi Arabia, Trust Fund 117, EA:TF/68; Rome: FAO.   -Documentation Section, Centre for Middle Eastern and Islamic Studies, University of Durham-

2o62 McNiel, J.R., 1963: Gastroenteritis in eastern Saudi Arabia. In: Texas State Journal of Medicine 59(8).   -38 M-

2o63 ———, 1965: Meningitis in eastern Saudi Arabia. In: Middle Eastern Medical Journal 2(1): 13-17.

2o64   McNiel, J.R., 1966a: Pediatric practice in eastern Saudi
       Arabia. In: Clinical Pediatrics 5: 385-39o.   -1a, 21,
       289-

2o65   ——, 1966b: Acute bacterial meningitis as seen in children
       of east Saudi Arabia. In: Clinical Pediatrics 5: 437-438.
       -1a, 21, 289-

2o66   ——, 1967: Family studies of thalassemia in Arabia. In:
       American Journal of Human Genetics 19: 1oo-111.   -7,
       1oo, 352-

2o67   Mecca pilgrimage. In: Aramco World 14.1963(9): 17-21.
       -H 223-

2o68   Mecci, M.S., 1975: Aspects of the urban geography of Medina,
       Saudi Arabia. M.A., University of Durham.

2o69   ——, 1979a: An analysis of the effects of modern pilgrimage
       on the urban geography of Medina. Ph.D., University of
       Durham.

2o7o   ——, 1979b: Mutawifs in Mecca. In: The Geographical Magazine
       51: 439.   -24, 18o, 464-

2o71   Mech, P., 1945: Au pays d'Ibn Sa'ûd: Le Wahhabisme. In: En
       Terre d'Islam III, 2o: 229-238.   -21, 3o-

2o72   Medawar, G.S., 1965a: The date problem in Saudi Arabia. N.p.
       (Beirut?). 12 p.

2o73   ——, 1965b: A report to the Ministry of Agriculture on agri-
       culture in Saudi Arabia. Beirut: Economic Research Insti-
       tute, American University of Beirut. 42 p.

2o74   ——, 1966: Agricultural expansion fulfills great hopes. In:
       Emergent Nations 2(2): 5o-52, 61.   -H 223-

2o75   Medical oasis in Saudi Arabia. In: Hospitals (Chicago) 39.
       1965: 51-56.   -1a, 289-

2o76   Meeker, M.E., 1979: Literature and violence in North Arabia.
       Cambridge Studies in Cultural Systems 3; Cambridge/Eng-
       land, London, New York, Melbourne: Cambridge University
       Press. XVI, 272 p.   -24, H 223-

2o77   Meglio, R. di, 197o: Il problema dei nomadi in Arabia Saudi-
       na e le sue soluzioni. In: Oriente Moderno 5o: 273-279.
       -5, 7, 21-

2078 Meglio, R. di, 1975: La donna in Arabia Saudita. In: Levante 23(1): 21-26.   -21-

2079 ———, 1978: Banū Khālid. In: EI^2 4: 925.   -21, 24, 212-

2080 Meigs, P., 1966: Geography of coastal deserts. Arid Zone Research 28; Paris: UNESCO, 1966. 140 p.   -7, 100-

2081 Meinertzhagen, R., 1954: Birds of Arabia. Edinburgh, London: Oliver & Boyd. XIII, 624 p.   -24-

2082 Mejcher, H., 1980: Die Politik und das Öl im Nahen Osten. Bd. 1: Der Kampf der Mächte und Konzerne vor dem Zweiten Weltkrieg. Stuttgart: Klett-Cotta. 278 S.   -21, 24-

2083 ———, 1982: Saudi Arabia's 'Vital Link to the West': some political, strategic and tribal aspects of the Transarabian Pipeline (TAP) in the stage of planning 1942-1950. In: Middle Eastern Studies 18: 359-377.   -1a, 21, 30-

2084 Melamid, A., 1955: The economic geography of neutral territories. In: The Geographical Review 45: 359-374.   -21, 24, 206-

2085 ———, 1956: The Buraimi oasis dispute. In: Middle Eastern Affairs 7(2): 56-63.   -1a, 21, 25-

2086 ———, 1957: The political geography of the Gulf of Aqaba. In: Annals of the Association of American Geographers 47: 231-240.   -1a, 7-

2087 ———, 1965: Political boundaries and nomadic grazing. In: The Geographical Review 55: 287-290.   -21, 24, 206-

2088 ———, 1980: Urban planning in eastern Arabia. In: The Geographical Review 70: 473-477.   -21, 24, 206-

2089 Melibary, A.R., 1980: Saudi Arabia, a technically developing country and the question of introducing nuclear power during 1980-2000. Diss., Fakultät für Maschinenbau, Universität Karlsruhe (zugleich: Bericht KfK 2997, Kernforschungszentrum Karlsruhe). VIII, 271 S.   -21, 93, H 3-

2090 Melikian, L.H., 1971: Actual and desired occupational status of acculturated Saudi youth. In: Al-Abhath 24: 125-132.   -21-

2091 Melikian, L.H., A. Ginsberg, D. Cüceloglu & R. Lynn, 1971: Achievement motivation in Afghanistan, Brazil, Saudi Arabia and Turkey. In: The Journal of Social Psychology 83: 183-184. -12, 18o, 352-

2092 ——, 1972: First drawn picture and modernization. In: Journal of Personality Assessment 36: 576-58o. -12, 291, 352-

2093 ——, 1977: The modal personality of Saudi college students: a study in national character. In: L.C. Brown & N. Itzkowitz (eds.): Psychological dimensions of Near Eastern studies. Princeton Studies on the Middle East; Princeton: The Darwin Press, 1977: 166-2o9. -21-

2094 Melka, R.L., 1966: The Axis and the Arab Middle East: 193o-1945. Ph.D., Faculty of the Graduate School, University of Minnesota, Minneapolis. 419 p. -21-

2095 Memon, A.F., 1967(?): Oil and the faith. Karachi: Inter Services Press. II, 15o p.

2096 Memun Abul Fadl, S., 1916: Zur Kultur und Verwertung der Dattelpalme im Gebiete von Medina. In: Archiv für Wirtschaftsforschung im Orient 1: 296-3o5. -1a, 25, 18o-

2097 Mengin, F., 1823: Histoire de l'Égypte sous le gouvernement de Mohammed-Aly, ou, récit des événemens politiques et militaires qui ont eu lieu depuis le départ des Français jusqu'en 1823. 2 vols. Paris: Bertrand. -16-

2098 Merriam, J.L., 1957: Aḥmad Zayny farm development program. Riyadh: Ministry of Agriculture.

2099 Mertens, R., 1949: Eduard Rüppell. Leben und Werk eines Forschungsreisenden. Senckenberg-Buch 24; zugleich: Frankfurter Lebensbilder 14; Frankfurt/M.: Kramer. 388 S.

2100 Mertz, R.A., 1971: Saudi Arabia. In: L.C. Deighton (ed.): The encyclopedia of education. Vol. 8. New York: Macmillan, The Free Press, 1971: 1o-13. -21-

2101 Messer, E., 1958: Een troon van luipaardvel; rondreis door het land van Koning Saoed. Amsterdam: Arbeiderspers. 112

2102 Messerschmidt, E.A., 1968: German economic delegation to Kuwait and Saudi Arabia. In: Orient (Opladen) 9: 187-188, 19o-191. -21, 1oo, H 223-

2103 Messiha, S.A., 1980: The export of Egyptian school teachers. Cairo Papers in Social Science 3, Monograph 4; Cairo: American University of Cairo. VI, 92, VI p. -2o6, H 223-

**** Metcalf, J.E.: author of no. 989.

2104 Metra Consulting Group, 1975: Saudi Arabia - business opportunities. London: The Financial Times. 189 p. -21, H 3-

2105 Metta, V.B., 1936: A decade of progress in the Hijaz. In: The Contemporary Review 15o: 473-477. -5, 25, 18o-

2106 Meulen, D. van der & H. von Wissmann, 1940: Harry St. John Bridger Philby C.I.E., I.C.S. In: Geographische Zeitschrift 46: 23-26. -16, 21, 24-

2107 ——, 1940: The Mecca pilgrimage and its importance to the Netherlands East Indies. In: Asiatic Review N.S. 36: 588-597 (reprinted in: The Moslem World 31.1941: 48-6o). -1a,21-

2108 ——, 1953: Een onderzoekingsreis in NO- en ZW-Arabië. I. Sa'ūdī Arabië, het land van het Wahhabisme wordt, land van olie. In: Tijdschrift van het Koninklijk Nederlandsch Aardrijkskundig Genootschaap Amsterdam II, 7o: 44-53, 56-57, 6o-69. -21, 3o, 291-

2109 ——, 1954: Mijn weg naar Arabië en de Islaam. Facetten der West-Europese Cultuur 5; Amsterdam: Meulenhoff. 68 S. -Koninklijke Bibliotheek 's-Gravenhage-

2110 ——, 1957: The wells of Ibn Saud. London: Murray; New York: Praeger. 27o p. -21, 24, H 223-

2111 ——, 1958: Ontwakend Arabië. Koning Ibn Sa'ūd, de laatste bedoeïnenvorst van Arabië. Meulenhoff's Flamingo-Reeks 3; Amsterdam: Meulenhoff (^{1}1953). 183 S. -21-

2112 ——, 1959: Verdwijnend Arabië. Ontmoetingen en avonturen in een woestijnland. Amsterdam: Meulenhoff. 24o S. -12-

2113 ——, 1961: Faces in Shem. London: Murray. XII, 194 p. -12, 21-

2114 ——, 1966: Changing Arabia: an observer's impressions - 1926-'64. In: Nederlands-Arabische Kring, 1955-1965. Leiden: Brill, 1966: 112-129.

2115 Meulen, D. van der, 1967: Memoirs of old Jiddah. In: Aramco World 18(2): 3o-33. -24-

2116 Meyer, A., 198o: Saudi-Arabien - ein Markt wird eng. In: Übersee-Rundschau 32(3): 8. -24, 3o, 18o-

2117 Meyer, W.C., 1962: Landwirtschaft in Saudi-Arabien. In: Der Deutsche Tropenlandwirt 63: 45-51. -7, 1oo, H 3-

2118 Meyer-Ranke, P., 197o: Die arabischen Staaten Vorderasiens. Edition Zeitgeschehen; Hannover: Verlag für Literatur und Zeitgeschehen. 144 S. -212-

2119 Middle East Airlines Airliban, Marketing Section, 1969: MEA's businessman's guide to Saudi Arabia. Beirut. 28 p.

212o Middle East countries - United States: joint statements and communiques on future cooperation and Middle East settlement (June 8 - August 19, 1974). In: International Legal Materials 13.1974: 1265-1274. -12, 24, 2o6-

2121 Middle East Supply Centre, Food Division, 1943: Note on the food situation in central and north-east Saudi Arabia. Cairo.

2122 ——, 1944a: Conditions of life and work of the agricultural worker in Saudi Arabia. N.p. (Cairo?). 7 p.

2123 ——, 1944b: The future development of the state farms at El Kharg, March 1944. N.p. (Cairo?). 16 p.

2124 ——, 1944c: Report on transport, distribution and roads in Saudi Arabia. N.p. (Jeddah?). 2o p.

2125 Middleton, D., 198o: Early moves in the 'Great Game'. In: The Geographical Magazine 52: 463-469. -7, 18o, 464-

2126 Midhat, A.H., 19o3: The life of Midhat Pasha: a record of his services, political reforms, banishment, and judicial murder, derived from private documents and reminiscences. London: Murray (französische Ausgabe: Midhat-Pacha. Sa vie - son œuvre. Paris, 19o8). XII, 292 p. -16-

2127 Midland Bank, International Division, 1976: Spotlight on overseas trade - selling to Saudi Arabia. London.

2128 Mihdar, A. al-, 1977: Business law in Saudi Arabia. London: Graham & Trotman.

2129   Mikesell, R.F., 1947: Monetary problems of Saudi Arabia. In: The Middle East Journal 1: 169-179.   -21, 38, 206-

2130   —— & H.B. Chenery, 1949: Arabian oil: America's stake in the Middle East. Chapel Hill: The University of North Carolina Press. XI, 201 p.   -21-

2131   Mikusch, D. von, 1942: König Ibn Sa'ud. Das Werden eines Staates. Leipzig: List. 380 S.   -12, 24-

2132   ——, 1944: König Ibn Sa'ud. In: W. Björkman, R. Hüber, E. Klingmüller, D. von Mikusch & H.H. Schaeder: Arabische Führergestalten. Arabische Welt 5; Heidelberg, Berlin, Magdeburg: Vowinckel, 1944: 115-136.   -Tü 17-

2133   ——, 1953: König Ibn Sa'ud. Mekka, Öl und Politik. Neubearb. & erw. Ausgabe von H. Ziock. List-Bücher 20; München: List. 238 S.   -12-

2134   Milburn, W., 1813: Oriental commerce; containing a geographical description of the principal places in the East Indies, China, and Japan, with their produce, manufactures, and trade, including the coasting or country trade from port to port; also the rise and progress of the trade of the various European nations with the eastern world, particularly that of the English East India Company, from the discovery of the passage round the Cape of Good Hope to the present period; with an account of the company's establishments, revenues, debts, assets, &c. at home and abroad. Deduced from authentic documents, and founded upon practical experience obtained in the course of seven voyages to India and China. Vol. 1. London: Black, Parry. CIII, 413 p.   -7-

2135   Miles, S.B., 1919: The countries and tribes of the Persian Gulf. 2 vols. London: Harrison (reprint: 2 vols. in one. London: Cass, 1966). 643 p.   -21, 212-

2136   'Milk run'. In: Aramco World 12.1961(8): 12-15.   -H 223-

2137   Miller, A.D., 1980: Search for security: Saudi Arabian oil and American foreign policy, 1939-1949. Chapel Hill: The University of North Carolina Press. XVIII, 320 p.   -12, 21-

2138 Mineau, W., 1958: The go devils: the story of Arabian oil. London: Cassell. XI, 243 p.   -38-

2139 Ministerial statement of 6 November 1962 by Prime Minister Amir Faysal of Saudi Arabia. In: The Middle East Journal 17.1963: 161-162.   -21, 38, 2o6-

214o Mishari, H., 1965: Statement on present situation of agriculture in Saudi Arabia. In: Levante 12(2): 3-6.   -21-

2141 ——, 1967: Towards full water utilization in Saudi Arabia. In: International Conference on Water for Peace. Vol. 2: Water supply technology. Washington: US Government Printing Office, 1967: 832-841.

2142 Mishlawi, T., 1978: The empire builders: Saudi Arabia's entrepreneurs. In: The Middle East (48): 86.   -1a, 12, 3o-

2143 ——, 1979: A new direction. - Oil for peace. In: The Middle East (55): 25-28.   -1a, H 3, H 223-

2144 Mission to Jiddah. In: Aramco World 13.1962(1o): 19-21. -H 223-

2145 Mittelmann, G., 1973: Über das Anlaufen einer Grossbaustelle im Ausland unter besonderer Berücksichtigung des Geräteeinsatzes. In: Baumaschine und Bautechnik 2o: 85-92, 149-156.   -89, 93, 18o-

2146 —— & W. Montada, 1974: Herstellung von Beton in einer Wüste. In: Der Tiefbau 16: 155-158, 26o-264.   -89, 93, 468-

2147 ——, 1977: Wasser für Al Hassa. Die Realisierung eines Bauvorhabens in der dritten Welt. In: Umschau in Wissenschaft und Technik 77: 1o1-1o8.   -21, 1oo, 18o-

2148 ——, 1981: Das Projekt Dammam. Deutscher Hafenbau im Ausland. In: Hansa (Hamburg) 118: 743-747.   -89, 9o, 93-

2149 Mittwoch, E., 1912: Burchardt, Hermann. In: A. Bettelheim (Hg.): Biographisches Jahrbuch und deutscher Nekrolog 14; Berlin: Reimer, 1912: 3o1-3o4.   -7-

215o Mixon, J.W., 1982: Saudi Arabia, OPEC, and the price of crude oil. In: Resources and Energy 4: 195-2o1.   -3o, 89, 2o6-

2151 Mizjaji, A.D. al-, 1982: The public attitudes toward the bureaucracy in Saudi Arabia. Ph.D., Florida State University, Tallahassee. 22o p.

2152 Moberg, A., 1936: Nadjrān. In: EI[1] 3: 890-891.   -21, 24-

2153 Moe, L.E., 1966: Saudi Arabia: supply and demand projections for farm products to 1975 with implications for U.S. exports. ERS-Foreign 168; Washington: Economic Research Service, US Department of Agriculture. VI, 25 p.   -H 3-

2154 Mohammed, S.K. el-, 1966: Die Entwicklung der saudi-arabischen Medizin in den letzten 15o Jahren unter besonderer Berücksichtigung der Chirurgie und Orthopädie. Diss., Univèrsität Düsseldorf. 64 S.   -61-

2155 Moliver, D.M., 1978: Oil and money in Saudi Arabia. Ph.D., Virginia Polytechnic Institute and State University, Blacksburg. IX, 11o p.

2156 ── & P.J. Abbondante, 198o: The economy of Saudi Arabia. Praeger Special Studies, Praeger Scientific; New York: Praeger. XV, 167 p.   -21, 1oo, H 3-

2157 Momberger, M., [2]1977: Arab airports: a background study of an expansive aviation market. Airport Profiles; Rutesheim ([1]1976). Var.pag.   -24-

2158 Monde arabe. Les fonds arabes de développement. Stratégie et moyens d'action. In: Syrie et Monde Arabe 28.1981(324): 25-34.   -2o6, H 3, H 223-

2159 Moneef, A.A., 1981: The attitude of the Saudi citizens toward the imposition of individual income tax. Ph.D., University of South Carolina, Columbia. 228 p.

216o Moneef, I.A. al-, 198o: Transfer of management technology to developing nations: the role of multinational oil firms in Saudi Arabia. New York: Arno Press (zugleich: D.B.A., Graduate School of Business, Indiana University, Bloomington, 1977). 5o6 p.   -21-

2161 Monfreid, H. de, 1932: Aventures de mer. Paris: Grasset (englische Ausgabe: Sea adventures. London: Methuen, 1937). 296 p.   -1o7-

2162 Monheim, C., 1934: Guerre en Arabie. In: Bulletin de la Société Royale de Géographie d'Anvers 54: 3o9-333.   -3o, 291-

2163 Monroe, E., 1973: Philby of Arabia. London: Faber & Faber (reprint: London: Quartet Books, 198o). 332 p. -21, Tü 17-

2164 Montagne, R., 1932: Notes sur la vie sociale et politique de l'Arabie du Nord: les Šemmâr du Neğd. In: Revue des Études Islamiques 6: 61-79.   -16, 21, 291-

2165 ——, 1947: La civilisation du désert. Nomades d'Orient et d'Afrique. Paris: Hachette. 271 p.

2166 Montgomery, J.A., 1927: Arabia to-day: an essay in contemporary history. In: Journal of the American Oriental Society 47: 97-132.   -21, 24, 352-

2167 Montgomery, P.A. & L.E.H. Tarkanian, c. 1980: Selected organization development interventions as related to Japan, Mexico, People's Republic of China, and Saudi Arabia. Ph.D., United States International University, San Diego. 22o p.

2168 Moore, F.L.,Jr., c. 1951: Origin of American oil concessions in Bahrein, Kuwait and Saudi Arabia. A thesis presented to The School of Politics and International Affairs, Princeton University, 1948; appendices and additions 1951. N.p.   -Library of Congress, Washington-

2169 Moran, T.H., 1981: Modeling OPEC behaviour: economic and political alternatives. In: International Organization 35: 241-272.   -18, 2o6-

217o Morano, L., 1979: Multinationals and nation-states: the case of Aramco. In: Orbis (Philadelphia) 23: 447-468.   -24/ 213, 2o6, 281-

2171 Mordtmann, J.H., 1927a: Ibn Rashīd. In: EI[1] 2: 433-435.   -21, 24-

2172 ——, 1927b: Ibn Sa'ūd. In: EI[1] 2: 441-444.   -21, 24-

2173 Moritz, B., 19o8: Ausflüge in der Arabia Petraea. In: Université Saint-Joseph, Beyrouth: Mélanges de la Faculté orientale 3, Fasc. 1; Paris: Geuthner; London: Luzac; Leipzig: Harrassowitz, 19o8: 387-436.   -7, 16, 21-

2174 ——, 1923: Arabien. Studien zur physikalischen und historischen Geographie des Landes. Hannover: Lafaire (Nachdruck: Osnabrück: Biblio, 1972). 133 S.   -7, 24, Tü 17-

2175 Morris, J., 1959: The Hashemite kings. London: Faber & Faber (New York: Pantheon, 1959). 231 p.   -21, 24-

2176 Morrison, O.F., 1964: Labor law and practice in Saudi Arabia. US Department of Labor, Bureau of Labor Statistics Report 269; Washington: US Government Printing Office. III, 44 p.   -H 223-

2177 Morsey, K., 1976: T.E. Lawrence und der arabische Aufstand 1916/18. Studien zur Militärgeschichte, Militärwissenschaft und Konfliktforschung 7; Osnabrück: Biblio (zugleich: Diss., Philosophische Fakultät, Universität Münster, 1975). XI, 457 S.   -352-

2178 Morsi Abbas, A., 1977: New lights on the Dariya war!! Did the Egyptians really take part in the invasion of Dariya? In: Addarah 3(3): 5-7.   -21-

2179 Morsy Abdullah, M., 1971: The first Sa'udi dynasty and 'Oman, 1795-1818. In: Proceedings of the Seminar for Arabian Studies 1: 34-41.   -1a, 21-

2180 Mortimer, E., 1982: Faith and power: the politics of Islam. London: Faber & Faber. 432 p.

2181 Moshaikeh, M.S.H., 1982: Patterns of instructional media utilization in preparation of elementary school teachers in Saudi Arabian junior colleges. Ph.D., University of Pittsburgh. 164 p.

2182 Mostra di dipinti di giovani alunni della Scuola Modello di Gedda. In: Levante 12.1965(3/4): 82-84.   -21-

2183 Mostyn, T., 1978: A desert journey. In: Middle East International (79): 28-3o.   -12, 3o-

2184 —— (ed.), 1981: Saudi Arabia. A MEED Practical Guide; London: Middle East Economic Digest. VIII, 28o p. -21, B 211-

2185 Moulla, E.A. al-, 198o: A system for evaluating the administration and effectiveness of vocational education programs in Saudi Arabia. Ed.D., University of Wyoming, Laramie. 172 p.

2186 Mourad, F., 1971: The attitude of village population in Saudi Arabia toward community development: a study in the social psychology of development. Ph.D., Department of Sociology, University of Southern California, Los Angeles. IV, 149 p.   -21-

2187 Mousa, S., 1966: T.E. Lawrence: an Arab view. London, New York, Toronto: Oxford University Press. X, 3o1 p. -12, 21-

2188 ――, 1967: The role of Syrians and Iraqis in the Arab Revolt. In: Middle East Forum 43(1): 5-17. -188/812-

2189 Moussali, M.S., Farid A. Shaker & O.A. Mandily, 1977: An introduction to urban patterns in Saudi Arabia: the central region. Art and Archaeology Research Papers; London. 7, 57 p. -21, Tü 17-

2190 ――, 1981: Development planning for higher education facilities in a developing country: a case study of Saudi Arabia. Ph.D., University of Cambridge.

2191 Moustapha, A.F. & F.J. Costa, 1981: Al Jarudiyah: a model for low rise/high density development in Saudi Arabia. In: Ekistics 48: 1oo-1o8. -1a, 2o6, Tü 17-

2192 Muarik, S.A. al-, 1982: Error analysis and English learning strategies among intermediate and secondary school students in Saudi Arabia. Ph.D., Indiana University, Bloomington. 149 p.

2193 Mücke, H. von, 1927: Ayesha. Neubearb. & erw. Ausgabe. Berlin: Scherl ([1]1915) (amerikanische Ausgabe: Emden. Boston: Ritter, 1917). 169 S. -7, 24/213-

2194 Mueller, J.H., 1974: Beduinen und Computer. Quer durch Saudi-Arabien. Zürich: Schweizer Verlagshaus. 253 S. -12, 21-

2195 ――, 1977: Verrat in schwarzen Zelten. Von letzten Stammeskriegern in Arabiens Wüsten. Zürich: Schweizer Verlagshaus. 3o3 S. -12, 21-

2196 Müller, V., 1931: En Syrie avec les Bédouins. Les tribus du désert. Paris: Leroux. XII, 347 p. -1a-

2197 Mughram, A.A., 1973: Assarah, Saudi Arabia: change and development in a rural context. Ph.D., Department of Geography, University of Durham. -Main Library, University of Durham-

2198 Muir, J.D., 1974: The boycott in international law. In: The Journal of International Law and Economics 9: 187-2o4. -2o6, B 2o8, B 212-

2199 Mulla, M.A., 1979: Aptitude, attitude, motivation, anxiety, intolerance of ambiguity, and other biographical variables as predictors of achievement in EFL by high school science major seniors in Saudi Arabia. Ph.D., University of Michigan, Ann Arbor. 374 p.

2200 Mulligan, W.E., 1960: Abḳayḳ. In: EI^2 1: 100.   -21, 24, 212-

2201 ——, 1973: Aramco's Bedouins. In: Aramco World Magazine 24 (4): 16-19.   -24-

2202 ——, 1976: Air raid! A sequel. In: Aramco World Magazine 27 (4): 2-3.   -21, 24-

2203 Munro, J., 1974: On campus in Saudi Arabia. In: Aramco World Magazine 25(4): 2-9.   -24-

2204 Murphy, C.J.V., 1944: Mr. Ickes' Arabian Nights. His pipeline faded away and American diplomacy got a black eye. But the desert oil remains. Do we want it - and under what terms? In: Fortune 29(6): 123-129, 273-274, 277-278, 280.   -30, 89, 180-

2205 Murr, K., 1969: Hydrologische Forschungsstation in Saudi-Arabien. In: Naturwissenschaftliche Rundschau 22: 357.   -16, 21, 100-

2206 Murshid, T.A., 1978: Saudi Arabia: administrative aspects of development. Ph.D., Claremont Graduate School. XII, 211 p.   -21, H 223-

2207 Musil, A., 1911: Vorbericht über seine letzte Reise nach Arabien. In: Anzeiger der kaiserlichen Akademie der Wissenschaften, Philosophisch-historische Klasse 48: 139-159.   -21, 24, 30-

2208 ——, 1914: Kulturpolitische Berichte aus Arabien. In: Österreichische Monatsschrift für den Orient 40: 49-51, 161-162, 245-246.   -1a, 16-

2209 ——, 1918: Zur Zeitgeschichte von Arabien. Leipzig: Hirzel; Wien: Manz. V, 102 S.   -21-

2210 ——, 1926: The northern Ḥeǧâz: a topographical itinerary. Oriental Explorations and Studies 1; New York: American Geographical Society. XII, 374 p.   -21, 24-

2211 Musil, A., 1927: Arabia Deserta: a topographical itinerary. Oriental Explorations and Studies 2; New York: American Geographical Society. XVII, 631 p. -21, 24-

2212 ——, 1928a: Northern Neğd: a topographical itinerary. Oriental Explorations and Studies 5; New York: American Geographical Society. XIII, 368 p. -21, 24-

2213 ——, 1928b: The manners and customs of the Rwala Bedouins. Oriental Explorations and Studies 6; New York: American Geographical Society. XIV, 712 p. -21, 24-

2214 ——, 1928c: Religion and politics in Arabia. In: Foreign Affairs (New York) 6: 675-681. -24, 25, 18o-

2215 ——, 193o: In the Arabian desert. Ed. by K.M. Wright. New York: Liveright (London, Toronto: Cape, 1931). XIV, 339 p. -21-

2216 Mussa, S.I., 1975: Problems of human resources development in Saudi Arabia. Ph.D., University of Southern California, Los Angeles. VII, 29o p. -21-

2217 Mustafa, Z., 1977: Forms of doing business in Saudi Arabia. In: W.G. Wickersham & B.P. Fishburne, III (eds.): Current legal aspects of doing business in the Middle East: Saudi Arabia, Egypt and Iran. Chicago: Section of International Law, American Bar Association, 1977: 123-134.

2218 ——, 1978: Legal aspects of doing business in Saudi Arabia. In: Corporate law and practice. Course handbook 266; New York, 1978: 13-95.

2219 Mutawia, H.H., 1979: Faisal and the fidelity of history. N.p. (Mecca): Mecca Cultural Club. 84 p. -212-

222o Myers, R.J., 1976: An instance of reverse heaping of ages. In: Demography (Washington) 13: 577-58o. -12, 1oo, 2o6-

2221 Mylrea, C.S.G., 1919: The politico-religious situation in Arabia today. In: The Moslem World 9: 3oo-3o5. -21, 291, 352-

2222 ——, 1933: The conquest of the desert. In: World Dominion 11: 117-121. -21-

N

2223 Nabti, F.G., 1980: Manpower, education, and economic development in the Kingdom of Saudi Arabia. Ph.D., School of Education, Stanford University. XIII, 273 p. -Tü 17-

2224 Nader, A., 1978: Special education in the Kingdom of Saudi Arabia. N.p.

2225 Nafa, M.A., 1977: Saudi Arabia: company and business law. London: Arab Consultants. 441 p. -B 212-

2226 Nagel, T., 1976: König Faişal von Saudi-Arabien und die 'islamische Solidarität'. In: Orient (Opladen) 17(1): 52-71. -21, 1oo, H 223-

2227 Nah- und Mittelost-Verein e.V., o.J. (1969): Bericht über die Reise der 1. Deutschen Wirtschaftsdelegation nach Kuwait und Saudi-Arabien, 8.-2o. November 1968. Hamburg. 115 S. -21, 212, Gö 153-

2228 ——, 1976: Saudi-Arabien - Verstärkte Kooperation mit der Bundesrepublik. In: Übersee-Rundschau 28(3): 18, 2o-21. -24, 3o, 18o-

2229 —— (Hg.), o.J. (1981): Saudi-Arabien, Partner für die Bundesrepublik Deutschland. Hamburg. 6o S. -21, 2o6, H 3-

223o Nahari, A.M. al-, 1982: The national library: an analysis of the critical factors in promoting library and information services in developing countries: the case of Saudi Arabia. Ph.D., University of California, Los Angeles. 365 p.

2231 Naiem, A.M. el-, 198o: International trade and balance of payments adjustments: a case study of the Saudi external payments. Ph.D., University of Colorado, Boulder. 169 p.

2232 Nairab, M.M., 1978: Petroleum in Saudi-American relations: the formative period, 1932-1948. Ph.D., North Texas State University, Denton. 235 p.

2233 Najai, A.M., 1982: Television and youth in the Kingdom of Saudi Arabia: an empirical analysis of the uses of television among young Saudi Arabian viewers. Ph.D., University of Wisconsin, Madison. 229 p.

2234   Nakhleh, E.A., 1975a: Arab-American relations in the Persian
        Gulf. Foreign Affairs Studies 17; Washington: American
        Enterprise Institute for Public Policy Research. 82 p.
        -21-

2235   ——, 1975b: The United States and Saudi Arabia: a policy
        analysis. Foreign Affairs Studies 26; Washington: American Enterprise Institute for Public Policy Research. 69
        p.   -21, 24/213, H 223-

2236   Nallino, C.A., 1926: Sulle dottrine dei Wahhabiti. In: Oriente Moderno 6: 337-338.   -1a, 5, 21-

2237   ——, 1939: Raccolta di scritti, editi e inediti. Bd. 1:
        L'Arabia Sa'ūdiana (1938), a cura di M. Nallino. Roma:
        Istituto per l'Oriente. IV, 472 S.   -21-

2238   Nallino, M., 1940: Importante progetto di bonifica e di sistemazione idraulica di terre dell'Arabia centrale (Arabia Sa'ūdiana). In: Oriente Moderno 2o: 1-8.   -5, 7, 21-

2239   ——, 1941a: L'Arabia Saudiana. Roma: Reale Accademia d'Italia. 31 S.   -Biblioteca Nazionale Centrale, Roma-

2240   ——, 1941b: La politica estera dell'Arabia Saudiana. Firenze: Sansoni. 28 S.   -Biblioteca Nationale Centrale, Roma-

2241   ——, 1958: Intorno ai recenti mutamenti nell'Arabia Saudiana. In: Oriente Moderno 38: 293-299.   -7, 18, 21-

2242   ——, s.d. (1961): L'évolution sociale en Arabie Séoudienne.
        In: Colloque sur la Sociologie Musulmane. Actes, 11-14
        Septembre 1961. Correspondance d'Orient 5, Publications
        du Centre pour l'Étude des Problèmes du Monde Musulman
        Contemporain; Bruxelles s.d. (1961): 431-451.   -21, 38-

2243   ——, 1973: Momenti essenziali nella vita e nella carriera
        scientifica di mio padre. In: Levante 2o(1): 11-23.   -21-

2244   Narayanan, R., 1965/66: A review of oil contract negotiations by Saudi Arabia with ARAMCO. In: International
        Studies (New Delhi) 7: 568-588.   -1a, 7, 18-

2245   ——, 1971: U.S. postures in West Asia: defence arrangements
        with Saudi Arabia - a case study. In: M.S. Rajan (ed.):
        Studies in politics: national and international. New
        Delhi: Vikas Publications, 1971: 298-323.

2246 Naseer, A.A., 1983: Managerial attitudes and needs in developing countries: an empirical study of managerial views in the Kingdom of Saudi Arabia. Ph.D., Claremont Graduate School. 172 p.

2247 Nasr, K.S., 1979: Business laws and taxation in Saudi Arabia. Riyadh. XI, 498 p.   -Esb 2ooo-

2248 Nasri, H.Y., 1979: Ibn 'Abd al-Wahhab's philosophy of society: an alternative to the tribal mentality. Ph.D., Fordham University, New York. VIII, 189 p.

2249 Nassar, F.M. al-, 1982: Saudi Arabian educational mission to the United States: assessing perceptions of student satisfaction with services rendered. Ph.D., University of Oklahoma, Norman. 219 p.

225o Nasser, S.A., 1976: The importance of community development in the development program of the Southwest Region of Saudi Arabia. Ph.D., Michigan State University, East Lansing. 245 p.

2251 National Commercial Bank, 1976-: Annual report. 1.1975-; Jeddah. -H 3-

2252 Natto, I.A. & Mohammad S. Khan, 1976: Growth of elementary education in the Kingdom of Saudi Arabia 196o to 1975 (with projections up to 1985). Riyadh.

2253 Natural gas - valuable by-product. In: The Arab World 5.1959 (4/5): 14-15.   -2o6-

2254 Nawwab, I.I., P. Lunde, M. Amin & M.E. Jansen, 1974: The Hajj: a special issue. In: Aramco World Magazine 25(6): 2-45.   -24-

2255 Nazer, I.S., 1977: Doing business in Saudi Arabia. In: W.G. Wickersham & B.P. Fishburne, III (eds.): Current legal aspects of doing business in the Middle East: Saudi Arabia, Egypt and Iran. Chicago: Section of International Law, American Bar Association, 1977: 116-122.

2256 Nazer: Our aim is to diversify Saudi economy. In: The Arab Economist 14.1982(152): 1o-11.   -H 3-

2257 Neaim, H.A. al-, 198o: An analysis of the recruitment of foreign employees in the civil service of Saudi Arabia. Ph.D., North Texas State University, Denton. 211 p.

2258 Neetix, H.W., 1982: Staatliche Politik zur Anlage von Petrodollars: das Beispiel Saudi Arabien. Diss., Rechts- und Staatswissenschaftliche Fakultät, Universität Bonn. XI, 232 S.   -5, 21-

2259 Nehme, M.G., 1983: Saudi Arabia: political implications of the development plans. Ph.D., Rutgers University, The State University of New Jersey, New Brunswick. 232 p.

2260 Neimans, R. von, 1858: Das rothe Meer und die Küstenländer im Jahre 1857 in handelspolitischer Beziehung, beleuchtet nach eigener Anschauung und Forschung während der Monate Juni bis November 1857 an der Küste von Hedjaz. In: Zeitschrift der Deutschen Morgenländischen Gesellschaft 12: 391-441.   -19, 21, 291-

2261 Neue Bestimmungen über Eheschliessung von Ausländern in Saudi-Arabien. In: Das Standesamt 27.1974: 276.   -1a, 3o, 352-

2262 Neumann, R.G., 1982: The internal stability of Saudi Arabia. In: Vierteljahresberichte - Probleme der Entwicklungsländer 1982(9): 2o9-215.   -24, 2o6, H 3-

2263 New education for Saudi girls. In: The Arab World 1o.1964 (5): 6-7.   -2o6-

2264 New trap for Shedgum. In: Aramco World 18.1967(2): 1o-11. -24-

2265 Niblock, T., 1982: Introduction. - Social structure and the development of the Saudi Arabian political system. In: T. Niblock (ed.): State, society and economy in Saudi Arabia. London: Croom Helm; Exeter: Centre for Arab Gulf Studies, 1982: 11-22 resp. 75-1o5.   -21-

2266 —— (ed.), 1982: State, society and economy in Saudi Arabia. London: Croom Helm; Exeter: Centre for Arab Gulf Studies. 314 p.   -21-

2267 Nichols, R.L., D.E. McComb, N.A. Haddad & E.S. Murray, 1963: Studies on trachoma. II. Comparison of fluorescent antibody, Giemsa, and egg isolation methods for detection of trachoma virus in human conjunctival scrapings. In: The American Journal of Tropical Medicine and Hygiene 12: 223-229.   -16, 18, 1oo-

2268 Nichols, R.L., S.D. Bell, Jr., E.S. Murray, N.A. Haddad & A.
A. Bobb, 1966: Studies on trachoma. V. Clinical observations in a field trial of bivalent trachoma vaccine at three dosage levels in Saudi Arabia. In: The American Journal of Tropical Medicine and Hygiene 15: 639-647. -1oo, 188, 289-

2269 ——, A.A. Bobb, N.A. Haddad & D.E. McComb, 1967: Immunofluorescent studies of the microbiologic epidemiology of trachoma in Saudi Arabia. In: American Journal of Ophthalmology 63: 1372-14o8. -1a, 16, 25-

227o ——, S.D. Bell, Jr., N.A. Haddad & A.A. Bobb, 1969: Studies on trachoma. VI. Microbiological observations in a field trial in Saudi Arabia of bivalent trachoma vaccine at three dosage levels. In: The American Journal of Tropical Medicine and Hygiene 18: 723-73o. -1oo, 188, 289-

2271 Niebuhr, B.G., 1816: Carsten Niebuhrs Leben. In: Kieler Blätter 3: 1-86. -1a, 21-

2272 Niebuhr, C., 1772: Beschreibung von Arabien. Aus eigenen Beobachtungen und im Lande selbst gesammelten Nachrichten. Kopenhagen: Möller (Nachdruck: Graz: Akademische Druck- und Verlagsanstalt, 1969). XLVII, 431 S. -21, 24-

2273 ——, 1774: Reisebeschreibung nach Arabien und andern umliegenden Ländern. Bd. 1. Kopenhagen: Möller (Nachdruck: Graz: Akademische Druck- und Verlagsanstalt, 1968. Auszugsweise Ausgabe: Entdeckungen im Orient. Reise nach Arabien und anderen Ländern 1761-1767. Hgg. & bearb. von R. und E. Grün. Alte abenteuerliche Reise- und Entdeckungsberichte; Tübingen, Basel: Erdmann, 1973. Französischsprachige Ausgabe: Voyage en Arabie et en autres pays circonvoisins. T. 1. Amsterdam, Utrecht, 1776. Englische Ausgabe: Travels through Arabia and other countries in the East. Vol. 1. Edinburgh, 1792; reprint: Beirut: Librairie du Liban, 1969). XVI, 5o5 S. -21, 24, 212-

2274 Nieuwenhuijze, C.A.O. van, 1971: Sociology of the Middle East. A stocktaking and interpretation. Études Sociales, Économiques et Politiques du Moyen-Orient / Social, Economic and Political Studies of the Middle East 1; Leiden: Brill. XIV, 819 p. -21-

2275 Nifay, A.M. al-, 1981: An assessment to redesign in-service training programs for paraprofessionals employed in military hospitals under the jurisdiction of the Saudi Arabian Ministry of Defense. Ph.D., University of Pittsburgh. 846 p.

2276 Nimatallah, Y.A.-W., 1967: Coordination of monetary and fiscal policies in Saudi Arabia. Ph.D., University of Massachusetts, Amherst. 341 p.

2277 Nixon, R.W., 1954a: Date culture in Saudi Arabia. In: Date Growers Institute Report (31): 15-2o.

2278 ——, 1954b: Date varieties of the Eastern Province of Saudi Arabia in relation to cultural practices. USA Operations Mission to Saudi Arabia; Washington: Foreign Operations Administration. 38 p.

2279 Noaim, A., 1975/76: Production of alfalfa under irrigation in Saudi Arabia. M.Sc., School of Agriculture, University College of North Wales, Bangor.

2280 Noel-Brown, S.J., 1931: Trading in Arabia: possibilities and difficulties of trading in the Hijaz and Nejd. In: International Business Men's Magazine 1931(Oct.): 1-4.

2281 Nolde, E., 1895: Reise nach Innerarabien, Kurdistan und Armenien 1893. Braunschweig: Vieweg. XV, 272 S. -24-

2282 Nollet, R., 1978: Regard sur le clan des al-Saoud. In: L'Afrique et l'Asie Modernes (118): 17-28. -1a, 21, 2o6-

2283 Nolte, R.H., 1958: Faisal takes over in Saudi Arabia. In: The Reporter (New York) 18(May 1, 1958): 7-1o. -18o-

2284 ——, 1963: From nomad society to new nation: Saudi Arabia. In: K.H. Silvert (ed.): Expectant peoples: nationalism and development. New York: Random House, 1963: 77-94. -12-

2285 Noris, J., J. Schulze & J. Stöhr, 1979: Ansätze für eine regionale Strukturverbesserung der Eastern Province in Saudi-Arabien. Approach towards regional structural improvement of the Eastern Province, Saudi Arabia. Diplomarbeit, Städtebauliches Institut, Universität Stuttgart. 8, 176 S.

2286 Norton, M., 1972: Sayhat: the town that became a family. In: Aramco World Magazine 23(1): 2-5.   -24-

2287 Nota del Governo Arabo Sa'ūdiano ai Governi italiano e britannico a proposito dell'Accordo Italo-Britannico del 16 aprile 1938 e risposte dei due Governi. In: Oriente Moderno 19.1939: 3o4-3o6.   -5, 7, 21-

2288 Novik, N., 1981: Weapons to Riyadh. U.S. policy and regional security. CSS Memorandum 4; Tel Aviv: Center for Strategic Studies, Tel Aviv University. 39 p.   -H 223-

2289 Nunè, E., 1941: L'Inghilterra nella Penisola Arabica. In: Oriente Moderno 21: 2o9-232.   -7, 18, 21-

229o Nursing comes naturally. In: Aramco World 13.1962(2): 3-6. -H 223-

2291 Nyrop, R.F., B.L. Benderly, L.N. Carter, D.R. Eglin & R.A. Kirchner, ³1977: Area handbook for Saudi Arabia. DA Pam. 55o-51; Washington: US Government Printing Office. XIV, 389 p.   -12, 24, 352-

2292 ——, 1982: General character of the society. In: R.F. Nyrop et al.: Saudi Arabia: a country study. Area Handbook Series DA Pam. 55o-51; Washington: US Government Printing Office, 1982: 1-9.   -H 223-

2293 —— et al., 1982: Saudi Arabia: a country study. Area Handbook Series DA Pam. 55o-51; Washington: US Government Printing Office. XIV, 39o p.   -H 223-

O

2294 Oasis fruit. In: Aramco World 13.1962(3): 18-2o.   -H 223-

2295 Obaid, A.S. al-, 1979: Human resources development in Saudi Arabia: case of technical manpower programs and needs. Ed.D., Oklahoma State University, Stillwater. 167 p.

2296 Oberkommando der Kriegsmarine, ³1937: Handbuch für das Rote Meer und den Golf von Aden. Seehandbücher 2o34; Berlin: Mittler. XXIII, 6o1 S.   -H 2-

2297 ——, ³1942: Handbuch für den Persischen Golf. Seehandbücher 2o35; Berlin: Mittler. XXIV, 3o9 S.   -1a, H 2-

2298 Obojski, R., 1979: Stamps and the history of the Hijaz. In: Aramco World Magazine 3o(5): 6-7. -21, 24-

2299 Ochel, W., 1978: Die Industrialisierung der arabischen OPEC-Länder und des Iran. Erdöl und Erdgas im Industrialisierungsprozess. Ifo-Studien zur Entwicklungsforschung 5; München, London: Weltforum. XIV, 192 S. -21, 212-

2300 ——, 1979: Länderstudie Saudi-Arabien. Strukturveränderungen der deutschen Wirtschaft, Die Industrialisierung der Entwicklungsländer und ihre Rückwirkungen auf die deutsche Wirtschaft - Perspektiven bis 199o 1o; Frankfurt/M.: Rationalisierungs-Kuratorium der Deutschen Wirtschaft. XVI, 115 S. -21, 2o6, H 3-

2301 Ochsenwald, W.L., 1973a: The financing of the Hijaz Railroad. In: Die Welt des Islams N.S. 14: 129-149. -16, 21, 2o6-

2302 ——, 1973b: Opposition to political centralization in South Jordan and the Hijaz, 19oo-1914. In: The Muslim World 63: 297-3o6. -21, 291, 352-

2303 ——, 1975: Ottoman subsidies to the Hijaz, 1877-1886. In: International Journal of Middle East Studies 6: 3oo-3o7. -1a, 21, 212-

2304 ——, 1976: A modern waqf: the Hijaz Railway, 19oo-48. In: Arabian Studies 3: 1-12. -12, 21, 212-

2305 ——, 1977a: The financial basis of Ottoman rule in the Hijaz, 184o-1877. In: W.W. Haddad & W.L. Ochsenwald (eds.): Nationalism in a non-national state: the dissolution of the Ottoman Empire. Columbus: Ohio State University Press, 1977: 129-149. -12, 352-

2306 ——, 1977b: The Jidda massacre of 1858. In: Middle Eastern Studies 13: 314-326. -1a, 21, 3o-

2307 ——, 198oa: The Hijaz Railroad. Charlottesville: The University Press of Virginia (zugleich: The Hijaz Railroad: a study in Ottoman political capacity and autonomy. Ph.D., University of Chicago, 1972). XVI, 169 p. -21-

2308 ——, 198ob: Muslim-European conflict in the Hijaz: the slave trade controversy, 184o-1895. In: Middle Eastern Studies 16: 115-126. -1a, 21, 3o-

2309　Ochsenwald, W.L., 1981: Saudi Arabia and the Islamic revival. In: International Journal of Middle East Studies 13: 271-286.　-12, 18, 21-

2310　———, 1982: The commercial history of the Hijaz Vilayet, 1840-1908. In: Arabian Studies 6: 57-76.　-21, Tü 17-

2311　Ocwieja, F.A., 1975: Marketing in Saudi Arabia. Overseas Business Reports 57; Washington. 25 p.　-2o6-

2312　Odell, R.M., 1913: Cotton goods in the Red Sea markets. US Department of Commerce 71; Washington: US Government Printing Office. 64 p.　-1a-

2313　O'Donnell, P.D., 1980: The Saudi story. In: Irish Arab News 6(1): 15-18.

2314　OECD, 1983: Aid from OPEC countries. Paris. 163 p.

2315　Önder, Z., 1980a: Die Entscheidungsprozesse im saudi-arabischen System. Die Rolle der 'Königlichen Familie'. In: Die Neue Gesellschaft 27: 349-353.　-1a, 12, 2o6-

2316　———, 1980b: Saudi-Arabien. Zwischen islamischer Ideologie und westlicher Ökonomie. Stuttgart: Klett-Cotta. 4o5 S.　-21, 212-

2317　Office Belge du Commerce Extérieur, 1966: L'Arabie Séoudite et les émirats du Golfe Persique. Mémento à l'usage de l'homme d'affaires belge. Informations du Commerce Extérieur, Suppl. D, 2; Bruxelles. 38 p.　-H 3-

2318　———, 1978: Arabie Séoudite. Collection 'Un Marché' 46; Bruxelles. 69 p.　-21, 2o6, H 3-

2319　O'Hali, A.A., 1974: Saudi Arabia in the United Nations General Assembly: 1946-1970. Ph.D., Graduate Faculty of Government, Claremont Graduate School. VI, 494 p.　-21-

2320　Oil show. In: Aramco World 14.1963(9): 2-7.　-H 223-

2321　Oil wells amid desert dunes. In: Standard (of California) Oil Bulletin 1946(Autumn): 4-5.

2322　Olsen, G.R., 1981: Saudi Arabia in the 1970s: state, oil and development policy. In: European Consortium for Political Research (ed.): The Middle East in the 1980s. Lancaster. 38 p.　-H 223-

2323 Omair, S.A., 1976: A study of the association between absorptive capacity and development strategy in Saudi Arabia. Ph.D., Texas Tech University, Lubbock. 292 p.

2324 Omran, A.N. al-, 1981: Perceptions of present and preferred student personnel services by administrators, faculty, and students in junior colleges in Saudi Arabia. Ed.D., Indiana University, Bloomington. 178 p.

2325 OPEC, Information Department, n.d.(1976): Selected documents of the international petroleum industry: Saudi Arabia, pre-1966. Vienna. 227 p. -21, 2o6-

2326 Opening up the Saudi west. In: Middle East International 1978(88): 2o-21. -12, 3o-

2327 Oppenheim, M. von, unter Mitbearbeitung von E. Bräunlich & W. Caskel, 1939-67: Die Beduinen. 4 Bde. Leipzig / Wiesbaden: Harrassowitz. -21, 24, 1o7-

2328 Os, J. van, 1967: Desert road. In: Aramco World 18(2): 22-29. -24-

2329 Osaimi, M.S.M.J. al-, 1981: The feasibility of modifying the grade-organization of the educational system in Saudi Arabia. Ph.D., Indiana University, Bloomington. 132 p.

233o Osman (auch: Othman), O.A., 1978: Formalism vs. realism: the Saudi Arabian experience with position classification. In: Public Personnel Management 7: 177-18o. -12, 291, 352-

2331 ——, 1979: Saudi Arabia: an unprecedented growth of wealth with an unparalleled growth of bureaucracy. In: International Review of Administrative Sciences 45: 234-24o. -188, 465-

2332 Owens, M.V., 1962: A health text for the fifth and sixth grades in the Saudi Arab government schools. Ed.D., Columbia University, New York. 217 p.

2333 Ownby, P., 1969: The development of human resources. In: T. C. Young (ed.): Middle East focus: the Persian Gulf. Princeton: Princeton University Press, 1969: 111-118.

2334 Ozoling, V., 1981: The prospects for the development of the social and economic structure of Saudi Arabia. In: Fourth

International Symposium, 29-31 March 1981: The Future of the Arab Gulf and the Strategy of Joint Arab Action. Basrah: Centre for Arab Gulf Studies, University of Basrah, 1981. 23 p.   -H 223-

2335 Ozoling, V., 1982: Oil and dollars of Arabian monarchies. In: Asia and Africa Today 1982(5): 49-51.   -3o, H 221-

P

2336 Page, S., 1971: The USSR and Arabia: the development of Soviet policies and attitudes towards the countries of the Arabian peninsula, 1955-197o. London: Central Asian Research Centre in association with Canadian Institute of International Affairs. 149 p.   -212-

2337 Palgrave, W.G., 1864a: Notes of a journey from Gaza, through the interior of Arabia, to El Khatif on the Persian Gulf, and thence to Omàn, in 1862-63. In: Proceedings of the Royal Geographical Society 8: 63-82.   -1a, 21, 9o-

2338 ——, 1864b: Observations made in Central, Eastern, and Southern Arabia during a journey through that country in 1862 and 1863. In: The Journal of the Royal Geographical Society 34: 111-154.   -1a, 21, 291-

2339 ——, 1865: Herrn Gifford Palgrave's Bericht über seine Reise durch das Innere Arabiens in den Jahren 1862 und 1863. In: Zeitschrift für allgemeine Erdkunde N.F. 18: 219-227.   -21, 24, 25-

234o ——, 1867, 1868: Reise in Arabien. 2 Bde. Leipzig: Dyk (Originalausgabe: Narrative of a year's journey through Central and Eastern Arabia. 2 vols. London, Cambridge/England: Macmillan, 1865; reprint: Farnborough: Gregg, 1969). VI, 354 bzw. 292 S.   -7, 21, 212-

2341 Pape, H., 1977: Er Riad: Stadtgeographie und Stadtkartographie der Hauptstadt Saudi-Arabiens. Bochumer Geographische Arbeiten, Sonderreihe 7; Paderborn: Schöningh (zugleich: Diss., Abteilung für Geowissenschaften, Universität Bochum, 1975). 1o1 S.   -21, 24, 93-

2342 Paréja, F., 1946: Le pèlerinage de La Mecque. In: En Terre d'Islam III, 21: 235-25o.   -21, 3o-

2343 Paréja, F., 1948a: Le pèlerinage musulman, des origines à nos jours. In: En Terre d'Islam IV, 23: 3-11.   -21, 3o-

2344 ——, 1948b: Le pèlerinage musulman, routes et organisation. In: En Terre d'Islam IV, 23: 166-177.   -21, 3o-

2345 Paret, R., 194o: Ḥāfiẓ Wahba's Arabienbuch. In: Die Welt des Islams 22: 67-1o1.   -16, 21, 2o6-

2346 Park, T.W. & M.D. Ward, 1979: Petroleum-related foreign policy: analytic and empirical analyses of Iranian and Saudi behavior (1948-1974). In: The Journal of Conflict Resolution 23: 481-512.   -24/213, 25, 18o-

2347 Parker, J.B., Jr., 1975: U.S. agricultural exports to the Arabian Peninsula soar. In: Foreign Agricultural Trade of the United States 1975(Oct.): 54-81.   -1a, 2o6-

2348 ——, 1976: Saudi Arabia - a flourishing market for U.S. farm products. In: Foreign Agricultural Trade of the United States 1976(July): 4-2o.   -1a, 2o6-

2349 ——, 1977a: Saudi Arabia: an expanding market for U.S. agricultural exports. In: Foreign Agricultural Trade of the United States 1977(May): 4-19.   -2o6-

235o ——, 1977b: The Mideast: again a growing market for U.S. farm products. In: Foreign Agricultural Trade of the United States 1977(Oct.): 4o-52.   -2o6-

2351 ——, 1979: Saudi Arabia: farm imports still soaring - $3 billion seen for 1979. In: Foreign Agriculture 17(19): 3o-31.   -1a, 98, 2o6-

2352 ——, 198o: Saudi farm imports head toward $4.5 billion. In: Foreign Agriculture 18(9): 14-16.   -1a, 98, 2o6-

2353 ——, 1981: Saudi Arabia now ranks among top three importers of poultry meat. In: Foreign Agriculture 19(8): 14-15.   -12, 98, 2o6-

2354 Parry, M.S., 197o: Research on date palms in Saudi Arabia. Dhahran: ARAMCO. 23 p.

2355 Parssinen, C., 198o: The changing role of women. In: W.A. Beling (ed.): King Faisal and the modernisation of Saudi Arabia. London: Croom Helm; Boulder: Westview Press, 198o: 145-17o.   -21, H 223-

2356 Pascal, A., E.M. Kennedy & S.J. Rosen, with P. Jabber et al., 1979: Men and arms in the Middle East: the human factor in military modernization. A report prepared for the Director of Net Assessment, Office of the Secretary of Defense. Rand Corporation Report R-2460-NA; Santa Monica: Rand Corporation. XV, 54 p.

2357 Passama, J., 1843: Notice géographique sur quelques parties de l'Yemen. In: Bulletin de la Société de Géographie (Paris) II, 19: 219-236. -291-

2358 Pastner, C.M., 1978: Englishmen in Arabia: encounters with Middle Eastern women. In: Signs 4: 3o9-323. -21, 24/ 213, 25-

2359 Patai, R., 1969: Golden river to golden road: society, culture, and change in the Middle East. 3rd, enl. ed. Philadelphia: University of Pennsylvania Press (^{1}1962, ^{2}1967). 560 p. -21-

2360 Pauling, N.G., 1961: Experience with an industrial research program in the social sciences. In: The Journal of Business (Chicago) 34: 14o-152. -38, 18o, 352-

2361 ——, 1964: Labor separations in an underdeveloped area: a case study of worker adjustment to change. In: The American Journal of Economics and Sociology 23: 419-434. -7, 18, 18o-

2362 Pechel, J., 1961: Zwischen Mekka und Teheran. Herrenalb: Erdmann. 319 S. -21, 24-

2363 Peck, M.C., 1970: Saudi Arabia in United States foreign policy to 1958: a study in the sources and determinants of American policy. Ph.D., Fletcher School of Law and Diplomacy, Tufts University, Medford.

2364 ——, 1972: Saudi Arabia's oil wealth: a two-edged sword. In: New Middle East (4o): 5-7. -1a, 352-

2365 ——, 1980: The Saudi-American relationship and King Faisal. In: W.A. Beling (ed.): King Faisal and the modernisation of Saudi Arabia. London: Croom Helm; Boulder: Westview Press, 1980: 230-247. -21, H 223-

2366 ——, 1982: The Arab states of the Gulf: a political and security assessment. In: R.A. Kilmarx & Y. Alexander (eds.):

Business and the Middle East: threats and prospects. Pergamon Policy Studies on Business and Economics; New York, Oxford, Toronto, Sydney, Paris, Frankfurt/M.: Pergamon Press, 1982: 91-116. -21-

2367 Pelly, L., 1865a: Visit to the Wahabee capital of Central Arabia. In: Proceedings of the Royal Geographical Society 9: 293-296. -1a, 21, 9o-

2368 ——, 1865b: A visit to the Wahabee capital, Central Arabia. In: The Journal of the Royal Geographical Society 35: 169-191. -1a, 7, 21-

2369 ——, 1978: Report on a journey to Riyadh in central Arabia (1865). Arabia Past and Present 6; Cambridge/England, New York: The Oleander Press; Naples: The Falcon Press (Originalausgabe: Report on a journey to the Wahabee capital of Riyadh in Central Arabia. Bombay: Bombay Government, 1866). XIV, 1oo p. -12, 21, Tü 17-

237o Pendleton, M., D.L. Davies, M.S. Davies & F.O. Snodgrass, ^{3}1980: The green book: guide for living in Saudi Arabia. Washington: Middle East Editorial Associates (^{1}1976, ^{2}1978). VII, 178 p. -21-

2371 Penisola Arabica: Arabia Saudita, Bahrein, Emirati Arabi Uniti, Kuwait, Oman, Qatar, Yemen, Yemen (R.D.P.), a cura di G. Schiavone. Guide per l'Esportazione 5/6; Roma: SIPI, 1978.

2372 Penrose, E., 1972: Oil and state in Arabia. In: D. Hopwood (ed.): The Arabian Peninsula: society and politics. Studies on Modern Asia and Africa 8; London: Allen & Unwin, 1972: 271-285 (reprinted from: E. Penrose: The growth of firms, Middle East oil and other essays. London: Cass, 1971: 281-295). -21, 24-

2373 Peppelenbosch, P.G.N., 1968: Nomadism on the Arabian Peninsula: a general appraisal. In: Tijdschrift voor Economische en Sociale Geografie 59: 335-346. -12, 21, 291-

2374 Perera, J., 1977: Three Arab states opt to 'go nuclear'. In: The Middle East (London) (34): 31-33. -12, 3o, H 223-

2375 ——, 1978: Together against the red peril: Iran and Saudi Arabia rivals for superpower role. In: The Middle East (London) (43): 16-31. -12, 3o-

2376 Peretz, D., 1977: Foreign policies of the Persian Gulf states. Part 3. In: New Outlook 2o(3): 48-53. -21-

2377 Pershits, A.I., 1966: The economic life of the nomads of Saudi Arabia. Reprinted in: C.P. Issawi (ed.): The economic history of the Middle East, 18oo-1914. A book of readings. Chicago, London: The University of Chicago Press, 1966: 343-349. -21-

2378 Pesce, A., 1972: Colours of the Arab fatherland. Riyadh (reprint: Arabia Past and Present 1; Cambridge/England, New York: The Oleander Press; Naples: The Falcon Press, 1978). 144 p. -21-

2379 ——, 1974: Jiddah: portrait of an Arabian city. Arabia Past and Present 2; Cambridge/England, New York: The Oleander Press; Naples: The Falcon Press (^{2}1977). IX, 239 p. -21, 212, H 223-

2380 Pesenti, G., 1917: La situazione politico-militare nell'Arabia e gli interessi dell'Italia. In: Rivista Coloniale 12(4): 165-171.

2381 Petit-Laurent, J., 1976: Arabie Séoudite: un plan difficile à réaliser. In: Industries et Travaux d'Outremer 24: 169-172. -1a, 89, 2o6-

2382 Peucker, K., 1910: Musil's explorations in northern Arabia, 19o8-9. In: The Geographical Journal 35: 579-581. -21, 291, 352-

2383 Peursem, G.D. van, 1936: Guests of Ibn Sa'ud. In: The Moslem World 26: 113-118. -21, 291, 352-

2384 ——, 1948: The Arabian Mission and Saudi Arabia. In: The Muslim World 38: 6-1o. -21, 291, 352-

2385 ——, 1965: Homage to Dr. Paul Wilberforce Harrison. In: The Muslim World 55: 17o-171. -21, 291, 352-

2386 Philby, H.St.J.B., 192oa: Southern Najd. In: The Geographical Journal 55: 161-191. -21, 291, 352-

2387 ——, 192ob: Across Arabia: from the Persian Gulf to the Red Sea. In: The Geographical Journal 56: 446-468. -21, 291, 352-

2388 ——, 192oc: The highways of Central Arabia. In: Journal of the Central Asian Society 7: 112-125. -21, 9o, 291-

2389 Philby, H.St.J.B., 1923: Jauf and the north Arabian desert. In: The Geographical Journal 62: 241-259.   -21, 291, 352-

2390 ——, 1925a: Das geheimnisvolle Arabien; Entdeckungen und Abenteuer. 2 Bde. Leipzig: Brockhaus (Originalausgabe: The heart of Arabia. A record of travel and exploration. 2 vols. London: Constable, 1922). 365 bzw. 320 S.   -21, 24-

2391 ——, 1925b: The Dead Sea to 'Aqaba. In: The Geographical Journal 66: 134-160.   -21, 291, 352-

2392 ——, 1925c: The recent history of the Hijaz. In: Journal of the Central Asian Society 12: 333-348.   -21, 90, 291-

2393 ——, 1926: The triumph of the Wahhabis. In: Journal of the Central Asian Society 13: 293-319.   -21, 90, 291-

2394 ——, 1928a: Arabia of the Wahhabis. London: Constable (reprints: The Middle East Collection; New York: Arno Press, 1973; London: Cass, 1977). XIV, 422 p.   -21, 24/213, Tü 17-

2395 ——, 1928b: The trouble in Arabia. In: The Contemporary Review 135: 705-715.   -25, 180, 465-

2396 ——, 1929a: Arabia - 1926-1929: three years of Wahhabi rule. In: The Contemporary Review 137: 714-719.   -25, 180, 465-

2397 ——, 1929b: A survey of Wahhabi Arabia, 1929: In: Journal of the Central Asian Society 16: 468-481.   -21, 90, 291-

2398 ——, 1930: Arabia. The Modern World, a Survey of Historical Forces; London: Benn. XIX, 387 p.   -24/213-

2399 ——, 1933a: Rub' al-Khali: an account of exploration in the Great South Desert of Arabia under the auspices and patronage of His Majesty 'Abdul 'Aziz ibn Sa'ud, King of the Hejaz and Nejd and its Dependencies. In: The Geographical Journal 81: 1-26.   -21, 291, 352-

2400 ——, 1933b: The Empty Quarter, being a description of the Great South Desert of Arabia known as Rub' al Khali. London: Constable; New York: Holt. XXIV, 433 p.   -21-

2401 ——, 1933c: Mecca and Madina. In: Journal of the Royal Central Asian Society 20: 504-518.   -21, 90, 291-

2402 Philby, H.St.J.B., 1935: Arabia to-day. In: International Affairs (London) 14: 619-634.   -18o, 291, 352-

2403 ——, 1936: Pax Wahhabica. In: The English Review (London) 62: 3o9-322.   -21, 18o, 352-

2404 ——, 1937: African contacts with Arabia. In: Journal of the Royal Society of Arts 86: 9o-1o2.   -1a-

2405 ——, 1938a: The land of Sheba. In: The Geographical Journal 92: 1-21, 1o7-132.   -21, 291, 352-

2406 ——, 1938b: Intervista con Sua Maestà il Re Ibn Sa'ūd sulla questione della Palestina. In: Oriente Moderno 18: 583-587.   -1a, 5, 21-

2407 ——, 1939: Sheba's daughters, being a record of travel in southern Arabia. London: Methuen. XIX, 485 p.   -21-

2408 ——, 1943: Halévy in the Yaman. In: The Geographical Journal 1o2: 116-124.   -21, 291, 352-

2409 ——, 1946: A pilgrim in Arabia. London: Hale ([1]1943). 198 p.   -21-

2410 ——, 1947: Palgrave in Arabia. In: The Geographical Journal 1o9: 282-285.   -21, 291, 352-

2411 ——, 1948: Arabian days: an autobiography. London: Hale. XVI, 336 p.   -21, 212-

2412 ——, 1949: Two notes from Central Arabia. In: The Geographical Journal 113: 86-93.   -21, 291, 352-

2413 ——, 195oa: The golden jubilee in Sa'udi Arabia. In: Journal of the Royal Central Asian Society 37: 112-123.   -21, 9o, 291-

2414 ——, 195ob: Motor tracks and Sabaean inscriptions in Najd. In: The Geographical Journal 116: 211-215.   -21, 291, 352-

2415 ——, 1952a: Arabian highlands. Ithaca: Cornell University Press (reprint: The Middle East in the Twentieth Century; New York: Da Capo Press, 1976). XVI, 771 p.   -7, 21, Tü 17-

2416 ——, 1952b: Arabian jubilee. London: Hale. XIV, 28o p.   -7, 21, 24-

2417 Philby, H.St.J.B., 1954: The new reign in Sa'udi Arabia. In: Foreign Affairs (New York) 32: 446-458.   -21, 24, 25-

2418 ——, 1955a: Sa'udi Arabia. Nations of the Modern World; London: Benn; New York: Praeger (reprints: Arab Background Series; Beirut: Librairie du Liban, 1968. World Affairs: National and International Viewpoints; New York: Arno Press, 1972). XIX, 393 p.   -12, 21, 212-

2419 ——, 1955b: The land of Midian. In: The Middle East Journal 9: 116-129.   -12, 21, 206-

2420 ——, 1957a: Forty years in the wilderness. London: Hale. XVI, 272 p.   -24-

2421 ——, 1957b: The land of Midian. London: Benn. XI, 286 p. -12, 21-

2422 ——, 1958a: Arabia in retrospect. In: Middle East Forum 33(1): 14-17.   -188/812-

2423 ——, 1958b: Saudi Arabia: the new Statute of the Council of Ministers. In: The Middle East Journal 12: 318-323.   -12, 21, 25-

2424 ——, 1958c: The exploration of Arabia. In: The Geographical Journal 124: 535-540.   -21, 291, 352-

2425 ——, 1959: Riyadh: ancient and modern. In: The Middle East Journal 13: 129-141.   -12, 21, 25-

2426 ——, 1964: Arabian oil ventures. Washington: The Middle East Institute. XIII, 134 p.   -21, H 3-

2427 ——, 1981: The Queen of Sheba. London, Melbourne, New York: Quartet Books. 141 p.   -21-

2428 Philipp, H.-J., 1976: Geschichte und Entwicklung der Oase al-Hasa (Saudi-Arabien). Bd. 1: Historischer Verlauf und traditionelles Bild. Sozialökonomische Schriften zur Agrarentwicklung 23; Saarbrücken: Breitenbach. VIII, 362 S.   -12, 21, 100-

2429 ——, 1979: Modernes Be- und Entwässerungssystem in einer Grossoase. Terra, Diareihen Geographie; Stuttgart: Klett. 12 Dias mit Beiheft, 8 S.

2430 ——, 1980: Die Dattelpalme und ihre Verwertung. Terra, Diareihen Geographie; Stuttgart: Klett. 12 Dias mit Beiheft, 12 S.

2431 Philipp, H.-J., 1983: Arnold Heims erfolglose Erdölsuche und erfolgreiche Wassersuche 1924 im nordöstlichen Arabien. In: Vierteljahrsschrift der Naturforschenden Gesellschaft in Zürich 128: 43-73. -16, 24, 25-

2432 Philipp Holzmann AG, 1974: Ausbildungszentrum Tabuk in Saudi-Arabien. Technischer Bericht (englischsprachige Ausgabe: Training Center Tabuk in Saudi Arabia. Technical Report); Frankfurt/M. 29 S.

2433 ——, 1978a: Fünf schlüsselfertige Krankenhäuser in Saudi Arabien (englischsprachige Ausgabe: Five turnkey hospitals in Saudi Arabia). Frankfurt/M. O.Pag.

2434 ——, 1978b: Hafen Dammam, Saudi Arabien (englischsprachige Ausgabe: Port of Dammam, Saudi Arabia). Frankfurt/M. O.Pag.

2435 ——, 1980: Medizinisches Zentrum in Riyadh, Saudi-Arabien. Schlüsselfertige Planung und Ausführung. Technischer Bericht (englischsprachige Ausgabe: Medical Center in Riyadh, Saudi Arabia. Turnkey planning and construction. Technical Report); Frankfurt/M. 15 S.

2436 ——, 1981a: Sportzentrum in Tabuk, Saudi Arabien. Schlüsselfertige Ausführung (englischsprachige Ausgabe: Sports center in Tabuk, Saudi Arabia. A turnkey project). Frankfurt/M. 8 S.

2437 ——, 1981b: Auditorium in Tabuk, Saudi Arabien. Schlüsselfertige Ausführung (englischsprachige Ausgabe: Auditorium in Tabuk, Saudi Arabia. A turnkey project). Frankfurt/M. O.Pag.

2438 ——, 1981c: Saudi Fund for Development, Hauptverwaltung in Riyadh. Schlüsselfertige Ausführung (englischsprachige Ausgabe: Saudi Fund for Development, administrative headquarters in Riyadh. A turnkey project). Frankfurt/M. 8 S.

2439 ——, 1981d: Residenz für den Emir von Tabuk/Saudi Arabien. Planung und schlüsselfertige Ausführung (englischsprachige Ausgabe: Residence for the Emir of Tabuk, Saudi Arabia. Planning and turnkey construction). Frankfurt/M. O.Pag.

2440 ——, 1982a: Verwaltungsgebäude für das medizinische Zentrum in Riyadh. Planung und schlüsselfertige Ausführung (englischsprachige Ausgabe: Administrative building for Medical Center in Riyadh. Planning and turnkey construction). Frankfurt/M. O.Pag.

2441 Philipp Holzmann AG, 1982b: Recreation-Center im medizinischen Zentrum in Riyadh. Planung und schlüsselfertige Ausführung (englischsprachige Ausgabe: The recreation center at the Riyadh Medical Center. Planning and turnkey construction). Frankfurt/M. 12 S.

2442 ——, 1982c: Schlüsselfertiger Bau einer Wohnstadt in Saudi-Arabien. Technischer Bericht (englischsprachige Ausgabe: Turnkey construction of a residential estate in Saudi Arabia. Technical Report); Frankfurt/M. 25 S. -H 3-

2443 Phoenix (pseud.), 1930: A brief outline of the Wahabi movement. In: Journal of the Central Asian Society 17: 401-416. -21, 90, 291-

2444 Pieper, W., 1923: Der Pariastamm der Ṣlêb. In: Le Monde Oriental 17: 1-75. -16, 21, 25-

2445 ——, 1934: Ṣulaib. In: EI[1] 4: 552-557. -21, 24-

2446 Pierre, A.J., 1982: The global politics of arms sales. A Council on Foreign Relations Book; Princeton: Princeton University Press. XVI, 352 p. -21-

**** Pilar Serrano de Lababidy, M. del: s. Serrano de Lababidy, M. del Pilar.

2447 Pilgrims' progress to Mecca. In: The National Geographic Magazine 72.1937: 627-642. -21, 24, 93-

2448 Pinto, O., 1935: Il viaggio a Hail di Anna Blunt e Gertrude Bell. In: Rivista Geografica Italiana 42: 15-23. -30, 206-

2449 Pirenne, J., 1958: À la découverte de l'Arabie. Cinq siècles de science et d'aventure. Le Livre Contemporain; Paris: Amiot-Dumont. 327 p. -21, 24, Tü 17-

2450 Piscatori, J.P., 1976: Islam and the international legal order: the case of Saudi Arabia. Ph.D., University of Virginia, Charlottesville. 433 p.

2451 ——, 1980: The roles of Islam in Saudi Arabia's political development. In: J.L. Esposito (ed.): Islam and development: religion and sociopolitical change. Contemporary Issues in the Middle East; Syracuse: Syracuse University Press, 1980: 123-138.   -21-

2452 Pivka, O. von, 1979: Armies of the Middle East. Cambridge/ England: Stephens. 168 p.   -21, 24/213-

2453 Planagan, A.J., 1970: Establishing a business in Saudi Arabia. Overseas Business Reports; Washington: Bureau of International Commerce, US Department of Commerce. 13 p. -206-

2454 ——, 1971: Basic data on the economy of Saudi Arabia. Overseas Business Reports 25; Washington: Bureau of International Commerce, US Department of Commerce. 13 p.   -206-

2455 Plans set for King Saud University. In: The Arab World 7.1961(5): 8-9.   -206-

**** Plant, M. & D. Cotton: authors of no. 2104.

2456 Plas, C.O. van der, 1953: Report to the Government of Saudi Arabia on the state of agriculture in the South Tihama area. Report 80; Rome: FAO.

2457 Plass, J.B., 1966: England zwischen Russland und Deutschland. Der Persische Golf in der britischen Vorkriegspolitik, 1899-1907, dargestellt nach englischem Archivmaterial. Schriftenreihe des Instituts für Auswärtige Politik 3; Hamburg: Hamburger Gesellschaft für Völkerrecht und Auswärtige Politik. VIII, 507 S.   -21, 24/213-

2458 Plüddemann, M., 1901: Kohlenstationen und die Farisan-Inseln. In: Asien (Berlin) 1(2): 31-33.   -1a, 90-

2459 Pönicke, H., 1956: Heinrich August Meissner-Pascha und der Bau der Hedschas- und Bagdadbahn. In: Die Welt als Geschichte 16: 196-210.   -16, 21, 30-

2460 ——, 1958: Die Hedschas- und Bagdadbahn, erbaut von Heinrich August Meissner-Pascha. Beiträge zur Technikgeschichte; Düsseldorf: VDI. 35 S.   -212-

2461 Point Four aid to the Middle East 1951-1953. In: Middle Eastern Affairs 4.1953: 62-67.   -1a, 21, 180-

2462 Polk, W.R. & W.J. Mares, 1973: Passing brave. New York: Knopf. 206 p.

2463   Pollitzer, R., 1959: Cholera. Geneva: WHO. 1o19 p.

2464   Pollog, C.H., 1934: Das 'Leere Viertel' von Arabien. In: Geographische Zeitschrift 4o: 41-49, 9o-99, 134-141. -16, 21, 24-

2465   Porath, Y., 1972: The Palestinians and the negotiations for the British-Hijazi Treaty, 192o-1925. In: Asian and African Studies 8: 2o-48.   -21, 3o, 2o6-

2466   Portrait of an editor. In: Aramco World 15.1964(1): 3-7. -H 223-

2467   Prax, 1841a: Un voyage de Suez à Médine. In: Bulletin de la Société de Géographie (Paris) II, 15: 129-152.   -4, 291-

2468   ——, 1841b: Notice sur la Mekke et sur les femmes musulmanes. In: Bulletin de la Société de Géographie (Paris) II, 15: 245-259.   -4, 291-

2469   Preece, R.M., 1979: The future role of Saudi Arabia. In: The U.S. role in a changing world political economy. Major issues for the 96th Congress. Washington, 1979: 526-539. -2o6-

247o   Presley, J.R., 198o: Saudi Arabia: a decade of economic progress. In: The Three Banks Review (127): 25-4o.   -38, 2o6, H 3-

2471   —— & M. Kebell, 1982: Saudi Arabia's financial system. In: The Banker 132(672): 43-47.   -1oo, 18o, H 3-

2472   ——, 1983: Trade and foreign aid: the Saudi Arabian experience. In: The Arab Gulf Journal 3(1): 61-73.   -21-

2473   Prest, M., 1981: Saudi Arabia faces the challenge of change. In: Middle East International (164): 12.   -3o-

2474   Price Waterhouse & Company, 1975: Doing business in Saudi Arabia. Price Waterhouse Information Guide; New York. 62 p

2475   Prisse d'Avennes, E., 19o9: Les Wahhabis. In: Société de Géographie de l'Est, Bulletin Trimestriel 3o: 41-47. -1a, 291-

2476   Pritzke, H., ^{4}1959: Nach Hause kommst du nie... Abenteuer des Lebens; Wien: Ullstein (^{1}1956) (amerikanische Ausgabe Bedouin doctor: my adventurous years with the Arabs. New York: Dutton, 1957). 325 S.   -Stadtbücherei Stuttgart-

2477   Problème du logement dans le monde Arabe. In: L'Économie des
       Pays Arabes 18.1975(214): 12-16. -2o6-

2478   Problems and progress in Saudi Arabia. In: Petroleum Press
       Service 21.1954: 288-291. -18o, 2o6-

2479   Prochazka, T., Jr., 1977: The architecture of the Saudi Arabian south-west. In: Proceedings of the Seminar for Arabian Studies 7: 12o-133. -1a, 21-

248o   Pröbster, E., 1935: Die Wahhabiten und der Maġrib. In: Islamica 7: 65-112. -16, 21, 291-

2481   Professional Business Reports on Market Development N.V.,
       1979: Der Saudi-Markt für Industrie und Bauwesen. Eine
       kommerzielle Analyse. Research for Industry and Construction; Königstein/Ts. O.Pag. -H 3-

2482   Programme d'industrialisation en Arabie Séoudite. In: Syrie
       et Monde Arabe 25.1978(298): 12-25. -2o6, H 3, H 223-

2483   Projets séoudiens dans la Mer Rouge. In: L'Économiste Arabe
       23.198o(258): 26-27. -2o6, H 223-

2484   Public execution of Saudi princess and husband. In: Women's
       International Network News 4.1978(2): 9.

2485   Puin, G.-R., 1973: Aspekte der wahhabitischen Reform auf der
       Grundlage von Ibn Ġannāms 'Rauḍat al-afkār'. In: T. Nagel,
       G.-R. Puin, C.-U. Spuler, W. Schmucker & A. Noth: Studien
       zum Minderheitenproblem im Islam 1. Bonner Orientalistische Studien N.S. 27,1; Bonn: Orientalistisches Seminar
       der Universität Bonn, 1973: 45-99. -21, H 223-

2486   ——, 1977: Der moderne Alltag im Spiegel ḥanbalitischer Fetwas aus ar-Riyāḍ. In: W. Voigt (Hg.): XIX. Deutscher
       Orientalistentag vom 28. September bis 4. Oktober 1975 in
       Freiburg im Breisgau, Vorträge, Bd. 1. Zeitschrift der
       Deutschen Morgenländischen Gesellschaft, Suppl. III; Wiesbaden: Steiner, 1977: 589-597. -21-

2487   Purdy, A., (ed.), 1976: The businessman's guide to Saudi Arabia. London: Arlington Books. 33o p. -21-

2488   Putman, J.J., 1975: The Arab World, Inc. In: National Geographic 148: 494-533. -21, 24, 93-

2489　Pye, B., 1976: The civil engineer in Saudi Arabia. In: Civil Engineering (London) 1976(4): 23, 25, 27, 29.　-83, 89, 93-

Q

2490　Qadi, S.H., 1980: Perception of adequacy of, and planning for educational facilities in Saudi Arabia. Ph.D., University of Colorado, Boulder. 153 p.

2491　Quandt, W.B., 1981a: Riyadh between the superpowers. In: Foreign Policy (44): 37-56.　-21, 2o6, H 223-

2492　——, 1981b: Saudi Arabia in the 1980s: foreign policy, security, and oil. Washington: The Brookings Institution. X, 190 p.　-12, 21-

2493　Qubain, F.I., 1966: Education and science in the Arab world. Baltimore: The Johns Hopkins Press. XXII, 539 p.　-12, 21-

2494　Quest for knowledge: a special issue. Aramco World Magazine 2o.1969(6).

2495　Quinlan, T., et al., 1974: Saudi Arabia: special report. In: Commerce International 1o5(1417): 39-57.　-H 3-

2496　Quota, B.M.N. (oder: B.N.M.), 1977: Accountability and audit in the Saudi Arabian government sector. Ph.D., University of Kent.

2497　Quotah, M.M.N., 1979: Mathematical model for food demand in Saudi Arabia. Ph.D., Case Western Reserve University, Cleveland. 259 p.

2498　——, 1981: On the use of simulation in national agricultural production planning model. In: Journal, Economics and Administration (12): 27-42.

2499　Quraishi, H.E.A. al-, 1980: Managing Saudi Arabia's reserve funds. In: Journal of Contemporary Business 9(3): 31-39. -12, 18o, 2o6-

R

2500 Raddady, M.M. al-, 1977: Transformation of agriculture in western Saudi Arabia: problems and prospects. Ph.D., University of Durham.

2501 Rafei, N., 1969: Income tax in Saudi Arabia. In: Bulletin for International Fiscal Documentation 23: 482-487. -3o, 2o6, 352-

2502 ———, 1976: The tax system in Saudi Arabia. In: Bulletin for International Fiscal Documentation 3o: 2-7. -3o, 2o6, 352-

2503 Ragette, F., 1971: Building on tradition. Modern architecture with an Arab flavor. In: Aramco World Magazine 22(3): 14-23. -24-

2504 Rahnama, S., 1980: Comparative analysis of the general management knowledge, leadership style, and effectiveness between Saudi Arabian and American business graduate students. Ph.D., United States International University, San Diego. 217 p.

2505 Raieky, M.I. al-, 1981: Construction and validation of an achievement evaluation instrument for twelfth grade physics in Saudi Arabia. Ed.D., University of Northern Colorado, Greeley. 127 p.

2506 Railroad opened. Trackage completed from Dammam to Riyadh. In: Aramco World 2.195o/51(11): 1, 12.

2507 Rajehi, M.O.R., 1981: The impact of social change on police development in Saudi Arabia: a case study of Riyadh Police Department. Ph.D., Department of Sociology, Michigan State University, East Lansing. XIII, 2o7 p.

2508 Ralli, A., 1909: Christians at Mecca. London: Heinemann. X, 283 p. -21-

2509 Ramm, H., 1971: Be- und Entwässerung für die Al-Hassa-Oase in Saudi-Arabien. Technische Berichte (englischsprachige Ausgabe: Irrigation and drainage of the Al Hassa oasis in Saudi Arabia. Technical Reports); Frankfurt/M.: Philipp Holzmann AG. 31 S.

2510 Ramm, H., 1972a: Be- und Entwässerungsprojekt Al Hassa in Saudi-Arabien. In: ICID-Nachrichten, verbunden mit Mitteilungen des KFK 1972(1): 1-4 (Beilage zu: Wasser und Boden 24.1972). -26, 1oo, 385-

2511 ——, 1972b: Be- und Entwässerungsprojekt Al Hassa in Saudi-Arabien. In: Strassen- und Tiefbau 26: 38-4o. -9o, 93, 2o6-

2512 Rand, C.T., 1975: Making democracy safe for oil: oilmen and the Islamic East. An Atlantic Monthly Press Book; Boston, Toronto: Little, Brown. X, 422 p. -21-

2513 Raoof, A.H., 197o: The Kingdom of Saudi Arabia. In: T.Y. Ismael (ed.): Governments and politics of the contemporary Middle East. The Dorsey Series in Political Science; Homewood: The Dorsey Press, 197o: 353-379. -21-

2514 Rasheed, M.S. al-, 1972/73: Criminal procedure in Saudi Arabian judicial institutions. Ph.D., University of Durham.

2515 Rashid, I. al- (ed.), 1976: Documents on the history of Saudi Arabia, 19o9-1935. 3 vols. (Documents on the History of Saudi Arabia 1-3;) Salisbury/N.C.: Documentary Publications. XII, 233; III, 246 resp. IV, 238 p. -21-

2516 —— (ed.), 198o: Saudi Arabia enters the modern world. Secret U.S. documents on the emergence of the Kingdom of Saudi Arabia as a world power, 1936-1949. Documents on the History of Saudi Arabia 4-5; Salisbury/N.C.: Documentary Publications. VII, 264 resp. V, 234 p. -21-

2517 Rashid, S.A.A. al-, 1977: A critical study of the pilgrim road between Kufa and Mecca (Darb Zubaydah) with the aid of fieldwork. Ph.D., University of Leeds.

2518 Rásky, L., 19o5: Die Wehrmacht der Türkei. Wien: Seidel. VIII, 187 S. -21-

2519 Ras Tanura refinery, a major plant. In: Standard (of California) Oil Bulletin 1946(Autumn): 6-7.

252o Raswan, C.R., 193o: Tribal areas and migration lines of the north Arabian Bedouins. In: The Geographical Review 2o: 494-5o2. -21, 24, 2o6-

2521 Raswan, C.R., 1934: Im Land der schwarzen Zelte. Mein Leben unter den Beduinen. Berlin: Ullstein (englische Ausgabe: The black tents of Arabia: my life amongst the Bedouins. London: Hutchinson, 1935. Französische Ausgabe: Au pays des tentes noires; moeurs et coutumes des Bédouins. Collection d'Études, de Documents et de Témoignages pour servir à l'Histoire de notre Temps; Paris: Payot, 1936. Amerikanische Ausgabe: Black tents of Arabia: my life among the Bedouins. New York: Creative Age Press, 1947). 156 S. -24-

2522 ——, 1952: Trinker der Lüfte. Auf der Suche nach Ismaels Pferden zwischen Euphrat und Nil. Zürich-Rüschlikon: Müller (^{2}1960) (Originalausgabe: Drinkers of the wind. London: Hutchinson, 1938; amerikanische Ausgabe: New York: Creative Age Press, 1942). 155 S. -21-

2523 Rathjens, C., 1941: Djiddah. In: Atlantis 13: 114-120. -1a, 21, 24-

2524 —— & H. von Wissmann, 1947: Landschaftskundliche Beobachtungen im südlichen Hedjaz. In: Erdkunde 1: 61-89, 200-205. -16, 21, 24-

2525 ——, 1948: Die Pilgerfahrt nach Mekka. Von der Weihrauchstrasse zur Ölwirtschaft. Hamburgische Abhandlungen zur Weltwirtschaft; Hamburg: Mölich. 144 S. -21-

2526 ——, 1960: Abdallah H.St. John B. Philby. In: Orient (Opladen) 1: 61-64. -21, 212, H 223-

2527 ——, 1963: Entwicklungen der Pilgerfahrt. In: Orient (Opladen) 4: 102-103, 106-107. -21, 100, H 223-

2528 Raunkiaer, B., 1912a: Beretning om min Rejse i Central-Arabien. In: Geografisk Tidsskrift 21: 283-289. -1a, 61, 206-

2529 ——, 1912b: Die Expedition der Kgl. Dänischen Geographischen Gesellschaft nach Arabien. Vorläufige Übersicht. In: A. Petermanns Mitteilungen aus Justus Perthes' Geographischer Anstalt 58, 2. Halbbd.: 84-85 (amerikanische Ausgabe: The expedition of the Danish Geographical Society to Arabia. Preliminary report. In: The Geographical Review 44: 657-660). -16, 21, 24-

2530 Raunkiaer, B., 1914: Viaggio nell'Arabia di Nord-Est. In: Bollettino della Reale Società Geografica Italiana V, 3 (zugleich: 51): 12o1-1213. -24, 3o-

2531 ——, 1969: Through Wahhabiland on camelback. Travellers and Explorers; London: Routledge & Kegan Paul (Originalausgabe: Gennem Wahhabiternes Land paa Kamelryg, 1912. Kopenhagen: Gyldendal, 1912; auszugsweise deutsche Ausgabe: Auf dem Kamelrücken durch das Land der Wahabiten. I. Geographische Übersicht von Nordostarabien. II. Kuwait. In: Mitteilungen der Geographischen Gesellschaft in Hamburg 3o; Hamburg: Friederichsen, 1917: 187-243). XII, 156 p. -21, Frei 119, Tü 17-

2532 Rautenbach, L., 1958: Fatime. Als Hofärztin im Harem König Ibn Saud's. Hamburg, Uelzen: Blume. 219 S. -24-

2533 Rawaf, O.Y. al-, 198o: The concept of the five crises in political development: relevance to the Kingdom of Saudi Arabia. Ph.D., Department of Political Science, Duke University, Durham/N.C. XIII, VII, 575 p. -21-

2534 Rayess, F., 1965: The face of Medina. In: Aramco World Magazine 16(6): 12-15. -24-

2535 Raymond, J., 1925: Mémoire sur l'origine des Wahabys, sur la naissance de leur puissance et sur l'influence dont ils jouissent comme nation. Rapport de J. Raymond daté de 18o6. Document inédit extrait des Archives du Ministère des Affaires étrangères de France. Société Royale de Géographie d'Égypte, Publications spéciales; Le Caire: Institut Français d'Archéologie Orientale pour la Société Royale de Géographie d'Égypte. VIII, 4o p. -21-

2536 Razvi, M., 1981a: The Mecca Summit. In: Pakistan Horizon 34(3): 44-55. -1a, 2o6, H 3-

2537 ——, 1981b: The Fahd Peace Plan. In: Pakistan Horizon 34(4): 48-61. -1a, 2o6, H 3-

2538 Reeves Palmer, M., 1917: The Kibla: a Mecca newspaper. In: The Moslem World 7: 185-19o. -21, 291, 352-

**** Regli, E.: Verfasser von Nr. 589.

2539 Regolamento del Consiglio dei Ministri e dei suoi 'Uffici dipendenti' nell'Arabia Saudiana (17 marzo 1954). In: Oriente Moderno 34.1954: 258-263.   -5, 7, 21-

2540 Rehatsek, E., 1880: The history of the Wahhábys in Arabia and in India. In: Journal of the Bombay Branch of the Royal Asiatic Society 14: 274-401.   -1a, 21-

2541 Reichert, H., 1978: Die Verstädterung der Eastern Province von Saudi Arabien und ihre Konsequenzen für die Regional- und Stadtentwicklung. Bamberg (zugleich: Diss., Fachbereich Orts-, Regional- und Landesplanung, Technische Universität Stuttgart). 5, 247 S.   -21, 24, 93-

2542 Reichs-Marine-Amt, 1906: Segelhandbuch für das Rote Meer und den Golf von Aden. Berlin: Mittler. XII, 588 S.   -21, H 2-

2543 ——, 1907: Segelhandbuch für den Persischen Golf. Berlin: Mittler. XIII, 277 S.   -77, H 2-

2544 Reintjens, H., 1975: Die soziale Stellung der Frau bei den nordarabischen Beduinen unter besonderer Berücksichtigung ihrer Ehe- und Familienverhältnisse. Bonner Orientalistische Studien N.S. 30; Bonn: Orientalisches Seminar der Universität Bonn. 236 S.   -21-

2545 Reisch, M., ^{2}1961: König im Morgenland. Im Auto durch Saudi-Arabien. Berlin, Frankfurt/M., Wien: Ullstein (^{1}1954). 156 S.   -21, 212-

2546 Reissner, J., 1980: Die Besetzung der grossen Moschee in Mekka 1979. Zum Verhältnis von Staat und Religion in Saudi-Arabien. In: Orient (Opladen) 21: 194-203.   -21, 100, H 223-

2547 ——, 1981: Die Idrīsīden in 'Asīr. Ein historischer Überblick. In: Universität Tübingen, Sonderforschungsbereich 19, Tübinger Atlas des Vorderen Orients, Arbeitsheft 11/2: Wissenschaftliche Beiträge einzelner Fächer; Arbeitsbericht 1979/80. Tübingen, 1981: 262-299.   -Tü 17-

2548 ——, 1983: Die Innenpolitik. In: T. Koszinowski (Hg.): Saudi-Arabien: Ölmacht und Entwicklungsland. Beiträge zur Geschichte, Politik, Wirtschaft und Gesellschaft. Mitteilungen des Deutschen Orient-Instituts 20; Hamburg: Deutsches Orient-Institut, 1983: 83-120.   -21, H 223-

2549 Rendel, George, 1957: The sword and the olive. Recollections of diplomacy and the foreign service, 1913-1954. London: Murray. 348 p.   -24/213-

2550 Rendel, Geraldine, 1937/38: Across Saudi Arabia. In: The Geographical Magazine 6: 163-180.   -7, 180, 464-

2551 Rentz, G., 1948: Muḥammad ibn 'Abd al-Wahhâb (1703/04-1792) and the beginnings of the Unitarian empire in Arabia. Ph.D., Graduate Division, University of California, Berkeley. IX, 326 p.   -21-

2552 ——, 1951a: Notes on Dickson's 'The Arab of the desert'. In: The Muslim World 41: 49-64.   -21, 291, 352-

2553 ——, 1951b: Pearling in the Persian Gulf. In: W.J. Fischel (ed.): Semitic and Oriental studies. A volume presented to William Popper, Professor of Semitic Languages, Emeritus, on the occasion of his seventy-fifth birthday, October 29, 1949. Berkeley, Los Angeles: University of California Press, 1951: 397-402 (reprinted in: C.P. Issawi (ed.): The economic history of the Middle East, 1800-1914. A book of readings. Chicago, London: The University of Chicago Press, 1966: 313-316).   -21-

2554 ——, 1957: Notes on Oppenheim's Die Beduinen. In: Oriens 10: 77-89.   -1a, 12, 21-

2555 —— & W.E. Mulligan, 1960a: Al-Aflādj (Aflādj al-Dawāsir). In: EI^2 1: 233-234.   -21, 24, 212-

2556 ——, 1960b: Djazīrat al-'Arab. In: EI^2 1: 533-556 (slightly modified version: A sketch of the geography, people, and history of the Arabian Peninsula; printed in: Society of Petroleum Engineers of AIME, Saudi Arabia Section: Fall field trip, November 14-16, 1962, Riyadh, Saudi Arabia. N.p., n.d.: 44-91).   -21, 24, 212-

2557 ——, 1960c: Al-'Āriḍ. In: EI^2 1: 628-629.   -21, 24, 212-

2558 ——, 1960d: Bi'r. II. Modern Arabia. In: EI^2 1: 1230-1231.   -21, 24, 212-

2559 —— & W.E. Mulligan, 1960e: Al-Buraymī. In: EI^2 1: 1313-1314.   -21, 24, 212-

2560 ——, 1965a: Al-Dawāsir. In: EI^2 2: 175-177.   -21, 24, 212-

2561 Rentz, G., 1965b: Al-Dir'iyya. In: EI² 2: 32o-322 (slightly modified version: Al-Dir'iyah: an oasis northwest of Riyadh; printed in: Society of Petroleum Engineers of AIME, Saudi Arabia Section: Fall field trip, November 14-16, 1962, Riyadh, Saudi Arabia. N.p., n.d.: 92-1oo). -21, 24, 212-

2562 ——, 1965c: Djayzān. In: EI² 2: 516-518. -21, 24, 212-

2563 ——, 1965d: Saudi Arabia: the Islamic island. In: Journal of International Affairs 19: 77-86 (reprinted in: J.H. Thompson & R.D. Reischauer (eds.): Modernization of the Arab world. New Perspectives in Political Science Series 11; Princeton, Toronto, New York, London: Van Nostrand, 1966: 115-125). -24, 25, 2o6-

2564 ——, 1969: The Wahhābīs. In: A.J. Arberry (ed.): Religion in the Middle East: three religions in concord and conflict. Vol. 2: Islam. Cambridge/England: Cambridge University Press, 1969: 27o-284. -21-

2565 —— & James Mandaville, 1971a: Banū Hādjir. In: EI² 3: 49-5o. -21, 24, 212-

2566 ——, 1971b: Hāshimids (al-Hawāshim). In: EI² 3: 262-263. -21, 24, 212-

2567 ——, 1971c: Al-Ḥawṭa. In: EI² 3: 294-295. -21, 24, 212-

2568 ——, 1971d: Al-Ḥidjāz. In: EI² 3: 362-364. -21, 24, 212-

2569 ——, 1971e: Hudhayl. In: EI² 3: 54o-541. -21, 24, 212-

257o ——, 1971f: Hutaym. - Al-Ḥuwayṭāt. In: EI² 3: 641-644. -21, 24, 212-

2571 ——, 1971g: Al-Ikhwān. In: EI² 3: 1o64-1o68. -21, 24, 212-

2572 ——, 1972: Wahhabism and Saudi Arabia. In: D. Hopwood (ed.): The Arabian Peninsula: society and politics. Studies on Modern Asia and Africa 8; London: Allen & Unwin, 1972: 54-66. -21, 24-

2573 ——, 1978a: Al-Kasīm. In: EI² 4: 717. -21, 24, 212-

2574 ——, 1978b: Al-Katīf. In: EI² 4: 763-765. -21, 24, 212-

2575 ——, 1978c: Khadīr, Banū. In: EI² 4: 9o5-9o6. -21, 24, 212-

2576 ——, 1978d: Al-Khardj. In: EI² 4: 1o72-1o73. -21, 24, 212-

2577 Rentz, G., 1978e: Khāwa. In: EI^2 4: 1133.   -21, 24, 212-

2578 ——, 1979a: Al-Khurma. In: EI^2 5, Fasc. 79-8o: 62-63.   -21, 24-

2579 ——, 1979b: Philby as a historian of Saudi Arabia. In: Studies in the history of Arabia. Vol. 1: Sources for the history of Arabia, pt. 2. Riyadh: Riyad University Press, 1979: 25-35.   -21-

258o ——, 198o: The Saudi monarchy. In: W.A. Beling (ed.): King Faisal and the modernisation of Saudi Arabia. London: Croom Helm; Boulder: Westview Press, 198o: 15-34.   -21, H 223-

2581 Reynolds, B., 1979a: A walk through history. In: Aramco World Magazine 3o(2): 12-17.   -21, 24-

2582 ——, 1979b: All that glitters is. In: The Middle East (London) (6o): 7o-71.   -1a, 12, 3o-

2583 ——, 198o: Their fathers' sons. In: Aramco World Magazine 31(1): 2-11.   -21, 24-

2584 Ricci, C., 1954: Note sulle comunicazioni stradali e ferroviarie nella provincia di el-Ḥasā' (Arabia Saudiana). In: Oriente Moderno 34: 293-3o3.   -5, 7, 21-

2585 Rice, A.G., 1973: Oil-rich Saudi Arabia's growth points to heightened trade. In: Overseas Trading 25: 162-163.   -2o6, H 3-

2586 Richards, J.M., 1947: Gateway to the Hedjaz. In: The Architectural Review 1o2(6o8): 47-53.   -25, 89, 93-

2587 ——, 1949: Desert city: an account of Hail in central Arabia. In: The Architectural Review 1o5(625): 35-41.   -25, 89, 93-

2588 Ridda, M., 1974: King Faisal's attitude. In: Journal of Palestine Studies 3(2): 226-228.   -12, 21, 25-

2589 Rihani, A.F., 1926: With the kingliest king of Arabia. In: Asia (New York) 26: 668-674, 76o-767, 864-871, 974-981.   -1a-

259o ——, 1928: Ibn Sa'oud of Arabia, his people and his land. London: Constable (amerikanische Ausgabe: Ibn Sa'oud of

Arabia: maker of modern Arabia. Boston, New York: Houghton, Mifflin, 1928). XVII, 370 p.   -12, 21-

2591 Rihani, A.F., 1929a: Arabia: an unbiased survey. In: Journal of the Central Asian Society 16: 35-55.   -21, 9o, 291-

2592 ——, 1929b: In the land of Wallah. We'll slay him. In: Asia (New York) 29: 716-721.   -1a-

2593 ——, 193oa: Around the coasts of Arabia. London: Constable. X, 364 p.   -12, 21-

2594 ——, 193ob: Arabian peak and desert: travels in Al-Yaman. London: Constable. IX, 28o p.   -12-

2595 ——, 1932: The political situation in Arabia. In: The Open Court 46: 806-827.   -1a-

2596 ——, 1937a: In Ibn Saud's palace. Glimpses of life as subject or guest of an Arab Sultan. In: Asia (New York) 37: 215-218.   -1a, 21-

2597 ——, 1937b: Ibn Sa'ud of Arabia. 25 years ago he believed in conquests; today, his first concern is for peace. In: Current History (New York) 46(Sept.): 62-64.   -21, 18o-

2598 ——, 1938: Conquest of the desert. In: Asia (New York) 38: 192-197.   -1a-

2599 Riley, C.L., 1972: Historical and cultural dictionary of Saudi Arabia. Historical and Cultural Dictionaries of Asia 1; Metuchen: The Scarecrow Press. VI, 133 p.   -12, 21, Tü 17-

26oo Ritter, C., 1846-47: Die Erdkunde von Asien. Bd. 1, 1. Abtheilung: Die Halbinsel Arabien (zugleich: Vergleichende Erdkunde von Arabien. 2 Bde.). Berlin: Reimer. XXVIII, 1o35 bzw. XIV, 1o57 S.   -24-

26o1 Ritter, W., 1971/72: Sandberge und Oasen in der Nafud Thuwayrat - Saudi Arabien. In: Geographischer Jahresbericht aus Österreich 34: 65-76.   -1a, 25, 3o-

26o2 ——, 1975: Central Saudi Arabia. In: Wiener Geographische Schriften 43/44/45: Beiträge zur Wirtschaftsgeographie, I. Teil. Hgg. von E. Winkler & H. Lechleitner; Wien: Hirt, 1975: 2o5-228.   -1a-

2603 Ritter, W., 1977: A note on the sedentarization of nomads in eastern Saudi Arabia. In: G. Gruber, H. Lamping, W. Lutz, R. Müller & K. Vorlaufer (Hgg.): Studien zur allgemeinen und regionalen Geographie. Josef Matznetter zum 6o. Geburtstag. Frankfurter Wirtschafts- und Sozialgeographische Schriften 26; Frankfurt/M.: Seminar für Wirtschaftsgeographie der Johann Wolfgang Goethe-Universität, 1977: 4o7-434. -12, 89, 291-

2604 ———, 1978: Die Arabische Halbinsel (Saudi-Arabien, Yemen, Südyemen, Kuwait, Bahrain, Qatar, Vereinigte Arabische Emirate, Oman). Reiseführer mit Landeskunde mit 2o Karten. Mai's Auslandtaschenbuch 34; Buchenhain: Volk und Heimat. 278 S. -21, 1oo, 212-

2605 ———, 1979: Die Anfänge eines Tourismus auf der Arabischen Halbinsel. In: Der Tourismus als Entwicklungsfaktor in Tropenländern. 2. Frankfurter Wirtschaftsgeographisches Symposium (27./28. Jan. 1978). Frankfurter Wirtschafts- und Sozialgeographische Schriften 3o; Frankfurt/M.: Institut für Wirtschafts- und Sozialgeographie der Johann Wolfgang Goethe-Universität, 1979: 87-1o3. -Tü 17-

2606 ———, 198o: The Arabian oases run dry? In: W. Meckelein (ed.): Desertification in extremely arid environments. Stuttgarter Geographische Studien 95; Stuttgart: Geographisches Institut der Universität Stuttgart, 198o: 73-92. -21, 24, 93-

2607 ———, 1983: Der Erdölgolf. Struktur- und Entwicklungsprobleme der Länder am Arabisch-persischen Golf. Problemräume der Welt 1; Köln: Aulis Verlag Deubner. 4o S.

2608 Rivoyre, D. de, 188o: Mer Rouge et Abyssinie. Paris: Plon. 3o8 p. -Bibliothèque Nationale, Paris-

2609 Riyadh, and five year development plan for municipalities. In: Doxiadis Associates Review 6.197o(67): 6-9.

2610 Riyadh Chamber of Commerce and Industry, n.d.: Riyadh Chamber of Commerce and Industry, trade directory. 1975/76 ed. Riyadh.

2611 Robbins, R.R., 1942: The legal background of Arabia and present tendency toward state development. Ph.D., Ohio State University, Columbus. XIV, 4o2, VIII p. -21-

2612   Robertson, N. (ed.), 1979: Origins of the Saudi Arabian oil empire. Secret U.S. documents, 1923-1944. Salisbury/N.C.: Documentary Publications. VI, 196 p.   -21-

2613   Robson, J., 1960: Bid'a. In: EI^2 1: 1199.   -21, 24, 212-

2614   Roches, L., 1885: Trente-deux ans à travers l'Islam (1832-1864). T. 2: Mission à la Mecque. - Le Maréchal Bugeaud en Afrique. Paris: Firmin-Didot. 503 p.   -1a-

2615   Rochet d'Héricourt, C.E.X., 1841a: Considérations géographiques et commerciales sur le golfe Arabique, le pays d'Adel et le royaume de Choa (Abyssinie méridionale). In: Bulletin de la Société de Géographie (Paris) II, 15: 269-293. -291-

2616   ——, 1841b: Voyage sur la côte orientale de la mer Rouge, dans le pays d'Adel et le royaume de Choa. Paris: Bertrand. XXIII, 432 p.   -12, 24-

2617   ——, 1843: Lettre de M. Rochet d'Héricourt à M. d'Avezac. In: Bulletin de la Société de Géographie (Paris) II, 19: 118-127.   -4, 291-

2618   ——, 1846: Second voyage sur les deux rives de la mer Rouge, dans le pays des Adels et le royaume de Choa. Paris: Bertrand (auszugsweise deutsche Ausgabe: Reise in das Königreich Schoa im mittäglichen Abyssinien während der Jahre 1842, 1843 und 1844. Stuttgart: Franckh, 1847). XLVIII, 406 p.   -12-

2619   Rock, A., 1935: Ibn Saud gründet das Gottesreich Arabien. Die kleine Geschichtsbücherei 10; Berlin: Hobbing. 75 S. -24/213-

2620   Roeder, A., 1963: Die arabischen Sekten und Bruderschaften. Eine soziologische Untersuchung. Diss., Philosophische Fakultät, Universität Heidelberg. 135 S.   -24-

2621   Rörig, H., 1955: Die arabische Welt. Dalp-Taschenbücher 313; München: Lehnen. 140 S.   -21, 93-

2622   Rösel, W., et al., 1979: Arabex '78. Exkursionsbericht des Fachgebiets Baubetrieb/Projektmanagement, Fachbereich Architektur, Universität Kassel, Gesamthochschule. Kassel. 114, XV S.

2623 Roff, W.R., 1982: Sanitation and security: the imperial powers and the nineteenth century Ḥajj. In: Arabian Studies 6: 143-16o. -21, Tü 17-

2624 Rohlfs, G., 1883: Meine Mission nach Abessinien. Auf Befehl Sr. Maj. des Deutschen Kaisers im Winter 188o/81 unternommen. Leipzig: Brockhaus. XX, 348 S. -21, 24-

2625 Roloff, M., 1915: Arabien und seine Bedeutung für die Erstarkung des Osmanenreiches. In: H. Grothe (Hg.): Länder und Völker der Türkei. Schriftensammlung des Deutschen Vorderasienkomitees; Leipzig: Veit, 1915: 111-136. -21, 212-

2626 Ronall, J.O., 1967a: Die wirtschaftliche und soziale Bedeutung des Hadjdj für Saudi-Arabien. In: Bustan 8(3): 33-34. -21, 212-

2627 ———, 1967b: Banking regulations in Saudi Arabia. In: The Middle East Journal 21: 399-4o2. -21, 2o6, 352-

2628 Rondot, P., 1973: Arabie Saoudite: Un puissant voisin pour l'Afrique. In: Revue Française d'Études Politiques Africaines 8(9o): 12-16. -18, 21, 2o6-

2629 ———, 198o: Les hommes au pouvoir en Arabie saoudite. In: Maghreb-Machrek (89): 16-26. -H 3-

263o ———, 1981: Whither Saudi Arabia? In: Nato's Fifteen Nations 26(1): 3o-35. -2o6-

2631 ———, 1982: L'Islam dans la péninsule Arabique. In: Centre d'Études et de Recherches sur l'Orient Arabe Contemporain (éd.): La péninsule Arabique d'aujourd'hui. Sous la direction de P. Bonnenfant. T. 1. Paris: Centre National de la Recherche Scientifique, 1982: 39-58. -21-

2632 Roosevelt, K., 1949: Arabs, oil and history; the story of the Middle East. New York: Harper & Row (reprint: Port Washington: Kennikat Press, 1969). 271 p. -21, 24-

2633 Rosen, G., 1865: Guarmani's Reise nach dem Neğd. Ein Beitrag zur geographischen Kenntniss Arabiens. In: Zeitschrift für allgemeine Erdkunde N.F. 18: 2o1-218. -21, 24, 25-

2634 Rosen, S.J. & H. Shaked, 1978: Arms and the Saudi connection. In: Commentary (65): 33-38. -7, 12, 18o-

2635 Rosenfeld, H., 1965: The social composition of the military in the process of state formation in the Arabian desert. In: The Journal of the Royal Anthropological Institute of Great Britain and Ireland 95: 75-86, 174-194. -16, 26, 352-

2636 Rosenthal, E., 1931: From Drury Lane to Mecca. Being an account of the strange life and adventures of Hedley Churchward (also known as Mahmoud Mobarek Churchward), an English convert to Islam. London: Low, Marston. 248 p. -12, 21-

2637 Ross, H.C., 1978: Bedouin jewellery in Saudi Arabia. London: Stacey International. 128 p. -21, H 223-

2638 ——, 1981: The art of Arabian costume: a Saudi Arabian profile. Fribourg: Arabesque Commercial. 188 p. -212-

2639 Rossi, E., 1939: La produzione di petrolio nel territorio arabo sa'udiano di el-Ahsa' (Golfo Persico). In: Oriente Moderno 19: 172. -5, 7, 21-

2640 ——, 1940: Completamento della 'Ferrovia di Baghdād' o 'Bosforo-Golfo Persico' e problemi ferroviari del Vicino Oriente. In: Oriente Moderno 2o: 513-52o. -5, 7, 21-

2641 ——, 1944: Documenti sull'origine e gli sviluppi della Questione Araba (1875-1944). Con introduzione storica. Pubblicazioni dell'Istituto per l'Oriente; Roma: Istituto per l'Oriente. LVI, 251 S. -21-

2642 Rousseau, J.B.L.J., 18o9: Description du Pachalik de Bagdad, suivie d'une notice historique sur les Wahabis, et de quelques autres pièces relatives à l'histoire et à la littérature de l'Orient. Paris: Treuttel & Würtz. VII, 261 p. -21-

2643 ——, 1811: Nouveaux renseignements sur les opérations militaires des Wahabis, depuis l'année 18o7 jusqu'au milieu de 181o. In: Annales des Voyages, de la Géographie et de l'Histoire 14: 1o2-112. -24-

2644 ——, 1818: Mémoire sur les trois plus fameuses sectes du musulmanisme, les Wahabis, les Nosaïris et les Ismaélis. Marseille: Masvert; Paris: Nève. 84 p.

2645 Rowley, G. & S.A. el-Hamdan, 1977: Once a year in Mecca. In: The Geographical Magazine 49: 753-759. -18o, 464-

2646 Royal Institute of International Affairs, ^{2}1954: The Middle East: a political and economic survey. London, New York (^{1}195o). XVIII, 59o p. -21-

2647 Royaume d'Arabie Séoudite, Ministère de l'Information, s.d.: Progrès de l'enseignement religieux en Arabie Séoudite. Riyâd. 46 p. -Bo 133-

2648 ——, Ministère du Travail et des Affaires Sociales, env. 1974: Aperçu sur les affaires sociales. Riyâd. 78 p.

2649 Rubelli, L., 1885: Djeddah. Wirthschaftliche Verhältnisse im Jahre 1884. In: Jahresberichte der k. und k. österreichisch-ungarischen Consulats-Behörden 13: 439-444. -35-

265o Rubin, B., 1979: Anglo-American relations in Saudi Arabia, 1941-45. In: Journal of Contemporary History 14: 253-267. -21, 3o, 352-

2651 ——, 198o: The great powers in the Middle East, 1941-1947: the road to the Cold War. London, Totowa: Cass. XIV, 254 p. -1a, 21-

2652 Rüppell, E., 1829: Reisen in Nubien, Kordofan und dem peträischen Arabien vorzüglich in geographisch-statistischer Hinsicht. Frankfurt/M.: Wilmans. XXVI, 388 S. -21, 24-

2653 ——, 1838: Reise in Abyssinien. Bd. 1. Frankfurt/M.: Schmerber. XVI, 434 S. -21, 24-

2654 Rugh, W.A., 1969: Riyadh, history and guide; a brief historical survey of the development of Riyadh from a small desert town into the capital of Saudi Arabia - and a guide book to the city and its environs. Dammam: Al-Mutawa Press. VIII, 112 p. -212-

2655 ——, 1973: Emergence of a new middle class in Saudi Arabia. In: The Middle East Journal 27: 7-2o. -12, 21, 2o6-

2656 ——, 1979a: The Arab press. News media and political process in the Arab world. Contemporary Issues in the Middle East. Syracuse: Syracuse University Press. XVIII, 2o5 p. -21-

2657 ——, 1979b: A tale of two houses. In: The Wilson Quarterly 3(1): 59-72. -7, 12, 3o-

2658 Rugh, W.A., 1979c: Islam: the Saudi Arabian way. In: Across the Board 16: 4-13. -2o6-

2659 ——, 198o: Saudi mass media and society in the Faisal era. In: W.A. Beling (ed.): King Faisal and the modernisation of Saudi Arabia. London: Croom Helm; Boulder: Westview Press, 198o: 125-144. -21-

266o Russel, S., 1884: Une mission en Abyssinie et dans la mer Rouge, 23 Octobre 1859 - 7 Mai 186o. Paris: Plon, Nourrit. XXVIII, 3o6 p. -12-

2661 Rustow, D.A., 1977: U.S.-Saudi relations and the oil crises of the 1980s. In: Foreign Affairs (New York) 55: 494-516. -21, 24, 25-

2662 ——, 1982: Oil and turmoil. America faces OPEC and the Middle East. New York, London: Norton. 32o p. -21-

2663 Rutter, E., 1929: The Muslim pilgrimage. In: The Geographical Journal 74: 271-273. -21, 291, 352-

2664 ——, 193oa: The Holy Cities of Arabia. London, New York: Putnam (^{1}1928, 2 vols.). XV, 593 p. -21-

2665 ——, 193ob: The habitability of the Arabian Desert. In: The Geographical Journal 76: 512-515. -21, 291, 352-

2666 ——, 1931a: Damascus and Hâil. In: Journal of the Central Asian Society 18: 61-73. -21, 9o, 291-

2667 ——, 1931b: The Hejaz. In: The Geographical Journal 77: 97-1o9. -21, 291, 352-

2668 ——, 1932: A journey to Hail. In: The Geographical Journal 8o: 325-331. -21, 291, 352-

2669 ——, 1933: Slavery in Arabia. In: Journal of the Royal Central Asian Society 2o: 315-332. -21, 9o, 291-

267o Rutter, O., 1937: Triumphant pilgrimage: an English Muslim's journey from Sarawak to Mecca. London, Bombay, Sydney: Harrap. 278 p. -12-

2671 Ryan, A., 1951: The last of the dragomans. London: Bles. 351 p. -12-

2672 Ryckmans, G., 1952: Prospectietocht door Saoedi-Arabië. In: Mededelingen van de Koninklijke Vlaamse Academie voor Wetenschappen, Letteren en Schone Kunsten van België, Klasse der Letteren 14(5). -Koninklijke Bibliotheek 's-Gravenhage-

2673 Ryckmans, G., 1954: Through Sheba's kingdom. In: The Geographical Magazine 27(3): 129-137. -7, 18o, 464-

2674 ——, 1961: H. Saint John B. Philby, le 'Sheikh 'Abdallah', 3 avril 1885 - 3o septembre 196o. Uitgaven van het Nederlands Historisch-Archaeologisch Instituut te Istanbul 1o; Istanbul: Nederlands Historisch-Archaeologisch Instituut in het Nabije Oosten. 24 p. -12, 21, Tü 17-

2675 Ryder, W., 198o: Saudi Arabia. A MEED Special Report; London: Middle East Economic Digest. 96 p.

2676 Rypka, J., 1938: Alois Musil, June 3oth, 1868 - June 3oth, 1938. In: Archiv Orientální 1o: 1-34. -1a, 16, 21-

S

2677 S., E., 1918: Ein Beduinenüberfall. In: Das heilige Land 62: 38-4o. -1a, 21, 25-

2678 Saab, R., 1974: Saudi agriculture. In: The Arab Economist 6(65): 51-53. -H 3-

2679 Saad, E.O. al-, 198o: The role of public school teachers as curriculum innovators in Riyadh, Saudi Arabia. Ed.D., University of Northern Colorado, Greeley. 149 p.

268o Sa'ad, N.M., n.d.: Centre for training and applied research in community development, Diriyah-Riyadh, Saudi Arabia. N.p.

2681 Saade, R.F., 1969: Réalités et problèmes de l'agriculture séoudienne. In: Revue de la Société d'Études et d'Expansion (235): 236-242. -2o6, 212, 282-

2682 Saadûn, M., 1957: Arabia Saudita: 'Arabia Feliz'? In: Cuadernos Africanos y Orientales (4o): 43-55. -12, 18, 2o6-

2683 Saaty, M.A., 1982: The constitutional development in Saudi Arabia. Ph.D., Claremont Graduate School. 264 p.

2684 Sabab, A.A.A. al-, 1973: An inquiry into the development of the current planning institutions for economic and socia development in Saudi Arabia. To those who are dedicated to education, progress, and welfare of the Kingdom of Saudi Arabia. Ph.D., Graduate School of Business Administration, New York University. 244 p.

2685 Sabbagh, G., 1972: Co-operative development in the Kingdom of Saudi Arabia. In: Plunkett Foundation for Co-operative Studies (ed.): Yearbook of agricultural co-operation, 1972. Oxford: Blackwell, 1972: 1o9-115.   -H 3-

2686 Sabban, A.A.S. al-, 1982: The municipal system in the Kingdom of Saudi Arabia: a case study of Makkah. Ph.D., Claremont Graduate School. 163 p.

2687 Sabini, J., 1973: Sea Island 4. In: Aramco World Magazine 24(2): 6-7.   -24-

2688 ——, 1981: Armies in the sand: the struggle for Mecca and Medina. London: Thames & Hudson. 223 p.   -21-

2689 Sablier, É., 1958: L'Arabie à l'âge du pétrole. In: La Table Ronde (126): 145-16o.   -1a, 21, 24-

269o Sabra, N., 198o: Arab financial assistance to Red Sea Arab countries. In: Istituto Affari Internazionali (ed.): Red Sea conflicts and cooperation. Regional balance and strategic implications. Rome. XIII, 87 p.   -H 223-

2691 Sabri, M.M., 1978: A demographic study in some selected villages in Kasseem area, Saudi Arabia. In: Addarah 4(3): 5-7.   -21-

2692 Sabry, Z.I., 196o: Processing of dates in the Eastern Province of Saudi Arabia. Beirut: Agricultural Science Faculty, American University of Beirut. 11 p.

2693 Sachar, H.M., 1969: The emergence of the Middle East: 1914-1926. New York: Knopf. XIII, 518, XXVII p.   -21, 24/213-

2694 ——, 1974: Europe leaves the Middle East, 1936-1954. London: Lane. XVIII, 687, XXXVIII p.   -21, 24/213-

2695 Sacher, R., 1976: Problemanalyse und Verfahrensvorschläge für ein Projekt zur Ansiedlung von Beduinen (Faisal Settlement Project Haradh, Saudi Arabien). Ein Beispiel für die Anwendung des 'situationsfunktionalen Ansatzes' in der Entwicklungshilfe. Diplomarbeit, Fachgebiet Kommunikationswissenschaft und landwirtschaftliches Beratungswesen, Institut für Agrarsoziologie, landwirtschaftliche Beratung und angewandte Psychologie, Universität Hohenheim. VIII, 91, 17 S.

2696 Sadek, F.M., 1972: Education of the mentally retarded: Saudi Arabia (Mission, December 1971 - August 1972). Paris: UNESCO. 58 p.

2697 Sadhan, A. al-, 1980: The modernisation of the Saudi bureaucracy. In: W.A. Beling (ed.): King Faisal and the modernisation of Saudi Arabia. London: Croom Helm; Boulder: Westview Press, 1980: 75-89. -21, H 223-

2698 Sadik, F.A., 1970: In-service training and development in Saudi Arabia with particular reference to the role of the IPA. M.A., Graduate Program of Development Administration, American University of Beirut. -Jafet Library, American University of Beirut-

2699 Sadlier (auch: Sadleir), G.F., 1823: Account of a journey from Katif on the Persian Gulf to Yanboo on the Red Sea. With a route. In: Transactions of the Literary Society of Bombay 3: 449-493. -1a, 21-

2700 ——, 1977: Diary of a journey across Arabia (1819). Arabia Past and Present 5; Cambridge/England, New York: The Oleander Press; Naples: The Falcon Press (^{1}1866). 161 p. -21, Tü 17-

2701 Safadi, A.I. al-, 1975: An evolving typology of personal constructs of critical thinking, curriculum planning and decision-making in teacher education programs based on the Islamic ideology. The case of Saudi Arabia. Ed.D., State University of New York, Buffalo. 347 p.

2702 Saggaf, A.A., 1981: An investigation of the English program at the Department of English, College of Education, King Abdul-Aziz University, in Mecca, Saudi Arabia. Ph.D., University of Kansas, Lawrence. 202 p.

2703 Saha, N., R.A. Bayoumi, F.S. el-Sheikh, A.P.W. Samuel, I. el-Fadil, I.S. el-Houri, Z.A. Sebai & H.M.A. Sabaa, 1980: Some blood genetic markers of selected tribes in western Saudi Arabia. In: American Journal of Physical Anthropology 52: 595-600. -26, 289, 352-

2704   Sahwell, 'A.S., 1956: The Buraimi dispute: the British armed aggression. In: The Islamic Review 44(4): 13-17.   -1a, 21-

2705   Said, A.H., 1979: Saudi Arabia: the transition from a tribal society to a nation-state. Ph.D., University of Missouri, Columbia. V, 199 p.   -206-

2706   Saif, J.A., 1973: An examination of the knowledge of traffic regulations and defensive driving among a selected sample of Saudi Arabian private car owners. Ph.D., Michigan State University, East Lansing. 240 p.

2707   Saif, S.M. al-, 1981: Recommended guidelines for the science education program in the public secondary schools of Saudi Arabia. Ph.D., University of Wyoming, Laramie. 231 p.

2708   Sakkar, S. al-, 1976: A Saudi-Iraqi family link. In: Arabian Studies 3: 189-190.   -12, 21, 212-

2709   Salah, S., n.d.: Brief pictorial guide to Saudi Arabia. Al-Khobar: The International Publication Agencies. N.pag.

2710   ——, 1975: Panorama of Saudi Arabia. A Souvenir Book Edition; Beirut: Systeco. 416 p.   -212-

2711   Salah, Y.S., 1969: Siedlung und Wirtschaft der Oase Al-Hofūf in Al Hasā (Saudi-Arabien). Diss., Philosophische Fakultät, Universität Münster. 65 S.   -6-

2712   Salam, Y.M., c. 1980: Transportation planning study for the Eastern Province of Saudi Arabia. In: Efficient transport in the Arab states: (conference held in) Amman, 1979. London: Contex, c. 1980: 113-129.

2713   Salamé (auch: Salameh), G., 1979a: Développement et dépendance: quelques remarques dérivées du cas saoudien. In: Oriente Moderno 58: 447-461.   -5, 7, 21-

2714   ——, 1979b: Développement du rôle régional et international de l'Arabie Saoudite depuis 1945. Thèse de doctorat, Université Paris I (Université Panthéon-Sorbonne).

2715   ——, 1980a: Political power and the Saudi state. MERIP Reports 91.   -1a-

2716   ——, 1980b: Les monarchies arabes du Golfe. Quel avenir? In: Politique Étrangère 45: 849-865.   -206, H 3, H 223-

2717 Salamé (auch: Salameh), G., 1980c: L'Islam en Arabie Saoudite. In: Pouvoirs 1980(12): 125-130 (nouvelle éd. 1983). -12, 21, H 223-

2718 ——, 1981: Arabie Saoudite. Une vocation de puissance régionale servie par l'alliance avec l'Amérique. In: Le Monde Diplomatique 28(331): 14.   -21, 24/213, H 223-

2719 Saleem Khan, M.A., 1971: Saudi Arabia: Wahabism and oil. In: Islam and the Modern Age 2: 87-101.   -21-

2720 ——, 1974: Foreign policy of Saudi Arabia: an introductory outline. In: Indian Journal of Politics 8(1/2): 79-90. -206-

2721 ——, 1981: Saudi Arabia in the 1980s: a preferred future. Fourth International Symposium, 29-31 March 1981: The Future of the Arab Gulf and the Strategy of Joint Arab Action. Basrah: Centre for Arab Gulf Studies, University of Basrah. 68 p.   -H 223-

2722 Saleh, F.S. al-, 1980: A case study and evaluation of a technological delivery system: construction management in Saudi Arabia. Ph.D., University of Washington, Seattle. 273 p.

2723 Saleh, M.A. eben, 1980: The development of energy-efficient building systems and techniques for housing the masses in hot dry climates, with special emphasis on Saudi Arabia. Arch.D., University of Michigan, Ann Arbor. 359 p.

2724 Saleh, Nabil A., 1981: The general principles of Saudi Arabian and Omani company laws (statutes and shari'a). London: Namara Publications. XI, 338 p.   -21, H 3-

2725 Saleh, Nasser O. al-, 1976: Some problems and development possibilities of the livestock sector in Saudi Arabia: a case study in livestock development in arid areas. Ph.D., University of Durham. 386 p.

2726 —— & B.N. Floyd, 1980: Livestock production in Saudi Arabia: some spatial considerations. In: H.K. Barth & H. Wilhelmy (Hgg.): Trockengebiete: Natur und Mensch im ariden Lebensraum; Festschrift zum 60. Geburtstag von Helmut Blume. Tübinger Geographische Studien 80 (Sbd. 13); Tübingen: Geographisches Institut der Universität Tübingen, 1980: 263-294.   -21, Tü 17-

2727 Saleh, Nassir A., 1975: The emergence of Saudi Arabian administrative areas: a study in political geography. Ph.D., University of Durham.

2728 ——, 1981: Provincial and district delimitation in the Kingdom of Saudi Arabia. In: J.I. Clarke & H. Bowen-Jones (eds.): Change and development in the Middle East. Essays in honour of W.B. Fisher. London, New York: Methuen, 1981: 3o5-317. -21, 1oo, H 3-

2729 Salem, F. al-, 1981: The issue of identity in selected Arab Gulf states. In: Journal of South Asian and Middle Eastern Studies 4(4): 3-2o. -1a, 12, 21-

273o Salem, M.S. al-, 1981: The interplay of tradition and modernity, a field study of Saudi policy and educational development. Ph.D., University of California, Santa Barbara. 226 p.

2731 Salibi, K.S., 198o: A history of Arabia. Delmar: Caravan Books. IX, 247 p. -21-

2732 Salloom, H.I. al-, 1974: A study of the relationship of school district size and administrative practices in schools in Saudi Arabia. Ph.D., University of Oklahoma, Norman. 268 p.

2733 Sammak, A. & H. Mantynen, 1975: Quantitative and qualitative aspects of educational wastage. A report on the UNEDBAS mission to Saudi Arabia 1975. N.p. 18 p. -H 1o8-

2734 Samman, N.H., 1982: Saudi Arabia and the role of the imarates in regional development. Ph.D., Claremont Graduate School. 8o1 p.

2735 Sampson, A., 1976: Die Sieben Schwestern. Die Ölkonzerne und die Verwandlung der Welt. Reinbek: Rowohlt (Originalausgabe: The Seven Sisters: the great oil companies and the world they made. London: Hodder & Stoughton (New York: Viking Press), 1975). 33o S. -24/213-

2736 ——, 1977: Die Waffenhändler: von Krupp bis Lockheed; die Geschichte eines tödlichen Geschäfts. Reinbek: Rowohlt (Originalausgabe: The arms bazaar. The companies, the dealers, the bribes: from Vickers to Lockheed. London, Sydney, Auckland, Toronto: Hodder & Stoughton, 1977). 346 S. -21, 24/213-

2737 Sams, T.A., 1980: A guide to development in Saudi Arabia. In: Business America 3(June 30, 1980): 2-13. -1a, 206, H 3-

2738 Sands of Saudi Arabia promise rich oil flow. In: World Petroleum 9.1938(7): 70-71. -1a, 90-

2739 Sanger, R.H., 1947: Ibn Saud's program for Arabia. In: The Middle East Journal 1: 180-190. -12, 21, 206-

2740 ———, 1954: The Arabian Peninsula. Ithaca: Cornell University Press. XIV, 295 p. -21, 212, Tü 17-

2741 Santani, A., 1966: Saudi Arabia: educational development in 1965-1966. In: International Yearbook of Education 28: 303-306. -1a, 24, 352-

2742 Saoedi-Arabië. Landendocumentatie 183/184; Amsterdam, 1974. 60 S. -Koninklijke Bibliotheek 's-Gravenhage-

2743 Sardar, Z., 1978: The Information Unit of the Hajj Research Centre. In: Aslib Proceedings 30: 158-164. -21, 352-

2744 ——— & Z. Badawi (eds.), 1978: Hajj studies 1. Hajj Research Centre Studies 2; London: Croom Helm for The Hajj Research Centre, King Abdul Aziz University, Jeddah. 164 p. -21, Tü 17-

2745 ———, 1979: Saudi Arabia: indigenous sources of information. In: Aslib Proceedings 31: 237-244. -21, 352-

2746 Sarhan, S., [2]1978: Who's who in Saudi Arabia 1978-79. Jeddah: Tihama; London: Europa Publications (1st ed.: Who's who in Saudi Arabia 1976-77. Jeddah: Tihama; London: Europa Publications, 1977). XVI, 309 p. -H 223-

2747 Saud, M.A.T. al-, 1982: Permanence and change: an analysis of the Islamic political culture of Saudi Arabia with special reference to the Royal Family. Ph.D., Claremont Graduate School. 206 p.

2748 Saudi and OPEC direct investment in the United States. In: International Currency Review 12.1980(1): 21-24. -180, 206, H 3-

2749 Saudi Arab driller. In: Aramco World 15.1964(3): 2-7. -H 223-

**** Saudi Arabia: s. auch Kingdom of Saudi Arabia.

2750 Saudi Arabia. In: The Arab World 11.1965(3): 83-9o. -2o6-

2751 Saudi Arabia. In: Afro-Asian Economic Review 13.1971(138/139): 28-33. -H 3-

2752 Saudi Arabia. In: Middle East Annual Review 1974: 137, 139, 143-145, 147. -21, 1oo, 2o6-

2753 Saudi Arabia. In: Middle East Annual Review 1975-76: 215, 217-22o, 225-227. -21, 1oo, 2o6-

2754 Saudi Arabia. In: OPEC Bulletin 13.1982(2): 28-35. -2o6, H 3-

2755 Saudi Arabia. In: Business America 5.1982(13): 18-22. -1a, 89, H 3-

2756 Saudi Arabia. In: The Middle East and North Africa, 1982-83. 29th ed. London: Europa Publications, 1982: 677-7o5. -12, 21, 93-

2757 Saudi Arabia. A construction boom. In: The Arab Economist 7.1975(75): 3o-35. -2o6-

2758 Saudi Arabia after Ibn Saud. In: The World Today N.S. 9.1953: 5o5-5o7. -18, 18o, H 3-

2759 Saudi Arabia. Agricultural sector registers slow growth. In: The Arab Economist 12.198o(135): 25-27. -2o6, H 3, H 223-

276o Saudi Arabia. American business just manages to hold firm. In: The Arab Economist 14.1982(152): 21-23. -2o6, H 3, H 223-

2761 Saudi Arabia and its place in the world. Lausanne: Three Continents; Jeddah: Dar Al Shorouq, 1979. 195 p. -21-

2762 Saudi Arabia. Central Committee for the Population Census. In: Population Bulletin of the United Nations Economic and Social Office in Beirut 1973(4): 62. -2o6, Tü 17-

2763 Saudi Arabia. Dangerous times for the West's oil reservoir. In: Lloyd's Shipping Economist 2.198o(2): 8-13. -H 3-

2764 Saudi Arabia: decree regulating service agents. In: International Legal Materials 17.1978: 1456-1458. -12, 24, 352-

2765　Saudi Arabia. Demographic sample survey. In: Population Bulletin of the United Nations Economic and Social Office in Beirut 1972(3): 85.　-2o6, Tü 17-

2766　Saudi Arabia. Demographic survey, 1392/93 AH (1972-1973 AD). Population and housing census of Saudi Arabia, 1394 AH (1974 AD). In: Population Bulletin of the United Nations Economic and Social Office in Beirut 1973(5): 78-8o. -2o6, Tü 17-

2767　Saudi Arabia: desert kingdom holds power over world-wide economies. In: Australian Foreign Affairs Record 45.1974: 82o-823.　-1a, 2o6, 291-

2768　Saudi Arabia - economic review, 1951. World Trade Series 24o; Washington, 1952. 8 p.

2769　Saudi Arabia - economic review, 1953. World Trade Series 612; Washington, 1954. 8 p.

277o　Saudi Arabia. Industrialization main policy objective. In: The Arab Economist 13.1981(145): 14-16.　-2o6, H 3, H 223-

2771　Saudi Arabia. Intensive industrialisation drive. In: The Arab Economist 14.1982(153): 25-26.　-2o6, H 3, H 223-

2772　Saudi Arabia - Iran: agreement concerning sovereignty over Al-'Arabiyah and Farsi Islands and delimitation of boundary-line separating submarine areas between the Kingdom of Saudi Arabia and Iran (signed at Tehran October 24, 1968; entered into force January 29, 1969). In: International Legal Materials 8.1969: 493-496.　-1a, 12, 291-

2773　Saudi Arabia, Kingdom of. In: A.S. Knowles (ed.): The international encyclopedia of higher education. Vol. 8. San Francisco, Washington: Jossey-Bass, 1977: 3673-368o. -21-

2774　Saudi Arabia (Kingdom of Saudi Arabia). In: Who's who in the Arab world. 6th ed. (thoroughly rev. & completed). Beirut: Publitec Publications, 1981/82: 429-456.　-93-

2775　Sa'udi Arabia. King Sa'ud I. The Crown Prince. In: The Islamic Review 43.1955(9): 22-24.　-1a, 21-

2776　Saudi Arabia - master plan for the city of Riyadh. In: Doxiadis Associates Review 4.1968(43): 1-8.

2777   Saudi Arabia. More importance to the non-oil sector. In: The Arab Economist 14.1982(148): 21-22.   -2o6, H 3, H 223-

2778   Saudi Arabian Airlines (SAUDIA), Passenger Sales Department, Marketing Division, 1973: The commercial guide to Saudi Arabia. Business Advisory Service; Jeddah. 28 p.   -212-

2779   ——, Public Affairs Division, c. 1977: Across Saudi Arabia. Jeddah. 143 p.   -H 3-

2780   Saudi Arabian budget for the year 1386/87 A.H. In: The Middle East Journal 21.1967: 86-91.   -21, 38, 2o6-

2781   Saudi Arabian decision limiting use of arbitration clauses in government contracts. In: International Legal Materials 3.1964: 45.   -1a, 3o, 291-

2782   Saudi Arabian decree relating to ownership of Red Sea resources (September 7, 1968). In: International Legal Materials 8.1969: 6o6.   -1a, 12, 291-

2783   Saudi Arabian foreign capital investment code. In: International Legal Materials 3.1964: 561-563.   -1a, 3o, 291-

2784   Saudi Arabian military purchases from the United States. In: International Currency Review 12.198o(5): 48-54.   -18o, 2o6, H 3-

2785   Saudi Arabian nomads and their settlement. In: The Muslim World 45.1955: 386-387.   -21, 291, 352-

2786   Saudi Arabian Royal decree on devaluation of the Riyal, and the establishment of a paper currency. In: The Middle East Journal 14.196o: 2o3-2o5.   -21, 38, 2o6-

2787   Saudi Arabia. Port development ensures foreign trade capability. In: The Arab Economist 14.1982(151): 28, 33.   -2o6, H 3, H 223-

2788   Saudi Arabia. Port plans geared to higher efficiency. In: The Arab Economist 12.198o(133): 21-23.   -2o6, H 3, H 223-

2789   Saudi Arabia: preliminary plan for Riyadh approved. In: Doxiadis Associates Review 5.1969(57): 7-8.

279o   Saudi Arabia. Prepared under the direction of the Commander, Second Air Division, Dhahran Airfield, Saudi Arabia. N.p., 1954. 54 p.   -24-

2791 Saudi Arabia. Record budget launched gigantomania or realism? In: The Arab Economist 7.1975(79): 56-61.   -2o6, H 3-

2792 Saudi Arabia: Riyadh master plan. In: Doxiadis Associates Review 4.1968(45): 7-9.

2793 Saudi Arabia: Riyadh preliminary plan. In: Doxiadis Associates Review 5.1969(53): 7-8.

2794 Saudi Arabia's boom in education. In: The Arab World 7.1961(9): 8-11.   -2o6-

2795 Sa'udi Arabia's finance. The achievements of the Ministry of Finance. In: The Islamic Review 43.1955(12): 33-34. -1a, 21-

2796 Saudi Arabia's industrial future. In: Middle East International 1978(85): 18-2o.   -3o-

2797 Saudi Arabia's port congestion. In: The Arab Economist 9.1977(9o): 3o-33.   -2o6-

2798 Saudi Arabia's port of Dammam. In: The Arab World 7.1961(8): 3-4.   -2o6-

2799 Saudi Arabia's rising resources. In: Petroleum Press Service 25.1958: 341-344.   -18o, 2o6-

28oo Saudi Arabia. Steady growth in economy. In: The Arab Economist 12.198o(129): 24-28.   -2o6, H 3, H 223-

28o1 Saudi Arabia. The awakening giant. In: The Shipping World and Shipbuilder 17o.1977: 267, 269, 271-272.   -83, H 3-

28o2 Sa'udi Arabia. The Buraimi dispute. An explanatory statement by the Sa'udi Arabian Embassy, London. In: The Islamic Review 44.1956(1): 35-36.   -1a, 21-

28o3 Saudi Arabia. The 1393/1394 A.H. budget. In: The Arab Economist 5.1973(59): 28-35.   -H 3-

28o4 Sa'udi Arabia today: Jiddah quarantine station. In: The Islamic Review 44.1956(1o): 24-25.   -1a, 21-

28o5 Saudi Arabia. Towards self-sufficiency in dairy products. In: The Arab Economist 14.1982(149): 22-23.   -2o6, H 3, H 223-

28o6 Saudi Arabia - U.A.R. agreement on self-determination in Yemen (signed at Jeddah, Saudi Arabia, August 24, 1965).

In: International Legal Materials 4.1965: 1139-1140.
-1a, 3o, 291-

**** Saudi-Arabien: s. auch Königreich Saudi-Arabien.

2807 Saudi-Arabien. In: Übersee-Nachrichten 14.1960(1): 5-7.
-3o, 18o, 212-

2808 Saudi-Arabien. In: Weltreise. Alles über alle Länder unserer Erde. Bd. 6: Vorderasien. München: Novaria; Basel: Kister, 1972: 299-326. -212-

2809 Saudi-Arabien. Merkblätter für Auslandtätige und Auswanderer 39; Köln: Bundesverwaltungsamt, 1979. 12 S. -212-

2810 Saudi Arabien bliver et stadig mere interessant marked. In: Udenrigsministeriets Tidsskrift 1981(Feb.): 4-15. -H 3-

2811 Saudi-Arabien. Gasnutzung für den Industrieaufbau. In: The Petroleum Economist 42.1975(5): 19o-191. -89, 2o6, H 3-

2812 Saudiarabien. Inkrafttreten des Rentensystems. In: Internationale Revue für Soziale Sicherheit 27.1974: 411-412. -21, 24, H 3-

2813 Saudi-Arabien, Jemen. In: Die Weltwirtschaft 1957(2): 74-77. -21, 1oo, 2o6-

2814 Saudi-Arabien, Königreich Saudi-Arabien. In: Handbuch der Weltpresse. Hg. vom Institut für Publizistik der Universität Münster unter Leitung von H. Prakke, W.B. Lerg & M. Schmolke. Köln, Opladen: Westdeutscher Verlag, [5]1970 ([1]1931), Bd. 1: 469-471, Bd. 2: 170-171. -21-

2815 Saudi Arabien. Økonomi og udenrigshandel 197o-71. Samhandelen med Danmark. In: Udenrigsministeriets Tidsskrift 53.1972(9): 129-134. -2o6-

2816 Saudi-Arabiens Chemie wird Realität. In: Chemische Industrie (Düsseldorf) 1o4.1981: 147-149, 167. -24, 2o6, H 3-

2817 Saudiarabische Zukunftsplanung mit ambitiösen Chemieprojekten. In: Chemische Industrie (Düsseldorf) 1o1.1978: 258-263. -24, 2o6, H 3-

2818 Saudia Williams pace setters. In: The Middle East (London) 1979(59): 54. -1a, 12, 3o-

2819   Saudi cement. In: Aramco World 13.1962(8): 2-7.   -H 223-

2820   Saudi Consulting House & Arthur D. Little International Inc., n.d.: Consulting for Saudi Arabia in the 1980s. N.p. 8 p.

2821   Saudi Plastics Products Company Ltd., Publicity Department, c. 1975: SAPPCO. Riyadh. 30 p.

2822   Saudi ports vital to development plans. In: Middle East Construction 2.1977(12): 82-84.   -89-

2823   Saudi railroad. In: Aramco World 3.1952(3): 3.

2824   Saudisch-Arabien. Bulletin International des Douanes / Internationaler Anzeiger für Zollwesen (62). ^{1}1955, ^{2}1970, ^{3}1978.   -H 3-

2825   Saudisch-Arabien, Königreich Saudisch-Arabien. In: Institut für Publizistik der Freien Universität Berlin unter Leitung von E. Dovifat (Hg.): Handbuch der Auslandspresse. Bonn: Athenäum; Köln, Opladen: Westdeutscher Verlag, 1960: 628-631.   -21, 24, 212-

2826   Saudis look for long-term solutions. In: The Middle East (London) 1980(63): 10-11.   -1a, 12, 30-

2827   Saudis now biggest US arms customer. In: Armed Forces Journal International 117.1979(10): 6.   -89-

2828   Saudi - UAR agreement on Yemen. In: The Middle East Journal 20.1966: 93-94.   -21, 38, 206-

2829   Sauer, G., 1969: Alois Musil's Reisen nach Arabien im ersten Weltkrieg. Ein Beitrag zu seinem Lebensbild aus Anlass seines 100. Geburtstages am 30. Juni 1968. In: Archiv Orientální 37: 243-263.   -16, 21, 291-

2830   Sauer, H.D., 1976: Der Aufbau der Stickstoffdüngerindustrie in den arabischen Golfstaaten. Ansätze zur Kooperation zwischen Industriestaaten, Ölproduzenten und Entwicklungsländern. Schriften des Deutschen Instituts für Entwicklungspolitik (DIE) 39; Berlin: Deutsches Institut für Entwicklungspolitik (DIE). V, 36 S.   -1a, 180-

2831   Saunders, H.H., 1979: Military equipment programs for Egypt and Saudi Arabia. In: Department of State Bulletin 79(2031): 52.   -21, 180, 206-

2832   Saussey, E., 1935/45: Un pèlerin d'aujourd'hui au Hijâz. In:
        Mélanges Gaudefroy-Demombynes. Mélanges offerts à Gaude-
        froy-Demombynes par ses amis et anciens élèves. Le Caire:
        Institut Français d'Archéologie Orientale, 1935/45: 91-
        1o2.   -21-

2833   Saxen, A., 1967: Situation der bewässerten Landwirtschaft in
        der Ostprovinz Saudi-Arabiens. In: Zeitschrift für Kultur-
        technik und Flurbereinigung 8: 321-344.   -1oo, Gö 153,
        H 223-

2834   Sayed, A.M.M. el-, 1982: An investigation into the syntactic
        errors of Saudi freshmen's English compositions. Ph.D.,
        Indiana University, Bloomington. 234 p.

2835   Sayigh, A.A.M. & E.M.A. el-Salam, 1978: Preliminary design
        data for a solar house in Riyadh. In: UNIDO (ed.): Tech-
        nology for solar energy utilization. Development and
        Transfer of Technology Series 5; New York, 1978: 1o1-
        1o6.   -Bo 149-

2836   Sayigh, Y.A., 1971: Problems and prospects of development in
        the Arabian Peninsula. In: International Journal of Mid-
        dle East Studies 2: 4o-58 (reprinted in: D. Hopwood (ed.):
        The Arabian Peninsula: society and politics. Studies on
        Modern Asia and Africa 8; London: Allen & Unwin, 1972:
        286-31o).   -12, 18, 21-

2837   ——, 1978: The economies of the Arab world: development
        since 1945. London: Croom Helm. 726 p.   -21- -

2838   Sayrafi, Y.H., 1981: Islam versus Planung? Situation der
        staatlichen und örtlichen Planung in Saudi-Arabien. Diss.,
        Fakultät für Bauwesen, Technische Hochschule Aachen.
        2o4 S.   -Bo 149-

2839   Scambio di lettere tra l''Iraq e il Regno Arabo Sa'ūdiano
        per i lavori di delimitazione dei confini. In: Oriente
        Moderno 18.1938: 154-155.   -1a, 5, 21-

284o   Scasso, C., 198o: Arabie saoudite. Les chemins de fer du
        désert. In: Le Rail et le Monde (13/298): 23-32.   -89,
        2o6-

2841   Schacht, J., 1931: Der Islām mit Ausschluss des Qor'āns. Re-
        ligionsgeschichtliches Lesebuch 16; 2., erw. Aufl. Tübin-
        gen: Mohr (Siebeck). XII, 196 S.   -21-

2842 Scharabi, M., 1974: Civic-Center in Djeddah, Saudi-Arabien. In: Deutsche Bauzeitung 1o8: 653-656.   -24, 25, 93-

2843 ——, 1979: Das traditionelle Wohnhaus der arabischen Halbinsel. In: Architectura 9: 77-9o.   -1a, 12, 24-

2844 ——, 198o: The new town of Jubail and the civic centre at Jedda. In: M. Meinecke (ed.): Islamic Cairo: architectural conservation and urban development of the historic centre. Proceedings of a seminar organised by the Goethe-Institute, Cairo (October 1-5, 1978). German Institute of Archaeology, Art and Archaeology Research Papers 18; London, 198o: 1oo-1o4.   -21-

2845 Schechterman, B., 1981/82: Political instability in Saudi Arabia and its implications. In: Middle East Review 14(1/2): 15-25, 75.   -21, H 223-

2846 Scheltema, J.F., 1917: Arabs and Turks. In: Journal of the American Oriental Society 37: 153-161.   -1a, 16, 21-

2847 Schemeil, Y., 1982: Du Cadi au Caddie: attitudes envers la modernisation dans les pays arabes du Golfe. In: Centre d'Études et de Recherches sur l'Orient Arabe Contemporain (éd.): La péninsule Arabique d'aujourd'hui. Sous la direction de P. Bonnenfant. T. 1. Paris: Centre National de la Recherche Scientifique, 1982: 245-276.   -21-

2848 Schleifer, J., 1913: Al-Dar'īya. In: EI[1] 1: 964.   -21, 24-

2849 ——, 1927a: Farasān (Farsān). In: EI[1] 2: 62.   -21, 24-

285o ——, 1927b: Hofhūf. In: EI[1] 2: 344-345.   -21, 24-

2851 ——, 1927c: Hutaim. In: EI[1] 2: 37o.   -21, 24-

2852 Schliephake, K., 1981: Die Arabische Halbinsel. In: Wissen heute 4; Niedernhausen, 1981: 128-171.

2853 —— & G. Schulz, 1982: Die Arabische Halbinsel. Landeskundliche Strukturanalyse mit didaktisch-schulpädagogischer Umsetzung. Würzburger Geographische Manuskripte 14; Würzburg: Geographisches Institut und Institut für Pädagogik der Universität Würzburg, 1982. 91 S.   -2o-

2854 ——, 1982: Regionalplanung in Saudi Arabien. Infrastruktur und Städtebau. In: Geographie im Unterricht 7: 462-464 (zugleich: Themenheft 13: Orient: 44-46).   -1a, 24, 3o-

2855 Schliephake, K., 1983: Wirtschafts- und Planungspolitik. In: T. Koszinowski (Hg.): Saudi-Arabien: Ölmacht und Entwicklungsland. Beiträge zur Geschichte, Politik, Wirtschaft und Gesellschaft. Mitteilungen des Deutschen Orient-Instituts 2o; Hamburg: Deutsches Orient-Institut, 1983: 251-297. -21, H 223-

2856 Schmid, P., 1974: Letter from Saudi Arabia: in the name of Allah. In: Encounter (London) 42(6): 48-51. -21, 24, 291-

2857 Schmidt, W., 1917: Der Kampf um Arabien zwischen der Türkei und England. In: Geographische Zeitschrift 23: 197-215. -21, 24, 2o6-

2858 Schmidt-Pathmann, W., 1979: Forecast of seaborne imports to Saudi Arabia and corresponding port capacities. In: Arab ports in the 7o's: a survey of ports, shipping and maritime trade in the Middle East, based on the proceedings of the Arab Ports Conference, London. London: International Communications, 1979: 99-1o6.

2859 Schmitz-Kairo, P., 1939: Politiker und Propheten am Roten Meer. Leipzig: Goldmann. 233 S. -21, 24/213-

286o ——, 1942: Die Arabische Revolution. Leipzig: Goldmann. 221 S. -21, 24/213-

2861 Schnabel, A. & K. Breidenbach, 1979: Einrichtung einer Ausbildungsabteilung für Kälte- und Klimatechnik an den SVS Saudi-Arabiens. Ermittlung von Planungsdaten. Studie im Auftrag der GTZ. Ludwigshafen. 237 S. -Esb 2ooo-

2862 Schnepp, B., 1865: Le Pèlerinage de la Mecque. Infidèles qui ont visité la Mecque. Djedda. Le Tombeau d'Ève. La Mecque. Le Kaaba. La Vallée de Menaa. Le Mont Arafat. Sacrifices. Dispersion des pèlerins. Conséquences pour la santé publique. Paris: Leclerc.

2863 Schoedl, P.F., 1965: Saudiarabien. In: Confrontation 5(2): 3-7. -2o6, 212, 291-

2864 Scholars on the job. In: Aramco World 14.1963(3): 3-6. -H 223-

2865 Scholz, F., 198o: Erdölfördergebiete und Oasenlandwirtschaft im arabischen Trockenraum. Die Provinz al-Hasa/Saudi-Arabien. In: Geographische Rundschau 32: 523-526.   -2o6, 212, Tü 17-

2866 Schools complex at Riyadh. In: Journal of the Royal Institute of British Architects 83.1976: 247-248.

2867 Schott, G., 1918: Geographie des Persischen Golfes und seiner Randgebiete. In: Mitteilungen der Geographischen Gesellschaft in Hamburg 31; Hamburg: Friederichsen, 1918: 1-11o.   -7-

2868 Schreiber, F., 1981: Die Saudis. Macht und Ohnmacht der Herrscher Arabiens. Wien, München, Zürich, New York: Molden. 352 S.   -12, 21, 212-

2869 Schütze, R.A., 1977: Gesellschaftsgründung in Saudi-Arabien. In: Internationale Wirtschaftsbriefe 1977: 2o5-2o8.   -282, 7o8-

287o ——, 1978: Die GmbH im saudischen Recht. In: GmbH-Rundschau mit Sonderfragen der GmbH & Co 69: 82-85.   -1a, 24, 18o-

2871 ——, 1979: GmbH-Recht in Saudi-Arabien. In: Internationale Wirtschaftsbriefe 1979: 3o3-3o4.   -282, 7o8-

2872 Schulz, G., 1983: Saudi-Arabien. Erdöl wandelt das Gesicht eines Landes (UE Sek. I/Hauptschule Kl. 8). In: Praxis Geographie 13(5): 25-31.   -1a, 5, 465-

2873 Schuster, W., 1979: Wirtschaftsgeographie Saudi Arabiens mit besonderer Berücksichtigung der staatlichen Wirtschaftslenkung. Dissertationen der Wirtschaftsuniversität Wien 27; Wien: Verband der wissenschaftlichen Gesellschaften Österreichs. 222 S.   -21, 24, 1oo-

2874 Schweizer, G., 1976: Bevölkerung, traditionelle Lebens- und Wirtschaftsformen. In: H. Blume (Hg.): Saudi-Arabien. Natur, Geschichte, Mensch und Wirtschaft. Ländermonographien 7; Tübingen, Basel: Erdmann, 1976: 167-252.   -12, 21, 93-

2875 ——, 198o: Gastarbeiter in Saudi-Arabien. In: H.K. Barth & H. Wilhelmy (Hgg.): Trockengebiete: Natur und Mensch im ariden Lebensraum; Festschrift zum 6o. Geburtstag von Helmut Blume. Tübinger Geographische Studien 8o (Sbd. 13); Tübingen: Geographisches Institut der Universität Tübingen, 198o: 353-365.   -21, Tü 17-

2876 Schweizerische Bankgesellschaft, 1976: Länderbericht für Saudi-Arabien. Zürich (englischsprachige Ausgabe: Union Bank of Switzerland: Country report for Saudi Arabia. Zurich, London, 1976). 12 S.

2877 Scott, R.W., 1974: Saudi Arabia. Where the oil is. I. In: World Oil 179(1): 145-159.   -89, 2o6, H 3-

2878 ——, 1976: Saudi Arabia's gas program. A big job for Aramco. In: World Oil 183(2): 25-3o.   -89, 2o6, H 3-

2879 Scouting with a difference. In: Aramco World 12.1961(4): 3-6.   -H 223-

288o Scoville, S.A. (ed.), 1979: Gazetteer of Arabia. A geographical and tribal history of the Arabian Peninsula. Vol. 1: A-E. Graz: Akademische Druck- und Verlagsanstalt. VII, 733 p.   -21, Tü 17-

2881 Seabrook, W.B., 1928: Adventures in Arabia: among the Bedouins, Druses, whirling dervishes and Yezidee devil-worshippers. London, Bombay, Sydney: Harrap. 293 p.   -7, 21, 93-

2882 Sea-going drilling rig. In: Aramco World 15.1964(2): 3-7. -H 223-

2883 Seaman, B.W., 198o: Islamic law and modern government: Saudi Arabia supplements the Shari'a to regulate development. In: Columbia Journal of Transnational Law 18: 413-481. -7, 2o6, 352-

2884 Search beneath the sands. In: Aramco World 12.1961(9): 3-7. -H 223-

2885 Sebai, Z.A., 1974: Knowledge, attitudes and practice of family planning: profile of a Bedouin community in Saudi Arabia. In: Journal of Biosocial Science 6: 453-461. -1a, 21, 2o6-

2886 —— & T.D. Baker, 1976: Projected needs of health manpower in Saudi Arabia, 1974-9o. In: Medical Education 1o: 359-361.   -26, 289, 291-

2887 —— & M.H. Shehata, 1978: Letters from a random sample of television viewers provide health education planning data in Saudi Arabia. In: International Journal of Health Education 21(1): 53-55.   -83-

2888 Seering, R., 1974: König Faisal - Koran und Öl. Bergisch Gladbach: Lübbe. 24o S.   -2o6, 212-

2889 Seetzen, U.J., 18o5: Fortgesetzte Reise-Nachrichten des Dr. U.J. Seetzen. In: Monatliche Correspondenz zur Beförderung der Erd- und Himmels-Kunde 12: 234-241.   -4, 21, 9o-

2890 ——, 18o8: Beyträge zur Geographie Arabiens. In: Monatliche Correspondenz zur Beförderung der Erd- und Himmels-Kunde 18: 371-393.   -4, 21, 9o-

2891 ——, 18o9: Beyträge zur Kenntniss der arabischen Stämme in Syrien und im wüsten und peträischen Arabien. In: Monatliche Correspondenz zur Beförderung der Erd- und Himmels-Kunde 19: 1o5-133, 213-233 (französische Ausgabe: Mémoire pour servir à la connoissance des tribus arabes en Syrie et dans l'Arabie déserte et pétrée. In: Annales des Voyages, de la Géographie et de l'Histoire 8.18o9: 281-324).   -4, 21, 9o-

2892 ——, 1813: Auszug aus einem Schreiben des Russ. Kais. Kammer-Assessors Dr. U.J. Seetzen. In: Monatliche Correspondenz zur Beförderung der Erd- und Himmels-Kunde 27: 61-79, 16o-182 (auszugsweise französische Ausgabe: Voyage de M. Seetzen sur la mer Rouge et dans l'Arabie. In: Annales des Voyages, de la Géographie et de l'Histoire 22.1813: 3o9-333).   -4, 21, 9o-

2893 ——, 1854: Ulrich Jasper Seetzen's Reisen durch Syrien, Palästina, Phönicien, die Transjordan-Länder, Arabia Petraea und Unter-Aegypten. Hgg. und commentirt von F. Kruse et al. Bd. 1. Berlin: Reimer. LXXV, 432 S.   -24-

2894 Seflan, A.M. al-, 1981: The essence of tribal leaders' participation, responsibilities, and decisions in some local government activities in Saudi Arabia: a case study of the Ghamid and Zahran tribes. Ph.D., Claremont Graduate School. 217 p.

2895 Seifert, W.W., M.A. Bakr & M.A. Kettani (eds.), 1973: Energy and development: a case study. Based upon an interdepartmental student project in systems engineering at the Massachusetts Institute of Technology, spring term, 1971, an M.I.T. sea grant project, and a study at the College of

Petroleum and Minerals, Dhahran, Saudi Arabia. M.I.T. Report 25; Cambridge/Mass., London: The MIT Press. XIX, 3oo p.   -21, 2o6, H 3-

2896 Sékaly, A., 1926: Les deux congrès musulmans de 1926. In: Revue du Monde Musulman 64: 3-222 (zugleich: Le Congrès du Khalifat (Le Caire, 13-19 Mai 1926) et le Congrès du Monde Musulman (La Mekke, 7 juin - 5 juillet 1926). Collection de la Revue du Monde Musulman; Paris: Leroux, 1926.. 219 p.).   -12, 21-

2897 Selow-Serman, K.E., 1917: Kapitänleutnant v. Möllers letzte Fahrt. Berlin: Scherl. 125 S.   -1a-

2898 Seoudi, M.A., 1917: Voyages au Hédjaz et en Arabie. In: Bulletin de la Société Sultanieh de Géographie N.S. 8: 356-36o.   -18o-

2899 Sergeant, R.B. & G.M. Wickens, 1949: The Wahhābis in western Arabia in 18o3-4 A.D. In: Islamic Culture 23: 3o8-3o9. -21-

29oo Serrano de Lababidy, M. del Pilar, 1956: La evolución contemporánea de los países árabes. Tetuán: Majzen. 225 S. -24-

29o1 Sertoli Salis, R., 194o: Italia, Europa, Arabia. Manuali di Politica Internazionale 24; Milano: Istituto per gli Studi Politica Internazionale. 41o S.

29o2 Seton-Williams, M.V., 1948: Britain and the Arab states: a survey of Anglo-Arab relations, 192o-1948. London: Luzac (reprint: Westport: Hyperion Press, 1981). IX, 33o p. -21, 24-

29o3 Seven Arabian Markets Ltd., Marketing Division, n.d.: Business guide to Saudi Arabia. N.p. 12o p.

29o4 Seven wells of Dammam. In: Aramco World 14.1963(1): 18-21. -H 223-

29o5 Severino, D., 1972: Golf in the Arab world. In: Aramco World Magazine 23(Special): 16-25.   -24-

29o6 Sha'afy, M.S. al-, 1973/74: Notes on the economic history of Juddah in the first half of the nineteenth century. In: Bulletin of the Faculty of Arts, University of Riyad 3: 23-39.   -21-

2907 Shaafy, M.S.M. el-, 1966/67: The first Sa'udi State in Arabia (with special reference to its administrative, military and economic features) in the light of unpublished materials from Arabic and European sources. Ph.D., University of Leeds.

2908 ——, 1969/73: The military organisation of the first Sa'udi state. In: The Annual of Leeds University Oriental Society 7: 61-74.   -12, 16, 21-

2909 Shadly, R.A., 1978: A study of the teacher training program at Riyadh University: by survey of opinions of teacher education graduates with in-service experiences. Ed.D., University of Northern Colorado, Greeley. 152 p.

2910 Shadukhi, S.M., 1981: Application of organization development (OD) in Saudi Arabia's public organizations: a feasibility study. Ph.D., Florida State University, Tallahassee. 235 p.

2911 Shaffer, R., 1952: Tents and towers of Arabia. New York: Dodd, Mead. XI, 276 p.   -18o, Tü 17-

2912 Shah, I.A., 1928: Westward to Mecca: a journey of adventure through Afghanistan, Bolshevik Asia, Persia, Iraq and Hijaz to the cradle of Islam. London: Witherby. 224 p.   -1a-

2913 ——, 1933: Alone in Arabian nights. London: Wright & Brown. 282 p.   -1a-

2914 Shah, S.A., 1977: The political and strategic foundations of international arms transfers: a case study of American arms supplies to, and purchases by, Iran and Saudi Arabia; 1968-76. Ph.D., University of Virginia, Charlottesville. XI, 337 p.   -H 223-

2915 Shair, I.M., 1977: Spatial pattern of Muslim pilgrim circulation. Ph.D., University of Kentucky, Lexington. 17o p.

2916 Shair Management Services, 1979: Business laws and practices of Saudi Arabia. Arabian Business Laws and Practices Series; London: Arabian Information. 159 p.

2917 Shaked, H. & T. Yegnes, 1977: Saudi Arabia. In: Middle East Record 5: 1o27-1o39.   -12-

2918　Shaked, H. & T. Yegnes, 1978: The Saudi Arabian Kingdom. In: Middle East Contemporary Survey 1: 565-585.   -21, 1oo, H 223-

2919　—— & ——, 1979: The Saudi Arabian Kingdom. In: Middle East Contemporary Survey 2: 675-694.   -21, 1oo, 2o6-

292o　Shaker, Fatina A., 1972: Modernization of the developing nations: the case of Saudi Arabia. Ph.D., Purdue University, Lafayette. XIII, 395 p.   -21-

2921　Shamekh, A.A., 1975: Spatial patterns of Bedouin settlement in Al-Qasim region, Saudi Arabia. Ph.D., Department of Geography, University of Kentucky, Lexington. XIII, 316 p. -21-

2922　——, 1975/76: Bedouin sedentarization in al-Qasim region, Saudi Arabia. In: Bulletin of the Faculty of Arts, University of Riyad 4: 21-29.

2923　——, 1977: Bedouin settlements. In: Ekistics 43: 249-259. -1a, 89, 2o6-

2924　——, 1979: Bedouin settlement in the Mihmal area of Saudi Arabia. In: Addarah 5(2): 3-17.   -21, 3o-

2925　——, 198o: Drainage system in the Qasim region of Saudi Arabia. In: Addarah 5(4): 5-16.   -3o-

2926　Shamekh, M.A.R. al-, 1978: The emergence of printing houses in Arabia. In: Addarah 4(4): 3-4.   -21, 3o-

2927　Shami, I.A. al-, 1977: Tradition and technology in the developmental education of Saudi Arabia and Egypt. Ph.D., University of Michigan, Ann Arbor. V, 22o p.   -21-

2928　Shamma, S. & W.D. Morrison, 1977a: The use of local representatives in Saudi Arabia. In: The International Lawyer 11: 453-465.   -1a, 4, 291-

2929　—— & ——, 1977b: Qualification, licensing and registration of foreign companies in Saudi Arabia. In: The International Lawyer 11: 693-699.   -1a, 4, 291-

293o　Shanneik, G., 1978: Einblick in das System der sozialen Sicherheit in Saudi-Arabien. In: Recht der Internationalen Wirtschaft 24: 6oo-6o2.   -12, 24, H 3-

2931 Shanneik, G., 198oa: Die Modernisierung des traditionellen politischen Systems in Saudi-Arabien. In: Orient (Opladen) 21: 3oo-319. -21, 1oo, H 223-

2932 ——, 198ob: Politischer Wandel und die Entwicklung der öffentlichen Verwaltung in Saudi-Arabien seit 1953. In: Orient (Opladen) 21: 486-5o1. -21, 1oo, H 223-

2933 ——, 1983: Die Modernisierung des traditionellen politischen Systems. - Voraussetzungen und Folgen der Arbeitskräftewanderung. In: T. Koszinowski (Hg.): Saudi-Arabien: Ölmacht und Entwicklungsland. Beiträge zur Geschichte, Politik, Wirtschaft und Gesellschaft. Mitteilungen des Deutschen Orient-Instituts 2o; Hamburg: Deutsches Orient-Institut, 1983: 151-176 bzw. 327-349. -21, H 223-

2934 Sharabi, H.B., 1962: Governments and politics of the Middle East in the twentieth century. Van Nostrand Political Science Series; Princeton, Toronto, London, New York: Van Nostrand. XIII, 296 p. -21-

2935 Sharbiny, S.U.T., 197o: Credit and education in the development of agriculture: the potential role of the Saudi Arabian Agricultural Bank. M.A., Graduate Program of Development Administration, American University of Beirut. IV, 1o3 p. -Jafet Library, American University of Beirut-

2936 Sharshar, A.M., 1977: Oil, religion, and mercantilism: a study of Saudi Arabia's economic system. In: Studies in Comparative International Development 12(3): 46-64. -3o, 352, H 3-

2937 ——, 1978: Trade policy and economic development in Saudi Arabia. In: Virginia Social Science Journal 13(Apr.): 5o-54.

2938 Shaw, J.A. & D.E. Long, 1982: Saudi Arabian modernization: the impact of change on stability. The Washington Papers 1o(89); New York: Praeger. XII, 111 p. -21-

2939 Shawly, A.T., 198o: Toward an educational-vocational guidance model in the Kingdom of Saudi Arabia. Ph.D., University of Wisconsin, Madison. 254 p.

294o Shea, T.W., 1969a: The Riyal: a miracle in money. In: Aramco World Magazine 2o(1): 26-33. -24-

2941 Shea, T.W., 1969b: Measuring the changing family consumption patterns of Aramco's Saudi Arab employees - 1962 and 1968. Dhahran: ARAMCO (reprinted in: D. Hopwood (ed.): The Arabian Peninsula: society and politics. Studies on Modern Asia and Africa 8; London: Allen & Unwin, 1972: 231-254). -21, 24-

2942 Shebl, A.H., 1978: Development of a model for a mathematics laboratory in Saudi Arabia. Ed.D., University of Northern Colorado, Greeley. 127 p.

2943 Sheean, V., 1966: King Faisal's first year. In: Foreign Affairs (New York) 44: 3o4-313. -16, 21, 18o-

2944 ——, 1975: Faisal: the king and his kingdom. Tavistock: University Press of Arabia. X, 161 p. -12, 24/213-

2945 Sheehan, E.R.F., 1974: Der Chefplaner der Erdölkrise. In: Das Beste aus Reader's Digest 1974(11): 35-4o. -12, 24, 18o-

2946 Sheikh, Abdulrahman Abdulaziz A.H. al-, 197o: Agriculture and economic development, with special emphasis on a strategy for Saudi Arabian economic development. Ph.D., University of Edinburgh. XII, 233 p.

2947 Sheikh, Abdul Ghafur, 1953: From America to Mecca on airborne pilgrimage. A Moslem student at Harvard Business School records Islam's sacred rites in color in the interest of world understanding. In: The National Geographic Magazine 1o4: 1-6o. -21, 24, 291-

2948 Sheikh, Abid M.S., 1965: Le développement industriel en Arabie Saoudite. In: Revue de la Société d'Études et d'Expansion 64: 5o5-5o9. -2o6, 212, 282-

2949 Sherbini, A.A. el-, 1978: Agricultural development in the Gulf States and the contribution of European technology. Beirut: Economic Commission for Western Asia. 14 p.

295o ——, 198o: Environmental adversity and food policy in the Arab Gulf states. In: Food Policy 5(2): 97-1o4. -12, 89, 2o6-

2951 Sherbiny, N.A., 1981: Sectoral employment projections with minimum data: the case of Saudi Arabia. In: N.A. Sherbiny (ed.): Manpower planning in the oil countries. Research in Human Capital and Development, Suppl. 1; Greenwich/ Conn.: Jai Press, 1981: 173-206. -H 223-

2952 Sherwood, M.A. (ed.), 1970: Guide to world science. Vol. 9: Near East. Guernsey: Hodgson. 209 p. -212-

2953 Sheshsha, J.A., 1982: The qualifications of a competent teacher of English in Saudi Arabia as perceived by successful EFL teachers and selected TESOL specialists. Ph.D., Indiana University, Bloomington. 166 p.

2954 Shiber, S.G., 1961: City growth in the Eastern Province of Saudi Arabia. In: Mid-East Commerce (54): 48-52, 54.

2955 Shilling, N.A., 1975: Doing business in Saudi Arabia and the Arab Gulf states. Doing Business in the Middle East 1; New York: Inter-Crescent Publishing and Information Corporation. 455 p.

2956 ——, 1978a: Commercial Regulations applying in Arabian Peninsula oil states. In: Middle East Annual Review 4: 43-50, 52-54 (reprinted in: Middle East Annual Review 5.1979: 43-46, 48-51, 53-55). -21-

2957 ——, 1978b: A practical guide to living and travel in the Arab world. New York: Inter-Crescent Publishing and Information Corporation. XII, 295 p.

2958 Shinawi, A.A.K., 1970: The role of accounting and accountant in the developing economy of Saudi Arabia. D.B.A., University of Southern California, Los Angeles. 294 p.

2959 Shirreff, D., 1976: Saudi Arabia. A MEED Special Report; London: Middle East Economic Digest. 34 p. -H 223-

2960 ——, 1978: Saudi Arabia. A MEED Special Report; London: Middle East Economic Digest. 84 p. -H 223-

2961 ——, 1979a: Saudi Arabia. A MEED Special Report; London: Middle East Economic Digest. 68 p.

2962 ——, 1979b: Saudi Arabia. In: Middle East Annual Review 5: 315, 317, 319, 321, 323, 325-326, 331, 335, 337-338. -1a, 21, 100-

2963 —— & W. Ryder (eds.), 1979: Saudi Arabia. A MEED Special Report; London: Middle East Economic Digest. 68 p. -H 223-

2964 Shoaib, M.S., 1980: Development of social studies education in Saudi Arabia since 1926. Ph.D., University of Missouri, Columbia. 157 p.

2965 Shobaili, A.S., 1971: An historical and analytical study of broadcasting and press in Saudi Arabia. Ph.D., Ohio State University, Columbus. VI, 349 p. -21-

2966 Shobokshi, S., 1965: Training for community development in Saudi Arabia, 1964-65. New York: UNO.

2967 Shomrany, S.A. al-, 1980: Types, distribution, and significance of agricultural terraces in Assarah, south-western Saudi Arabia. M.A., Michigan State University, East Lansing. 197 p.

2968 Shuaiby, A.M. al-, 1976: The development of the Eastern Province, with particular reference to urban settlement and evolution in eastern Saudi Arabia. Ph.D., Faculty of Social Science, University of Durham.

2969 Shukri, A.I., 1965/66: The development of formal education in Saudi Arabia, with particular reference to social and economic changes. M.A., Institute of Education, University of London.

2970 ——, 1972/73: Educational manpower needs and socio-economic development in Saudi Arabia. Ph.D., Institute of Education, University of London.

2971 Shurrab, S., 1972: Agricultural conditions in Al Hassa and their effect on agricultural production. Paper presented at the Hofuf Agricultural Research Seminar, 24-30 March 1972. 5 p.

2972 Shwadran, B., 1973: The Middle East, oil and the great powers. 3rd ed., rev. & enl. The Shiloah Center for Middle Eastern and African Studies, The Monograph Series; Jerusalem: Israel Universities Press (^{1}1955, ^{2}1959). XVIII, 630 p. -21, 93, 100-

2973 ——, 1978: Middle East oil developments. In: Middle East Contemporary Survey 1: 263-275. -21, 100, H 223-

2974 Sikandar Begam, 1906: Pilgrimage to Mecca, by the Nawab Sikandar Begam of Bhopal, G.C.S.I. Ed. by Mrs. Willoughby-Osborne. Calcutta: Thacker, Spink. XI, 205 p.

2975 Silverfarb, D.N., 1972: British relations with Ibn Saud of Najd, 1914-1919. Ph.D., Graduate School, University of Wisconsin, Madison. IV, 222 p.   -21-

2976 ——, 1979a: The British Government and the Khurmah dispute, 1918-1919. In: Arabian Studies 5: 37-6o.   -1a, 12, 21-

2977 ——, 1979b: The Philby mission to Ibn Sa'ud, 1917-18. In: Journal of Contemporary History 14: 269-285.   -21, 18o, 352-

2978 ——, 198o: The Anglo-Najd treaty of December 1915. In: Middle Eastern Studies 16: 167-177.   -21, 188, 2o6-

2979 ——, 1982: The Treaty of Jiddah of May 1927. In: Middle Eastern Studies 18: 276-285.   -1a, 21, 3o-

298o Sim, K., 1969: Desert traveller: the life of Jean Louis Burckhardt. London: Gollancz. 447 p.   -25-

2981 Simansky, N., 1955: Report to the Government of Saudi Arabia on the preliminary project for land and water use development in Wadi Jizan. Report 41o; Rome: FAO. 38 p.

2982 Simmersbach, B., 19o7: Die Hedschasbahn. In: Asien (Berlin) 6: 139-141.   -1a, 25, 9o-

2983 ——, 1919: Arabien. In: Asien (Berlin) 16: 157-163, 176-178.   -1a, 21, 2o6-

2984 Simmons, A., 1981: Arab foreign aid. London, Toronto: Associated University Presses. 196 p.   -H 223-

2985 Simmons, J.S., T.F. Whayne, G.W. Anderson, M.M. Horack, R.A. Thomas & collaborators, 1954: Global epidemiology: a geography of disease and sanitation. Vol. 3: The Near and Middle East. Philadelphia, London, Montreal: Lippincott. XXIV, 357 p.   -21, 25-

2986 Simon, K., 1929: Slavery. London: Hodder & Stoughton (^{2}193o) XIII, 284 p.   -21-

2987 Simpich, F., 1919: The rise of the new Arab nation. In: The National Geographic Magazine 36: 369-393.   -16, 21, 93-

2988 Sinclair, R.W. (ed.), 1976: Documents on the history of southwest Arabia. Tribal warfare and foreign policy in Yemen, Aden and adjacent tribal kingdoms, 192o-1929. 2 vol

Salisbury/N.C.: Documentary Publications. XI, 224 resp.
III, 225-479 p. -21-

2989 Sindi, A.M., 1980: King Faisal and Pan-Islamism. In: W.A. Beling (ed.): King Faisal and the modernisation of Saudi Arabia. London: Croom Helm; Boulder: Westview Press, 1980: 184-2o1. -21, H 223-

2990 Sindi, S.B., 1975: A rationale and comprehensive traffic safety education program for Saudi Arabia. Ph.D., Michigan State University, East Lansing. 351 p.

2991 Singer, S.F., 1982: Saudi Arabia's oil crisis. In: Policy Review 21: 87-1oo. -2o6-

2992 SIPRI. Stockholm International Peace Research Institute, 1971: The arms trade with the third world. Stockholm: Almqvist & Wiksell; New York: Humanities Press. XXXI, 91o p. -24/213-

2993 ——, 1972-: World armaments and disarmament. SIPRI yearbook 1972-. London, New York: Taylor & Francis. -21-

2994 Siradj, A., 1925: La question du Califat. Le Hedjaz et le comité Hindou. In: Renseignements Coloniaux et Documents, Suppl. à l'Afrique Française 35.1925: 355-364. -3o, 291-

2995 Sisley, T., 1980: Saudi Arabia: the political future. In: Middle East International (127): 7-8. -12, 3o-

2996 Situation économique de l'Arabie Séoudite en 1974. In: Syrie et Monde Arabe 22.1975(261): 11-18. -2o6, H 3-

2997 Skornia, V., 1978: Stadtentwässerung für Hofuf und Mubarraz in Saudi-Arabien. In: Bauen macht Freude, Heitkamp-Mitteilungen 1978(1): 7-9.

2998 Slaugh, F.S., 1968: Arabian fires. New York: Carlton Press. 3o6 p.

2999 Slemman (pseud. de Lammens), H., 19oo: Le chemin de fer de Damas-La Mecque. In: Revue de l'Orient Chrétien 1o: 5o7-534. -21, 291, 352-

3ooo Slow growth in Saudi agricultural sector; water storage remains major obstacle. In: The Arab Economist 9.1977(91): 33-35. -2o6-

3oo1 Sluglett, P. & M. Farouk-Sluglett, 1982: The precarious monarchy: Britain, Abd al-Aziz ibn Saud and the establishment of the Kingdom of Hijaz, Najd and its Dependencies, 1925-1932. In: T. Niblock (ed.): State, society and economy in Saudi Arabia. London: Croom Helm; Exeter: Centre for Arab Gulf Studies, 1982: 36-57.   -21-

3oo2 Smalley, W.F., 1932: The Wahhabis and Ibn Sa'ud. In: The Moslem World 22: 227-246.   -21, 291, 352-

3oo3 Smeaton, B.H., 1973: Lexical expansion due to technical change as illustrated by the Arabic of Al Hasa, Saudi Arabia. Indiana University Publications, Language Science Monographs 1o; Bloomington: Indiana University Press (zugleich: Ph.D., Columbia University, New York, 1959). XV, 19o p.   -12, 21-

3oo4 Smith, Adam, 1978: The Saudi connection. In: Atlantic Monthly 242(6): 44-47.   -12, 18o, 352-

3oo5 Smith, Alan M., 1982a: Saudi ports prosper under SPA policies. In: Containerisation International 16(2): 45-49. -H 3-

3oo6 ——, 1982b: Saudi Arabia's 'national' lines reveal their colours. In: Containerisation International 16(3): 15-17. -H 3-

3oo7 Smith, J.T., 1951: Development of irrigation farms with special reference to irrigation and crop production under desert conditions as observed in Saudi Arabia. In: Proceedings of the United Nations Scientific Conference on the Conservation and Utilization of Resources 4: 385-388. -2o, 188, 2o6-

3oo8 Smith, W.R., 1912: A journey in the Hejâz. In: J.S. Black & G. Chrystal (eds.): Lectures and essays of William Robertson Smith. London: Black, 1912: 484-597.   -21-

3oo9 Smithers, R., 1966: Bedouin development in Saudi Arabia: the Haradh Project. Beirut: Ford Foundation.

3o1o Snouck Hurgronje, C., 1888-89: Mekka. 2 Bde. Den Haag: Nijhoff (auszugsweise englische Ausgabe: Mekka in the latter part of the 19th century. Daily life, customs and learning. The Moslims of the East-Indian-Archipelago. Leiden:

Brill; London: Luzac, 1931 (reprint: Leiden: Brill, 197o). VI, 3o9 p.). XXIII, 228 bzw. XVIII, 397 S. -21, 24-

3o11 Snouck Hurgronje, C., 1923-27: Verspreide Geschriften (Gesammelte Schriften) van C. Snouck Hurgronje. 6 Bde. Bonn, Leipzig: Schroeder. -21-

3o12 ——, 1977: Some of my experiences with the Muftis of Mecca (1885). In: Asian Affairs (London) 64/N.S. 8: 25-37. -1a, 21, 291-

3o13 Snyder, H.R., 1963: Community college education for Saudi Arabia. A report of a Type A project. Ph.D., Teachers College, Columbia University, New York. 18o p. -21-

3o14 Snyder, J.C., R.L. Nichols, S.D. Bell, Jr., N.A. Haddad, E. S. Murray & D.E. McComb, 1964: Vaccination against trachoma in Saudi Arabia: design of field trials and initial results. In: Harvard School of Public Health: Industry and Tropical Health 5: 65-73.

3o15 Sobaihi, M.A., 1982: The use of television to meet social needs in Saudi Arabia: an historical analytical study with a plan for fuller utilization of the medium. Ph.D., University of Southern California, Los Angeles.

3o16 Sobaihi, S.M. al-, 1976: Water resources of Wadi Hanifah, Saudi Arabia: a case study. M.A., University of Durham.

3o17 Soccer/baseball of Saudi Arabia. In: Aramco World 14.1963(8): 2-7. -H 223-

3o18 Solaim, S.A., 197o: Constitutional and judicial organization in Saudi Arabia. Ph.D., Johns Hopkins University, Baltimore. 3, VI, 231 p. -21-

3o19 ——, 1971: Saudi Arabia's judicial system. In: The Middle East Journal 25: 4o3-4o7. -12, 21, 38-

3o2o Soliman, T.M.A. al-, 1981: School buildings for boys' general education in Saudi Arabia: present functioning, future demands and proposed alternatives under conditions of social change. 2 vols. Arch.D., University of Michigan, Ann Arbor. 529 p.

3o21 Soubhy, S., 1894: Pèlerinage à la Mecque et à Médine, précédé d'un aperçu sur l'Islamisme et suivi de considérations générales au point de vue sanitaire et d'un appendice sur la circoncision. Le Caire: Imprimerie Nationale. 129 p. -21-

3022 Soulié, G.J.-L. & L. Champenois, 1966: Le royaume d'Arabie saoudite face à l'Islam révolutionnaire 1953-1964. Cahiers de la Fondation Nationale des Sciences Politiques 145; Paris: Colin. 135 p.   -24, 212-

3023 ——, 1966: Formes et action actuelle du Wahabisme. In: L'Afrique et l'Asie (74): 3-1o.   -1a, 21, 2o6-

3024 —— & L. Champenois, 1977: La politique extérieure de l'Arabie Saoudite. In: Politique Étrangère 42: 6o1-622. -2o6, 212, H 3-

3025 —— & ——, 1978: Le Royaume d'Arabie Saoudite à l'épreuve des temps modernes. Un homme providentiel: Fayçal. Présence du Monde Arabe; Paris: Michel. 248 p.   -21-

3026 South Asian Student Association & Arab Support Committee, 1972: Struggle, oppression and counter-revolution in Saudi Arabia. Berkeley: Arab Support Committee.

3027 Soviet and American military bases in Iraq and Saudi Arabia. In: International Currency Review 12.198o(1): 4o-44. -18o, 2o6, H 3-

3028 Sowaygh, I.A. al-, M.E. el-Sayed & M.E. Abdullah, 1977: Designing a new curriculum for undergraduate pharmacy in Saudi Arabia. In: American Journal of Pharmaceutical Education 41(Feb.): 38-44.   -38 M-

3029 Sowayyegh, A.H. al-, 198o: Saudi oil policy during King Faisal's era. In: W.A. Beling (ed.): King Faisal and the modernisation of Saudi Arabia. London: Croom Helm; Boulder: Westview Press, 198o: 2o2-229.   -21, H 223-

3o3o Sowygh, H. ibn Z. el-, 1981: Performance of a Piagetian test by Saudi Arab students in Colorado colleges and universities in relation to selected sociodemographic and academic data. Ph.D., University of New Mexico, Albuquerque. 166 p.

3031 Spärck, R., 1965: Peter Forsskål and the Arabian Expedition. In: The American-Scandinavian Review 53: 161-17o.   -212-

3032 Sparrow, G., 197o: Modern Saudi Arabia. London: Knightly Vernon. VIII, 124 p.   -H 223-

3o33 Speetzen, H., 1974: Land settlement projects and agricultural development. An analysis of development factors and processes based on four case studies in Ghana, Libya and Saudi Arabia. Ph.D., Faculty of Sciences, University of Durham. -Main Library, University of Durham-

3o34 Spiegel, S.L., 1982: Saudi Arabia and Israel: the potential for conflict. In: Middle East Review (New York) 14(3/4): 33-43. -21, H 223-

3o35 Spies, O., 1936: Die Bibliotheken des Hidschas. In: Zeitschrift der Deutschen Morgenländischen Gesellschaft 9o: 83-12o. -12, 16, 21-

3o36 Spinks, W., 1976: Doing business in Saudi Arabia. In: Accountancy 87(994): 88, 9o-91. -38, 282-

3o37 Spiro, S., 1932: The Moslem pilgrimage; an authentic account of the journey from Egypt to the Holy Land of Islam, and a detailed description of Mecca and Medina and all the religious ceremonies performed there by the pilgrims from all parts of the Mohammedan world. Alexandria: Whitehead Morris. 7o p. -21-

3o38 Sprenger, A., 1863: Ein Beitrag zur Statistik von Arabien. In: Zeitschrift der Deutschen Morgenländischen Gesellschaft 17: 214-226. -12, 16, 21-

3o39 Sprengling, M.J., 1932: Mysterious Arabia modernizes. In: The Open Court 46: 793-8o5. -1a-

3o4o Squire, C.B., 1956: Charles Richard Crane: an American philanthropist of the Middle East. In: The Islamic Review 44(6): 23-27. -1a, 21-

3o41 Stachels, R., 1977: Das Gesellschaftsrecht des Königreichs Saudi-Arabien. In: Recht der Internationalen Wirtschaft 23: 396-399. -24, 18o, H 3-

3o42 Städtebauliches Institut im Fachbereich ORL der Universität Stuttgart (Hg.), 1977: Grundzüge des islamisch-arabischen Städtebaus. Seminarbericht Planen in Entwicklungsländern. Arbeitsbericht 28 des Städtebaulichen Instituts der Universität Stuttgart; Stuttgart. 144 S. -24-

3o43 Staehelin, W., o.J. (1968): Saudi-Arabien - Land ohne Gegenwart. O.O. (Basel?). 46 S. -21-

3044　Stanford Research Institute, 1975: Business in Saudi Arabia. Menlo Park. 5o p.

3045　Stanton, H.U.W. & C.L. Pickens, 1934: The Mecca Pilgrimage. In: The Moslem World 24: 229-235.   -21, 291, 352-

3046　Statistisches Bundesamt, 1957: Saudisch-Arabien. Der Aussenhandel der Bundesrepublik Deutschland, Ergänzungsreihe: Der Aussenhandel des Auslandes 41; Stuttgart: Kohlhammer. 86 S.   -24-

3047　——, 1958: Saudisch-Arabien. Der Aussenhandel der Bundesrepublik Deutschland, Ergänzungsreihe: Der Aussenhandel des Auslandes 72; Stuttgart: Kohlhammer. 72 S.   -24-

3048　——, 1959: Saudisch-Arabien und Aden. Allgemeine Statistik des Auslandes, Länderberichte; Stuttgart, Mainz: Kohlhammer. 32 S.   -24, 212-

3049　——: Saudi-Arabien. Allgemeine Statistik des Auslandes, Länderkurzberichte; Stuttgart, Mainz: Kohlhammer, [1]1968: 24 S., [2]1970: 21 S., [3]1973: 26 S., [4]1975: 27 S., [5]1979: 28 S., [6]1982: 3o S.   -24, 1oo, H 3-

3050　Status report. Arab tanker business. In: The Arab Economist 7.1975(82): 38-39.   -2o6, H 3-

3051　Statut de la Nationalité Arabe Saoudienne. In: La Revue Marocaine de Droit 12.196o: 49-52.   -B 2o8, B 212-

3052　Stauffer, T.B., 1955: The industrial worker. In: S.N. Fisher (ed.): Social forces in the Middle East. Ithaca: Cornell University Press, 1955: 83-98.   -24, Tü 17-

3053　Steffen, H., 1928: Arabien tritt in die Weltpolitik ein. In: Zeitschrift für Geopolitik 5: 1o24-1o36.   -21, 24, Tü 17-

3054　Steffen, W.G., 1944: Ibn Saud. In: Der Nahe Osten (Berlin) 5(1/1o): 48-49.   -21-

3055　Stegar, W., 1969: Always bells. Sydney: Angus & Robertson. 171 p.

3056　Stegner, W., 1971: Discovery! The search for Arabian oil. As abridged for Aramco World Magazine. Beirut: Middle East Export Press. XII, 19o p.   -21-

3057　Stein, L., 1965: Wirtschaftsgrundlagen und Wirtschaftswandel der Šammar-Gerba unter Berücksichtigung der Veränderung der sozialen Verhältnisse. Diss., Philosophische Fakultät, Universität Leipzig. VI, 248 S. -15-

3o58 Steinbach, U., 1974: Saudi Arabiens neue Rolle im Nahen Osten. In: Aussenpolitik 25: 2o2-213. -21, 1oo, 2o6-

3o59 Steineke, M. & M.P. Yackel, 195o: Saudi Arabia and Bahrein. In: W.E. Pratt & D. Good (eds.): World geography of petroleum. American Geographical Society, Special Publication 31; Princeton: Princeton University Press, 195o: 2o3-229. -25-

3o6o Stevens, G.P., Jr., 1949: Saudi Arabia's petroleum resources. In: Economic Geography 25: 216-225. -93, 188, 2o6-

3o61 Stevens, J.H. & E. Cresswell, 1972: The future of date cultivation in the Arabian Peninsula. In: Asian Affairs (London) 59/N.S. 3: 191-197. -1a, 21, 291-

3o62 ——, 1972: Oasis agriculture in the central and eastern Arabian Peninsula. In: Geography 57: 321-326. -1a, 25, 3o-

3o63 ——, 1974a: Stabilization of aeolian sands in Saudi Arabia's Al Hasa Oasis. In: Journal of Soil and Water Conservation 29: 129-133. -89, 98-

3o64 ——, 1974b: Man and environment in Eastern Saudi Arabia. In: Arabian Studies 1: 135-145. -12, 21, 212-

3o65 ——, 1974c: The role of major agricultural projects in the economic development of Arabian Peninsula countries. In: Proceedings of the Seminar for Arabian Studies 4: 14o-144. -1a, 21-

3o66 Stevens, P., 1982: Saudi Arabia's oil policy in the 197os - its origins, implementation and implications. In: T. Niblock (ed.): State, society and economy of Saudi Arabia. London: Croom Helm; Exeter: Centre for Arab Gulf Studies, 1982: 214-234. -21-

3o67 Stewart, D.S., 198o: Mecca. New York: Newsweek. 172 p. -21-

3o68 Stifani, E.I., 1976: Civil aviation in the Gulf. The role of commercial interests in the issue of traffic rights. In: Arabian Studies 3: 29-49. -12, 21, 212-

3o69 Stilgoe, E., 198o: Commercial regulations in Algeria, Libya, Egypt, Iraq, Saudi Arabia and the Gulf oil states. In: Middle East Annual Review 6: 125-131. -1a, 21, 24/213-

3o7o   Stitt, G.M.S., 1948: A prince of Arabia: the Emir Shereef
       Ali Haidar. London: Allen & Unwin. 314 p.   -21-

3o71   Stocking, G.W., 1971: Middle East oil: a study in political
       and economic controversy. London: Lane (Originalausgabe:
       Nashville: Vanderbilt University Press, 197o). XII, 485
       p.   -21, 24-

3o72   Stokes, B.R., 1981: Introducing public transit in Saudi Ara-
       bia. In: Traffic Quarterly 35: 489-5o5.   -1a, 93, 2o6-

3o73   Stone, R.A. (ed.), 1977: OPEC and the Middle East; the im-
       pact of oil on societal development. Praeger Special
       Studies in International Politics and Government; New
       York, London: Praeger. XVIII, 267 p.   -21-

3o74   Storm, I.P., 195o: Highways in the desert. Nashville: Broad-
       man Press. 135 p.   -Library of Congress, Washington-

3o75   Storm, W.H., 1937a: Leper survey of the Arabian Peninsula.
       In: Leprosy Review 8: 5-11.   -38 M-

3o76   ———, 1937b: A doctor's tour in neglected Arabia. In: The
       Missionary Review of the World 6o(1o): 2o-25.   -1a, 4, 21-

3o77   ———, 1938: Whither Arabia? A survey of missionary opportu-
       nity. World Dominion Survey Series; London, New York:
       World Dominion Press. XVI, 132 p.   -1a-

3o78   Storrs, R., 1937: Orientations. London: Nicholson & Watson
       (^{2}1939). XVII, 624 p.   -21, 3o-

3o79   Stratégies du nouveau plan de développement. In: L'Économiste
       Arabe 24.1981(266): 23-26.   -2o6, H 223-

3o8o   Stresemann, E., 1954: Hemprich und Ehrenberg. Reisen zweier
       naturforschender Freunde im Orient, geschildert in ihren
       Briefen aus den Jahren 1819-1826. Abhandlungen der Deut-
       schen Akademie der Wissenschaften zu Berlin, Klasse für
       Mathematik und allgemeine Naturwissenschaften 1954(1);
       Berlin(-Ost): Akademie-Verlag. 177 S.   -12, 24-

3o81   Strika, V., 1974a: Due capitali saudiane: Gedda e Riàd. In:
       Levante 21(3): 25-32.   -21-

3o82   ———, 1974b: Istruzione e ideologia islamica nell'Arabia Sau-
       diana. In: Annali dell'Istituto Orientale di Napoli
       34/N.S. 24: 437-456.   -1a, 21, 291-

3083 Strika, V., 1975: Studi saudiani. In: Annali dell'Istituto Orientale di Napoli 35/N.S. 25: 555-585. -1a, 21, 291-

3084 ——, 1979a: Ideologia e madīḥ politico in Arabia Saudiana. In: Annali dell'Istituto Orientale di Napoli 39/N.S. 29: 1o7-118. -1a, 21, 291-

3085 ——, 1979b: Note sulla qiṣṣah e riwāyah saudiana. In: Oriente Moderno 59: 677-683. -5, 7, 21-

3086 Strippelmann, W.-D., 1977: Errichtung von Ausbildungswerkstätten für Schweissfachkräfte in Saudi-Arabien. Gutachten im Auftrag der GTZ. Ketsch. 73 S. -Esb 2ooo-

3087 Strohm, J.L., 1949: The king's county agents. In: Country Gentleman 119(Jan.): 22-23, 71-74.

3088 Stross, L., 1886: Sclaverei und Sclavenhandel in Ost-Afrika und im Rothen Meere. In: Österreichische Monatsschrift für den Orient 12: 211-215. -1a, 3o-

3089 Stuhlmann, F., 1916: Der Kampf um Arabien zwischen der Türkei und England. Hamburgische Forschungen, Wirtschaftliche und politische Studien aus hanseatischem Interessengebiet 1; Hamburg, Braunschweig, Berlin: Westermann. XVI, 277, 72* S. -21, 212, Tü 17-

3o9o Subait, A.S., 1976: The educational psychological preparation of elementary school teachers in Saudi Arabia. Ed.D., University of Northern Colorado, Greeley. 15o p.

3091 Succession assured in Saudi Arabia. In: The Middle East (London) 1975(8): 4-6. -12, 3o, 188-

3092 Sudairy, M.A. el-, 1976: Grundprobleme der Infrastrukturentwicklung Saudi Arabiens unter besonderer Berücksichtigung des Erziehungswesens. Diss., Rechts-, wirtschafts- und sozialwissenschaftliche Fakultät, Universität Freiburg/ Schweiz. XI, 152 S. -21, 2o6-

3093 Sullivan, R.R., 197o: Saudi Arabia in international politics. In: The Review of Politics 32: 436-46o. -16, 21, 24-

3094 Sultan Jahan Begam, 19o9: The story of a pilgrimage to Hijaz. Calcutta: Thacker, Spink. 36o p. -21-

3095 Suraisry, J.E., 1979: Development of a dualistic economy: a case study of Saudi Arabia. Ph.D., University of Colorado, Boulder. 2o4 p.

3096 Surur, R.S., 1981: Survey of students', teachers', and administrators' attitudes toward English as a foreign language in the Saudi Arabian public schools. Ph.D., University of Kansas, Lawrence. 160 p.

3097 Sweet, L.E., 1965a: Camel pastoralism in north Arabia and the minimal camping unit. In: A. Leeds & A.P. Vayda (eds.): Man, culture, and animals: the role of animals in human ecological adjustments. Publication 78; Washington: American Association for the Advancement of Science, 1965: 129-152. -21-

3098 ——, 1965b: Camel raiding of North Arabian Bedouin: a mechanism of ecological adaptation. In: American Anthropologist 67: 1132-1150 (reprinted in: L.E. Sweet (ed.): Peoples and cultures of the Middle East: an anthropological reader. Vol. 1: Cultural depth and diversity. Garden City: The Natural History Press for the American Museum of Natural History, 1970: 265-289). -21, 212-

3099 —— (ed.), 1971: The Central Middle East. A handbook of anthropology and published research on the Nile valley, the Arab Levant, Southern Mesopotamia, the Arabian Peninsula and Israel. New Haven: HRAF Press. XI, 323 p. -21, 25-

3100 Sykes, P.M., 1939: Charles R. Crane. In: Journal of the Royal Central Asian Society 26: 361-362. -21, 90, 291-

3101 ——, [3]1949: A history of exploration from the earliest times to the present day. London: Routledge & Kegan Paul ([1]1934) (reprint: Westport: Greenwood Press, 1975). XIV, 426 p. -12, 21-

3102 Szyliowicz, J.S., 1973: Education and modernization in the Middle East. Ithaca, London: Cornell University Press. XIII, 480 p. -12, 21-

3103 ——, 1979: The prospects for scientific and technological development in Saudi Arabia. In: International Journal of Middle East Studies 10: 355-372. -18, 21, 206-

T

3104 Taher, A.H., 1966: Petromin and its role in the international oil industry. Riyadh: General Petroleum and Mineral Organization. 12 p. -H 223-

3105 ——, 1980: The role of state petroleum enterprises in developing countries: the case of Saudi Arabia. In: UNO, Centre for Natural Resources, Energy and Transport (ed.): State petroleum enterprises in developing countries. New York, Frankfurt/M., Oxford, Toronto, Sydney, Paris: Pergamon Press, 1980: 23-27. -H 3-

3106 Tahir-Kheli, S. & W.O. Staudenmaier, 1982: The Saudi-Pakistani military relationship: implications for U.S. policy. In: Orbis (Philadelphia) 26: 155-171. -18, 21, 206-

3107 Tahtinen, D.R., 1974: Arms in the Persian Gulf. Foreign Affairs Studies 10; Washington: American Enterprise Institute for Public Policy Research. 31 p. -21-

3108 ——, 1978: National security challenges to Saudi Arabia. Studies in Defense Policy, American Enterprise Institute Studies 194; Washington: American Enterprise Institute for Public Policy Research. 45 p. -21, 206-

3109 Takroni, M.H., 1980: Evaluating loan repayment in the Saudi Arabian agricultural sector by means of a farm credit interdependent system. Ph.D., Oklahoma State University, Stillwater. 224 p.

3110 Talhouk, A.M.S., 1957: Report on the diseases and insect pests of crops found in the Eastern Province of Saudi Arabia, with recommendations for their control. Beirut: ARAMCO. 86 p.

3111 ——, 1970: Pests among the palms. In: Aramco World Magazine 21(5): 16-17. -24-

3112 Tallon, P., 1979: Saudias zurückhaltende Expansionspolitik zahlt sich aus. In: Interavia 34: 773-775. -206, H 3-

3113 Tallqvist, K., 1905: Bref och Dagboksanteckningar af Georg August Wallin. Svenska Litteratursällskapet i Finland, Skrifter 70; Helsingfors. CXXXIV, 366 S. -46-

3114 Tamisier, M., 1840: Voyage en Arabie. Séjour dans le Hedjaz.-Campagne d'Assir. 2 vols. Paris: Desessart (reproduction: Graz: Akademische Druck- und Verlagsanstalt, 1976). 399 resp. 4o2 p.   -21, 212, Tü 17-

3115 Tannous, A.I., 1948: The Middle East challenges modern agricultural technology. In: Foreign Agriculture 12: 1o1-1o5.   -1a, 2o6-

3116 ———, 195o: Land ownership in the Middle East. In: Foreign Agriculture 14: 263-269.   -1a, 98, 2o6-

3117 ———, 1951: Land reform: key to the development and stability of the Arab world. In: The Middle East Journal 5: 1-2o.   -12, 21, 2o6-

3118 Tapline operations: Trans-Arabian line delivers 17o million barrels of crude across desert with less than 24 hours shutdown time for repairs. In: World Petroleum 23.1952 (Sept.): 84-89.   -9o-

3119 Tarabzune, M.R., 1975: An analysis of the effect of capitalizing exploration and developing costs in the petroleum industry - with emphasis on possible economic consequences in Saudi Arabia. Ph.D., University of Arkansas, Fayetteville. 367 p.

312o Tarbush, S., 1981: Saudis weigh manpower prospects, companies. In: The Middle East (81): 62-63.   -12, H 3, H 223-

3121 Tarizzo, M.L., 1957: Saudi Arabia - epidemiological notes on the Eastern Province. In: The American Journal of Tropical Medicine and Hygiene 6: 786-8o4.   -12, 16, 1oo-

3122 Tashkandi, M.O., 1981: Effects of instruction and personal traits of Saudi pre-service science teachers on the use of higher cognitive questions. Ph.D., Indiana University, Bloomington. 214 p.

3123 Tashkandy, A.-J., 1977: Bibliographical control in Saudi Arabia: an inquiry into the printing and distribution of government publications, with recommendations for improvement Ph.D., University of Pittsburgh. V, 1o5 p.   -H 223-

3124 Tawail, M.A. al-, 1974: Institute of Public Administration in Saudi Arabia: a case study in institution building. Ph.D., University of Pittsburgh. 414 p.

3125 Tawati, A.M., 1976: The civil service of Saudi Arabia: problems and prospects. Ph.D., West Virginia University, Morgantown. 318 p.

3126 Taylor, B., 1872: Travels in Arabia. Compiled and arranged by B. Taylor. Illustrated Library of Travel, Exploration and Adventure; New York: Scribner, Armstrong. IV, 325 p.

3127 Taylor, D.C., 1968: Research on agricultural development in selected Middle Eastern countries. New York: The Agricultural Development Council. X, 166 p.   -21-

3128 Taylor, F.W., 1978: Changing circumstances: Saudi labor law and recent developments. In: The International Lawyer 12: 661-669.   -1a, 291, 352-

3129 Taylor, J.W., 1963: Cancer in Saudi Arabia. In: Cancer 16: 153o-1536.   -16, 21, 24-

313o Tayyeb, M.A. al-, 1982: Information technology transfer to Saudi Arabia. Ph.D., University of Pittsburgh. 3o2 p.

3131 Taxation and other commercial legislation in Saudi Arabia. Kuwait: Peat; Beirut: Marwick; Dubai: Mitchell, 1976. 35 p.   -H 223-

3132 Team from the University College of North Wales, 1971-74: Research and development at the Hofuf Agricultural Research Centre, Saudi Arabia. Reports for 197o-71, 1971-72, 1972-73, 1973-74. N.p.(Bangor?).

3133 Telecommunications all set for spectacular growth. In: The Arab Economist 13.1981(137): 12-14.   -2o6, H 3, H 223-

3134 Territorial waters of Saudi Arabia. Decree No. 6/4/5/3711, May 28, 1949. In: Supplement to the American Journal of International Law 43.1949: 154-157.   -21, 24, 25-

3135 That secret agreement. In: International Currency Review 12.198o(1): 7-2o.   -18o, 2o6, H 3-

3136 The Arabian Peninsula. In: The British Survey, Main Series 28.1966(2o5): 1-11.   -212-

3137 The boundaries of the Nejd: a note on special conditions. In: The Geographical Review 17.1927: 128-134.   -21, 24, 2o6-

3138 The changing face of Sa'udi Arabia. In: The Islamic Review 41.1953(1): 22.   -1a, 21-

3139   The dawn of a new era of public education in Sa'udi Arabia.
        In: The Islamic Review 43.1955(1): 16-19.   -1a, 21-

3140   The development of the Su'udi Arabian economy. In: The Is-
        lamic Review and Arab Affairs 58.1970(7/8): 11-19.   -1a,
        21-

3141   The fleet's out... In: Aramco World 18.1967(3): 1-9.   -24-

3142   The Hajj by motor bus: comfort for modern pilgrims. In: Great
        Britain and the East 67.1951(1824): 23-24.   -21, 2o6-

3143   The iron camel. In: Aramco World 1.1949/5o(7): 4.

3144   The Kingdom of Saudi Arabia. 6th, rev. ed. London: Stacey
        International, 1982 (^{1}1977) (französischsprachige Ausgabe:
        Le Royaume de l'Arabie Saoudite. London: Stacey Interna-
        tional, 1979; spanischsprachige Ausgabe: El Reino de
        l'Arabia Saudita. London: Stacey International, 1979).
        256 p.   -12, 21, 93-

3145   The labor situation in Saudi Arabia. In: Labor Development
        Abroad 14.1969(8): 1-1o, (9): 1-7.   -2o6, 281-

3146   The modernisation of Mecca. British group's contract for
        electrification. In: Great Britain and the East 68.1952
        (1827): 2o.   -1a, 21, 2o6-

3147   The motor vehicle market in Saudi Arabia. In: Motor Business
        1979(1oo): 44-64.   -H 3-

3148   The oil fields of Saudi Arabia. Operations of Arabian Ameri-
        can Oil Company, an affiliate. - The people of Saudi Ara-
        bia. In: Standard (of California) Oil Bulletin 1946(Autumn):
        1-2 resp. 1o-11.

3149   The potters of al-Qarah. In: Aramco World 12.1961(8): 3-5.
        -H 223-

3150   The re-opening of the Hedjaz Railway. In: The Islamic Review
        43.1955(4): 32.   -1a, 21-

3151   The Riyadh dairy. In: Aramco World 12.1961(1o): 18-2o.
        -H 223-

3152   The Safaniya Field. In: Aramco World 13.1962(7): 3-7.
        -H 223-

3153 The Sa'udi Arabian budget. Total expenditure 1,355,ooo,ooo Sa'udi Riyals of which the sum of 231,519,35o is earmarked for development projects. In: The Islamic Review 43.1955 (11): 12-15.   -1a, 21-

3154 The Saudi Arabian Kingdom. In: Middle East Record 1.196o: 372-381, 2.1961: 417-431.   -1a, 12, 21-

3155 The search began in 1933. In: Aramco World 14.1963(2): 17-19.   -H 223-

3156 Thesiger, W.P., 1947: A journey through the Tihama, the 'Asir, and the Hijaz mountains. In: The Geographical Journal 11o: 188-2oo.   -21, 291, 352-

3157 ——, 1948a: Across the Empty Quarter. In: The Geographical Journal 111: 1-21.   -21, 291, 352-

3158 ——, 1948b: Two years in the Arabian desert. In: Synopsis 1948(3): 52-55.

3159 ——, 1949: A further journey across the Empty Quarter. In: The Geographical Journal 113: 21-46.   -21, 291, 352-

316o ——, 195o: The Badu of southern Arabia. In: Journal of the Royal Central Asian Society 37: 53-61.   -21, 9o, 291-

3161 ——, 1959: Die Brunnen der Wüste. Mit den Beduinen durch das unbekannte Arabien. München: Piper (Originalausgabe: Arabian sands. New York: Dutton, 1959). 353 S.   -24, Tü 17-

3162 ——, 198o: The last nomad. One man's forty year adventure in the world's most remote deserts, mountains and marshes. New York: Dutton (Originalausgabe: Desert, marsh and mountain; the world of a nomad. London: Collins, 1979). 3o4 p.   -21-

3163 The technological progress of Su'udi Arabia: the Petroleum and Minerals College. In: The Islamic Review 54.1966(4): 19-22.   -1a, 21, 212-

3164 The Turko-Egyptian conflict over the right of possessing the ports of northern Hijāz and Sinai and the British interference: a study based mainly on the British and Turkish archives. In: Addarah 5.1979(1): 3-6.   -21, 3o-

3165 The voice with a smile - Dhahran style. In: Aramco World 9.1958(7): 3-5.   -H 223-

3166 Thomas, A., 1965: The educational renaissance of the Kingdom of Saudi Arabia; a report of research travel grant 1964, American Friends of the Middle East, American Association of Collegiate Registrars and Admissions Officers. 2 vols. Tempe.

3167 ———, 1967: Saudi Arabia; a study of the educational system of the Kingdom of Saudi Arabia and guide to the academic placement of students from the Kingdom of Saudi Arabia in United States educational institutions. Washington: American Association of Collegiate Registrars and Admissions Officers. 138 p.

3168 Thomas, B.S., 1931a: A journey into Rub' al Khali - the southern Arabian desert. In: The Geographical Journal 77: 1-37. -21, 24, 291-

3169 ———, 1931b: Alarms and excursions in Arabia. London: Allen & Unwin. 296 p. -12-

3170 ———, 1931c: A camel journey across the Rub' al Khali. In: The Geographical Journal 78: 2o9-242. -21, 24, 291-

3171 ———, 1931d: Across the Rub' al Khali. In: Journal of the Royal Central Asian Society 18: 489-5o4. -21, 9o, 291-

3172 ———, 1932: Arabia Felix: across the Empty Quarter of Arabia. London: Cape. XXIX, 396 p. -1a, 21, Tü 17-

3173 Thomas, K. & W. Tracy, 1979: America as alma mater. The students and the States. In: Aramco World Magazine 3o(3): 2-11, 26-33. -21, 24-

3174 Thomas, R.H. (ed.), 1856: Selections from the records of the Bombay Government. N.S. 24: Historical and other information, connected with the province of Oman, Muskat, Bahrai and other places in the Persian Gulf. Bombay: East India Company.

3175 Thubaity, A.M. al-, 1981: Department chairpersons' perception of their position regarding the requirements and process of selection, the major responsibilities and the requirements for job satisfaction in Saudi Arabian univer sities. Ph.D., Michigan State University, East Lansing. 197 p.

3176 Thubaity, K.K.M. al-, 1981: Rural migrants in Taif: their migration and residential mobility. Ph.D., Michigan State University, East Lansing. 297 p.

3177 Thum, W., 1976: Saudi-Arabien, Eldorado für westliche Unternehmer. In: Information der Internationalen Treuhand AG (52): 25-3o.   -2o6, 212-

3178 Tibi, B., 198o: Zum Verhältnis von Politik, Religion und Staat in islamisch legitimierten Monarchien. Eine komparative Studie über Marokko und Saudi-Arabien. In: Orient (Opladen) 21: 158-174.   -21, 1oo, H 223-

3179 Tidrick, K., 1981: Heart-beguiling Araby. Cambridge/England, London, New York, New Rochelle, Melbourne, Sydney: Cambridge University Press. XII, 244 p.   -21-

318o Tietjen, W.V., 1965: Rig ahoy! In: Aramco World 16.1965(2): 1-7.   -24-

3181 Time of the hundred men. In: Aramco World 13.1962(8): 22-24.   -H 223-

3182 Timm, W., 1977: Das Steuerrecht Saudi-Arabiens. In: Internationale Wirtschaftsbriefe 1977: 199-2o4.   -282, 7o8-

3183 Titi, C., 1972: Progretti idrici e risorse minerarie nell' Arabia Saudita. In: Rivista Geografica Italiana 79: 438.   -7, 12, 18-

3184 Tomiche, F.-J., 1979: L'Arabie séoudite. 3e éd., mise à jour. Que sais-je? 1o25; Paris: Presses Universitaires de France (^{1}1962, ^{2}1969). 128 p.   -12, 24-

3185 Tomkinson, M., 1969: Seaside city for Mecca pilgrims. Jedda - second largest town of Saudi Arabia. In: The Geographical Magazine 42: 95-1o4.   -84, 464-

3186 ——, 197o: Requiem for the Empty Quarter. In: The Geographical Magazine 43: 183-193.   -84, 464-

3187 Tončić von Sorinj, D., 1915: Handelsverhältnisse in Oman und im Persischen Golf. In: Österreichische Monatsschrift für den Orient 41: 2o1-211.   -1a, 16-

3188 Tončić von Sorinj, M., 193o: Eine Pilgerfahrt nach Mekka. In: Atlantis 2: 134-141.   -21, 24, 18o-

3189 Topf, E., 1929: Die Staatenbildungen in den arabischen Teilen der Türkei seit dem Weltkriege nach Entstehung, Bedeutung und Lebensfähigkeit. Hamburgische Universität, Abhandlungen aus dem Gebiet der Auslandskunde A, 31, Rechts- und Staatswissenschaften 3; Hamburg: Friederichsen, de Gruyter. X, 26o S.   -21, 24-

319o Top priority for projects at Yanbu. In: The Middle East 198o(63): 56-57.   -1a, 12, 3o-

**** Torki, S. al-: s. Altorki, S.

3191 Tothill, J.D., 1952: Report to the Government of Saudi Arabia on agricultural development. Report 76; Rome: FAO. 88 p.

3192 Totten, D.E., 1959: Erdöl in Sa'ūdī-Arabien. Heidelberger Geographische Arbeiten 4; Heidelberg, München: Keyser (zugleich: Sozial- und wirtschaftsgeographische Auswirkungen der Ölindustrie im östlichen Sa'ūdī-Arabien. Diss., Philosophische Fakultät, Universität Heidelberg, 1957). 174 S.   -21, 24/213, 1oo-

3193 Towagry, A.M., 1973: Organization analysis and proposed reorganization of the Ministry of Education of the Kingdom of Saudi Arabia. Ed.D., University of Arkansas, Fayetteville. 145 p.

3194 Towards an integrated Arab rail network. In: The Arab Economist 14.1982(148): 12-13.   -2o6, H 3, H 223-

3195 Townsend, J., 198oa: Saudis review joint venture policy. In: The Middle East (64): 57-58.   -1a, 12, 3o-

3196 ——198ob: L'industrie en Arabie saoudite. In: Maghreb-Machrek (89): 41-53.   -2o6, H 3, H 223-

3197 Toy, B., 1957: A fool strikes oil: across Saudi Arabia. London: Murray. XII, 2o7 p.

3198 ——, 1968: The highway of the three kings: Arabia - from south to north. London: Murray. IX, 188 p.   -21-

3199 Toynbee, A.J., 1927: Survey of international affairs 1925. Vol. 1: The Islamic World since the Peace Settlement. London: Oxford University Press. XVIII, 611 p.   -21, 24-

32oo ——, 1929: A problem of Arabian statesmanship. In: Journal of the Royal Institute of International Affairs 8: 367-375.   -18o-

3201   Toynbee, A.J., assisted by V.M. Boulter, 1929, 1931, 1935, 1937: Survey of international affairs 1928 (resp. 1930, 1934, 1936). London: Oxford University Press.   -21, 24-

3202   ——, 1970: The McMahon-Hussein Correspondence: comments and a reply. In: Journal of Contemporary History 5(4): 185-193.   -21, 25, 188-

3203   Tracks in the sand. $50,000,000 railroad nears finish; Arabia's biggest public works project. In: Aramco World 2.1950/51(6): 1, 7.

3204   Tracy, W., 1965a: A path to progress. In: Aramco World 16(1): 18-23.   -24-

3205   ——, 1965b: The restless sands. In: Aramco World 16(3): 2-9.   -24-

3206   ——, 1965c: The big shop. In: Aramco World 16(4): 26-33.   -24-

3207   ——, 1965d: Through the reefs. In: Aramco World 16(6): 28-33.   -24-

3208   ——, 1966a: Island of steel. In: Aramco World 17(3): 1-7.   -24-

3209   ——, 1966b: Pink gold. In: Aramco World 17(5): 21-27.   -24-

3210   ——, 1966c: Scribe. In: Aramco World 17(6): 6-7.   -24-

3211   ——, 1967: Fly the desert sky. In: Aramco World 18(6): 1-7.   -24-

3212   ——, 1974: Made in...Saudi Arabia. In: Aramco World Magazine 25(3): 2-15.   -24-

3213   Traini, R., 1962: Arabia Saudiana. Richiamo dell'Università islamica all'osservanza dell'uso del velo. In: Oriente Moderno 42: 460-461.   -5, 7, 21-

3214   ——, 1963: Arabia Saudiana. 'Associazione per l'elevazione della donna'. - La prima donna giornalista saudiana. In: Oriente Moderno 43: 492-493.   -5, 7, 21-

3215   ——, 1964: Arabia Saudiana. Circa la libertà della donna nella scelta del marito. In: Oriente Moderno 44: 588-589.   -5, 7, 21-

3216 Transfer of powers from H.M. King Sa'ūd to H.R.H. Amīr Faysal. In: The Middle East Journal 18.1964: 351-354. -12, 21, 2o6-

3217 Trattato di amicizia del 6 maggio 1932 tra il Regno del Ḥigiāz e Neǵd e l'Afghānistān. In: Oriente Moderno 14.1934: 196-197. -1a, 5, 21-

3218 Trattato di amicizia e Trattato di commercio del 1o febbraio 1932 fra il Regno d'Italia e il Regno del Ḥigiāz, Neǵd e dipendenze. In: Oriente Moderno 12.1932: 23o-233. -1a, 5, 21-

3219 Trattato di amicizia fra l'Arabia Saudiana e il Pakistan (2o novembre 1951). In: Oriente Moderno 33.1953: 233-234. -7, 18, 21-

322o Trautz, M., 193o: G.A. Wallin and 'The penetration of Arabia'. In: The Geographical Journal 76: 248-252. -21, 291, 352-

3221 Treskow, C. von, 1962: Beregnungsprojekt in der saudi-arabischen Wüste. In: Entwicklungsländer 4: 197-199. -21, 1oo, 291-

3222 Tresse, R., 1937: Le pèlerinage syrien aux villes saintes de l'Islam. Thèse pour le Doctorat d'Université présentée à la Faculté des Lettres de l'Université de Paris. Paris: Chaumette. VIII, 383 p. -16, 21-

3223 Trial, G.T. & R.B. Winder, 195o: Modern education in Saudi Arabia. In: History of Education Journal 1(3): 121-133.

3224 Trietsch, D. (Hg.), 19o9: Levante-Handbuch. Eine Übersicht über die wirtschaftlichen Verhältnisse der Europäischen und Asiatischen Türkei, der christlichen Balkanstaaten, Ägyptens und Tripolitaniens. Berlin: Gea. V S., 224 Sp. -24-

3225 Troeller, G., 1971: Ibn Sa'ud and Sharif Husain: a comparison in importance in the early years of the First World War. In: Historical Journal (Cambridge/England) 14: 627-633. -21, 24, 25-

3226 ——, 1976: The birth of Saudi Arabia. Britain and the rise of the house of Sa'ud. London: Cass (zugleich: British policy towards Ibn Sa'ud, 191o-1926. Ph.D., University of Cambridge, 1971/72). XXII, 287 p.   -12, 21, Tü 17-

3227 Trunec, H., 197o: Mit Mekkapilgern unterwegs. Ein Reisebericht. Rosenheim: Rosenheimer Verlagshaus. 2o8 S.   -12-

3228 Turki, Abdel M. & R. Souami, 1979: Récits de pèlerinage à la Mekke. Étude analytique. Journal d'un pèlerin. Paris: Maisonneuve & Larose. 118 p.   -21-

3229 Turki, Abdulaziz M.I. al-, 1978: Forecasting model for cement demand in Saudi Arabia. Ph.D., University of Arizona, Tucson. XIV, 17o p.   -21, H 223-

323o Turki, M.I. al-, 1971: Accelerating agricultural production in Saudi Arabia. Ph.D., Colorado State University, Fort Collins. XV, 11o p.   -21-

3231 Turky, H., 1974: Die wirtschaftliche Bedeutung der Erdölproduktion für die Industrialisierung Saudi-Arabiens. Diss., Rechts-, wirtschafts- und sozialwissenschaftliche Fakultät, Universität Freiburg/Schweiz. Reinheim. XVIII, 338 S. -212, H 3-

3232 Turner, L., 1977: Die Rolle des Öls im Nord-Süd-Dialog. In: Europa-Archiv, Beiträge und Berichte 32: 1o2-112.   -16, 21, 1oo-

3233 —— & J. Bedore, 1977: Saudi and Iranian petrochemicals and oil refining: trade warfare in the 1980s? In: International Affairs (London) 53: 572-586.   -21, 18o, 352-

3234 —— & ——, 1978a: Saudi Arabien - eine Geldmacht. In: Europa-Archiv, Beiträge und Berichte 33: 397-41o (zugleich: Saudi Arabia: the power of the purse-strings. In: International Affairs (London) 54.1978: 4o5-42o).   -21, 2o6, H 223-

3235 —— & ——, 1978b: The trade politics of Middle Eastern industrialization. In: Foreign Affairs (New York) 57: 3o6-322.   -21, 24, 18o-

3236 —— & ——, 1979: Middle East industrialisation: a study of Saudi and Iranian downstream investments. Farnborough: Saxon House. X, 219 p.   -12, H 3, H 223-

3237 Turner, W.O., Jr., 1982: U.S. arms sales to Saudi Arabia: implications for American foreign policy. Ph.D., George Washington University, Washington. 342 p.

3238 Turpin, R.E., 1982: Classroom climate, general ability, and anxiety in a basic skills program for Saudi Arabian naval trainees. Ph.D., Stanford University. 266 p.

3239 Tuson, P., 1979: Lieutenant Wyburd's journal of an excursion into Arabia. In: Arabian Studies 5: 21-36.   -1a, 12, 21-

3240 Tweedie, W., 1894: The Arabian horse. His country and people. Edinburgh: Blackwood (reprint: Los Angeles: Borden, 1961). XIX, 411 p.   -12-

3241 Tweedy, O., 1934: An unbeliever joins the Hadj. On the age-old pilgrimage to Mecca, babies are born, elders die, and families may halt a year to earn funds in distant lands. In: The National Geographic Magazine 65: 761-789.   -21, 93, 291-

3242 Twitchell, K.S., 1941: American ideas for Arabia. In: Asia and the Americas 41: 631-636.

3243 ——, A.L. Wathen & J.G. Hamilton, 1943: Report of the United States Agricultural Mission to Saudi Arabia. Cairo: Misr Printing Office. 147 p.

3244 ——, 1944: Water resources of Saudi Arabia. In: The Geographical Review 34: 365-386.   -21, 24, 206-

3245 ——, with the collaboration of E.J. Jurji & R.B. Winder, ^{3}1958: Saudi Arabia; with an account of the development of its natural resources. Princeton: Princeton University Press (^{1}1947, ^{2}1953; reprint: New York: Greenwood Press, 1969). XIV, 281 p.   -7, 12, 1oo-

3246 ——, 1959: Nationalism in Saudi Arabia. In: Current History (Philadelphia) 36: 92-96.   -21, 18o, 352-

3247 Twitchell, N.G., 1955: Reminiscences of the Yemen and Sa'udi Arabia. King Ibn Sa'ud's interest in the development of his country. Mr. Crane's philanthropic interest in the Yemen and Sa'udi Arabia. In: The Islamic Review 43(3): 32-35.   -1a, 21-

U

3248 Udhailiyah - a town is born. In: Aramco World 8.1957(8): 1o-13.   -H 223-

3249 Uhlig, D., 1965: Das Be- und Entwässerungsprojekt der Al Hassa-Oasen in Saudi-Arabien. In: Die Wasserwirtschaft (Stuttgart) 55: 53-57. -89, 9o, 1oo-

3250 ——, 1969a: King Faisal Settlement Project Haradh/Saudi Arabia. N.p.: WAKUTI, Karl Erich Gall KG. 15 p.

3251 ——, 1969b: Das König-Faisal-Projekt Haradh - Ein Siedlungsvorhaben in der Wüste Arabiens. In: Die Wasserwirtschaft (Stuttgart) 59: 171-176. -89, 9o, 1oo-

3252 ——, 1971a: Anwendung von Stahlbetonfertigteilen für das Bewässerungssystem des Al Hassa Projektes. In: Die Bautechnik 48: 1o9-113. -89, 9o, 93-

3253 ——, 1971b: Al Hassa Irrigation and Drainage Project, Kingdom of Saudi Arabia. N.p.: WAKUTI Karl Erich Gall KG.

3254 —— & P. Klein, 1972: Der Bau des Bewässerungssystems des Faisal-Siedlungsprojekts Haradh in Saudi Arabien. In: Zeitschrift für Kulturtechnik und Flurbereinigung 13: 2o-33. -7, 9o, 1oo-

3255 Un budget de développement. In: L'Économie des Pays Arabes 18.1975(211): 18-21. -2o6, H 223-

3256 UN, Department of Economic Affairs, 1951: Review of economic conditions in the Middle East. Supplement to World Economic Report, 1949-5o. New York. VIII, 84 p. -21-

3257 ——, ——, 1952: Summary of recent economic developments in the Middle East. Supplement to World Economic Report, 195o-51. New York. -2o6-

3258 ——, ——, 1953: Review of economic conditions in the Middle East 1951-52. Supplement to World Economic Report. New York. IX, 161 p. -21-

3259 ——, Department of Economic and Social Affairs, 1956-64: Economic developments in the Middle East 1954-1955 (resp. 1955-1956, 1956-1957, 1957-1958, 1958-1959, 1959-1961, 1961-1963). New York. -1a-

326o ——, Commissioner for Technical Co-operation, Department of Economic and Social Affairs, 1966: Desalination in the Eastern Province of Saudi Arabia. Prepared for the Government of Saudi Arabia by the Bechtel Corporation, appointed under the United Nations Programme of Technical Assistance. Report TAO/SAU/6; New York. VIII, 134 p.   -2o6-

3261 UN, 1967a: Industrial development in the Arab countries. Selected documents presented to the Symposium on Industrial Development in the Arab Countries, Kuwait, 1-1o March 1966. New York. IV, 14o p.   -H 3-

3262 ——, 1967b: Studies on selected development problems in various countries in the Middle East. New York. 87 p.

3263 ——, Economic and Social Office in Beirut (at head of title: Department of Economic and Social Affairs), 1967c: Foreign trade statistics of Saudi Arabia, 196o-63, reclassified according to the United Nations Standard International Trade Classification, 196o revision (SITC, rev.). New York. XIV, 126 p.   -12-

3264 ——, ——, 1969: Étude de certains problèmes que pose le développement dans divers pays du Moyen-Orient, 1968. New York. VI, 94 p.   -H 223-

3265 ——, 197o: Perspectives de croissance et de développement en Arabie Séoudite. In: Le Commerce du Levant (116): 23-29.   -H 223-

3266 —— (ed.), 197o: Growth and development perspectives in Saudi Arabia. Expert Group Meeting on Planning the Agricultural Sector in Relation to Overall Planning and Sectoral Programming, 1-5 June 197o, Beirut, Lebanon. Beirut. 9 p.   -H 223-

3267 UN, Department of Economic and Social Affairs, 1978: A manual and resource book for popular participation training. Vol. 2: Selected examples of innovative training activities. New York. IV, 21 p.

3268 ——, Department of International Economic and Social Affairs, 1981: National experience in the formulation and implementation of population policy, 1958-1979. Saudi Arabia. ST/ESA/SER.R/35; New York. IV, 55 p.   -H 3-

3269 UNCTAD/GATT, International Trade Centre, 1969: Saudi Arabia as a market for manufactured products from developing countries. ITC/MR/1o1/P; n.p. (Geneva?). VIII, 1o9 p.   -H 3-

3270 UNCTAD/GATT, International Trade Centre, (Pakistan) Export Promotion Bureau (ed.), 1970: Saudi Arabia. The market for selected manufactured products from developing countries. An UNCTAD-GATT study. Karachi. XII, 1o3 p. -2o6-

3271 ——, ——, 1976: The oil-exporting developing countries. New market opportunities for other developing countries. 1. Saudi Arabia. Geneva. XII, 133 p. -2o6-

3272 ——, ——, 1977: Focus on Saudi Arabia. Geneva. 133 p.

3273 Une polémique saoudo-libyenne. In: Maghreb-Machrek 1981(93): 1o2-1o7. -21, 3o, H 3-

3274 UNESCO, 1955: World survey of education: handbook of educational organization and statistics. Paris. 943 p. -21-

3275 ——, 1958: World survey of education II: primary education. Paris. 1387 p. -21-

3276 ——, 1961: World survey of education III: secondary education. Paris. 1482 p. -21-

3277 ——, 1966: World survey of education IV: higher education. Paris. 1433 p. -21-

3278 ——, 1969: Special education. Éducation spéciale. International Directories of Education 5; Paris. 142 p.

3279 ——, 1971a: A study of the present situation of special education. Paris. 157 p.

3280 ——, 1971b: World survey of education V: educational policy, legislation and administration. Paris. 1418 p. -21-

3281 United States aid to the Middle East 194o-1951. In: Middle Eastern Affairs 4.1953: 58-62. -1a, 21, 25-

3282 Université Catholique de Louvain, Institut des Pays en Développement, Centre de Recherches sur le Monde Arabe Contemporain (C.R.M.A.C.) (éd.), env. 1981: L'Arabie Saoudite d'hier à demain. Université Catholique de Louvain, Institut des Pays en Développement, Centre de Recherches sur le Monde Arabe Contemporain, Cahiers 8-9-1o-11-12; Louvain-la-Neuve. 298 p. -12, 21-

3283 University co-operation: co-operation between Saudi Arabia and the United States. In: International Association of Universities Bulletin 28.198o: 132-135. -16, 24, 188-

3284 University Securities Ltd., Business Aids Division, 1974: Business directory of Saudi Arabia. One of the first and most definitive guides to one of the most promising markets in the world for your products and services. London. 163 p. -21-

3285 US Department of Commerce, 1948: Saudi Arabia; summary of current economic information. International Reference Service 5(8o); Washington. 6 p.

3286 ———, Bureau of Foreign and Domestic Commerce, Office of International Trade, 1951: Foreign trade of the United States 1936-49. International Trade Series 7; Washington: US Government Printing Office. -18o-

3287 ———, Office of Economic Affairs, Near Eastern and African Division, 1955: Economic developments in Saudi Arabia 1954. In: World Trade Information Service, Economic Reports 1(55-58); Washington. 6 p.

3288 ———, 1975a: Marketing in Saudi Arabia. Overseas Business Report 75-57; Washington: US Government Printing Office. 25 p.

3289 ———, 1975b: Major projects: Saudi Arabian Government Desalination Program. Commerce Action Group for the Near East; Washington: US Government Printing Office.

3290 ———, 1976: Construction costs in Saudi Arabia. Commerce Action Group for the Near East; Washington: US Government Printing Office.

3291 US Department of Housing and Urban Development, 1977: Housing and urban development in Saudi Arabia. Washington: US Government Printing Office.

3292 US Department of the Interior, Division of Geography, 1949: Preliminary NIS Gazetteer: Arabian Peninsula. Washington. III, 165 p.

3293 US Department of State, 1971: Kuwait - Saudi Arabia. Boundary Studies 1o3; Washington: Bureau of Intelligence.

3294 US Hydrographic Office, ^{2}1922: Red Sea and Gulf of Aden pilot; comprising the Suez Canal, the Gulfs of Suez and Akaba, the Red Sea and Strait of Bab el Mandeb, the Gulf of Aden with Sokotra and adjacent islands, and the southeast coast of Arabia to Ras al Hadd. H.O. 157; Washington: US Government Printing Office (^{1}1916). VIII, 668 p. -24, 9o-

3295 US Information Agency, 1964: Arabian Peninsula: a communication fact book. Washington: Research and Reference Service.

3296 US military activity in Saudi Arabia. In: International Currency Review 12.198o(6): 24-3o. -18o, 2o6, H 3-

3297 US-Saudi Arabian Joint Commission on Economic Cooperation, 1975: Summary of Saudi Arabian Five-Year Development Plan, 1975-198o. Washington: US Department of the Treasury.

3298 ——, 1977: Project agreement for technical cooperation in science and technology. Washington: US National Science Foundation.

3299 'Uthaymin, A.-A.S. al-, 1972: Muhammad Ibn 'Abd al-Wahhab: the man and his works. Ph.D., University of Edinburgh.

33oo Uvarov, B.P., 1945/46: War against locusts. In: The Geographical Magazine 18: 219-227. -7, 18o, 464-

33o1 ——, 1951: Locust research and control, 1929-195o. Colonial Research Publication 1o; London: HMSO. IV, 67 p.

33o2 Uzayzi, R.Z. & J. Chelhod, 1969: L'amour et le marriage dans le désert. In: Objets et Mondes 9: 269-278. -12, 2o4-

V

33o3 Vajda, A.J. de, 1953: Proposal for the Wadi Jizan Irrigation Development Scheme. With notes on the Jizan Water Supply by T.O. Smallwood. Based on work of Burningham, Ferris, Tothill, Van der Plas and Smallwood. Report 81; Rome: FAO.

33o4 Valancogne, A., 1937: Le maître de l'Arabie: Ibn Sa'ûd. In: En Terre d'Islam 12: 67-72. -5, 21, 3o-

33o5 Valentia, G., 18o9: Voyages and travels to India, Ceylon, the Red Sea, Abyssinia, and Egypt, in the years 18o2, 18o3, 18o4, 18o5, and 18o6. Vol. 3. London: Miller (französische Ausgabe: Voyages dans l'Indoustan, à Ceylan, sur les deux côtes de la mer Rouge, en Abyssinie et en Égypte, durant les années 1802, 1803, 1804, 1805 et 1806. T. 4. Paris: Lepetit, 1813). III, 506 p.   -21-

3306 Vander Clute, N.R., 1975: Doing business in Saudi Arabia. In: M. Maged (ed.): Legal aspects of doing business with Egypt, Iran, Saudi Arabia and the Gulf states. Corporate Law and Practice, Course Handbook Series 195; New York: Practising Law Institute, 1975: 255-266.   -21-

3307 Vander Werff, L.L., 1977: Christian mission to Muslims: the record. Anglican and Reformed approaches in India and the Near East, 1800-1938. The William Carey Library Series on Islamic Studies; South Pasadena: Carey Library. XI, 366 p.   -21-

3308 Vanguard at Jubail. In: Aramco World 13.1962(9): 17-19.   -H 223-

3309 Vasco, G., 1908: Le chemin de fer musulman du Hedjaz. In: Revue Française de l'Étranger et des Colonies 33: 505-516.   -F 1-

3310 Vasile, R.C., 1978: The impact of modernization in Saudi Arabia. Final report. Alexandria/Va.: Army Military Personnel Center. 112 p.

3311 Vaste plan de développement et de promotion économiques en Arabie Séoudite. In: Syrie et Monde Arabe 16.1969(185): 54-58.   -206, H 3-

3312 Veccia Vaglieri, L., 1934: Notizie aneddotiche su Ibn Sa'ūd, l'Imām Yaḥyà ed il Yemen. In: Oriente Moderno 14: 417-433.   -1a, 5, 21-

3313 ———, 1939: Congresso di capi dell'Arabia Sa'ūdiana a er-Riyāḍ (agosto 1939). In: Oriente Moderno 19: 556-565.   -5, 7, 21-

3314 Verbesserung der Wasserversorgung in Saudi-Arabien. In: Naturwissenschaftliche Rundschau 34.1981: 522-523.   -16, 21, 100-

3315 Verein Deutscher Maschinenbau-Anstalten (VDMA) e.V., 1976: Das Arabien-Geschäft. Daten, Fakten, Perspektiven. Frankfurt/M. 443 S.   -H 223-

3316 Vereinbarung zwischen der Bundesrepublik Deutschland und der Regierung des Königreichs Saudi-Arabien über die Rechtswahrung bei garantierten privaten Kapitalanlagen vom 2.2. 1979 (BGBl. 1980 II S. 693). Nachdruck in: K.-H. Böckstiegel (Hg.): Vertragspraxis und Streiterledigung im Wirtschaftsverkehr mit arabischen Staaten. Schriftenreihe des Deutschen Instituts für Schiedsgerichtswesen 2; Köln, Berlin, Bonn, München: Heymanns, 1981: 313-314.   -21-

3317 Vesey-Fitzgerald, D.F., 1951: From Hasa to Oman by car. In: The Geographical Review 41: 544-560.   -21, 24, 206-

3318 Vianney, J.J., 1966: Aspects of Saudi Arabia through the ages. With reference to the impact of the culture of this country on Europe. In: Levante 13(3/4): 54-60.   -21-

3319 ——, 1968: Oil, social change and economic development in the Arabian Peninsula. In: Levante 15(4): 45-48.   -21-

3320 Vicker, R., 1975: The kingdom of oil: the Middle East; its people and its power. London: Hale (Originalausgabe: New York: Scribner, 1974). 264 p.   -21-

3321 Vickery, C.E., 1922: A journey in Arabia. In: Blackwood's Magazine 211: 166-177.   -180, 188, 385-

3322 ——, 1923: Arabia and the Hedjaz. In: Journal of the Central Asian Society 10: 46-67.   -21, 90, 291-

3323 Vidal, F.S., 1954: Date culture in the Oasis of al-Hasa. In: The Middle East Journal 8: 417-428 (reprinted in: A.M. Lutfiyyah & C.W. Churchill (eds.): Readings in Arab Middle Eastern societies and cultures. The Hague, Paris: Mouton, 1970: 205-217).   -12, 21, 38-

3324 ——, 1955: The Oasis of al-Hasa. N.p. (New York): Arabian Research Division, Local Government Relations Department, ARAMCO (enl. ed.: Ph.D., Department of Anthropology, Harvard University, Cambridge/Mass., 1965). X, 216 p.   -12, H 3-

3325 ——, 1960: Bukūm, al-. In: EI^2 1: 1299.   -21, 24, 212-

3326 ——, 1965a: Notes on social change in Saudi Arabia. In: ARAMCO: Management Development Seminar. Sponsored by The Executive Management Development Committee. Dhahran, 1965: 3-11.

3327 Vidal, F.S., 1965b: Ghāmid. In: EI^2 2: 1oo1.   -21, 24, 212-

3328 ——, 1971a: Al-Ḥasā. In: EI^2 3: 237-238.   -21, 24, 212-

3329 ——, 1971b: Al-Hufūf. In: EI^2 3: 548-549.   -21, 24, 212-

3330 ——, 1975a: Anthropometric measurements of 71 al-Hasa males, Saudi Arabia. Coconut Grove: Field Research Projects. 19 p.

3331 ——, 1975b: Bedouin migrations in the Ghawar oil field, Saudi Arabia. Coconut Grove: Field Research Projects. 26, 9 p.

3332 ——, 1976a: The evolution of the 'Dirah' concept and Bedouin settlement in Arabia. Paper presented at the Middle East Studies Association Meetings, Los Angeles, 1976. 13 p.

3333 ——, 1976b: Mutayr: a tribe of Saudi Arabian pastoral nomads. In: D.E. Hunter & P. Whitten (eds.): The study of anthropology. New York, Hagerstown, San Francisco, London: Harper & Row, 1976: 486-5o5 (reprinted in: D.E. Hunter & P. Whitten (eds.): The study of cultural anthropology. New York, Hagerstown, San Francisco, London: Harper & Row, 1977).   -352, 7o8-

3334 ——, 1978a: Khamīs Mushayṭ. In: EI^2 4: 994.   -21, 24, 212-

3335 ——, 1978b: Utilization of surface water by northern Arabian Bedouins. In: N.L. Gonzalez (ed.): Social and technological management in dry lands: past and present, indigenous and imposed. AAAS Selected Symposium Series 1o; Boulder: Westview Press, 1978: 111-126.   -2o6-

3336 ——, 198o: Development of the Eastern Province: a case study of al-Hasa Oasis. In: W.A. Beling (ed.): King Faisal and the modernisation of Saudi Arabia. London: Croom Helm; Boulder: Westview Press, 198o: 9o-1o1.   -21, H 223-

3337 ——, 1981: Ḳurayyāt al-Milḥ. In: EI^2 5, Fasc. 85-86: 435-436.   -21, 24-

3338 Vieille, P., 1977: Pétrole et classe fonctionelle. Le cas de l'Arabie Séoudite. In: Peuples Méditerranéens (1): 153-193.   -21, 3o, 2o6-

3339 ——, 1978: I meccanismi sociali in Arabia Saudita. Il clan-classe dominante e la sua dipendenza dal sistema mondiale. In: Politica Internazionale 1978(5): 43-65.   -H 223-

3340 Villiers, A.J., 1940: Sons of Sinbad. An account of sailing with the Arabs in their dhows, in the Red Sea, around the coasts of Arabia, and to Zanzibar and Tanganyika: pearling in the Persian Gulf: and the life of the shipmasters, the mariners and merchants of Kuwait. New York: Scribner (London: Hodder & Stoughton, 1940). XV, 429 p.   -21-

3341 ——, 1948: Some aspects of the Arab dhow trade. In: The Middle East Journal 2: 399-416 (reprinted in: L.E. Sweet (ed.): Peoples and cultures of the Middle East. An anthropological reader. Vol. 1: Cultural depth and diversity. Garden City: The Natural History Press, 1970: 155-172). -12, 21, 38-

3342 Vincent-Barwood, A., 1980: A park for 'Asir. In: Aramco World Magazine 31(5): 22-23.   -21, 24-

3343 —— & A. Clark, 1981: Saudi Arabia and solar energy. In: Aramco World Magazine 32(5): 16-29.   -21, 24-

3344 Vitray—Meyerovitch, E. de, 1982: Mekka und Medina. Die Städte des Propheten in 18 Bänden 2; Freiburg/Br., Basel, Wien: Herder (Originalausgabe: L'Universo dello Spirito. La Mecca e Medina. Le città del Profeta. Milano: Mondadori, 1981). 138 S.   -21, 24-

3345 Voices over the sands. In: The Arab World 4.1958(9): 15-16. -206-

3346 Voll, J., 1975: Muḥammad Ḥayyā al-Sindī and Muḥammad ibn 'Abd al-Wahhāb: an analysis of an intellectual group in eighteenth-century Madīna. In: Bulletin of the School of Oriental and African Studies 38: 32-39.   -12, 21, 24-

3347 Voss, B., 1979: Ein Beitrag zur Hydrogeologie und Wasserwirtschaft der Oase Al Hassa in Saudi Arabien. Fortschritt-Berichte der VDI-Zeitschriften IV, 48; Düsseldorf: VDI (zugleich: Diss., Fakultät für Maschinenbau und Elektrotechnik, Technische Universität Braunschweig, 1978). X, 165 S.   -84, 93, Tü 17-

3348 Vredenbregt, J., 1962: The haddj: some of its features and functions in Indonesia. In: Bijdragen tot de Taal-, Landen Volkenkunde 118: 91-154.   -1a, 7, 21-

W

3349 Waagenaar, S., 1957: Länder am Roten Meer. München: Süddeutscher Verlag. 122 S. -12-

3350 Wabli, S.M. al-, 1982: An evaluation of selected aspects of the secondary teacher preparation program at the Umm Al-Qura University, Makkah, Saudi Arabia, based on a follow-up of 1978-79 graduates. Ph.D., Michigan State University, East Lansing. 256 p.

3351 Wagner, D., 1978: Saudi-Arabien als Zentrum der arabischen Reaktion und Verbündeter des Imperialismus in der Klassenauseinandersetzung im Nahen und Mittleren Osten (1970-1975). Diss., Sektion Afrika- und Nahostwissenschaften, Universität Leipzig. VI, 225 S. -15-

3352 ——, 1979: Aktuelle Aspekte der inneren Entwicklung Saudi-Arabiens. In: Asien, Afrika, Lateinamerika 7: 643-652. -21, 206, H 223-

3353 ——, 1980: Die Rolle des Islam in der Aussenpolitik Saudi-Arabiens. In: Asien, Afrika, Lateinamerika 8: 871-880. -21, 206, H 223-

3354 Wahba, H., 1929: Wahhabism in Arabia, past and present. In: Journal of the Central Asian Society 16: 458-467. -21, 90, 291-

3355 ——, 1964: Arabian days. London: Barker. 183 p. -21-

3356 WAKUTI Karl Erich Gall KG, 1964a: Studies for the project of improving irrigation and drainage in the region of Al Hassa, Saudi Arabia. 4 vols. Siegen.

3357 ——, 1964b: Final design for the project of improving irrigation and drainage in the region al Al Hassa, Saudi Arabia. Explanation report. Siegen. II, 99 p. -Gö 153-

3358 Walford, G.F., 1963: Arabian locust hunter. London: Hale. 176 p. -1a, 21-

3359 Wallace, D.M., 1976: Saudi Arabia's building costs. In: Hydrocarbon Processing 55(11): 189-196. -89-

3360 Wallin, G.A., 1979: Travels in Arabia (1845 and 1848). With introductory material by W.R. Mead and M. Trautz. Arabia Past and Present 9; Cambridge/England, New York: The Oleander Press; Naples: The Falcon Press. XLI, 146 p.   -21-

3361 Walpole, N.C., A.J. Bastos, F.R. Eisele, A.B. Herrick, H.J. John & T.K. Wieland, ^{2}1971: Area handbook for Saudi Arabia. Washington: US Government Printing Office (^{1}1966). XLVIII, 373 p.   -1oo, H 3-

3362 Walt, J.W., 196o: Saudi Arabia and the Americans: 1928-1951. Ph.D., Northwestern University, Evanston. IX, 5o4 p.   -21-

3363 Ward, P., 1983: Ha'il: oasis city of Saudi Arabia. Arabia Past and Present 11; Cambridge/England, New York: The Oleander Press. XIV, 747 p.   -21-

3364 Ward, R.J., 197o: The long run employment prospects for Middle East labor. In: The Middle East Journal 24: 147-162.   -12, 21, 38-

3365 Ward, T.E., 1965: Negotiations for oil concessions in Bahrain, El Hasa (Saudi Arabia), the Neutral Zone, Qatar and Kuwait; a story dedicated to those who worked with heart, hand, and mind in discovering and developing the petroleum resources of the Persian Gulf. New York. XIII, 296 p.   -2o6-

3366 Warren, P., 1966: Saudi Arabia attains leadership in planning. In: Emergent Nations 2(2): 57-58.   -H 223-

3367 Wasser für Al Hassa. In: Umschau in Wissenschaft und Technik 77.1977: 1o1-1o8.   -21, 1oo, 18o-

3368 Waterbury, J. & R. el-Mallakh, 1978: The Middle East in the coming decade: from wellhead to well-being? 1980s Project/Council on Foreign Relations; New York (etc.): McGraw-Hill. XII, 219 p.   -21-

3369 Watt, D.C., 1963: The foreign policy of Ibn Saud, 1936-39. In: Journal of the Royal Central Asian Society 5o: 152-16o.   -21, 9o, 291-

337o Waugh, T., 1937: The German counter to revolt in the desert. In: Journal of the Royal Central Asian Society 24: 313-317.   -21, 9o, 291-

3371 Wavell, A.J.B., 1912: A modern pilgrim in Mecca and a siege in Sanaa. London: Constable (^{2}1918). IX, 349 p.   -21-

3372 Wedin, W.E., 1981: Saudi Arabia builds for the future. In: The Futurist 15(June): 4-6.   -38, 2o6, 352-

3373 Weeks, A.E., 1982: The Dhahran Academy: a case study. Ed.D., University of Utah, Salt Lake City. 396 p.

3374 Weidauer, R., 1957: Arabienfahrt in die Länder des Nahen und Mittleren Ostens. Hg. von der Werbeabteilung der Klöckner-Humboldt-Deutz AG, Werk Ulm. O.O. (Ulm). 17 S.   -24-

3375 Weintraub, Sidney, 1978: Saudi Arabia's role in the international financial system. In: Middle East Review (New York) 1o(4): 16-2o.   -21-

3376 Weintraub, Stanley & R. Weintraub (eds.), 1968: Evolution of a revolt: early postwar writings of T.E. Lawrence. University Park, London: The Pennsylvania State University Press. 175 p.   -21-

3377 Weir, S., 1976: The Bedouin. Aspects of the material culture of the Bedouin of Jordan. World of Islam Festival, 1976. N.p. (London): World of Islam Festival. X, 89 p.   -21-

3378 Weisl, W. von, 1928: Zwischen dem Teufel und dem Roten Meer. Fahrten und Abenteuer in Westarabien. Leipzig: Brockhaus. 318 S.   -24-

**** Weiss, L.: s. auch Asad, Muhammad.

3379 ——, 193oa: Zwischen Nedschd und Irak. In: Zeitschrift für Geopolitik 7: 58-67, 135-143.   -1a, 21, 93-

338o ——, 193ob: In der Heiligen Stadt. In: Atlantis 2: 142-146.   -21, 24, 18o-

3381 ——, 193oc: Riadh, die Stadt des Königs Ibn Saud. In: Atlantis 2: 522-53o.   -21, 24, 18o-

3382 Weiss, W., 1975: Wirtschaftliche Zukunftsaspekte Saudi Arabiens. In: Zeitschrift für Wirtschaftsgeographie 19: 5o-55.   -16, 24, Tü 17-

3383 ——, 1977a: Bewässerung in Saudi-Arabien. In: Zeitschrift für Wirtschaftsgeographie 21: 91-92.   -16, 24, Tü 17-

3384 Weiss, W., 1977b: Saudiarabien. Bern: Kümmerly & Frey; München, Bern, Wien: BLV (französische Ausgabe: Arabie saoudite. Merveilles de notre Monde; Paris, Bruxelles: Elsevier-Sequoia, 1978). 223 S.   -12, 21, 93-

3385 ——, ^{2}1978: Das Ende von 1oo1 Nacht. Ein Saudi-Arabien-Buch. Wien, München: Jugend und Volk (^{1}1976). 2o6 S.   -12, 21, 212-

3386 Wells, D.A., 1971: Aramco: the evolution of an oil concession. In: R.F. Mikesell et al.: Foreign investment in the petroleum and mineral industries; case studies of investor-host country relations. Baltimore, London: The Johns Hopkins Press for Resources for the Future, 1971: 216-236.   -21-

3387 ——, 1974: Saudi Arabian revenues and expenditures. The potential for foreign exchange savings. Baltimore, London: Resources for the Future. IX, 34 p.   -21, 2o6, H 3-

3388 ——, 1976: Saudi Arabian development strategy. National Energy Study 12; Washington: American Enterprise Institute for Public Policy Research. 8o p.   -2o6, H 223-

3389 Wellsted, J.R., 1836: Observations on the coast of Arabia between Rás Mohammed and Jiddah. In: The Journal of the Royal Geographical Society 6: 51-96.   -1a, 21, 291-

339o ——, 1842: J.R. Wellsted's Reisen in Arabien. Deutsche Bearbeitung hg. von E. Rödiger. Bd. 2. Halle/S.: Buchhandlung des Waisenhauses (Originalausgabe: Travels in Arabia. Vol. 2: Sinai; survey of the Gulf of Akabah; coasts of Arabia and Nubia, etc. London: Murray, 1838). VI, 412 S.   -16, 212, Tü 17-

3391 Welzel, C.S., 1978: Treffpunkt Wüste, 5o° östliche Länge. Bericht über das moderne Königreich von Saudi Arabien. Mainz: von Hase & Koehler. 2o2 S.   -12, 212-

3392 Wenner, M.W., 1967: Modern Yemen: 1918-1966. The Johns Hopkins University Studies in Historical and Political Science LXXXV, 1; Baltimore: The Johns Hopkins Press. 257 p.   -21, 24/213-

3393 ——, 1975: Saudi Arabia: survival of traditional elites. In: F. Tachau (ed.): Political elites and political development in the Middle East. States and Societies of the Third World; New York, London, Sydney, Toronto: Wiley, 1975: 157-191. -21, 24/213-

3394 Wensinck, A.J., 1927: Ḥadjdj. In: EI[1] 2: 152-157. -21, 24-

3395 ——, 1936: Mekka. In: EI[1] 3: 514-525. -21, 24-

3396 ——, J. Jomier & B. Lewis, 1971: Ḥadjdj. In: EI[2] 3: 31-38. -21, 24, 212-

3397 Westlake, H.J.H., with contributions from G.J. Walters, 1977: Saudi Arabia. Guides to Multinational Business 1; San Francisco: Four Corners Group.

3398 Wetzstein, J.G., 1865: Nordarabien und die syrische Wüste nach den Angaben der Eingeborenen. In: Zeitschrift für allgemeine Erdkunde N.F. 18: 1-47, 241-282, 4o8-498. -4, 21, 24-

3399 What it's like to live and work in Saudi Arabia. In: World Oil 179.1974(1): 16o-162. -89, 2o6, H 3-

34oo Wheatcroft, A., 1982: Arabia and the Gulf: in original photographs, 188o-195o. London, Boston, Melbourne, Henley-on-Thames: Kegan Paul International. XIV, 184 p. -21-

34o1 Whelan, J. (ed.), 1981: Saudi Arabia. A MEED Special Report; London: Middle East Economic Digest. 12o p.

34o2 Wiedensohler, G., 198o: Zur Neuordnung der Justiz in Saudi-Arabien. In: Orient (Opladen) 21: 5o2-51o. -21, 1oo, H 223-

34o3 ——, 1982: Arbeitsrecht in Saudi-Arabien. In: Orient (Opladen) 23: 36-44. -21, 1oo, H 223-

34o4 Wieland, E., 1976: Secondary Agricultural School in Al Hassa/Saudi-Arabien. Planungsstudie im Auftrag des Ministry of Education des Königreiches Saudi-Arabien und der Deutschen Gesellschaft für Technische Zusammenarbeit (GTZ) GmbH. Riad. 58 S. -Esb 2ooo-

34o5 Wien, J., 198o: Saudi-Egyptian relations: the political and military dimensions of Saudi financial flows to Egypt. The Rand Paper Series, P-6327; Santa Monica: Rand Corporation. VII, 91 p.

3406 Wiese, E., 1968: 1o,ooo miles through Arabia. London: Hale. 191 p.   -21-

3407 Wiese, J., 19o7: Mekka-Pilger. In: Asien (Berlin) 6: 162-165.   -1a, 25, 9o-

3408 Wiet, G., 195o: Un rapport britannique sur la prise de la Mecque par les Wahabites (18o3). In: Mélanges offerts à William Marçais par l'Institut d'Études Islamiques de l'Université de Paris. Paris: Maisonneuve, 195o: 321-329.   -21-

3409 Wilkins, C.E., 1964: Learn, remember and know. In: Aramco World 15(6): 26-27.   -H 223-

341o Willert, L., 1973: Talsperre Wadi Jizan in Saudi-Arabien. In: Die Wasserwirtschaft (Stuttgart) 63: 148-161.   -89, 9o, 1oo-

3411 Williams, Keith, 1957: Commercial aviation in Arab states: the pattern of control. In: The Middle East Journal 11: 123-138.   -12, 21, 38-

3412 Williams, Kenneth, 1933: Ibn Sa'ud, the puritan King of Saudi Arabia. London: Cape. 299 p.   -Universitätsbibliothek Wien-

3413 Williams, Maurice J., 1976: The aid programs of the OPEC countries. In: Foreign Affairs (New York) 54: 3o8-324.   -21, 24, 291-

3414 Williams, Maynard O., 1945: Guest in Saudi Arabia. In: The National Geographic Magazine 88: 463-487.   -21, 24, 38-

3415 ——, 1948: Saudi Arabia, oil kingdom. In: The National Geographic Magazine 93: 497-512.   -21, 24, 291-

3416 Willson-Pepper, C.R., 197o/71: Office in Arabian Bedouin societies. B.Litt., University of Oxford.

3417 Wilmington, M.W., 1971: The Middle East Supply Centre. Ed. by L. Evans. Albany: State University of New York Press (englische Ausgabe: London: University of London Press, 1972). XIX, 248 p.   -21, 24/213-

3418 Wilson, A.T., [2]1954: The Persian Gulf; an historical sketch from the earliest times to the beginning of the twentieth century. London: Allen & Unwin (Oxford: Clarendon Press, [1]1928). XI, 327 p.   -7, 21, 24/213-

3419 Wilson, J.C., 1952: Apostle to Islam: a biography of Samuel M. Zwemer. Grand Rapids: Baker. 261 p.

3420 ——, 1967: The epic of Samuel Zwemer. In: The Muslim World 57: 79-93.   -21, 291, 352-

3421 Wilson, Rodney, 1979: The economies of the Middle East. London, Basingstoke: Macmillan. XIII, 209 p.   -21-

3422 ——, 1982: The evolution of the Saudi banking system and its relationship with Bahrain. In: T. Niblock (ed.): State, society and economy in Saudi Arabia. London: Croom Helm; Exeter: Centre for Arab Gulf Studies, 1982: 278-300.   -21-

3423 Wilson, Rodney J.A., 1980: The rising costs of industrial construction. In: M. Ziwar-Daftari (ed.): Issues and development: the Arab Gulf States. London: MD Research & Services, 1980: 65-75.   -21-

3424 Wilton, J., 1981: The pros and cons of the Saudi plan. In: Middle East International (163): 8-9.   -12, 30-

3425 ——, 1982: The Saudi succession. In: Middle East International (177): 12-13.   -12, 30-

3426 Winder, R.B., 1950: A history of the Su'ūdi state from 1233/1818 until 1308/1891. Ph.D., Department of Oriental Languages and Literatures, Princeton University. 328 p. -21-

3427 ——, 1965: Saudi Arabia in the nineteenth century. New York: St. Martin's Press; London, Melbourne, Toronto: Macmillan. XIV, 312 p.   -7, 18, 24-

3428 ——, 1980: 'Abd al-'Azīz b. 'Abd al-Raḥmān b. Fayṣal Āl Su'ūd. In: EI^2 Suppl., Fasc. 1-2: 3-4.   -21, 24-

3429 ——, 1982: Fayṣal b. 'Abd al-'Azīz b. 'Abd al-Raḥman Āl Su'ūd. In: EI^2 Suppl., Fasc. 5-6: 305-306.   -21, 24-

3430 Windfuhr, V., 1981: Saudi-Arabien: Vom Kamel zum Cadillac. In: Merian 34(7): 38-41.   -21, 24, 180-

3431 Windsor, G.W., Jr., 1981a: From suqs to supermarkets. In: Aramco World Magazine 32(1): 8-16.   -21, 24-

3432 ——, 1981b: In Saudi Arabia, another search for the other minerals. In: Aramco World Magazine 32(4): 24-29.   -21, 24-

3433 Wingerter, R.B., 1981: More behind Awacs than meets the eye. In: Middle East International (162): 7-8. -12, 3o-

3434 Winkler, E., 1952: Eisenbahnbau vom Persischen Golf zum Roten Meer. In: Mitteilungen der Geographischen Gesellschaft Wien 94: 1o5-1o6. -7, 16, Tü 17-

3435 Winstone, H.V.F. & Z. Freeth, 1972: Kuwait: prospect and reality. London: Allen & Unwin. 232 p. -21-

3436 ——, 1976: Captain Shakespear: a portrait. London: Cape. 236 p. -21-

3437 ——, 1978: Gertrude Bell. London: Cape. XII, 322 p. -21, 24-

3438 Wirth, E., 1969: Das Problem der Nomaden im heutigen Orient. In: Geographische Rundschau 21: 41-51. -12, 21, 1oo-

3439 ——, 1973: Vorderasien. In: H. Mensching & E. Wirth unter Mitarbeit von H. Schamp: Nordafrika und Vorderasien. Fischer Länderkunde 4; Frankfurt/M.: S. Fischer, 1973: 166-263. -21, 24, 93-

344o Wirthschaftliche Verhältnisse der wichtigsten Handelsplätze am Rothen Meer, dann von Gibraltar und Tanger. In: Jahresberichte der k. und k. österreichisch-ungarischen Consulats-Behörden 13.1885: 121-139. -35-

3441 Wirtschaftliches über Sa'udi-Arabien. In: Orient-Nachrichten 4.1938: 231-232. -2o6, B 2o8-

3442 Wissa-Wassef, C., 1974: L'Arabie Séoudite et le conflit israélo-arabe du mois d'octobre 1973. In: Politique Étrangère 39: 185-199. -21, 212, H 3-

3443 Wissmann, H. von, 1937: Arabien. In: F. Klute (Hg.): Handbuch der Geographischen Wissenschaft. Bd. 5: Vorder- und Südasien in Natur, Kultur und Wirtschaft. Potsdam: Akademische Verlagsgesellschaft Athenaion, 1937: 178-211. -21, 24, 212-

3444 ——, 1941: Arabien und seine kolonialen Ausstrahlungen. Eine geographisch-geschichtliche Skizze. In: K.H. Dietzel, O. Schmieder & H. Schmitthenner (Hgg.): Lebensraumfragen europäischer Völker. Bd. 2: Europas koloniale Ergänzungsräume. Leipzig: Quelle & Meyer, 1941: 374-488. -24/213-

3445 Wissmann, H. von, 1961: 'Abdallah H.St.J.B. Philby (1885-1960), sein Leben und Wirken. In: Die Welt des Islams N.S. 7: 1oo-141. -16, 21, 2o6-

3446 Wizarat, S., 1975: Saudi Arabia: an economic review. In: Pakistan Horizon 28(4): 131-141. -21, 2o6-

3447 Wohaibi, M.N., 1978: Cultural perspectives of the adult reading problem in Riyadh, Saudi Arabia. Ph.D., North Texas State University, Denton. IV, 225 p. -H 223-

3448 Wohlers-Scharf, T., 1982: Les fonds nationaux de développement: la coopération financière des pays arabes et la période post-pétrolière. In: Centre d'Études et de Recherches sur l'Orient Arabe Contemporain (éd.): La péninsule Arabique d'aujourd'hui. Sous la direction de P. Bonnenfant. T. 1. Paris: Centre National de la Recherche Scientifique, 1982: 3o7-341. -21-

3449 Wohlfahrt, E., 1977: Die Bildung von Planungsregionen in Gebieten mit vorwiegend semi-ariden und voll-ariden Klimabedingungen zum Zwecke der Strukturentwicklung: Abgrenzungskriterien und deren Anwendung, dargestellt am Beispiel des südwestlichen Saudi-Arabien (Südlicher Hedschas). Diss., Fakultät für Wirtschaftswissenschaften, Universität Karlsruhe (zugleich: Die Bildung von Planungsregionen in Trockengebieten zum Zwecke der Strukturentwicklung. International Land Development Institute Studies 1; München: International Land Development Institute, 1977). VII, 233 S. -9o, 212, H 3-

345o ——, 198o: Die Arabische Halbinsel. Länder zwischen Rotem Meer und Persischem Golf. Berlin, Frankfurt/M., Wien: Safari bei Ullstein. 1276 S. -21, 24, 1oo-

3451 Wolff, G., 195o: Eduard Rüppell. In: Passat 2(3): 66-67. -21, 2o6, 212-

3452 Women and travel: Saudi Arabia. In: Women's International Network News 3.1977(4): 63.

3453 Wong, S.S., 1981: Planning for a relevant medical education in the Kingdom of Saudi Arabia. Ph.D., University of Pittsburgh. 265 p.

3454 Wood, J.G., 1977: De l'eau pour l'industrie du Moyen-Orient. In: Industries et Travaux d'Outremer 25(280): 2o9-211. -1a, 89, H 3-

3455 Wood, Junius B., 1923: A visit to three Arab kingdoms. Transjordania, Iraq, and the Hedjaz present many problems to European powers. In: The National Geographic Magazine 43: 535-568. -21, 93-

3456 Woods, G., 1978: Saudi Arabia. A First Book; New York, London: Watts. 63 p.

3457 Wozelko, H., 1939: Handel am Roten Meer. In: Orient-Nachrichten 5: 178-18o, 182-183. -12, 2o6-

3458 Wray, T., 1976: UK project to develop fisheries of Saudi Arabia. In: Fishing News International 15(2): 2o-22. -H 3-

3459 ———, 1977: U.K. team helps develop Saudi Arabian fisheries. In: Fishing News International 16(9): 18-23. -H 3-

3460 Wren, G.R., 1976: The hospitals of Saudi Arabia. In: World Hospitals 12: 12o-125. -38 M, 83-

3461 Wright, C. & R. Manning, 1982: Setting the limits of power. In: The Middle East (London) (88): 13-16. -1a, 12, H 223-

3462 Wright, J.K., 1927: Northern Arabia: the explorations of Alois Musil. In: The Geographical Review 17: 177-2o6. -21, 24, 2o6-

3463 Wylde, A.B., 1887: The Red Sea trade. In: The Journal of the Manchester Geographical Society 3: 181-195. -3o, F 1-

Y

3464 Yamani, A.Z., 1975: Die Interessen der Erdöl-Export-Länder. In: Europa-Archiv, Beiträge und Berichte 3o: 693-698. -21, 24, 2o6-

3465 Yegnes, T., 1981: Saudi Arabia and the peace process. In: The Jerusalem Quarterly (18): 1o1-12o. -12, 21, 24/213-

3466 Yémen - Péninsule Arabique. Arabie Saoudite - Oman - Émirats Arabes Unis - Qatar - Bahrein - Koweit. Poche-Voyage Marcus, Collection Moderne de Guides de Voyage 32; Paris: Marcus, 1979. 64 p. -21-

3467 Yodfat, A. & Mordechai Abir, 1977: In the direction of the Persian Gulf: the Soviet Union and the Persian Gulf. London, Totowa: Cass. XII, 167 p.   -21-

3468 Young, A., 1976: Saudi Arabia: personal impressions. In: Irish Arab News 2(2): 17-19.

3469 Young, Arthur N., 1953: Saudi Arabian currency and finance. In: The Middle East Journal 7: 361-380, 539-556.   -12, 21, 38-

3470 ——, 1960: Financial reforms in Saudi Arabia. In: The Middle East Journal 14: 466-469.   -12, 21, 38-

3471 Young, G.B., et al., 1977: Marketing in the Middle East. In: Industrial Marketing 62(7): 50-106.   -38, 467, H 3-

3472 Young, H.W., 1933: The independent Arab. London: Murray (reprint: New York: AMS Press, 1974). XI, 344 p.   -12, 21-

3473 Young, P.L., 1980a: Saudi Arabia: a political and strategic assessment. (Part I). In: Asian Defence Journal 1980(July/Aug.): 36-44.   -1a-

3474 ——, 1980b: Saudi Arabia's defence situation. (Part II). In: Asian Defence Journal 1980(Sept./Oct.): 47-59.   -1a-

3475 Young, R., 1949: Saudi Arabian offshore legislation. In: The American Journal of International Law 43: 530-532.   -7, 12, 21-

3476 ——, 1970: Equitable solutions for offshore boundaries: the 1968 Saudi Arabia-Iran agreement. In: The American Journal of International Law 64: 152-157.   -7, 12, 21-

3477 ——, 1973: The law of the sea in the Persian Gulf: problems and progress. In: R. Churchill, K.R. Simmonds & J. Welch (eds.): New directions in the law of the sea. Vol. 3: Collected papers. Dobbs Ferry: Oceana Publications, 1973.

3478 Younger, S., 1977: The Saudis go for goal. In: Middle East International (74): 28-29.   -12, 30-

Z

3479 Zafer, M.I., 1972: An investigation of factors which are associated with enrollment and non-enrollment in teacher education programs of public secondary education in Saudi Arabia. Ph.D., Michigan State University, East Lansing. 388 p.

3480 Zahlan, R.S., 1979: The creation of Qatar. London: Croom Helm; New York: Harper & Row. 192 p.   -21, 25-

3481 ——, 1982: King Abd al-Aziz's changing relationship with the Gulf states during the 1930s. In: T. Niblock (ed.): State, society and economy in Saudi Arabia. London: Croom Helm; Exeter: Centre for Arab Gulf Studies, 1982: 58-74.   -21-

3482 Zaid, A.M., 1978: The law of bequest in traditional Islamic law and in the contemporary law of Saudi Arabia. Ph.D., School of Oriental and African Studies, University of London.

3483 Zaid, Abdulla Mohamed al-, 1982: Education in Saudi Arabia: a model with difference. 2nd ed. Jeddah: Tihama Publications. 115 p.

3484 Zaid, I.A. al-, 1975: Sozialversicherung in Saudiarabien. In: Internationale Revue für Soziale Sicherheit 28: 280-285 (englischsprachige Ausgabe: Social insurance in Saudi Arabia. In: International Social Security Review 28.1975: 256-261).   -1a, 12, 206-

3485 Zaidi, A.A. al-, 1982: Adequacies of curriculum and training in agriculture provided at three Saudi institutions as assessed by administrators, instructors, senior students, and regional directors. Ed.D., Oklahoma State University, Stillwater. 159 p.

3486 Zaidi, Z.H., 1971: Ḥidjāz Railway. In: EI^2 3: 364-365.   -21, 24, 212-

3487 Zaidy, A.K. al-, 1980: Food and agriculture in the arid environment of Saudi Arabia under human pressures and changes. M.A., Michigan State University, East Lansing. 138 p.

3488 Zakka, N.M., 1979: Amin ar-Rihani, penseur et homme de lettres libanais. Étude sur la littérature arabe contemporaine. Lille: Presses Universitaires de Lille (zugleich: Thèse de doctorat en Littérature Arabe Contemporaine, Université de Strasbourg, 1975). 271 p.   -21-

3489 Zalatimo, Y.N., 1981: Curriculum design for a developing society: a vocational business and office education program for women in Saudi Arabia. Ph.D., Southern Illinois University, Carbondale. 192 p.

3490 Zamel, A. & M. Tartir, 1975: Saudi Arabia plans to expand exploratory work. In: World Oil 181(1): 161, 165, 168, 172. -89, 9o, 2o6-

3491 Zawadzky, K., 1975: Nicht zu Lasten der unentgeltlichen Hilfe. Technische Zusammenarbeit gegen Entgelt. Grundsätze für Technische Zusammenarbeit in Kraft getreten. In: Auslandskurier 16(8): 37. -21, 38, 212-

3492 Zedan, F.M., 1981: Political development of the Kingdom of Saudi Arabia. Ph.D., Claremont Graduate School. 151 p.

3493 Zehme, A., 1875: Arabien und die Araber seit hundert Jahren. Eine geographische und geschichtliche Skizze. Halle/S.: Buchhandlung des Waisenhauses. VIII, 4o7 S. -212, Tü 17-

3494 Zein, Y.T. el-, et al., 1974: Saudi Arabia. In: The Arab Economist 6(65.), Suppl.: 3-78. -H 3-

3495 Zillmann, G., 1977: Exportmarkt Saudi-Arabien. Sulzberg: Export-Verlag. 256 S. -212, H 223, Tü 17-

3496 Zimmermann, F. & A. Saxen, 1964: Bewässerungsprojekt Oase Al Hasa, Hofuf, Saudi-Arabien. Siegen: WAKUTI Karl Erich Gall KG.

**** Zischka, A.: s. auch Donkan, R.

3497 ——, 1976: Europas bedrohte Hauptschlagader. Arabische Renaissance oder neue Grossmacht Iran? Bern: Kümmerly & Frey. 2oo S. -12, 24/213, 212-

3498 Zoli, C., 1934: Lievi modificazioni alla carta politica dell'Arabia. In: Bollettino della Società Geografica Italiana VI, 11 (zugleich: 71): 635-652. -188, Tü 17-

3499 ——, 1936: Il confine tra Arabia Saudita e Yemen. In: Bollettino della Società Geografica Italiana VII, 1 (zugleich: 73): 113-115. -21-

3500 Zughoul, M.R., 1978: Lexical interference of English in Eastern Province Saudi Arabic. In: Anthropological Linguistics 2o: 214-225. -1a, 7, 21-

3501 Zulfa, M. al-, 1979: Ibn Abd al-Wahhāb's call and its impact on the 'Asīr resistance to Turkish-Egyptian rule 1811-1840. M.A., University of Kansas, Lawrence.

3502 ——, 1982: Village communities in Bilād Rufaydah: their political and economic organisation. In: Arabian Studies 6: 77-96.   -21, Tü 17-

3503 Zwemer, S.M., 1901: The Wahābîs: their origin, history, tenets, and influence. In: Journal of the Transactions of the Victoria Institute, or, Philosophical Society of Great Britain 33: 311-333.   -291-

3504 ——, 1906: Islam in Arabia (the Wahabis). In: S.M. Zwemer, E.M. Wherry & J.L. Barton (eds.): The Mohammedan World of To-day: being papers read at the first Missionary Conference on behalf of the Mohammedan World held at Cairo April 4th-9th, 1906. New York, Chicago, Toronto, London, Edinburgh: Revell, 1906: 99-112.   -21-

3505 ——, 1907: Oman and Eastern Arabia. In: Bulletin of the American Geographical Society 39: 597-606.   -1a-

3506 ——, 1912: Arabia: the cradle of Islam. Studies in the geography, people and politics of the peninsula with an account of Islam and mission-work. 4th, rev.ed. New York, Chicago, Toronto, London, Edinburgh: Revell (^{1}1900). 437 p.   -1a-

3507 ——, 1917a: Mecca the mystic. A new kingdom within Arabia. In: The National Geographic Magazine 32: 157-172.   -7, 21, 93-

3508 ——, 1917b: Three visits to Jiddah. In: The Church Missionary Review 69: 222-228.   -4-

3509 ——, 1932: Snouck Hurgronje's 'Mekka'. In: The Moslem World 22: 219-226.   -1a, 21, 30-

3510 —— & J. Cantine, 1938: The Golden Milestone: reminiscences of pioneer days fifty years ago in Arabia. New York, London, Edinburgh: Revell. 157 p.   -1a-

3511 ——, 1943: Islam in 'Arabia Deserta'. In: The Moslem World 33: 157-164.   -21, 291, 352-

3512 ——, 1947: Al Haramain: Mecca and Medina. In: The Moslem World 37: 7-15.   -21, 291, 352-

III NACHTRÄGE / ADDITIONS

BIBLIOGRAPHIEN / BIBLIOGRAPHIES

3513 Bryson, T.A., 1979: United States/Middle East diplomatic relations, 1784-1978: an annotated bibliography. Metuchen, London: The Scarecrow Press. XIV, 2o5 p.

3514 Newman, D., E.W. Anderson & G.H. Blake, 1982: The security of Gulf oil: an introductory bibliography. Occasional Paper Series 13; Durham: Centre for Middle Eastern and Islamic Studies, University of Durham. II, 55 p.

LITERATUR / LITERATURE

3515 Abbas, A.M., n.d.: Report to the Government of Saudi Arabia on nutritional conditions in Saudi Arabia. Rome: FAO.

3516 Abdo, Albert N., 1962: Saudi Arabia, a market for U.S. products. Washington: Bureau of International Commerce, US Department of Commerce. VIII, 44 p.

3517 'Abdu, I., 1955: What I saw in Sa'udi Arabia. In: The Islamic Review 43(5): 24-26. -1a, 21-

3518 Abdul-Samad, M., 1953: Saudi-Arabien zwischen gestern und heute. In: Übersee-Rundschau 5: 65o. -3o, 18o, 212-

3519 Abu-Rizaiza, O.S., 1982: Municipal, irrigational and industrial future water requirements in Saudi Arabia. Ph.D., University of Oklahoma, Norman. 222 p.

352o Adwani, S.H., 1981: The relationship between teacher and supervisor as perceived by teachers, supervisors, and principals in secondary schools in Saudi Arabia. Ph.D., University of Oregon, Eugene. 321 p.

3521 Ajaji, A., 1980: A Ministry of Agriculture's perspective on agricultural education and development. In: J. Ryan & A. T. Saad (eds.): Agricultural education for development in the Middle East. Proceedings of the Conference on The Role of Agricultural Education in the Development of the Middle East, April 24-28, 1979. Beirut: Faculty of Agricultural and Food Sciences, American University of Beirut, 1980: 36-39.   -21-

3522 Aldoasary, F.S., 1983: Impact of the oil sector on the development of the non-oil economy of Saudi Arabia. Ph.D., American University, Washington. 413 p.

3523 Algosaibi, G.A. (wohl identisch mit dem Verfasser von Nr. 1120), 1982: Arabian essays. London, Boston, Melbourne: Kegan Paul International. VII, 117 p.   -21-

3524 Ali, I. al-, 1964: Past and present conditions of Bedouins. Riyadh: Ministry of Agriculture and Water.

3525 Alikattan, M., 1974: Le développement de l'enseignement des jeunes filles en Arabie Saoudite. Thèse de 3e cycle en Sociologie, Université Paris V (Université René Descartes). 347 p.

3526 Almana, A.M., 1981: Economic development and its impact on the status of women in Saudi Arabia. Ph.D., University of Colorado, Boulder. 297 p.

3527 Almas, H.M., 1983: Investigation of opinions and performance regarding physics instructional procedures in Saudi secondary schools. Ed.D., University of Northern Colorado, Greeley. 279 p.

3528 Alsaigh, M.N.H., 1981: A proposed model for evaluating teacher performance in the Saudi Arabian high schools. Ed.D., Indiana University, Bloomington. 230 p.

3529 Altwaijri, A.O., 1982: The adequacy of students' preparation in English as a foreign language in the Saudi schools. Ph.D., University of Oregon, Eugene. 196 p.

3530 Amaldi, D., 1979: Assalto alla Grande Moschea della Mecca. In: Oriente Moderno 59: 792-793.   -5, 7, 21-

3531 Arms sales to Saudi Arabia. In: International Defense Review 8.1975(1): 23.

3532 Athubaity, M.M., 1981: An exploratory study of the leadership behavior of deans and presidents in higher education institutions in Saudi Arabia. Ph.D., Michigan State University, East Lansing.

3533 Ayubi, N.N.M., 1982: Vulnerability of the rich: the political economy of defense and development in Saudi Arabia and the Gulf. The Gulf Project; Washington: Center for Strategic and International Studies, Georgetown University.

3534 Bahrain - Iran - Iraq - Kuwait - Oman - Qatar - Saudi Arabia - United Arab Emirates: agreements from the Kuwait Regional Conference on the Protection and Development of the Marine Environment and the Coastal Areas (held in Kuwait, April 15-23, 1978). In: International Legal Materials 17.1978: 5o1-54o.   -1a, 12, 24-

3535 Bakri, T.H., 1983: Factors influencing the use of instructional media by middle school teachers in two school districts in Saudi Arabia. Ph.D., University of Oklahoma, Norman. 2o8 p.

3536 Basabrain, A.A.A.A., 1983: Modernization of agriculture: an analysis of incentives, disincentives, and the economical, educational factors influencing the adoption of agricultural innovations in Saudi Arabia. Ph.D., University of Massachusetts, Amherst. 4o4 p.

3537 Bassam, A.A., 1981: An examination of absorptive capacity in Saudi Arabia. Ph.D., Michigan State University, East Lansing. 192 p.

3538 Bayoumi, R.A., A. Omer, A.P.W. Samuel, N. Saha, Z.A. Sebai & H.M.A. Sabaa, 1979: Haemoglobin and erythroctic glucose-6-phosphate dehydrogenase variants among selected tribes in western Saudi Arabia. In: Tropical and Geographical Medicine 31: 245-252.   -1a, 12, 61-

3539 Bericht über die Deutsche Schule Djidda, Juli 1981. O.O., 1981. 9 S.   -212-

354o Binder, F., 1939: Ġabrīn and Maqainama, zwei merkwürdige Ortsnamen in der Wüste ar-rub' al-ḫālī. In: Wiener Zeitschrift für die Kunde des Morgenlandes 46: 219-222. -4, 21, 188-

3541   Binder, F., 1941: Über die Altersbezeichnungen der Kamele bei den heutigen Beduinen Arabiens. In: Wiener Zeitschrift für die Kunde des Morgenlandes 48: 89-96.   -4, 21, 188-

3542   Bitter realities. In: Civil Engineering (London) 1976(4): 4o-41.   -83, 89, 93-

3543   Bleiber, F., 1939: Die völkerrechtliche Stellung der Staaten Arabiens. In: Zeitschrift für öffentliches Recht 19: 137-163.   -12, 21, 24-

3544   Blume, H. (Hg.), 1976: Saudi-Arabien. Natur, Geschichte, Mensch und Wirtschaft. Ländermonographien 7; Tübingen, Basel: Erdmann. 359 S.   -12, 21, 24-

3545   Brébant, É. & J. Arcache, 1978: Péninsule Arabique. Le Guide Pratique de l'Homme d'Affaires 79; Paris: Bréa. 19o p. -21-

3546   Buchan, J., 198o: Take counsel among yourselves. In: The Middle East (London) (71): 34-36.   -1a, 3o, 188-

3547   Budget of Saudi Arabia - 1962-1963. In: Middle Eastern Affairs 14.1963: 49-52.   -1a, 21, 18o-

3548   Burgoyne, J.A., Jr., 1977: Specific problems and unique aspects of doing business in Saudi Arabia. In: W.G. Wickersham & B.P. Fishburne, III (eds.): Current legal aspects of doing business in the Middle East: Saudi Arabia, Egypt and Iran. Chicago: Section of International Law, American Bar Association, 1977: 135-143.

3549   Campbell, R.B., 1976: Some indications of the literary movement in Saudi Arabia in 1976. In: CEMAM Reports 1976: 213-229.   -3o-

355o   Carlson, T., 1954: Report to the Government of Saudi Arabia on the establishment of date processing and packing plants. Report 266; Rome: FAO.

3551   Carver, J.L. (pseud.), 1948: Slavery's last stronghold. Handmaidens and field hands are still sold in Arabia - not economics but sexual-psychological factors keep the ancient trade alive. In: United Nations World 2(5): 24-27. -1a, 17, 26-

3552 Casillas, R.J., 1983: Oil and diplomacy: the evolution of American foreign policy in Saudi Arabia, 1933-1945. Ph.D., University of Utah, Salt Lake City. 244 p.

3553 Caso, R.G., 1977: Local tax considerations: Iran and Saudi Arabia. In: W.G. Wickersham & B.P. Fishburne, III (eds.): Current legal aspects of doing business in the Middle East: Saudi Arabia, Egypt and Iran. Chicago: Section of International Law, American Bar Association, 1977: 18-24.

3554 Celli, J.P., 1980: Special libraries of the Kingdom of Saudi Arabia. In: Special Libraries 71: 358-364.   -1a, 21, 24-

3555 Chaglassian, H.T., N. Bustani & H.H. Anderson, 1952: Endemic treponematosis (balash or bejel) in Saudi Arabia. In: The American Journal of Tropical Medicine and Hygiene 1: 826-830.   -21, 25, 1oo-

3556 Chamieh, J. (ed.), c. 1980: Sa'udi Arabia yearbook 1979/80. Beirut, Riyadh, London: The Research & Publishing House. XX, 260 p.   -212-

3557 Clerc, M., 1948: L'évolution dans la protection sanitaire du pèlerinage de la Mecque. In: Annales d'Hygiène Publique, Industrielle et Sociale N.S. 26: 194-199.

3558 Communiqué issued by Egypt, Saudi Arabia and Syria, March 12, 1956. In: Middle Eastern Affairs 7.1956: 143-145.   -1a, 21, 180-

3559 Consortium formed to house Mecca pilgrims. In: Civil Engineering (London) 1976(1): 15.   -83, 89, 93-

3560 Council for Middle East Trade, 1963: Report of the British Trade Mission to Saudi Arabia. London. 31 p.

3561 Crane, R.D., 1978: Some considerations on Saudi industrialization. In: N.A. Shilling (ed.): Arab markets - 1978. New York: Inter-Crescent Publishing and Information Corporation, 1978: 107-131.

3562 Daher, S.A., 1983: From crisis to crisis: Saudi-American relations - 1973-1979. Ph.D., Harvard University, Cambridge/Mass. 362 p.

3563 Daniels, R.J., 1977: Al Jubail industrial harbour. In: Civil Engineering (London) 1977(6): 27, 29, 31.   -83, 89, 93-

3564 Darrab, I.A., 1970: The present date situation and programs prolonged for date production and market development. Kingdom of Saudi Arabia, Ministry of Agriculture and Water, Special Report 5; n.p. (Riyadh?). 56 p.

3565 Descoudray, M.(?), 1829: Voyage à la Mekke, dans les années 1826-1827. In: Nouvelles Annales des Voyages et des Sciences Géographiques II, 11: 198-216. -4, 21-

3566 Deutsches Hydrographisches Institut, ^{6}1983: Handbuch für das Rote Meer und den Persischen Golf. Hamburg. 485 S. -H 2-

3567 Dhohayan, A.I., 1981: Islamic Resource Sharing Network: a feasibility study for its establishment among university libraries of Saudi Arabia and the Republic of Turkey as representative Islamic nations. D.L.S., University of Southern California, Los Angeles.

3568 Dossary, Saleh J., 1981: Guidance needs of secondary school students in Saudi Arabia. Ph.D., University of Denver. 378 p.

3569 Dowson, V.H.W., 1968: The improvement of the Saudi Arabian date industry. Riyadh: Ministry of Agriculture and Water.

3570 Draz, O., 1964: Range improvement in the Western Province in Saudi Arabia. Riyadh: Ministry of Agriculture and Water.

3571 Etibi, M.H. al-, 1974/75: Innovation and traditionalism in contemporary Hijazi poetry. M.A., University of Manchester.

3572 Exchange of letters between President John F. Kennedy and Amir Faisal, Premier and Foreign Minister of Saudi Arabia. In: Middle Eastern Affairs 14.1963: 47-48. -21, 25, 180-

3573 FAO, 1952: Saudi Arabia, country background information paper. Report 52/12/7970; Rome.

3574 ——, Human Resources, Institutions and Agrarian Reform Division, 1974: Directory of agricultural education and training institutions in the Near East region. Rome. IV, 111 p. -5-

3575 Farm Machinery Corporation, 1965: Haradh Project, Kingdom of Saudi Arabia. San José.

3576 Feng, M., R.S.-M. Chang, T.R. Smith & J.C. Snyder, 1959: Adenoviruses isolated from Saudi Arabia. II. Pathogenicity of certain strains for man. In: The American Journal of Tropical Medicine and Hygiene 8: 5o1-5o4. -1a, 16, 1oo-

3577 Filemban, S.N., 1982: Verbal classroom interaction in elementary school mathematics classes in Saudi Arabia. Ph.D., Oregon State University, Corvallis. 125 p.

3578 Forslund, B., 1954: Saudi-Arabia - en växande svensk exportmarknad. In: Utlandssvenskarna 16(11): 8-11. -212-

3579 Fougerouse, D., 1976: La Santé publique dans le Royaume d'Arabie Saoudite; l'Hôpital de spécialistes du Roi Fayçal à Riyadh et de son Centre de recherches. Thèse de doctorat en Médecine, Académie de Clermont-Ferrand. 69 p.

358o Gérard, I., C. Gérard & J. Gérard, 1967: Une famille belge en Arabie Saoudite. Bruxelles: Éditions 'Arts & Voyages', de Meyer. 124 p.

3581 Ghamdi, Abdulaziz S. al-, 1981: An approach to planning a primary health care delivery system in Jeddah, Saudi Arabia. Ph.D., Michigan State University, East Lansing. 1o7 p.

3582 Ghamdi, Mohammed S.D. al-, 1982: The impact of ecological factors upon the attitudes of Saudi students toward work values: a search for development approach. Ph.D., Florida State University, Tallahassee. 2o7 p.

3583 Giles, M.L. & C.E. Chatham, Jr., 1976: Design of Jubail Harbor, Saudi Arabia, Royal Saudi Naval Expansion Program; hydraulic model investigation. U.S. Army Engineer Waterways Experiment Station, Technical Report H-76-2; Vicksburg. Var.pag. -H 2-

3584 Gouin, É., 1847: L'Égypte au XIXe siècle. Histoire militaire et politique, anecdotique et pittoresque de Méhémet-Ali, Ibrahim-Pacha, Soliman-Pacha (Colonel Sèves). Paris: Boizard. IV, 47o p.

3585 Grayson, B.L., 1982: Saudi-American relations. Washington: University Press of America. III, 163 p. -21-

3586 Haass, R., 1981: Saudi Arabia and Iran: the twin pillars in revolutionary times. In: H. Amirsadeghi (ed.): The security of the Persian Gulf. London: Croom Helm, 1981: 151-169.   -21-

3587 Hacker, B., 1963: Sojourn in Saudi Arabia; the story of three years in the land of Cadillacs and walls. New York: Exposition Press. 277 p.

3588 Haddad, G.A., 1969: Schéma de croissance industrielle pour l'Arabie Séoudite. In: Annales de la Faculté de Droit de Beyrouth 59: 153-23o.

3589 Hagra, H.H. & A.A. al-Noty, 1962: Date situation in Saudi Arabia. Riyadh: Ministry of Agriculture and Water.

359o Hamouda, M., 1936: L'état sanitaire du Hedjaz, notamment au point de vue du pèlerinage. In: Office International d'Hygiène Publique, Bulletin Mensuel 28: 887-895.   -4, 17, 25-

3591 Harrison, P.W., 192o: Fanatical Moslems of Central Arabia. In: The Missionary Review of the World 192o(July): 597-6oo.   -4-

3592 Harsaghy, F.J., Jr., 1968: Report on a library consultative visit to King Abdulaziz University and other Jeddah libraries. Dhahran. 6 p.

3593 Heck, G. & M. Wöbcke, 1983: Arabische Halbinsel. Saudi-Arabien und Golfstaaten. Reise-Handbuch. Richtig reisen; Köln: DuMont. 294 S.   -21-

3594 Hötzl, H., C. Job & J. Zötl, 1977: Saudi-Arabien: Wann ist das Grundwasser erschöpft? In: Umschau in Wissenschaft und Technik 77: 518-519.   -21, 1oo, 18o-

3595 Hubert-Rodier, L., 1975: Survol de l'Arabie Séoudite. In: Europe Outremer (535): 42-44.   -3o, 2o6, H 3-

3596 Hudson, M.C., 1977: Arab politics: the search for legitimacy. New Haven, London: Yale University Press. XI, 434 p. -21-

3597 Husain, Z.A. al-, 1981: Development planning: a realistic approach for Saudi Arabia. Ed.D., University of Northern Colorado, Greeley. 443 p.

3598 Jabber, P., 1982: Oil, arms, and regional diplomacy: strategic dimensions of the Saudi-Egyptian relationship. In: M.H. Kerr & S. Yassin (eds.): Rich and poor states in the

Middle East. Egypt and the new Arab order. Westview's Special Studies on the Middle East; Boulder: Westview Press; Cairo: The American University in Cairo Press, 1982: 415-447. -21-

3599 Juynboll, T.W., 1913: 'Abd. In: EI[1] 1: 17-2o. -21, 24-

36oo Kamerbeek, G.J., 1978: Modernizing the Saudi Arabian telephone network. In: Philips' Telecommunication Review 36: 217-233. -1a, 89, 93-

36o1 Karabuda, B. & G. Karabuda, 1967: Till Mecka. Ett reportage från Saudi-Arabien. Stockholm: Bonnier. 11o S.

36o2 Kattan, R.A.S., 1982: The impact of government expenditures in the transformation process of a traditional economy: a case study of Saudi Arabia. Ph.D., George Washington University, Washington. 321 p.

36o3 Kazemzadeh, I. (identisch mit dem Verfasser von Nr. 1514), 1934-35: Meine Pilgerfahrt nach Mekka, der heiligen Stadt des Islams. In: Moslemische Revue 11: 75-87, 12: 36-46. -1a, 7, 21-

36o4 King, G.R.D., 1979: Some European travellers in Najd in the 19th and early 2oth centuries A.D. In: Studies in the history of Arabia. Vol. 1: Sources for the history of Arabia, pt. 2; Riyadh: Riyad University Press, 1979: 255-265. -21-

36o5 Kingdom of Saudi Arabia, Ministry of Agriculture and Water, Department of Agricultural Development, 1979: A guide to agricultural investment in Saudi Arabia. N.p. 146 p.

36o6 ——, Ministry of Information, n.d. (1973?): Royal speech at banquet given by H.M. King Faisal in Mecca for notable pilgrims on Dhul Hijjah 6, 1392 A.H., corresponding to January 1o, 1973 (Aussentitel: Address by H.M. King Faisal ibn Abdul Aziz to pilgrimage missions for 1392 A.H.-1973 A.D.; französischsprachige Ausgabe: Texte du discours prononcé par Sa Majesté le roi Fayçal le 6 Zul Hijjah 1392 - 1o Janvier 1973 devant les délégations venues au pélerinage à la Mecque). Jeddah: Dar al-Asfahani. 12 p. -212-

36o7 ——, Ministry of Planning, 1982: Saudi Arabia: achievements of the First and Second Development Plans, 139o-14oo (197o-198o). Facts & Figures; Jeddah: Tihama Publications. V, 146 p.

3608 Kisnawi, M.M., 1981: Attitudes of students and fathers toward vocational education: the role of vocational education in economic development in Saudi Arabia. Ph.D., University of Colorado, Boulder. 342 p.

3609 Kley, H.D., 1974: Saudi-Arabien: Mit dem Ölboom kam der Fortschritt. Trotz Wirtschaftswachstumsrate ist die Infrastruktur mangelhaft. In: Auslandskurier 15(7): 2o-21. -21, 38, 212-

3610 Knauerhase, R., 1983: Saudi Arabia: fifty years of economic change. In: Current History (Philadelphia) 82: 19-23, 35-36. -1a, 21, 18o-

3611 Kopf, W., 1982: Saudiarabien. Insel der Araber. Stuttgart: Seewald. 231 S. -21, 24, 212-

3612 Laoust, H., 1939: Essai sur les doctrines sociales et politiques de Taķī-d-Dīn Aḥmad b. Taimīya, canoniste ḥanbalite, né à Ḥarrān en 661/1262, mort à Damas en 728/1328. Thèse pour le Doctorat ès Lettres, présentée à la Faculté des Lettres de l'Université de Paris. Recherches d'Archéologie, de Philologie et d'Histoire 1o; Le Caire: Institut Français d'Archéologie Orientale. 755 p. -12, 21-

3613 Lawton, J., et al., 1983: The greening of the Arab East: an update. In: Aramco World Magazine 34(5): 2-4o. -21, 24-

3614 Leeson, H.S., 1948: Anopheline larvae collected in Arabia. In: Annals of Tropical Medicine and Parasitology 42: 253-255. -1a, 61, 291-

3615 Linde, G., 1978: Sowjetische Politik auf der arabischen Halbinsel. Berichte des Bundesinstituts für ostwissenschaftliche und vergleichende Studien 44-1978; Köln: Bundesinstitut für ostwissenschaftliche und vergleichende Studien. 36 S. -212-

3616 Long, D.E., 1979: International health aspects of the Ḥajj. In: Studies in the history of Arabia. Vol. 1: Sources for the history of Arabia, pt. 2; Riyadh: Riyad University Press, 1979: 183-195. -21-

3617 Lowth, G.T., 1855: The wanderer in Arabia; or Western footsteps in eastern tracks. 2 vols. London: Hurst & Blackett.

3618 Makari, J.G., 1952: Benadryl and sulfadiazine in the treatment of trachoma. In: The Journal of Tropical Medicine and Hygiene 55: 263-265.   -21, 61, 289-

3619 Martin, H.E., 1972: Report to the Government of Saudi Arabia on research in plant protection. SAU/TF 63; Rome: FAO. 38 p.

3620 Marx, E., 1974: The organization of nomadic groups in the Middle East. In: M. Milson (ed.): Society and political structure in the Arab world. The Van Leer Jerusalem Foundation Series; New York: Humanities Press, 1974: 3o5-336. -21-

3621 Maull, H.W., 1975: Ölmacht: Ursachen, Perspektiven, Grenzen. Frankfurt/M., Köln: Europäische Verlagsanstalt. 235 S. -21-

3622 Maurizi, V., 1819: History of Seyd Said, Sultan of Muscat; together with an account of the countries and people on the shores of the Persian Gulf, particularly of the Wahabees by Shaik Mansur (pseud., i.e. V. Maurizi), a native of Rome, who after having practised as a physician in many parts of the East, became commander of the forces of the Sultan of Muscat, against the Geovasseom and Wahabee pirates. Translated from the original Italian ms. hitherto not published. London: Booth. 174 p.

3623 McCarthy, J., 1979: Ottoman sources on Arabian population. In: Studies in the history of Arabia. Vol. 1: Sources for the history of Arabia, pt. 2; Riyadh: Riyad University Press, 1979: 113-133.   -21-

3624 McNiel, J.R., 1968: Variations in the response of childhood iron deficiency anemia to oral iron. In: Blood 31: 641-646.   -1a, 21, 1oo-

3625 Medawar, G.S., 1964: Dates in Saudi Arabia; production estimates and projections. Beirut: Economic Research Institute, American University of Beirut. 29 p.

3626 Merkel, G., 1979: Strassenplanung in Saudi-Arabien. Die verkehrstechnische Erschliessung des Azir-Gebirges. In: Auslandskurier - Diplomatischer Kurier 2o(6): 21.   -21, 38, 212-

3627 Milyani, H.M., 1983: The budgetary process in Saudi Arabia. Ph.D., Florida State University, Tallahassee. 165 p.

3628 Mofti, F.A., 1981: Urban housing design in the context of Saudi Arabia's cultural and physical conditions: potentials and constraints. Ph.D., Rensselaer Polytechnic Institute, Troy. 3o8 p.

3629 Mourad, F., 1966: Life in an Arabian village: a community study. M.Sc., Department of Sociology, University of Southern California, Los Angeles.

363o Murray, E.S., R.S.-M. Chang, S.D. Bell, Jr., M.L. Tarizzo & J.C. Snyder, 1957: Agents recovered from acute conjunctivitis cases in Saudi Arabia. In: American Journal of Ophthalmology 43: 32. -16, 25, 291- ◆

3631 Musil, A., 19o8: Arabia Petraea. Bd. 3: Ethnologischer Reisebericht. Wien: Hölder. XV, 55o S. -21-

3632 N., A. de, 1818: Notice sur les Arabes et sur les Wahabis. In: Annales Encyclopédiques 5: 5-3o. -7, 17, 25-

3633 Nettelbeck, J., 198o: Deutsche Wissenschaftler an das 'MIT des Nahen Ostens'. Kontaktstelle der University of Petroleum and Minerals beim DAAD eröffnet. DAAD-Pressenotiz 17/8o. 3 S. -212-

3634 Österreich, Bundeskammer der Gewerblichen Wirtschaft, 1966: Jordanien, Libanon, Saudi-Arabien, Syrien. Österreichs Handelspartner; Wien: Oberösterreichischer Wirtschaftsverlag. 62 S.

3635 Olayan, S.S., 198o: A Saudi view of oil. In: The Washington Quarterly 3(4): 175-179. -12, 24/213, 2o6-

3636 Omar, W., 1952: The Mecca Pilgrimage; its epidemiological significance and control. In: Postgraduate Medical Journal 28: 269-274. -38 M-

3637 Philipp, H.-J., 1982: Extremfälle der Ablehnung und Übernahme komplexer technischer Innovationen im ländlichen Raum Saudi-Arabiens. Manuskript eines auf der Frühjahrstagung der Sektion Entwicklungssoziologie in der Deutschen Gesellschaft für Soziologie am 26. Mai 1982 in Bonn gehaltenen Vortrags. 19 S.

3638 Phillips, T.O., 1963: Getting to know Saudi Arabia. New York: Coward-McCann (rev. ed.: 1971). 64 p.

3639 Rentz, G., 1966: Saudi Arabia is older than the United States. In: Emergent Nations 2(2): 8-9. -H 223-

3640 Saleh, M. al-, 1966/67: Modern Najdi poetry. Ph.D., University of Cambridge.

3641 Santamaria, R., 1977a: Arabia Saudita aumenta su influencia internacional. In: Africa (Madrid) (421): 24. -1a, 12, 89-

3642 ——, 1977b: ¿acuerdo entre Washington y Ryad? In: Africa (Madrid) (428/429): 32. -1a, 12, 89-

3643 Saudi Consulting House, 1981: A guide to industrial investment. 6th ed. Riyadh: Al-Mutawa Press. VI, 169 p.

3644 Seiz, G., 1977: Die Institute für die Ausbildung auf dem Gebiet des Fernmeldewesens - einschliesslich Rundfunk und Fernsehen - in Saudi-Arabien. In: Zeitschrift für das Post- und Fernmeldewesen 1977(11): 38-44. -1a, 21, 291-

3645 Shamikh, M.A.-H. (oder: M.A.-R.) el-, 1967/68: A survey of Hijāzī prose literature in the period 1908-1941, with some reference to the history of the press. Ph.D., School of Oriental and African Studies, University of London.

3646 Sherbiny, N.A. & I. Serageldin, 1982: Expatriate labor and economic growth: Saudi demand for Egyptian labor. In: M.H. Kerr & S. Yassin (eds.): Rich and poor states in the Middle East. Egypt and the new Arab order. Westview's Special Studies on the Middle East; Boulder: Westview Press; Cairo: The American University in Cairo Press, 1982: 225-257. -21-

3647 Shobaili, A.S., 1969: Saudi Arabian television; a historical and descriptive study. M.Sc., University of Kansas, Lawrence.

3648 Sudais, M.A.-S. al-, 1976: A critical and comparative study of modern Nejdi Arabic proverbs. Ph.D., University of Leeds.

3649 Talhouk, A.M.S., 1979: Über landwirtschaftliche und soziologische Einflüsse auf das Auftreten und die Bekämpfung von Schädlingen in Saudi-Arabien. In: Anzeiger für Schädlingskunde, Pflanzenschutz, Umweltschutz 52: 9o-92. -21, 1oo, 291-

3650 Taton, R., 1975: L'Arabie Séoudite: troisième producteur et premier exportateur mondial de pétrole brut. In: Europe Outremer (535): 45-46. -3o, 2o6, H 3-

3651 Wilson, Rodney, 1978: The great Saudi gusher. Saudi Arabia's mammoth drive to become an industrial power is coming under control. In: Management Today 1978(Sept.): 114-122. -38-

3652 Yaseen, N.H., 1983: A study of the factors relating to admission and academic achievement of female students in the College of Education, Mecca, Saudi Arabia. Ed.D., University of Northern Colorado, Greeley. 126 p.

WEITERE NACHTRÄGE / FURTHER ADDITIONS

3653 Abd al-Hay, A.A. al-, 1983: Contemporary women's participation in public activities: differences between ideal Islam and Muslim interpretation with emphasis on Saudi Arabia. Ph.D., University of Denver. 362 p.

3654 Abulfaraj, W.H., 1983: Development and application of a decision methodology for the planning of nuclear research and development in Saudi Arabia. Ph.D., Iowa State University, Ames. 262 p.

3655 Abul-Nasr, S.E., 1966: Report to the Government of Saudi Arabia on plant protection development. Report 2o77; Rome: FAO. 8 p. -Gö 153-

3656 Abu Yaman, I.K., 1966: Insect pests of Saudi Arabia. In: Zeitschrift für angewandte Entomologie 58: 266-278. -1a, 1oo, 352-

3657 Alkazaz, A., 1975: Der Aufbau moderner Forschungseinrichtungen in Saudi-Arabien. Ein einführender Bericht. In: Orient (Opladen) 16(1): 99-111. -1a, 21, 1oo-

3658 Baltow, A.M., 1983: A historical analysis of the Saudi Arabian Ministry of Education's policies regarding fathers' involvement in the schooling of their children. Ph.D., Michigan State University, East Lansing. 165 p.

3659 Beling, W.A., 1980: Introduction. In: W.A. Beling (ed.): King Faisal and the modernisation of Saudi Arabia. London: Croom Helm; Boulder: Westview Press, 1980: 9-14. -21-

3660 Blunt, W.S., 1880: A visit to Jebel Shammar (Nejd). New routes through Northern and Central Arabia. In: Proceedings of the Royal Geographical Society and Monthly Record of Geography 2: 81-102. -21, 24, 291-

3661 British Bank of the Middle East, 1979: Saudi Arabia. Business Profile Series; n.p. 28 p. -Tü 17-

3662 Crecelius, D., 1975: Sa'udi-Egyptian relations. In: International Studies 14: 563-585. -1a, 18, 30-

3663 Cuinet, V., 1894: La Turquie d'Asie. Géographie administrative, statistique, descriptive et raisonnée de chaque province de l'Asie Mineure. T. 3. Paris: Leroux.

3664 Dieckmann, P., 1944: Das Eisenbahnwesen im Nahen Osten und seine Bedeutung für den Handel, insbesondere mit Deutschland. In: Studien zur Auslandskunde; Vorderasien 1: 48-63. -7, 12, 212-

3665 Gamie, M.N., 1982: The status of rural development in Saudi Arabia. In: King Saud University, College of Agriculture, Research Center Bulletin 9; Riyadh: King Saud University Press, 1982: 5-32.

3666 Heim, A., 1941: Charakterbewertung bei den Wahabiten. In: Der Wendepunkt im Leben und im Leiden 18(10): 424-429. -1a, F 1-

3667 Heuglin, M.T. von, 1868: Reise nach Abessinien, den Gala-Ländern, Ost-Sudán und Chartúm in den Jahren 1861 und 1862. Jena: Costenoble. XII, 459 S. -21, 24-

3668 Ingenieurgemeinschaft Lässer-Feizlmayr Consulting Engineers, 1983: Water for Riyadh. Innsbruck: Tyrolia. N.pag.

3669 Kay, S., 1977: The landscape of the Asir. In: Ahlan Wasahlan 1(2): 14-15, 17.

3670 Khālid ibn 'Abd al-'Azîz Âl Sa'ûd. In: Orient (Opladen) 16.1975(2): 8-9. -1a, 21, 100-

3671 Kingdom of Saudi Arabia, Ministry of Finance and National Economy, Zakat and Income Tax Department, 1978: Regulations for income tax, road tax, and Zakat (up to the end of September 1978). Riyadh: Safir Bureau. 555, XV p.

3672 Knauerhase, R., 1976: Social factors and labor market structure in Saudi Arabia. Discussion Paper 247; New Haven: Yale University Economic Growth Center.

3673 Königreich Saudi-Arabien, Ministerium für Erziehung, Generaldirektion für berufliches Bildungswesen, Berufliche Sekundarschule Hofuf, o.J.(1974): Berufliche Ausbildung - Fakten und Bilder. (Ausschliesslich arabischsprachiger Aussen- und Innentitel, aber arabisch- und englischsprachiger Text.) Dschidda: Dar al-Asfahani. 44 S.

3674 Koszinowski, T., 1975: Fahd ibn 'Abd al-'Azîz Âl Sa'ûd. In: Orient (Opladen) 16(4): 9-11. -1a, 21, 1oo-

3675 Momenah, A.A., 1977: King Faisal Medical City. In: Ahlan Wasahlan 1(2): 6-7, 9.

3676 Murray, E.S., S.D. Bell, Jr., A.T. Hanna, R.L. Nichols & J.C. Snyder, 196o: Studies on trachoma. I. Isolation and identification of strains of elementary bodies from Saudi Arabia and Egypt. In: The American Journal of Tropical Medicine and Hygiene 9: 116-124. -1a, 16, 1oo-

3677 Rasch, B., 198o: Die Zeltstädte des Hadsch. Mitteilungen des Instituts für leichte Flächentragwerke (IL) der Universität Stuttgart 29; Stuttgart: Institut für leichte Flächentragwerke (IL) der Universität Stuttgart (zugleich: Diss., Fakultät für Architektur und Stadtplanung, Universität Stuttgart, 198o). 24o S. -21, 93-

3678 Rustow, D.A., 1979: U.S.-Saudi relations and the oil crises of the 198os. In: S.B. Hunt (ed.): The energy crisis: a critical analysis of the energy policy of the United States. Contemporary Issues Series; Skokie: National Textbook Co., 1979: 315-335. -H 3-

3679 Smith, J., 1977: More precious than oil. In: Ahlan Wasahlan 1(2): 1o-11, 13.

368o Todd, W.F., 1969: The impact of oil on Middle Eastern economies. Much remains to be done before development goals of producing countries are achieved, but oil revenues have already resulted in considerable progress. In: World Petroleum 4o(1).

REGISTERTEIL / INDEX SECTION

I SCHLAGWÖRTER

Anthropologische Kennzeichen   759, 97o, 972, 11o1, 1894, 2o16,
   236o, 27o3, 333o, 3538
Arabischer Aufstand: Vorgeschichte und Verlauf   196, 336, 426, 455,
   512, 515, 516, 628, 648, 784-786, 836, 91o, 931, 1o13, 1o28,
   1o29, 1o99, 1132, 1187, 1247, 1248, 1338, 1381, 144o, 1449,
   1471, 1477, 1519-1521, 1767, 1778, 1821, 1868-1871, 1921,
   1933, 1967, 2166, 2175, 2177, 2187, 2188, 2193, 22o9, 2465,
   2641, 2677, 2693, 286o, 2897, 3o11, 3o7o, 3o78, 32o2, 3225,
   3322, 337o, 3376, 3378, 3472
Arbeitsmarkt und Beschäftigungspolitik   54, 67, 89, 11o, 191, 225,
   275, 299, 323, 331, 333, 449-452, 52o, 548, 619, 633, 69o,
   694, 73o, 756, 8o8, 8o9, 884, 897, 9o4, 939, 94o, 984, 985,
   1o53, 1o74, 1185, 1214, 1394, 1399, 1444, 1456, 1495, 1497,
   1538, 1585, 16o1, 1642, 168o, 1691, 1695, 1727, 173o, 1763,
   18oo, 195o, 1972, 1991, 2o13, 2176, 2185, 2216, 2223, 2275,
   2295, 23oo, 2334, 236o, 2361, 2713, 2875, 2933, 2938, 2951,
   297o, 3o13, 3o52, 31o3, 312o, 3145, 3154, 323o, 3264, 3521,
   3526, 3672
Außenpolitik
— Überblick   192, 418, 443, 462, 483, 5o7, 543, 634, 646, 782,
   819, 87o, 975, 1oo5, 1o1o, 1o51, 1183, 1196, 1197, 1481,
   1542, 1759, 1793, 18o4, 18o5, 1836, 1862, 1946, 1957, 1958,
   2o12, 2o28, 2o44, 2143, 224o, 2283, 2319, 2396, 2513, 2721,
   2845, 2868, 2934, 3o22, 3o25, 3232, 3246
— Verhältnis zu arabischen und islamischen Staaten, Völkern und
   Organisationen   75, 97, 99, 113, 137, 149, 194, 283, 3o6,

— Verhältnis zu arabischen und islamischen Staaten, Völkern und
  Organisationen (Fortsetzung) 3o8, 319, 341, 371, 399, 416,
  425, 437, 485, 5o3, 5o5, 5o8-511, 514, 544, 547, 548, 673,
  684, 733, 745, 774, 781, 783, 82o, 822, 841, 842, 893, 898,
  9o9, 961, 962, 1o2o, 1o45, 1o95, 11o9-1111, 112o, 13o3, 1334,
  1349, 1356, 14oo-14o3, 1415, 1452, 1453, 1472, 15oo, 1522,
  1525, 1533, 1534, 1544, 1545, 1578, 17o8, 1758, 176o, 1773,
  1776, 1777, 1791, 1794, 1798, 1799, 1818, 182o, 1822, 1834,
  1866, 19o3, 19o7, 1912, 192o, 1964, 1971, 1994, 2o11, 2111,
  2118, 2162, 2219, 2226, 2316, 2366, 2375, 2376, 2492, 2536,
  2537, 2547, 2559, 2588, 2628, 2634, 2641, 269o, 2714, 2716,
  2718, 272o, 2752, 2753, 28o6, 2828, 2917-2919, 2975, 2989,
  3o24, 3o29, 3o58, 3o93, 31o6, 3154, 3199, 32o1, 3217, 3219,
  3226, 3234, 3273, 3313, 3351, 3353, 3369, 3392, 3424, 3442,
  3461, 3465, 348o, 3481, 3498, 3523, 3543, 3556, 3558, 3562,
  3598, 36o6, 3621, 3662

— Verhältnis zu anderen Entwicklungsländern 1579, 3o24, 3154,
  3556

— Verhältnis zu westlichen Industriestaaten 111, 137, 155, 185,
  186, 193, 283, 337, 34o, 341, 35o, 371, 416, 422, 423, 425,
  427, 442, 482, 5o8-511, 517, 546, 548, 627, 628, 647, 651,
  684, 688, 723, 774, 777, 781, 799, 834, 875-877, 893, 9o9,
  978, 1o83-1o85, 1o96, 11o8-1112, 1119, 1128, 1144, 115o, 1194,
  1199, 12o3, 1226, 1262, 1282, 1295, 13o3, 1314, 1316, 1325,
  1334, 1352, 1353, 1362, 1377, 1381, 1415, 1437, 1452, 1453,
  1462, 1519, 1521, 1522, 1533, 1534, 1538, 1545, 1568, 158o,
  17o8, 176o, 1762, 1789, 1791, 1799, 1818-182o, 1822, 1827,
  1861, 1863, 1879, 1885, 19o4, 1914, 1915, 1918-192o, 1963,
  197o, 1996, 1997, 2o14, 2o24, 2o3o, 2o83, 2o85, 2o94, 211o,
  2111, 2137, 217o, 2229, 2232, 2235, 2287, 2289, 23o6, 23o8,
  2316, 2346, 2363, 2365, 2366, 2376, 24o6, 2465, 2491, 2492,
  2516, 2549, 2588, 259o, 2591, 2634, 2641, 265o, 2651, 2661,
  2671, 2694, 2714, 274o, 286o, 29o2, 2911, 2917-2919, 2975-
  2979, 2988, 3oo1, 3o24, 3o27, 3o29, 3o53, 3o54, 31o7, 3154,
  3189, 3199, 32o1, 3218, 3225, 3226, 3234, 3237, 3298, 3316,
  3351, 3355, 3362, 3369, 3461, 3523, 3552, 3556, 3562, 3572,
  3585, 3611, 3621, 3642, 3678

— Verhältnis zu Ostblockstaaten 283, 51o, 511, 774, 781, 9o9,
  111o, 1111, 1194, 1414, 1452, 176o, 192o, 2336, 2375, 2491,
  2492, 3467

Bank- und Versicherungswesen 1o8, 2o6, 223, 347, 387, 515, 52o,
    521, 57o, 615, 633, 73o, 8o8, 884, 1o43, 1181, 1283, 164o,
    1691, 1695, 1749-1752, 1754-1756, 195o, 1985, 2156, 2251,
    2471, 2495, 2627, 2756, 2935, 3422, 3494
Bauern
— Materielle und immaterielle Kultur 66, 117, 413, 741, 8o1,
    855, 947, 1235, 15o4, 2122, 2935, 3323, 3324, 35o2, 3536,
    3629
— Sozialstruktur und Sozialorganisation 117, 741, 8o1, 1235,
    15o4, 165o, 3324, 35o2, 3536, 3629
Baugewerbe (s. auch Dörfer: Hausarchitektur und Wohnungseinrichtung sowie Wasserbau und -wirtschaft; Eisenbahnbau und -verkehr; Entwicklung der materiellen Infrastruktur: Überblick; Flughafenbau, Fluglinien und Luftverkehr; Hafenbeschreibungen, -bau und -verkehr; Gesundheitswesen: Hygienische und medizinische Einrichtungen und Beschäftigte; Landwirtschaft einschließlich Viehwirtschaft: Wasserbau und -wirtschaft; Meerwasserentsalzungsanlagen; Öl- und Gasindustrie: Förderung und Transport sowie Verarbeitung und Verschiffung; Städte: Planung und Entwicklung, Hausarchitektur und Wohnungseinrichtung sowie Wasserbau und -wirtschaft; Straßenbau und -verkehr; Verarbeitende Industrie und Industriepolitik; Verkehrswesen: Überblick) 11o, 279, 488, 557, 582, 588, 598, 669, 725, 73o,
    897, 986, 1o3o, 1o31, 1149, 1171, 124o, 1473, 1493, 16o7-
    161o, 1616, 1617, 1624, 1628-1632, 1751, 1757, 1939, 1973,
    2145-2147, 2194, 2229, 2432-2442, 2477, 2489, 25o9-2511,
    2622, 2722, 2723, 2757, 277o, 3229, 3252, 3254, 329o, 3359,
    341o, 3559
Beherbergungsgewerbe (s. auch Geschäftsratgeber und -praxis sowie Reiseführer und Fremdenverkehr) 994, 124o, 1695, 1893, 19o9,
    198o, 1981, 2532, 2744, 2779, 3197, 3677
Bevölkerung
— Methoden und Resultate von Volksschätzungen und -zählungen
    89, 9o, 1o4, 189, 289, 45o, 483, 631, 661, 732, 8o1, 8o6,
    874, 94o, 1127, 12o8, 1278, 13o9, 1311, 1511, 1588, 159o,
    169o, 1695, 1696, 18oo, 1832, 195o, 1951, 1968, 2o12, 2o42,
    2o54, 222o, 2341, 2369, 2428, 2691, 2728, 2762, 2765, 2766,
    2873-2875, 292o, 2921, 2941, 2968, 3224, 3324, 3378, 3623,
    3665
— Bevölkerungspolitik 661, 874, 2885, 3268

Bibliotheks- und Dokumentationswesen 51, 296, 4lo, 683, 1239, 1252, 1419, 1562, 1691, 2oo6, 2o41, 223o, 2581, 2743, 3o35, 3123, 3554, 3567, 3592

Bildungswesen

— Überblick 56, 88, 143, 144, 171, 333, 4o2, 413, 463, 478, 524, 673, 8o1, 864, 879, 883, 892, 9o6, 937, 1o51, 1o74, 1o77, 1177-118o, 1182, 14o4, 1534, 1543, 1559, 159o, 167o-1674, 1679, 1681, 1683-1686, 1691, 1695, 171o, 1751, 1957, 2oo6, 2oo9, 2o54, 21oo, 2216, 2223, 2242, 2329, 2333, 2455, 2493, 2494, 25o7, 2713, 2733, 2741, 2794, 2836, 2863, 2894, 292o, 2933, 3o82, 3o92, 31o2, 3139, 3223, 3274-3277, 328o, 3492, 3523

— Geschichtliche Entwicklung 54, 57, 87, 115, 123, 2o1, 37o, 65o, 839, 84o, 1o98, 1113, 1214, 1448, 22o8, 2216, 2223, 273o, 2773, 2927, 2969, 3o11-3o13, 3355, 3483

— Vor- und Grundschulwesen 57, 87, 95, 117, 18o, 2o1, 37o, 65o, 928, 1o79, 128o, 1448, 1883, 2o15, 2252, 2332, 2647, 2866, 3o13, 3o2o, 3166, 3167, 3275, 3324, 3483, 3539, 3577

— Mittel- und höheres Schulwesen 5o, 57, 76, 87, 115, 131, 153, 2o1, 3oo, 3o1, 322, 325, 333, 385, 65o, 767, 788, 793, 861, 926, 938, 1o77, 1o79, 1168, 13o5, 1423, 1424, 1436, 1448, 1492, 1982, 1992, 2o15, 2o48, 2o49, 2192, 25o5, 2647, 27o7, 2834, 2866, 2942, 2953, 2964, 3o13, 3o2o, 3o96, 3166, 3167, 3276, 3373, 34o4, 3479, 3483, 352o, 3527-3529, 3535, 3568, 36o8

— Hochschulwesen 54, 57, 67, 1o5, 114, 151, 161, 167, 191, 2o1, 296, 3o3, 3o4, 326, 327, 33o, 348, 369, 64o, 844, 9o5, 92o, 952, 998, 1o39, 1185, 119o, 12o6, 1214, 1228, 127o, 1276, 13o4, 13o5, 13o8, 1319, 1356, 1366, 1376, 1413, 1419, 1447, 1448, 1483, 15o1, 1515, 1539, 1557, 1558, 1563, 1599, 1645, 17oo, 17o1, 1719, 1747, 1992, 2oo5, 2o23, 2o29, 2o9o-2o93, 2181, 219o, 22o3, 2223, 2249, 2324, 2455, 2493, 25o3, 25o4, 2583, 2655, 27o2, 2745, 2773, 2834, 2856, 2864, 29o9, 2927, 2952, 3o28, 3o3o, 3o9o, 31o3, 3163, 3166, 3167, 3173, 3175, 3277, 3283, 34o9, 3453, 3479, 3483, 3485, 3529, 3532, 3567, 3574, 3582, 3633, 3652, 3657

— Sonderschulwesen 57, 2o1, 12o5, 1524, 1676, 2224, 2696, 3166, 3167, 3278, 3279

Bildungswesen (Fortsetzung)
— Berufliches Bildungswesen (ohne staatliche Erwachsenenbildung)
    57, 121, 141, 184, 2o1, 25o, 265, 331, 421, 422, 497, 619,
    645, 694, 758, 793, 8o2, 81o, 913, 982, 984, 985, 1o51,
    1o7o, 11o5, 1211, 1224, 1225, 1399, 1444, 1448, 1538, 1644,
    168o, 1728, 1748, 1864, 1881, 1883, 1972, 2o38, 2o5o, 2182,
    2185, 2216, 2227, 2229, 229o, 2295, 2432, 2697, 2698, 2861,
    2886, 2935, 2965, 2966, 3o13, 3o86, 3112, 3124, 3166, 3167,
    3192, 32o4, 32o6, 3238, 34o4, 3485, 3489, 3521, 3574, 36o8,
    3644, 3657, 3673
— Mädchenschulwesen 57, 2o1, 37o, 454, 881, 1195, 1448, 1563,
    1591-1593, 17o9, 1719, 2o78, 2263, 2355, 3166, 3167, 3483,
    3489, 3525, 3526, 3652
— Staatliche Erwachsenenbildung 57, 164, 2o1, 332, 1165, 12o9,
    1448, 1669, 1677, 1678, 1687, 1883, 2o78, 3166, 3167, 3447
— Lehr- und Verwaltungssystem und -personal 76, 78, 87, 95, 1o5,
    153, 2o1, 3oo, 3o1, 322, 325-327, 37o, 65o, 767, 788, 793,
    861, 88o, 928, 938, 1o76, 1o79, 11o7, 1165, 1168, 119o,
    12o5, 12o6, 12o9, 1227, 127o, 1276, 13o5, 1419, 1423, 1424,
    1436, 1448, 1492, 15o1, 1549, 1563, 1982, 1992, 2o15, 2o49,
    21o3, 2181, 2185, 2199, 2249, 2252, 2324, 2329, 2332, 249o,
    2679, 27o1, 27o2, 27o7, 273o, 2732, 29o9, 2942, 2953, 2964,
    3o28, 3o9o, 3o96, 3122, 3166, 3167, 3175, 3193, 328o, 335o,
    3373, 3447, 3453, 3479, 3483, 3485, 352o, 3527-3529, 3532,
    3535, 3568, 3577, 3658
— Bildungspolitik 54, 57, 168, 2o1, 332, 478, 65o, 9o5, 1448,
    1483, 1675, 1682, 249o, 273o, 2927, 3o13, 3166, 328o, 3283,
    3483, 3658, 3673
Biographien
— Mohammed ibn Abdalwahhab (17o3/o4-92) 86, 663, 729, 829, 1249,
    1343, 1856, 2248, 2418, 254o, 2551, 2572, 3299, 35o3, 3612
— König Abdalasis (Ibn Saud) (188o-1953) 149, 278, 415, 427, 438,
    533, 651, 843, 1o47, 1113, 1362, 1373, 1415, 1443, 1553,
    211o, 2111, 2131-2133, 2345, 239o, 2413, 2416, 2422, 2589,
    259o, 2597, 2619, 2621, 2632, 2868, 3192, 33o4, 3312, 3355,
    3412, 3428
— König Saud (19o2-69) 416, 274o, 2775
— König Faisal (19o5/o6-75) 418, 482, 924, 1o1o, 1o54, 139o,
    1541, 1577, 1862, 2219, 2226, 2283, 2775, 2943, 2944, 3429,
    3523, 3659

— König Khalid (1912-82)   441, 139o, 1542, 1892, 2629, 2746, 367o
— König Fahd (1922/23-)   441, 898, 139o, 1542, 2629, 2746, 3674
— Großscherif und König Hussain ibn Ali (1853/54-1931)   336, 455, 1128, 1332, 2175, 2593, 2636
— Andere Staatsmänner und sonstige Elitemitglieder   142, 149, 441, 455, 693, 866, 1132, 1235, 1236, 139o, 1559, 1892, 2175, 2512, 2629, 2746, 2945, 3o7o
— Bekannte westliche Arabienreisende   145, 149, 384, 4o5, 424, 444, 474, 513, 531, 532, 623, 642, 643, 657, 685, 882, 9o1, 981, 1ooo, 1159, 1212, 1254, 1326, 1331, 1341, 136o, 1368, 1534, 1778, 179o, 1811, 1911, 1928, 2o99, 21o6, 21o9, 2149, 2163, 2177, 2187, 2243, 2271, 2281, 2369, 2385, 2411, 242o, 2427, 2431, 2526, 2531, 2674, 2676, 27oo, 2829, 2893, 298o, 3o31, 3o4o, 31oo, 3113, 3162, 33o7, 336o, 3365, 3419, 342o, 3435-3437, 3445, 3451, 3488

Dörfer
— Geschichte und Beschreibung   79, 117, 617, 656, 658, 8o1, 852, 853, 915, 1o42, 1451, 15o2-15o4, 1664, 1694, 1951, 2o21, 2o22, 2191, 2197, 2211, 2212, 2394, 2421, 2541-2543, 26o1, 2711, 2874, 288o, 2921, 2924, 3o42, 3324, 336o, 3629, 366o
— Planung und Entwicklung   86o, 956-958, 1o42, 2191, 2197, 2428, 2954, 3o42, 3324
— Hausarchitektur und Wohnungseinrichtung   8o, 8o1, 85o, 856, 957, 958, 1o42, 1493, 157o-1574, 2191, 2479, 2524, 2541, 2622, 2723, 2843, 2874, 3o42, 3324, 3669
— Wasserbau und -wirtschaft   914, 2985, 3324, 3519
— Community Development   332, 661, 1255, 1556, 17o5, 1726, 2186, 225o, 268o, 292o, 2935, 2966, 3267

Druck- und Pressewesen   18, 486, 487, 9o5, 942, 1o72, 1268, 145o, 1691, 1759, 2oo6, 2242, 2466, 2538, 2656, 2659, 2745, 2756, 2814, 2825, 292o, 2926, 2965, 3o1o, 3123, 313o, 3345

Eisenbahnbau und -verkehr   61, 224, 3o9, 311, 336, 392, 396, 431, 457, 459, 46o, 47o, 516, 658, 662, 695, 752, 826, 827, 896, 931, 946, 1o31, 1o9o, 1118, 1143, 1166, 1242, 1243, 1277, 1284, 1332, 1336, 1371, 1373, 1375, 1382, 14o6, 1449, 1451, 1471, 1576, 1764, 1847, 1848, 1886, 1888, 1913, 1967, 2o32, 2o45, 2o46, 2o59, 2177, 2187, 22o7, 221o, 2242, 23o1, 23o2, 23o4, 23o7, 2341, 24o1, 2421, 2459, 246o, 25o6, 25o8, 2518, 2532, 2545, 2584, 264o, 2677, 2823, 284o, 2896, 2982, 2983, 2999, 3143, 315o, 3192, 3194, 32o3, 3222, 33o9, 3371, 3434, 3472, 3486, 3517, 3637, 3664

Eliten
— Königliche Familie 75, 124, 142, 194, 256, 280, 441, 549, 629,
    712, 773, 781, 822, 870, 902, 961, 976, 1111, 1317, 1352,
    1359, 1362, 1526, 1534, 1542, 1759, 1791, 1834, 1892, 1942,
    2014, 2143, 2282, 2283, 2292, 2315, 2316, 2492, 2516, 2532,
    2533, 2548, 2560, 2629, 2715, 2746, 2747, 2917-2919, 2931,
    2933, 2938, 3154, 3216, 3425, 3546, 3674
— Sonstige politische Elite 256, 441, 2265, 2316, 2492, 2533,
    2655, 2746, 3393
— Religiöse Autoritäten 256, 271, 452, 533, 549, 744, 877, 902,
    905, 1113, 2265, 2292, 2315, 2492, 2533, 2931, 2933, 2938,
    3011, 3012
— Wirtschaftliche Elite 139, 158, 159, 268, 441, 452, 461, 549,
    632, 659, 725, 945, 973, 1559, 2142, 2246, 2265, 2533, 2736,
    2746, 2938
— Sonstige Eliten 452, 2466, 2746, 2938
Energiewirtschaft und -politik 215, 463, 592, 884, 897, 1031,
    1038, 1691, 1695, 1751, 1756, 1827, 1874, 1917, 2089, 2374,
    2770, 2835, 2895, 3146, 3343, 3654
Entwicklung der materiellen Infrastruktur: Überblick 279, 477,
    480, 563, 564, 766, 815, 843, 986, 1038, 1162, 1278, 1446,
    1504, 1554, 1559, 1603, 1748, 1957, 1997, 2104, 2105, 2114,
    2118, 2130, 2228, 2300, 2402, 2409, 2413, 2416, 2418, 2420,
    2489, 2516, 2545, 2563, 2622, 2632, 2739, 2752, 2753, 2774,
    2777, 2837, 2854, 2863, 2911, 2938, 2968, 3039, 3049, 3059,
    3077, 3092, 3197, 3231, 3257, 3259, 3287, 3421, 3494, 3495,
    3508, 3518, 3607, 3643, 3665, 3677, 3680
Entwicklungshilfe: bi- und multilateral
— Erhaltene Entwicklungshilfe 350, 463, 535, 546, 614, 740, 777,
    810, 877, 934, 978, 1051, 1157, 1170, 1256, 1262, 1345, 1377,
    1538, 1587, 1760, 1789, 1878, 1997, 2108, 2110, 2120, 2129,
    2130, 2137, 2229, 2461, 2516, 2650, 2651, 2740, 2911, 2972,
    3071, 3087, 3258, 3281, 3362, 3417, 3458, 3459, 3491, 3552,
    3585
— Geleistete Entwicklungshilfe 140, 387, 463, 514, 668, 783, 813,
    841, 842, 917, 1008, 1038, 1422, 1754, 1794, 1867, 1932,
    1964, 1989-1991, 2012, 2158, 2258, 2314, 2438, 2472, 2690,
    2746, 2984, 3106, 3405, 3413, 3448, 3556
Erzlagerstätten und -bergbau 227, 291, 463, 625, 626, 661, 1038,
    1049, 1051, 1509, 1559, 1597, 1733-1735, 2012, 2174, 2426,
    2612, 2740, 3183, 3246, 3247, 3414, 3432

Fernmelde- und Postwesen 2o6, 463, 497, 715, 722, 757, 772, 1o34,
    1134, 1268, 14o6, 145o, 1691, 1695, 1751, 1874, 1997, 2227,
    2298, 231o, 2345, 2391, 2396, 24o3, 2593, 2624, 313o, 3133,
    3138, 3165, 3222, 3355, 35o7, 36oo, 3644
Finanzwirtschaft und -politik 1o8, 1o9, 12o, 128, 133, 134, 175,
    18o, 2o6, 2o9, 21o, 213, 216, 223, 237, 238, 284, 289, 291,
    344, 363, 414, 463, 484, 521, 57o, 617, 638, 676, 697, 73o,
    776, 841, 878, 884, 925, 1o38, 1o51, 1164, 1171, 1219, 1221,
    1257, 1283, 1298, 1315, 138o, 1422, 1442, 1445, 1495, 1496,
    1569, 1691, 1695, 1751, 1771, 1774, 1791, 1849, 19o5, 195o,
    1985, 199o, 1991, 2ooo, 2o13, 2o31, 21o4, 211o, 2129, 2155,
    2156, 2229, 2231, 2258, 2276, 23o1, 23o3, 23o5, 2323, 2334,
    2372, 2396, 24o3, 2413, 247o, 2471, 2516, 2525, 2746, 2748,
    2752, 2753, 2756, 278o, 2791, 2795, 28o3, 2896, 29o6, 2972,
    3135, 3153, 3234, 3255, 3259, 3287, 3362, 3369, 3375, 3387,
    3421, 3423, 3469, 347o, 3494, 3522, 3547, 36o2, 36o7, 3627
Fischereiwirtschaft einschließlich Perlenfischerei 1o7, 2o4, 491,
    492, 863, 9o8, 1235, 1889, 2296, 2457, 2553, 2867, 32o9,
    334o, 3458, 3459, 36o5
Flughafenbau, Fluglinien und Luftverkehr 6o, 61, 3o5, 54o, 7o2,
    7o3, 755, 758, 777, 778, 817, 877, 1o31, 1o34, 1119, 175o,
    1873, 1874, 2o18, 2136, 2157, 2341, 25o3, 2779, 279o, 3o68,
    3112, 3211, 3411
Forschungseinrichtungen und -politik 258, 383, 78o, 8o2, 948-95o,
    1188, 1319, 1384, 1411, 1895-1897, 1954, 2oo4, 2oo6, 2o41,
    22o5, 2354, 236o, 2583, 2745, 31o3, 3127, 3132, 3579, 3649,
    3654, 3657, 3677
Forstwirtschaft 7o, 77, 1oo, 112, 391, 3o63
Frauen: Status und Aktivitäten 127, 139, 158, 159, 277, 334, 37o,
    413, 454, 465, 655, 759, 773, 79o, 791, 821, 847, 852, 881,
    892, 944, 1o93, 113o, 1195, 1288, 13o2, 151o, 1516, 1518,
    1725, 1742, 178o, 1845, 1877, 1935, 2o27, 2o78, 2355, 2358,
    2468, 2484, 2486, 2532, 2544, 268o, 2938, 3o57, 3213-3215,
    33o2, 3452, 3526, 361o, 3653
Freizeitbeschäftigungen und -einrichtungen einschließlich Sportleben 146, 172, 173, 38o, 777, 1147, 1156, 1386, 1538, 1781,
    2o81, 2432, 2436, 2437, 2441, 26o5, 2818, 29o5, 3o17, 3478
Gastarbeiterzusammensetzung und -probleme 275, 299, 32o, 449, 451,
    452, 466, 472, 614, 62o, 73o, 939, 1297, 1311, 1356, 1393,
    1394, 1534, 1763, 18oo, 2o26, 2o54, 21o3, 2216, 2223, 2334,
    25o7, 2713, 2874, 2875, 2919, 2933, 2938, 3364, 3646

Geld und Währungspolitik   1o8, 223, 347, 776, 1o38, 1o69, 1145,
    1146, 1181, 1218-122o, 1283, 1383, 1691, 1751, 1774, 1849,
    1985, 199o, 1991, 1999, 2134, 2155, 2156, 2276, 228o, 231o,
    2499, 2531, 2786, 2941, 3324, 3469, 347o
Genossenschaftswesen   129, 1129, 1691, 1695, 2685
Geschäftsratgeber und -praxis   81, 82, 94, 182, 198, 388, 526,
    527, 551, 557, 56o, 562, 565, 566, 568, 57o, 574, 578, 581,
    582, 584, 587-591, 593-612, 614, 66o, 714, 718, 719, 73o,
    737, 78o, 789, 8o8, 929, 966, 989, 99o, 1o91, 1176, 12o1,
    1318, 14o7, 148o, 1487, 1488, 15o6, 1586, 16o3, 1613, 1663,
    18o9, 1852, 2o37, 2o52, 21o4, 2119, 2127, 2128, 2184, 2217,
    2218, 2247, 2255, 228o, 231o, 2317, 2318, 2371, 2453, 2474,
    2481, 2487, 2489, 2495, 2675, 2737, 2746, 2755, 2764, 2778,
    29o3, 2916, 2928, 2929, 2955, 2956, 3o36, 3o44, 3o69, 3177,
    3271, 3272, 3284, 3288, 33o6, 3315, 3397, 34o1, 3457, 3471,
    3495, 3516, 3542, 3545, 3548, 36o5, 3643, 3661
Gesamtgesellschaft (s. auch Bauern, Nomaden sowie Städter)
— Materielle und immaterielle Kultur   1o3, 132, 16o, 349, 37o,
    5o2, 661, 816, 838, 847, 85o, 1185, 1214, 1233, 1421, 1924,
    194o, 1978, 2o29, 2o52, 2o91, 2o93, 2167, 2223, 2272, 2359,
    2729, 28o8, 2998, 3o42, 3385, 3582, 36o8, 3648, 3658, 3665
— Sozialstruktur und Sozialorganisation   16o, 162, 2o2, 271, 318,
    328, 413, 5o9, 614, 664, 7o9, 712, 8o1, 8o5, 852, 866, 879,
    979, 1191, 1237, 1282, 1299, 131o, 1321, 1353, 1355, 1448,
    147o, 1958, 1964, 2o55, 2o78, 2o9o, 2242, 2259, 2265, 2274,
    2284, 2292, 2334, 2359, 2513, 2533, 2563, 2575, 2655, 27o5,
    2713, 28o8, 2874, 288o, 2933, 2938, 3oo3, 3o22, 3o99, 3338,
    3339, 3351, 3352, 3385, 3393, 3447, 3665
— Kultureller, sozialer und sozialökonomischer Wandel   117, 124,
    127, 132, 139, 162, 163, 183, 259, 334, 345, 348, 349, 413,
    44o, 461, 466, 471, 483, 485, 5o2, 614, 661, 667, 7o6-711,
    721, 724, 731, 768, 783, 8o5, 816, 866, 9o5, 9o6, 944, 968,
    1o5o, 113o, 1148, 1175, 1221, 125o 128o, 1297, 1299, 1317,
    1321, 1333, 1352, 1353, 1355, 1357, 1359, 1392-1394, 1421,
    1432, 1453, 15o4, 15o7, 15o9, 151o, 1533, 1539, 1558, 1776,
    18o2, 18o3, 1836, 1844, 1845, 19oo, 19o3, 1919, 1924, 1936,
    1942, 1957, 1958, 1987, 1991, 2o23, 2o42, 2o53, 2o55, 2o92,
    2o93, 21o8, 211o, 2114, 2115, 2151, 2194, 2242, 2259, 2262,
    2284, 2323, 2359, 2362, 2417, 242o, 2425, 25o7, 2513, 2546,
    2563, 2564, 26o4, 2655, 2659, 2689, 2719, 273o, 2832, 2847,

— Kultureller, sozialer und sozialökonomischer Wandel (Fortsetzung) 2923, 2933, 2934, 2938, 2941, 2943, 2969, 2970, 2998, 3004, 3043, 3057, 3064, 3073, 3077, 3192, 3213-3215, 3310, 3319, 3326, 3331-3333, 3338, 3339, 3362, 3382, 3385, 3393, 3517, 3610

Geschichte der westlichen Erforschung (Saudi-)Arabiens im 18.-20. Jahrhundert 170, 314, 384, 433, 500, 518, 624, 625, 658, 678, 914, 1007, 1021, 1050, 1071, 1073, 1131, 1159, 1213, 1322, 1324, 1325, 1327, 1455, 1479, 1532, 1564, 1570, 1839, 1951, 1963, 2034, 2035, 2125, 2163, 2254, 2369, 2379, 2392, 2408, 2410, 2413, 2424, 2425, 2428, 2448, 2449, 2464, 2508, 2531, 2587, 2600, 2604, 2740, 2862, 2898, 3039, 3101, 3126, 3179, 3220, 3540, 3604

Gesundheitswesen

— Verbreitung und Bekämpfung von Krankheiten und Seuchen 55, 148, 261, 307, 314, 408, 475, 555, 617, 624, 639, 661, 687, 725, 760, 761, 765, 768, 770-773, 799, 821, 865, 941, 1059-1065, 1130, 1158, 1192, 1275, 1321, 1339, 1416, 1433, 1434, 1559, 1576, 1924, 1943, 1953, 1984, 2018, 2034, 2039, 2062, 2063, 2065, 2066, 2154, 2222, 2267-2270, 2332, 2383, 2384, 2463, 2525, 2532, 2623, 2653, 2862, 2985, 3014, 3074-3077, 3121, 3129, 3222, 3324, 3555, 3557, 3576, 3590, 3614, 3616, 3618, 3624, 3630, 3636, 3676

— Hygienische und medizinische Einrichtungen und Beschäftigte 102, 148, 201, 265, 314, 434, 463, 475, 555, 631, 661, 682, 687, 695, 696, 773, 801, 844, 865, 1031, 1058, 1074, 1088, 1158, 1192, 1228, 1231, 1256, 1280, 1321, 1339, 1433, 1434, 1503, 1538, 1691, 1695, 1714, 1751, 1924, 1943, 1953, 1984, 2018, 2051, 2054, 2064, 2075, 2110, 2112, 2113, 2154, 2275, 2332, 2432, 2433, 2435, 2440, 2441, 2463, 2476, 2532, 2533, 2545, 2622, 2623, 2804, 2886, 2887, 2896, 2911, 2985, 3010, 3021, 3074, 3129, 3192, 3460, 3526, 3557, 3579, 3581, 3590, 3616, 3636, 3675

Grenzstreitigkeiten und -regelungen 98, 106, 111, 125, 194, 274, 287, 312, 468, 523, 534, 693, 822, 1080, 1084, 1096, 1097, 1124, 1133, 1140, 1142, 1203, 1282, 1347, 1361, 1381, 1401, 1449, 1527-1529, 1535, 1545, 1580, 1741, 1805, 1826, 1853, 1916, 1929, 1960, 1961, 1970, 2024, 2084-2087, 2289, 2391, 2395, 2476, 2559, 2571, 2590, 2611, 2667, 2671, 2689, 2704, 2772, 2802, 2839, 2860, 2902, 3001, 3053, 3134, 3137, 3189,

Grenzstreitigkeiten und -regelungen (Fortsetzung) 3199-3201,
  3293, 3379, 3435, 3475-3477, 3480, 3481, 3497-3499
Hafenbeschreibungen, -bau und -verkehr 269, 329, 617, 720, 811,
  812, 814, 946, 1015, 1016, 1031, 1044, 1134, 1309, 1418,
  1485, 1513, 1667, 1743-1746, 1874, 1888, 1951, 1965, 2021,
  2022, 2148, 2296, 2297, 2310, 2379, 2390, 2434, 2483, 2542,
  2543, 2660, 2763, 2787, 2788, 2797, 2798, 2801, 2822, 2858,
  3005, 3207, 3294, 3305, 3390, 3457, 3563, 3566, 3583
Handel und Handelspolitik
— Außenhandel (s. auch Agrarim- und -exporte) 96, 110, 206, 209,
  211, 233, 235, 272, 289, 313, 463, 482, 517, 520, 525, 552,
  565, 614, 617, 669, 728, 777, 809, 820, 875, 879, 884, 964,
  977, 1012, 1038, 1083, 1091, 1134, 1138, 1148, 1163, 1184,
  1240, 1397, 1408, 1414, 1426, 1445, 1472, 1497, 1513, 1532,
  1540, 1546, 1567, 1691-1693, 1695, 1743-1745, 1751, 1765,
  1807, 1816, 1840, 1880, 1888, 1893, 1899, 1901, 1930, 1938,
  1947, 1951, 1952, 1966, 1990, 1991, 1999, 2037, 2104, 2127,
  2129, 2134, 2156, 2227, 2229, 2231, 2260, 2265, 2280, 2306,
  2310-2312, 2379, 2457, 2472, 2478, 2495, 2515, 2585, 2615-
  2617, 2649, 2752, 2753, 2756, 2760, 2763, 2787, 2810, 2815-
  2817, 2824, 2858, 2867, 2873, 2880, 2906, 2937, 3046-3049,
  3147, 3187, 3218, 3224, 3229, 3259, 3263, 3269-3271, 3285-
  3288, 3305, 3390, 3431, 3435, 3440, 3441, 3457, 3463, 3494,
  3516, 3518, 3560, 3578, 3634
— Binnenhandel 110, 117, 169, 268, 361, 463, 561, 617, 637, 735,
  759, 825, 852, 853, 884, 914, 1018, 1131, 1235, 1426, 1485,
  1514, 1547, 1554, 1691, 1951, 1965, 2096, 2213, 2310, 2340,
  2341, 2508, 2531, 2842, 2844, 2880, 2906, 3010, 3057, 3222,
  3323, 3324, 3360, 3431, 3457
Handwerk und Handwerker 117, 617, 759, 850, 852, 897, 915, 1157,
  1235, 1285, 1291, 1508, 1509, 1605, 2017, 2096, 2213, 2341,
  2428, 2582, 3149, 3192, 3323, 3324
Innenpolitik 418, 443, 507, 508, 614, 655, 673, 782, 818, 821,
  822, 906, 909, 975, 1005, 1017, 1096, 1113, 1169, 1282,
  1358, 1359, 1415, 1491, 1526, 1708, 1773, 1776, 1805, 1833,
  1836, 1862, 1866, 1903, 1958, 1994, 2057, 2118, 2139, 2262,
  2345, 2492, 2548, 2564, 2571, 2716, 2719, 2721, 2758, 2845,
  2917-2920, 2995, 3025, 3154, 3201, 3352, 3355
Kindererziehung und -beschäftigungen 139, 821, 1023, 1516, 1518,
  2879

Kulturpolitik   9o5, 1681, 2oo6
Kunst   146, 413, 44o, 1o72, 113o, 1389, 2oo6, 2746, 28o8, 3o85,
    3523, 3549, 3571, 364o, 3645
Landeskunde   26o, 298, 649, 11o2, 1135, 1137-1139, 1266, 172o,
    1882, 1884, 19o8, 1923, 2237, 2291, 2556, 3o32, 3184, 3245,
    3282, 3361, 3384, 3493
Landeskundlicher Überblick (z.T. reich bebildert)   54, 11o, 2o3,
    29o, 324, 351, 366, 37o, 394, 417, 428, 437, 464, 497, 519,
    558, 56o, 562, 634, 635, 667, 67o, 698, 73o, 746, 764, 775,
    792, 796, 8o1, 8o5, 8o7, 89o, 899, 992, 993, 1o35, 11o8, 1117,
    1174, 1214, 1257, 126o, 1261, 1269, 1312, 1317, 1323, 1348,
    135o, 1372, 1374, 1419, 1493, 1523, 1546, 1551, 1565, 17o6,
    1711, 1715, 1718, 1746, 1774, 1782, 1783, 1824, 1855, 186o,
    189o, 1898, 1934, 1935, 1945, 197o, 2o36, 2o54, 2o6o, 21o4,
    213o, 2184, 2235, 2239, 2285, 237o, 2378, 2481, 2487, 25o7,
    2541, 2599, 26o4, 26o7, 2625, 2646, 2657, 27o9, 271o, 2734,
    2742, 275o, 2756, 2761, 2775, 2778, 2779, 28o7, 28o8, 2852,
    2853, 2872, 2873, 2888, 2895, 2955, 2959-2963, 2985, 2987,
    3o25, 3o43, 3o48, 3o49, 3136, 3144, 3284, 3292, 332o, 3338,
    3349, 3362, 3385, 3391, 34oo, 34o1, 343o, 3439, 3443, 3444,
    345o, 3456, 3457, 3466, 3495, 3497, 35o6, 3526, 3545, 3556,
    3578, 3593, 36o9, 3638
Landwirtschaft einschließlich Viehwirtschaft
— Überblick   11o, 228, 291, 575, 621, 622, 691, 716, 717, 792,
    8o2, 862, 884, 1o74, 1278, 1335, 1429, 143o, 15o5, 159o,
    1655, 1657, 1691, 1751, 1838, 1865, 2o12, 2o38, 2o54, 2o73,
    2o74, 214o, 2156, 2456, 25oo, 2695, 2759, 2837, 2855, 2935,
    2949, 2971, 3o49, 3191, 323o, 3243, 3487, 3494, 3573
— Agrarstruktur und -verfassung   84, 289, 352, 8o1, 8o2, 8o4,
    1189, 1235, 1456, 1457, 15o4, 165o, 1897, 2428, 2873, 2874,
    2946, 3o33, 3o62, 3116, 3117, 323o, 3324, 35o2
— Pflanzenproduktion   66, 79, 84, 1o7, 117, 2o4, 253, 258, 289,
    335, 352, 361, 419, 476, 493, 536, 537, 617, 678, 679, 74o,
    75o, 78o, 8oo, 8o1, 8o4-8o6, 855, 856, 872, 914, 947-95o,
    965, 983, 996, 997, 117o, 1235, 1253, 1274, 1319, 1378, 1384,
    1385, 1388, 1411, 1456, 1457, 15o4, 1552, 1554, 1559, 16o2,
    16o4, 1611, 1638, 1646-1652, 1655, 1659, 1695, 1766, 183o-
    1832, 1864, 1895-1897, 1951, 1954, 1965, 2o25, 2o61, 2o72,
    2o96, 21o8, 211o, 2117, 2153, 2277-2279, 2294, 234o, 2354,
    2357, 2428, 243o, 2498, 2524, 2545, 26oo, 26o1, 26o2, 274o,

— Pflanzenproduktion (Fortsetzung) 2848, 2873, 2874, 2921, 2946,
    2967, 2968, 2981, 3oo7, 3o33, 3o61, 3o64, 3o87, 311o, 3111,
    3132, 3192, 3221, 323o, 3242, 3249, 3251, 33oo, 33o1, 3314,
    3323, 3324, 3328, 3356, 3357, 3496, 35o2, 3536, 3564, 357o,
    3589, 36o5, 3613, 3619, 3625, 3637, 3649, 3655, 3656, 3669
— Tierproduktion (s. auch Nomaden: wirtschaftliche Grundlagen)
    1o4, 1o7, 117, 199, 2o4, 258, 289, 465, 538, 539, 75o, 78o,
    8o1, 8o4, 821, 85o, 867, 914, 947, 948, 1159, 1274, 1384,
    1411, 1428, 1456, 1457, 15o7, 1559, 1646-1652, 1655, 1659,
    1766, 183o, 1832, 1841, 1872, 1951, 1954, 2o61, 2153, 2428,
    2498, 26oo, 2665, 2725, 2726, 28o5, 2946, 3o33, 3o64, 3132,
    3151, 3181, 3192, 324o, 3324, 357o, 36o5, 3637, 366o
— Entwicklungsprojekte, Fördermaßnahmen und Agrarpolitik 7o, 77,
    91, 1oo, 1o4, 1o7, 112, 117-119, 15o, 2o4, 2o6, 218, 228,
    229, 258, 265, 289, 352, 367, 419, 463, 528, 573, 622, 644,
    7o4, 71o, 711, 721, 725, 74o, 754, 78o, 8o1-8o3, 8o5, 83o,
    831, 864, 869, 872, 873, 884, 889, 918, 919, 933, 942, 967,
    996, 997, 1o67, 1o68, 11o3, 11o4, 1129, 117o, 1188, 1189,
    1253, 1259, 128o, 13o7, 1378, 1384, 1395, 1411, 1457, 1524,
    1552, 1554, 1559, 1586, 16o2, 16o4, 1611, 1638, 1646-1649,
    1651-1662, 1695, 1712, 1713, 1716, 1749, 183o, 1864, 1872,
    1878, 1889, 1922, 193o, 1954, 2o25, 2o38, 2o61, 2o72, 21o8,
    211o, 2123, 214o, 2145-2147, 2227, 2238, 2354, 2412, 2429,
    25o9-2511, 2545, 26o3, 2678, 2681, 2695, 274o, 2759, 28o5,
    2833, 2865, 2935, 2946, 295o, 3o33, 3o63-3o65, 3o87, 31o9,
    3115, 314o, 3192, 32o5, 3249-3254, 3259, 3324, 3336, 3347,
    3356-3358, 3367, 3383, 3391, 341o, 3417, 3496, 3521, 3536,
    3575, 36o5, 3655, 3665
— Wasserbau und -wirtschaft 66, 92, 1o1, 1o7, 117, 2o6, 258,
    364, 365, 367, 39o, 395, 419, 463, 493, 5o4, 535, 621, 622,
    632, 679, 721, 75o, 8o1, 8o2, 8o4, 8o5, 852, 862, 864, 872,
    873, 889, 936, 95o, 997, 1o34, 1188, 1253, 1256, 1259, 1364,
    1378, 1384, 1411, 1443, 1456, 1457, 15o4, 1524, 1552, 1554,
    1646-1658, 1661, 1712, 1713, 1716, 1779, 183o, 1872, 1874,
    1895-1897, 1951, 1965, 2oo3, 2o25, 2o38, 2o4o, 2o61, 21o8,
    211o, 213o, 214o, 2141, 2145-2147, 2227, 2238, 2277, 2279,
    2386, 2412, 2428, 2429, 25o9-2511, 2524, 2558, 26o1-26o3,
    26o6, 2678, 2695, 274o, 2833, 2865, 2873, 2874, 2921, 2925,
    2946, 2967, 2981, 3ooo, 3oo7, 3o16, 3o33, 3o62, 3o64, 3115,
    314o, 3183, 3192, 3221, 3242-3244, 3249-3254, 33o3, 3314,

— Wasserbau und -wirtschaft (Fortsetzung) 3323, 3324, 3335, 3336, 3347, 3356, 3357, 3366, 3367, 3383, 3391, 3410, 3496, 3519, 3575, 3594, 3605, 3613, 3679

— Agrarim- und -exporte 90, 691, 780, 795, 804, 1445, 1611, 1691, 1695, 1841, 1857, 1930, 1956, 2153, 2231, 2347-2353, 2895, 2911, 3230

— Agroindustrie 253, 258, 361, 780, 897, 911, 1388, 1604, 1611, 1618, 1619, 1638, 1639, 1841, 1925, 2061, 2072, 2430, 2692, 3550, 3569, 3589, 3605

Lebens- und Arbeitsbedingungen für westliche Ausländer 110, 127, 139, 172, 173, 394, 494, 559, 684, 719, 730, 749, 769, 810, 834, 1006, 1032, 1128, 1566, 1855, 1890, 1934, 2018, 2052, 2060, 2184, 2308, 2370, 2489, 2622, 2671, 2689, 2790, 2809, 2911, 2955, 2957, 3399, 3580

Meerwasserentsalzungsanlagen 65, 181, 204, 206, 390, 393, 463, 777, 922, 1713, 1874, 2003, 2010, 2895, 3260, 3289, 3454, 3668

Nahrungsmittelzubereitung und Ernährungsweise 104, 253, 272, 289, 617, 661, 801, 821, 1302, 1451, 1512, 1604, 1638, 1832, 1841, 1924, 2096, 2121, 2213, 2430, 2497, 2941, 2985, 3008, 3057, 3487, 3515, 3631

Nomaden einschließlich teilweise seßhafte Stämme

— Wirtschaftliche Grundlagen 119, 246, 285, 309, 316, 471, 618, 656, 658, 704-711, 724, 726, 821, 822, 848, 849, 852, 853, 873, 914, 915, 918, 919, 951, 967, 1001, 1093, 1096, 1126, 1175, 1235, 1250, 1274, 1280, 1288, 1302, 1392, 1451, 1484, 1507, 1580, 1844-1846, 1951, 2042, 2165, 2196, 2210-2213, 2215, 2307, 2327, 2373, 2400, 2520, 2521, 2552, 2558, 2570, 2603, 2635, 2665, 2695, 2874, 2880, 2881, 2891, 2921, 2923, 3057, 3064, 3097, 3099, 3157-3161, 3168-3172, 3192, 3331-3333, 3360, 3389, 3390, 3435, 3438, 3524, 3541, 3620, 3631, 3637

— Materielle und immaterielle Kultur 246, 285, 479, 501, 618, 624, 626, 680, 708, 726, 821, 822, 845, 849, 852, 853, 914, 915, 951, 969, 971, 1001, 1026, 1093, 1096, 1097, 1125, 1175, 1235, 1250, 1273, 1286, 1288-1290, 1292-1294, 1302, 1320, 1357, 1392, 1435, 1451, 1484, 1502-1504, 1507, 1517, 1518, 1845, 1846, 1951, 1955, 2042, 2076, 2165, 2196, 2210-2213, 2215, 2274, 2327, 2373, 2377, 2400, 2521, 2522, 2531, 2544, 2552, 2577, 2637, 2638, 2695, 2880, 2881, 2885, 2891, 3008,

— Materielle und immaterielle Kultur (Fortsetzung) 3o57, 3o97-
  3o99, 3157-3161, 3168-3172, 3192, 3331-3333, 336o, 3377,
  3389, 339o, 3416, 3511, 3541, 3565, 3631, 3666
— Sozialstruktur und Sozialorganisation 119, 246, 293-295, 413,
  5o1, 665, 7o4, 7o5, 7o8, 71o, 711, 726, 821, 845, 848, 849,
  852, 853, 873, 951, 967, 1oo1, 1126, 1175, 1235, 125o, 1288,
  1289, 13o2, 1392, 1451, 15o2-15o4, 15o7, 1529, 158o, 1845,
  1846, 1986, 1991, 2164, 2165, 2196, 221o-2213, 2215, 2274,
  2327, 2388, 252o, 2544, 2552, 2554, 256o, 2565, 2566, 2569,
  257o, 2635, 2695, 27o5, 2874, 288o, 3o38, 3o57, 3o97, 3o99,
  3161, 3172, 33o2, 3325, 3327, 3333, 336o, 3393, 3416, 3435,
  362o, 3631
— Seßhaftmachung und -werdung 118, 119, 127, 176, 285, 316, 471,
  517, 651, 7o4-7o7, 71o, 721, 8o3, 8o5, 821, 822, 869, 873,
  879, 892, 918, 919, 951, 967, 979, 1o67, 1o68, 1o96, 1113,
  1123, 1169, 1175, 125o, 1282, 13oo, 1392, 1443, 1491, 15o7,
  151o, 1655, 1766, 1788, 1986, 2o77, 2212, 2345, 2373, 2377,
  24o3, 2443, 2548, 2554, 2565, 2571, 26o2, 26o3, 2695, 27o5,
  2785, 2874, 2894, 2921-2924, 2946, 3oo9, 3o33, 3o57, 325o,
  3251, 3254, 3313, 3326, 3331, 3332, 3355, 3438, 3492, 3524,
  3575, 3591
— Kennzeichen von Pariastämmen 686, 821, 845, 846, 849, 852,
  853, 1o14, 1o94, 1285, 1287, 1291, 1875, 1955, 221o, 2211,
  2213, 2327, 2369, 2444, 2445, 257o, 2635, 2851, 2921

Öl- und Gasindustrie
— Überblick 11o, 24o-244, 248, 254, 26o, 264, 266, 425, 653,
  667, 676, 734, 9oo, 932, 968, 985, 1o36-1o38, 1216, 1278,
  1295, 1296, 1334, 1346, 1425, 1445, 1466, 1467, 15o9, 1538,
  1559, 1691, 1695, 1732, 1737, 1751, 176o, 1774, 181o, 1833,
  1837, 186o, 1899, 19o2, 199o, 2o12, 2o24, 2o56, 2118, 2156,
  2322, 2362, 26o7, 2632, 2694, 2754, 2817, 2837, 2863, 2865,
  2868, 2873, 2888, 2911, 3o49, 3o6o, 3148, 3231, 3319, 3382,
  3393, 3497, 365o
— Konzessionen und Explorationen 116, 149, 2o6, 213, 219, 226,
  227, 31o, 337, 362, 372, 379, 381, 429, 463, 467, 468, 534,
  652, 666, 681, 685, 725, 794, 799, 822, 884, 1o34, 115o,
  1211, 1258, 1313, 1381, 1432, 1538, 1597, 1598, 1947, 1948,
  197o, 2oo8, 2o44, 2o82, 2o84, 21o8, 213o, 2137, 2138, 2144,
  2168, 2232, 2244, 2325, 2426, 2431, 2478, 2512, 259o, 2612,
  2877, 2884, 29o4, 2972, 3o4o, 3o56, 3o59, 3o71, 3155, 318o,

— Konzessionen und Explorationen (Fortsetzung)   3186, 3192, 33o8,
   3362, 3365, 3386, 349o
— Förderung und Transport   116, 12o, 166, 2o6, 213, 219, 251,
   252, 262, 31o, 312, 343, 373, 375, 381, 429, 463, 467, 534,
   681, 725, 751, 774, 835, 884, 987, 1211, 124o, 1422, 1431,
   1495, 1497, 1597, 1598, 1736, 1874, 1948, 197o, 1983, 2oo7,
   2o44, 2o83, 2o84, 21o8, 213o, 2137, 22o4, 2253, 2264, 2321,
   2326, 2426, 2478, 2512, 2527, 2612, 2639, 2738, 274o, 2749,
   2811, 2877, 2878, 2882, 2938, 2972, 3o56, 3o59, 3o71, 3118,
   3152, 3181, 3192, 3362
— Verarbeitung und Verschiffung   72, 2o6, 2o9, 213, 262, 263,
   31o, 312, 376, 377, 429, 463, 467, 534, 835, 884, 1o86,
   1211, 1495, 1581-1584, 1597, 1598, 1736, 1874, 1948, 197o,
   2o44, 213o, 22o4, 2478, 2512, 2519, 2687, 274o, 2817, 2878,
   2972, 3o56, 3o59, 3141, 3192, 32o8, 3233, 3235
— Kennzeichen der ARAMCO, ihrer beiden Vorgängerinnen und der
   Trans-Arabian Pipe Line Company (TAP Line) einschließlich
   ihres Personals   116, 148, 155, 176, 185, 186, 191, 197,
   2o2, 241, 242, 244, 245, 247, 248, 25o, 254, 257-26o, 265,
   266, 279, 331, 349, 35o, 37o, 372-379, 381, 41o, 425, 429,
   463, 635, 636, 653, 681, 699, 725, 75o, 866, 932, 945, 946,
   961, 984, 985, 1o27, 1o51, 1157, 1211, 1252, 1253, 1258,
   1295, 1296, 13o9, 1313, 1333, 1365, 1377, 1398, 14o4, 1478,
   1524, 1534, 1538, 1554, 1559, 1566, 1581-1584, 176o, 1789,
   1872, 1881, 1883, 19o2, 1924, 1926, 1948, 197o, 1996, 1997,
   2o41, 2o44, 2o83, 21o8, 211o, 213o, 2137, 2138, 216o, 217o,
   22o1, 22o2, 2253, 2264, 2265, 232o, 2333, 236o-2362, 2426,
   2478, 2512, 2548, 2612, 2632, 2655, 2662, 27o5, 2735, 2739,
   274o, 279o, 2877, 2878, 2911, 2954, 2972, 3oo3, 3o13, 3o52,
   3o56, 3o59, 3o6o, 3o71, 3118, 3119, 3145, 3181, 3192, 32o3-
   32o6, 32o8, 3211, 3231, 3246, 3358, 3362, 3386, 3393, 349o,
   3552, 368o
— Kennzeichen anderer Ölgesellschaften einschließlich ihres Personals   227, 27o, 463, 534, 1o81, 1167, 1597, 1598, 2o44,
   216o, 2265, 2516, 2972, 31o4, 31o5, 3119, 3231
— Staatliche Preis- und Mengenpolitik   116, 12o, 128, 137, 186,
   283, 319, 422, 437, 462, 463, 5o6, 548, 638, 783, 866, 884,
   961, 975, 11o8, 1121, 1144, 1162, 1199, 1215, 1216, 1334, 1351,
   141o, 1422, 146o, 1494-1497, 1534, 1535, 1736, 1762, 1777,
   1791, 1798, 1799, 1822, 1833, 1836, 1851, 1859, 19o4, 191o,

— Staatliche Preis- und Mengenpolitik (Fortsetzung) 1944, 1974,
    1983, 1991, 2o14, 2o31, 2o43, 2o44, 2143, 215o, 2169, 217o,
    2198, 2258, 2283, 2325, 2346, 2372, 2491, 2492, 2735, 2945,
    2972, 2973, 3o29, 3o58, 3o66, 3232, 3234, 332o, 3338, 3386,
    3464, 3494, 3562, 3585, 361o, 3621, 3635, 3678
Oppositionsgruppen und -aktivitäten 75, 124, 127, 162, 281, 282,
    319, 466, 5o7-5o9, 549, 681, 893, 9o5, 961, 976, 1oo5, 111o,
    1196-1198, 1334, 1352, 1354, 1359, 1394, 1533-1535, 1542,
    1569, 1759, 18o3, 1833, 1836, 1866, 19oo, 1914, 1959, 1964,
    2o14, 2242, 2265, 23o9, 2492, 2546-2548, 2826, 2868, 2917-
    292o, 2938, 3o26, 32o1, 353o, 3621
Pilgerwesen
— Pilgerreise und -riten 52, 53, 68, 69, 178, 354, 412, 481,
    617, 623, 624, 682, 7oo, 7o1, 766, 832, 865, 1o46, 1o58,
    1o66, 1o71, 1o89, 1122, 1161, 12oo, 121o, 1339, 1395, 142o,
    1465, 1486, 149o, 1514, 1516-1518, 1547, 1559, 1792, 18o9,
    1943, 1981, 1998, 2o67, 2254, 2342, 24o9, 2447, 25o8, 2636,
    2664, 267o, 2744, 2832, 2892, 2912, 2913, 2947, 2974, 3o11,
    3o37, 3o55, 3o67, 3o94, 3227, 3228, 3344, 3371, 3394, 3396,
    34o7, 3512, 36o3, 3677
— Pilgerverkehr und -administration 52, 68, 69, 97, 154, 169,
    177, 178, 189, 3o9, 314, 336, 389, 445-448, 456-459, 463,
    465, 47o, 475, 481, 495, 555, 617, 749, 765, 766, 777, 827,
    828, 832, 834, 851-854, 865, 868, 895, 941, 1o26, 1o46,
    1o57, 1o58, 1o66, 1122, 1128, 1131, 1161, 1166, 1218, 1219,
    1265, 1331, 1339, 134o, 137o, 1417, 142o, 1465, 1472, 1486,
    1514, 1516-1518, 1547, 1576, 1691, 1695, 1698, 1699, 1751,
    1792, 1842, 1873, 19o6, 1943, 198o, 1981, 1998, 2o12, 2o67,
    2o69, 2o7o, 21o5, 21o7, 211o, 2112, 2113, 2254, 228o, 23o7,
    2327, 2343, 2344, 2379, 2396, 24o3, 24o4, 24o9, 2447, 25o8,
    2515, 2517, 2523, 2525, 2527, 2623, 2626, 2645, 2663, 2664,
    267o, 2671, 2744, 286o, 2862, 2896, 2915, 2947, 2999, 3o1o,
    3o11, 3o21, 3o37, 3o45, 3142, 3199, 3222, 3227, 3241, 33o9,
    3344, 3348, 3371, 3378, 3389, 339o, 3394, 3396, 34o7, 35o7,
    35o9, 359o, 36o3, 36o7, 3616, 3636, 366o, 3677
Politische Geschichte
— Überblick 286, 425, 692, 766, 8o5, 921, 991, 1o5o, 1o71, 1o74,
    1115, 12o3, 1232, 1246, 1317, 1322, 1328, 1344, 1432, 1443,
    1476, 1481, 1542, 1559, 1761, 1765, 1776, 1787, 1797, 18o4,
    1828, 1856, 1942, 1945, 1971, 1972, 1995, 1997, 2o12, 2o24,

— Überblick (Fortsetzung)   2o71, 2111, 2118, 2166, 22o8, 2214,
2221, 2272, 2327, 2338, 238o, 24o2, 2443, 2533, 2566, 2576,
258o, 2591, 26o4, 2621, 2631, 2654, 27o5, 2719, 2731, 2746,
2846, 2857, 286o, 2863, 2868, 29oo, 29o2, 292o, 3oo2, 3o89,
3246, 3354, 3444, 3492, 35o6, 3556, 3611, 3612, 3639

— Erster saudischer Staat (1745-1818): Überblick   86, 162, 465,
542, 545, 617, 663, 882, 1244, 1415, 1452, 1534, 155o, 18o5,
18o8, 2178, 2216, 2265, 2369, 239o, 2398, 2611, 2657, 35o3,
3662

— Erster saudischer Staat (1745-1818): Details   53, 63, 73, 85,
382, 443, 618, 729, 98o, 1249, 1527, 1529, 1532, 158o, 1951,
2o97, 2135, 2172, 2179, 22o9, 2212, 2418, 2428, 2485, 2535,
254o, 2551, 2561, 2572, 26oo, 2642-2644, 2688, 27oo, 2848,
2899, 29o7, 29o8, 3299, 33o5, 3355, 34o8, 3418, 3426, 3427,
3584, 3622, 3632

— Zweiter saudischer Staat (1824-91): Überblick   64, 86, 111,
162, 315, 443, 465, 542, 1o14, 1249, 1415, 1452, 1534, 1773,
18o5, 2265, 239o, 2611, 3226, 35o3

— Zweiter saudischer Staat (1824-91): Details   382, 744, 822, 995,
1527, 1529, 153o, 1532, 158o, 1819, 1951, 2135, 2172, 22o9,
2212, 2337, 2339, 234o, 2369, 2398, 2418, 2428, 254o, 2572,
27o8, 3355, 3418, 3426, 3427, 3584

— Dritter saudischer Staat (seit 19o2): Überblick   111, 155, 162,
345, 371, 438, 512, 542, 627, 63o, 651, 663, 673, 797, 818,
87o, 879, 912, 1o2o, 1o47, 1196, 1197, 1281, 1329, 1359,
1545, 177o, 1773, 18o1, 18o3, 1861, 1876, 1919, 1958, 2o5o,
2o55, 2132, 2217, 24o3, 2413, 2422, 2513, 2564, 2572, 2595,
2619, 2657, 274o, 2774, 286o, 2927, 2934, 3o22, 3o25, 3o53,
3192, 3312, 3339, 3455, 3497, 3596

— Dritter saudischer Staat (seit 19o2): Details   149, 196, 278,
283, 337, 34o, 341, 35o, 415, 418, 423, 426, 427, 43o, 443,
485, 517, 533, 559, 628, 693, 771, 821, 822, 834, 843, 876, 1o58,
1o8o, 1o84, 1o96, 1o97, 11o9, 1112, 1113, 1128, 1169, 1183,
1187, 1229, 1235, 1244, 1249, 1282, 1334, 1362, 1363, 138o,
1415, 1452, 1472, 1525, 1527, 1529, 1534, 1553, 158o, 176o,
1786, 18o5, 18o8, 1818, 1819, 1833, 1836, 1914, 1918, 1951,
2o83, 2o94, 211o, 2131, 2133, 2162, 2163, 218o, 22o2, 22o6,
22o9, 2212, 2226, 2265, 2345, 2386, 2387, 2389, 239o, 2393-
2398, 2411, 2416, 2418, 242o, 2428, 2457, 2515, 2516, 2547,
2548, 2571, 2578, 2579, 259o, 2611, 2664, 2671, 2894, 2931,

— Dritter saudischer Staat (seit 1902): Details (Fortsetzung)
2933, 2944, 2975-2979, 3001, 3189, 3199-3201, 3217-3219,
3226, 3324, 3355, 3363, 3369, 3378, 3379, 3392, 3412, 3428,
3435, 3436, 3480, 3481, 3498, 3585, 3662
— Der Hedschas unter türkischer und ägyptischer Hoheit (bis 1916)
73, 74, 93, 179, 196, 309, 314, 315, 336, 382, 431, 455, 515,
516, 617, 624, 626, 786, 787, 825, 839, 852, 853, 980, 1024,
1053, 1132, 1187, 1243, 1247, 1281, 1332, 1451, 1465, 1532,
1567, 1797, 1999, 2030, 2097, 2126, 2175, 2193, 2207, 2209-
2211, 2302, 2303, 2305-2308, 2392, 2418, 2467, 2578, 2611,
2618, 2688, 2700, 3008, 3010, 3012, 3070, 3114, 3164, 3222,
3305, 3355, 3371, 3395, 3408, 3565, 3637
— Haschemitische Herrschaft im Hedschas (1916-25) (s. auch Arabischer Aufstand) 336, 426, 455, 515, 559, 628, 787, 839,
1028, 1029, 1053, 1082, 1106, 1128, 1132, 1187, 1249, 1329,
1440, 1449, 1468, 1519-1521, 1545, 1548, 1778, 1786, 1821,
1870, 2175, 2304, 2345, 2390, 2392, 2418, 2420, 2465, 2515,
2578, 2593, 2611, 2664, 2860, 2976, 2994, 3053, 3070, 3078,
3189, 3199, 3222, 3226, 3378, 3379, 3543
— Raschidische Dynastie in Nedschd (1835-1921) 315, 406, 430,
443, 465, 637, 851-853, 915, 1244, 1249, 1369, 1370, 1846,
1875, 1951, 2164, 2171, 2209, 2211, 2212, 2281, 2339, 2340,
2389, 2418, 2443, 2572, 2611, 2635, 2829, 3189, 3360, 3363,
3660
— Beni Khalid-Herrschaft in al-Hasa (1680-1793/94 und 1819-30)
85, 2079, 2418, 2428, 2551
— Al-Hasa unter türkischer Hoheit (1871-1913) 430, 431, 616,
627, 1112, 1235, 1244, 1532, 1796, 1819, 1951, 2126, 2212,
2418, 2428, 3324, 3505, 3506
— Asir im 18. und 19. Jahrhundert 297, 315, 337, 339, 431, 732,
1272, 2030, 2260, 2357, 2415, 2547, 2880, 3114, 3501, 3502
— Idrisiden-Herrschaft in Asir (1908-30) 337-341, 517, 628, 732,
742, 743, 999, 1272, 1437, 2415, 2515, 2547, 2593, 2594,
2612, 2880, 2988, 3189, 3199, 3201, 3378, 3392, 3498, 3502
Politisches System und einzelne politische Institutionen 142, 149,
162, 163, 255, 256, 281, 282, 318, 425, 432, 435, 438, 443,
483, 485, 507, 549, 553, 601, 655, 673, 681, 712, 713, 730,
774, 818, 879, 902, 905, 906, 909, 952, 1005, 1017, 1034,
1038, 1074, 1108, 1113, 1229, 1278, 1333, 1358, 1359, 1379,
1453, 1461, 1481, 1526, 1531, 1533, 1534, 1542, 1555, 1600,

Politisches System und einzelne politische Institutionen (Fortsetzung)   1721, 1759, 1793, 1801-1803, 1833, 1836, 1860, 1903, 1914, 1942, 1943, 1945, 1958, 1964, 1970, 2050, 2057, 2105, 2180, 2235, 2241, 2242, 2262, 2265, 2283, 2284, 2315, 2316, 2417, 2423, 2451, 2492, 2513, 2516, 2533, 2539, 2548, 2564, 2580, 2683, 2686, 2730, 2734, 2747, 2752, 2753, 2756, 2774, 2795, 2845, 2856, 2883, 2894, 2917-2920, 2927, 2931-2934, 2938, 2943, 2965, 3018, 3022, 3023, 3066, 3091, 3154, 3178, 3192, 3216, 3313, 3351, 3352, 3355, 3393, 3425, 3492, 3517, 3556, 3596, 3612, 3643

Radio und Fernsehen   117, 267, 496-498, 715, 753, 801, 892, 905, 942, 1185, 1450, 1556, 1717, 1759, 2006, 2114, 2233, 2242, 2656, 2659, 2887, 2920, 2965, 3015, 3130, 3345, 3644, 3647

Rechtswesen   83, 110, 135, 136, 168, 187, 195, 214, 230, 274, 275, 355-357, 359, 401, 530, 578, 584, 589, 593, 594, 601, 605, 633, 655, 673, 679, 680, 690, 708, 713, 730, 808, 809, 830, 902, 903, 916, 952, 1091, 1093, 1113, 1125, 1176, 1200, 1202, 1234, 1241, 1279, 1292, 1293, 1367, 1379, 1457, 1480, 1506, 1559, 1560, 1585, 1596, 1603, 1663, 1691, 1695, 1733, 1734, 1753, 1759, 1768, 1801, 1812, 1813, 1815, 1817, 1835, 1852, 1891, 1941, 1960, 1961, 1970, 1973, 2128, 2176, 2218, 2225, 2247, 2261, 2369, 2450, 2451, 2476, 2484, 2507, 2514, 2544, 2596, 2724, 2737, 2764, 2774, 2781-2783, 2856, 2869-2871, 2883, 2894, 2916, 2928, 2929, 2931, 2933, 3011, 3012, 3018, 3019, 3041, 3051, 3069, 3128, 3131, 3182, 3271, 3316, 3402, 3403, 3475-3477, 3482, 3492, 3495, 3643, 3653

Regionalanalysen und -entwicklung   66, 84, 89, 169, 257, 258, 297, 397, 725, 732, 761, 762, 765, 858, 859, 874, 891, 954, 959, 960, 974, 1014, 1136, 1152, 1154, 1217, 1267, 1272, 1306, 1332, 1391, 1412, 1429, 1430, 1456, 1457, 1505, 1517, 1518, 1652, 1664, 1723, 1727, 1796, 1951, 2080, 2174, 2250, 2285, 2404, 2541, 2555, 2557, 2560, 2562, 2567, 2568, 2573, 2574, 2576, 2578, 2602, 2667, 2712, 2734, 2849, 2865, 2880, 2921, 2968, 3038, 3064, 3099, 3192, 3240, 3324, 3327, 3328, 3336, 3337, 3398, 3449, 3663

Regionalplanung und -politik   152, 858, 859, 959, 960, 1391, 1412, 1429, 1430, 1483, 1652, 1665, 1716, 1724, 2285, 2854, 2954, 3342, 3449

Reiseberichte und Erlebnisschilderungen   53, 71, 127, 139, 145, 168, 170, 276, 280, 286, 309, 324, 354, 384, 386, 403, 405,

Reiseberichte und Erlebnisschilderungen (Fortsetzung)   4o6, 4o9,
    433, 435, 436, 454, 461, 465, 466, 481, 482, 512, 513, 516, 519,
    529, 541, 55o, 559, 616, 617, 623-626, 629, 635-637, 656-
    658, 677-679, 681, 684, 693, 695, 696, 7oo, 7o1, 722, 736,
    748, 749, 77o, 772-774, 821-825, 832, 834, 851-854, 856,
    882, 886, 887, 889-891, 9o1, 914, 915, 927, 964, 98o, 981,
    999, 1ooo, 1oo2-1oo4, 1o2o, 1o22, 1o25, 1o4o, 1o47, 1o48,
    1o5o, 1o52, 1o55, 1o57, 1o58, 1o71, 1o73, 1o95-1o97, 11oo,
    1128, 115o, 1159, 1172, 1193, 1222, 1223, 123o, 1312, 132o,
    133o, 1331, 1337-1339, 1356, 1369, 137o, 1395, 1418, 1451,
    1478, 1489, 1516-1518, 1523, 1524, 1547, 1559, 1769, 18o9,
    1854, 1875, 1876, 1893, 1922, 1926, 1933, 1936, 194o, 1978,
    1993, 1998, 1999, 2o2o, 2o52, 2o81, 2o95, 21o1, 21o2, 21o6,
    21o8, 211o, 2112, 2113, 2161, 2173, 2183, 2193, 2194, 2195,
    22o7, 221o-2213, 2215, 2227, 2237, 2254, 2273, 2281, 2337-
    234o, 2358, 2367-2369, 2382-2384, 2386, 2387, 2389-2391,
    2394, 2399, 24oo, 24o5, 24o7-2412, 2414, 2415, 2419, 2421,
    2422, 2431, 2462, 2464, 2467, 2476, 2488, 2521, 2522, 2528-
    2532, 2545, 2549, 255o, 2589-2594, 2596, 26o8, 2614, 2616-
    2618, 2622, 2624, 2632, 2633, 2636, 2652, 2653, 266o, 2664,
    2667, 2668, 267o-2673, 2699, 27oo, 274o, 2881, 2888-289o,
    2892, 2897, 2911-2913, 2944, 2974, 3oo8, 3o1o-3o12, 3o21,
    3o43, 3o55, 3o56, 3o74, 3o76, 3o78, 3o8o, 3o87, 3o94, 3113,
    3114, 3126, 3156-3162, 3168-3172, 3174, 3186, 3188, 3197,
    3198, 322o, 3227, 3239, 3241, 3247, 33o5, 3317, 3321, 3358,
    336o, 3371, 3374, 3378, 3389-3391, 34o6, 3414, 3415, 3436,
    3437, 345o, 3462, 3468, 35o5-35o8, 351o, 358o, 3587, 36o1,
    3611, 3617, 3638, 366o, 3666
Reiseführer und Fremdenverkehr   554, 1o35, 11o3, 11o4, 1454, 1565,
    26o4, 26o5, 2654, 3397, 3466, 3593
Religiöse Minderheiten: Status und Aktivitäten
— Moslems   87, 89, 549, 976, 1191, 1217, 1235, 1999, 2428, 2492,
    2548, 2938, 3oo3, 3324, 3355
— Nicht-Moslems   87, 314, 559, 623, 624, 7oo, 749, 825, 834,
    1o53, 1128, 1172, 1193, 1241, 1514, 1538, 1922, 194o, 1999,
    2o3o, 211o, 2112, 2113, 2115, 2163, 23o6, 23o8, 2379, 2411,
    26o8, 2617, 2623, 2624, 266o, 3114, 35o8
Schiffahrt   124, 2oo, 247, 489, 49o, 811, 812, 1o15, 1o16, 11oo,
    1134, 1143, 137o, 1418, 1598, 1816, 1965, 1999, 2o21, 2o22,
    226o, 2296, 2297, 23o6, 231o, 2379, 2458, 2542, 2543, 26o8,

Schiffahrt (Fortsetzung)   2623, 2624, 266o, 2763, 28o1, 3oo6,
    3o5o, 3294, 334o, 3341, 3389, 339o, 3423, 3566, 3667
Sicherheitskräfte, Sicherheitspolitik und Rüstung   75, 124, 142,
    162, 274, 282, 283, 42o, 453, 5o7, 5o8, 51o, 511, 684, 688,
    7o5, 712, 756, 781, 783, 87o, 888, 952, 961, 976, 991, 1oo5,
    1o45, 1o49, 111o, 1111, 1196, 1197, 1199, 1316, 1317, 1356,
    138o, 1393, 1394, 141o, 1462, 1534, 1537, 1544, 1545, 1559,
    1759, 1762, 1791, 1834, 1942, 2o14, 2o43, 2o44, 21o8, 211o,
    2229, 2234, 2245, 2258, 2288, 2292, 2316, 2356, 2366, 2446,
    2452, 2469, 2491, 2492, 25o7, 2516, 2518, 2532, 2634, 2635,
    2713, 2736, 2784, 279o, 2827, 2831, 2894, 2914, 2917-2919,
    2938, 2962, 2992, 2993, 3o27, 3o34, 31o6-31o8, 3237, 3296,
    3352, 3362, 3415, 3433, 3442, 3473, 3474, 3492, 3531, 3533,
    3562, 3585, 3598, 3611, 3615, 3621
Siedlungstypen und -struktur   66, 89, 413, 8o1, 2189, 2341, 2388,
    2428, 2541, 26o1-26o3, 2873, 2874, 2923, 2954, 3251, 3324
Sklavenhandel, -behandlung und -befreiung   317, 454, 5o1, 559,
    617, 773, 9o5, 961, 1141, 1469, 1516, 1518, 1547, 1914,
    2o3o, 2213, 23o8, 2388, 2392, 24o4, 2476, 2545, 2635, 2669,
    2986, 3oo8, 3o88, 3169, 3551, 3599
Sozialleistungen und -politik   136, 463, 633, 661, 8o8, 988, 1o74,
    12o2, 1398, 1585, 1596, 1691, 1695, 17o5, 1729, 1751, 1924,
    2o12, 2176, 22o6, 2227, 2258, 2286, 2533, 259o, 2596, 2648,
    2812, 2931, 3128, 3192, 3264, 34o3, 3484, 3526, 36o5, 36o7
Sprachliche Modernisierung   3oo3, 35oo
Städte
— Geschichte und Beschreibung   52, 53, 68, 9o, 139, 156, 157, 172,
    173, 189, 19o, 288, 354, 386, 465, 469, 473, 481, 499, 556,
    616, 617, 623, 624, 637, 679, 681, 7o8, 762, 763, 766, 8o7,
    811, 812, 821, 825, 834, 852, 853, 857, 91o, 915, 964, 1oo2-
    1oo4, 1o14, 1o51, 1o56, 1o57, 1o71, 1o73, 11oo, 11o3, 11o4,
    1128, 1134, 1151, 1153, 1155, 1161, 1215, 123o, 1245, 1251,
    1264, 1265, 1271, 1339, 1342, 1369, 1396, 1454, 1475, 1485,
    1514, 1547, 1559, 1664, 1694, 1843, 1876, 1927, 1951, 1965,
    1979, 1987, 1998, 2oo1, 2oo2, 2o21, 2o68, 2o69, 21o8, 211o,
    2112, 2113, 2152, 2164, 2184, 22oo, 2211, 2273, 2281, 2285,
    2296, 2297, 23o6, 2338, 234o, 2341, 2379, 2386, 2387, 239o,
    2394, 24o1, 24o9, 2415, 2421, 2425, 2428, 2467, 2468, 25o8,
    2524, 2541-2543, 2545, 255o, 2586, 2587, 259o, 26o4, 26o8,
    2616, 2635, 2636, 2652-2654, 2664, 2668, 2711, 2779, 28o8,

— Geschichte und Beschreibung (Fortsetzung) 2850, 2860, 2862,
    2874, 2880, 2906, 2921, 3008, 3010, 3021, 3037, 3042, 3067,
    3114, 3188, 3192, 3241, 3294, 3324, 3329, 3334, 3344, 3360,
    3363, 3371, 3378, 3380, 3381, 3389, 3390, 3395, 3506, 3508,
    3512, 3566, 3603, 3604, 3677
— Planung und Entwicklung 139, 189, 288, 292, 367, 393, 469,
    472, 473, 681, 727, 762, 763, 834, 860, 871, 1127, 1221,
    1263, 1311, 1342, 1352, 1356, 1387, 1391, 1395, 1396, 1459,
    1467, 1478, 1483, 1509, 1536, 1566, 1722, 1731, 1748, 1751,
    1785, 1823, 1874, 1919, 1979-1981, 1987, 2054, 2068, 2069,
    2088, 2108, 2189, 2285, 2341, 2362, 2379, 2409, 2413, 2425,
    2442, 2513, 2534, 2541, 2545, 2602, 2609, 2654, 2740, 2776,
    2789, 2792, 2793, 2838, 2842, 2844, 2854, 2855, 2874, 2911,
    2921, 2923, 2954, 2968, 3042, 3067, 3081, 3118, 3176, 3185,
    3192, 3248, 3260, 3291, 3324, 3363, 3372, 3628, 3677
— Hausarchitektur und Wohnungseinrichtung 53, 80, 90, 139, 149,
    342, 469, 473, 617, 727, 762, 763, 773, 777, 822, 823, 825,
    832, 915, 1047, 1387, 1395, 1485, 1493, 1516, 1518, 1536,
    1570-1575, 1893, 2189, 2285, 2341, 2378, 2379, 2390, 2479,
    2523, 2524, 2531, 2532, 2541, 2587, 2590, 2596, 2622, 2723,
    2843, 3008, 3010, 3042, 3083, 3197, 3321, 3324, 3604, 3628
— Wasserbau und -wirtschaft (s. auch Meerwasserentsalzungsanla-
    gen) 90, 188, 204, 365, 367, 390, 393, 556, 617, 762, 837,
    914, 923, 1031, 1057, 1426, 1458, 1504, 1509, 1516, 1518,
    1646-1649, 1713, 1999, 2010, 2108, 2141, 2227, 2379, 2524,
    2739, 2985, 2997, 3010, 3021, 3244, 3260, 3324, 3519, 3668
Städter
— Materielle und immaterielle Kultur 53, 342, 617, 624, 773,
    1071, 1965, 1998, 2340, 2394, 2508, 2531, 2638, 3008, 3010,
    3011, 3037, 3321, 3322, 3324, 3509, 3565
— Sozialstruktur und Sozialorganisation 53, 74, 89, 472, 473,
    617, 624, 964, 1056, 1235, 1342, 1516, 1518, 1951, 1987,
    1998, 2108, 2164, 2340, 2341, 2394, 2428, 2468, 2602, 2635,
    2636, 2744, 3008, 3010, 3114, 3176, 3324, 3395, 3509
Steuersystem und -erhebung 249, 346, 359, 520, 522, 589, 633, 808,
    809, 1091, 1113, 1235, 1257, 1282, 1480, 1527, 1529, 1580,
    1689, 1815, 1891, 1950, 1970, 2159, 2213, 2247, 2501, 2502,
    2635, 2958, 3131, 3182, 3495, 3553, 3605, 3671
Straßenbau und -verkehr 58, 59, 61, 62, 392, 466, 654, 748, 762,
    946, 1031, 1034, 1049, 1050, 1258, 1395, 1472, 1665-1668,

Straßenbau und -verkehr (Fortsetzung)   1814, 1874, 1978, 1979,
    2112, 2113, 2124, 2222, 2242, 228o, 2328, 2341, 2515, 2525,
    2545, 2584, 26o1, 26o2, 27o6, 2895, 299o, 3142, 3147, 3242,
    3374, 3378, 3391, 3414, 3611, 3626, 3664, 3669, 3677
Umweltverschmutzung und -schutz   778, 3534
Verarbeitende Industrie und Industriepolitik   138, 147, 165, 2o6,
    2o8, 222, 225, 227, 231, 252, 268, 279, 291, 3o2, 32o, 323,
    398, 414, 463, 477, 48o, 548, 567, 571, 572, 578, 583, 59o,
    591, 595, 596, 598-6oo, 6o2-6o4, 6o6-6o9, 611-613, 714, 777,
    8o5, 8o9, 864, 884, 897, 9o6, 977, 1o18, 1o31, 1127, 1157,
    116o, 1162, 1163, 1167, 1173, 1186, 12o4, 124o, 1446, 1473,
    1474, 1561, 1586, 1597, 1598, 16oo-16o3, 16o6, 16o7, 1612-
    1617, 162o-1623, 1625-1637, 1639-1643, 1691, 1695, 17o2-
    17o4, 1748, 1751, 1755-1757, 1784, 1795, 1829, 185o, 1851,
    1858, 1874, 195o, 1969, 1975, 1977, 199o, 1991, 2o12, 2o13,
    2o54, 2156, 216o, 2167, 2227, 2228, 2258, 2299, 23oo, 2316,
    2322, 2326, 2335, 2473, 2482, 261o, 277o, 2771, 2796, 2811,
    2816, 2817, 2819-2821, 283o, 2855, 2895, 2948, 2958, 2968,
    31o4, 314o, 319o, 3195, 3196, 3212, 3229, 3231, 3233, 3235,
    3236, 3261, 3271, 3382, 3423, 3494, 3519, 3522, 3561, 3588,
    36o2, 3643, 3651
Verkehrswesen: Überblick (s. auch Eisenbahnbau und -verkehr; Flug-
    hafenbau, Fluglinien und Luftverkehr; Hafenbeschreibungen,
    -bau und -verkehr; Schiffahrt sowie Straßenbau und -ver-
    kehr)   59, 2o6, 2o7, 367, 463, 661, 716, 717, 792, 1o74,
    1295, 1296, 1298, 13o9, 1576, 1691, 1695, 1751, 21o8, 2227,
    2254, 2392, 2712, 2744, 2756, 2855, 3o72, 3382
Verwaltungssystem und -personal   134, 149, 151, 249, 318, 439, 614,
    63o, 673, 674, 979, 1o34, 1o5o, 1o75, 12o7, 1439, 1551, 1644,
    1691, 1722, 1759, 1972, 2o47, 2o5o, 211o, 2151, 22o6, 2216,
    2257, 2292, 23o5, 233o, 2331, 2392, 2415-2417, 242o, 2428,
    2439, 2476, 2496, 2513, 2516, 2533, 2541, 26o2, 2655, 2671,
    2684, 2686, 2697, 2698, 27o5, 2727, 2728, 2734, 2756, 2774,
    2838, 291o, 2931-2933, 3o18, 31o3, 3124, 3125, 3145, 3324,
    3492, 3517, 3556, 3597, 3627, 3665, 3666
Wahhabiya: Dogmatik und Praxis   53, 87, 371, 411, 413, 485, 533,
    542, 641, 663, 729, 744, 783, 829, 879, 9o2, 9o5, 915, 952,
    955, 995, 1oo5, 1113, 1114, 1123, 1169, 12oo, 1249, 125o, 1343,
    1432, 1499, 1532, 1534, 1539, 155o, 18o5, 1856, 1936, 1962,

Wahhabiya: Dogmatik und Praxis (Fortsetzung)   1971, 2o19, 2o71,
    2212, 2214, 2236, 2248, 2272, 2292, 23o9, 2337, 2339, 234o,
    239o, 2443, 2451, 2475, 248o, 2485, 2486, 2546, 2551, 2564,
    2571, 2572, 2613, 262o, 2631, 2644, 2658, 2717, 2719, 2841,
    2927, 3oo2, 3o23, 3299, 3346, 3354, 336o, 3426, 3455, 35o1,
    35o3, 35o4, 35o6, 3591, 3612, 3632
Wirtschaftsentwicklung und -struktur   1o9, 11o, 116, 122, 13o, 174,
    175, 181, 198, 2o5, 2o6, 217, 219-221, 229, 232-237, 239,
    281, 282, 291, 299, 313, 318, 328, 344, 353, 358, 363, 368,
    4oo, 414, 425, 45o, 452, 463, 464, 483, 485, 552, 563, 564,
    569, 579, 58o, 586, 614, 619, 632, 659, 667, 672, 675, 676,
    697, 71o, 711, 725, 734, 764, 792, 798, 8o1, 8o5, 8o8, 8o9,
    815, 842, 864, 878, 884, 885, 897, 9o6, 9o7, 921, 952, 963,
    966, 973, 975, 989-991, 1oo5, 1o18, 1o19, 1o27, 1o38, 11o8,
    111o, 1111, 1121, 116o, 1171, 1173, 1182, 1257, 1278, 13o9,
    1315, 1334, 1362, 14o8, 14o9, 1427, 1461, 1482, 1495, 1589,
    159o, 16o1, 16o3, 1642, 1688, 1695, 1697, 1711, 1727, 1732,
    1751, 1752, 1756, 1772, 1774-1776, 1791, 1794, 18o4, 1822,
    1832, 1836, 1859, 1887, 1899, 19o1, 193o, 1931, 1937, 1938,
    1942, 1949, 195o, 1958, 1969, 1988, 199o, 1991, 1995, 1997,
    2o31, 2o37, 2o54, 2o56, 21o4, 211o, 2116, 2118, 2155, 2156,
    2223, 2227, 2242, 2251, 2256, 2259, 2276, 2285, 2292, 23oo,
    2318, 2323, 2334, 2335, 24o9, 2416, 2454, 2469, 247o, 2585,
    2622, 2675, 2737, 2746, 2751-2753, 2756, 2768, 2769, 2774,
    2777, 2778, 28oo, 281o, 2813, 2836, 2837, 2855, 2868, 2873,
    2876, 292o, 2933, 2937, 2946, 2951, 2959-2963, 2996, 3o22,
    3o42, 3o49, 3o95, 3147, 3154, 3192, 3231, 3256-3259, 3262,
    3266, 3269-3271, 3285, 3287, 3352, 3368, 34o1, 3421, 3446,
    3457, 3471, 3494, 3495, 3518, 3522, 3526, 3533, 3537, 3556,
    36o5, 36o7, 361o, 3621, 3643, 3646, 3661
Wirtschaftspolitik und Entwicklungsplanung   54, 1o9, 138, 152,
    193, 2o7, 21o, 212, 22o, 221, 237, 239, 273, 318, 328, 4oo,
    414, 422, 442, 463, 484, 485, 5o2, 52o, 548, 576, 585, 614,
    633, 638, 646, 672, 675, 697, 7o9-711, 737-739, 745, 764,
    78o, 783, 792, 815, 878, 884, 885, 911, 93o, 952, 953, 963,
    975, 1oo8, 1o41, 1127, 1162, 1171, 1176, 1214, 1257, 1315,
    1318, 1391, 14o9, 1422, 1438, 1442, 1462, 147o, 1473, 1482,
    15o5, 1551, 1561, 1586, 1589, 159o, 16o1, 16o3, 1627, 166o,
    1661, 1682, 17o6, 1715, 1722, 1728, 1738-174o, 1751, 1774,
    1775, 18oo, 1822, 1836, 1838, 1858, 186o, 1874, 1889, 1899,

Wirtschaftspolitik und Entwicklungsplanung (Fortsetzung) 19o1,
    1937, 1949, 195o, 1972, 1976, 199o, 1991, 2ooo, 2o12, 2o31,
    2o43, 2o5o, 2156, 22o6, 2216, 2227, 2235, 2256, 2258, 2318,
    2322, 2323, 2335, 2381, 247o, 2482, 2488, 2495, 2513, 2533,
    2675, 2684, 2737, 2745, 2754, 2755, 2759, 276o, 2787, 2788,
    2791, 28oo, 2822, 2855, 2873, 2933, 2936, 2938, 2955, 3o49,
    3o79, 3o92, 31o3, 313o, 3196, 3255, 3264-3266, 3297, 3311,
    3318, 3352, 3366, 3388, 34o1, 34o4, 3446, 3494, 3495, 3526,
    3537, 3597, 36o2, 36o7, 3643

I SUBJECTS

Accomodations (s. also business guides and practices as well as
    guidebooks and tourism)   994, 124o, 1695, 1893, 19o9, 198o,
    1981, 2532, 2744, 2779, 3197, 3677
Agriculture including animal production
— Synopsis   11o, 228, 291, 575, 621, 622, 691, 716, 717, 792,
    8o2, 862, 884, 1o74, 1278, 1335, 1429, 143o, 15o5, 159o,
    1655, 1657, 1691, 1751, 1838, 1865, 2o12, 2o38, 2o54, 2o73,
    2o74, 214o, 2156, 2456, 25oo, 2695, 2759, 2837, 2855, 2935,
    2949, 2971, 3o49, 3191, 323o, 3243, 3487, 3494, 3573
— Agrarian structure and land tenure   84, 289, 352, 8o1, 8o2, 8o4,
    1189, 1235, 1456, 1457, 15o4, 165o, 1897, 2428, 2873, 2874,
    2946, 3o33, 3o62, 3116, 3117, 323o, 3324, 35o2
— Plant production   66, 79, 84, 1o7, 117, 2o4, 253, 258, 289, 335,
    352, 361, 419, 476, 493, 536, 537, 617, 678, 679, 74o, 75o,
    78o, 8oo, 8o1, 8o4-8o6, 855, 856, 872, 914, 947-95o, 965,
    983, 996, 997, 117o, 1235, 1253, 1274, 1319, 1378, 1384, 1385,
    1388, 1411, 1456, 1457, 15o4, 1552, 1554, 1559, 16o2, 16o4,
    1611, 1638, 1646-1652, 1655, 1659, 1695, 1766, 183o-1832,
    1864, 1895-1897, 1951, 1954, 1965, 2o25, 2o61, 2o72, 2o96,
    2108, 211o, 2117, 2153, 2277-2279, 2294, 234o, 2354, 2357,
    2428, 243o, 2498, 2524, 2545, 26oo, 26o1, 26o2, 274o, 2848,
    2873, 2874, 2921, 2946, 2967, 2968, 2981, 3oo7, 3o33, 3o61,
    3o64, 3o87, 311o, 3111, 3132, 3192, 3221, 323o, 3242, 3249,
    3251, 33oo, 33o1, 3314, 3323, 3324, 3328, 3356, 3357, 3496,
    35o2, 3536, 3564, 357o, 3589, 36o5, 3613, 3619, 3625, 3637,
    3649, 3655, 3656, 3669
— Animal production (s. also nomads: economic bases)   1o4, 1o7,
    117, 199, 2o4, 258, 289, 465, 538, 539, 75o, 78o, 8o1, 8o4,

— Animal production (continuation) 821, 85o, 867, 914, 947, 948, 1159, 1274, 1384, 1411, 1428, 1456, 1457, 15o7, 1559, 1646-1652, 1655, 1659, 1766, 183o, 1832, 1841, 1872, 1951, 1954, 2o61, 2153, 2428, 2498, 26oo, 2665, 2725, 2726, 28o5, 2946, 3o33, 3o64, 3132, 3151, 3181, 3192, 324o, 3324, 357o, 36o5, 3637, 366o

— Development projects, assistance measures and agricultural policy 7o, 77, 91, 1oo, 1o4, 1o7, 112, 117-119, 15o, 2o4, 2o6, 218, 228, 229, 258, 265, 289, 352, 367, 419, 463, 528, 573, 622, 644, 7o4, 71o, 711, 721, 725, 74o, 754, 78o, 8o1-8o3, 8o5, 83o, 831, 864, 869, 872, 873, 884, 889, 918, 919, 933, 942, 967, 996, 997, 1o67, 1o68, 11o3, 11o4, 1129, 117o, 1188, 1189, 1253, 1259, 128o, 13o7, 1378, 1384, 1395, 1411, 1457, 1524, 1552, 1554, 1559, 1586, 16o2, 16o4, 1611, 1638, 1646-1649, 1651-1662, 1695, 1712, 1713, 1716, 1749, 183o, 1864, 1872, 1878, 1889, 1922, 193o, 1954, 2o25, 2o38, 2o61, 2o72, 21o8, 211o, 2123, 214o, 2145-2147, 2227, 2238, 2354, 2412, 2429, 25o9-2511, 2545, 26o3, 2678, 2681, 2695, 274o, 2759, 28o5, 2833, 2865, 2935, 2946, 295o, 3o33, 3o63-3o65, 3o87, 31o9, 3115, 314o, 3192, 32o5, 3249-3254, 3259, 3324, 3336, 3347, 3356-3358, 3367, 3383, 3391, 341o, 3417, 3496, 3521, 3536, 3575, 36o5, 3655, 3665

— Water resources development and water usage 66, 92, 1o1, 1o7, 117, 2o6, 258, 364, 365, 367, 39o, 395, 419, 463, 493, 5o4, 535, 621, 622, 632, 679, 721, 75o, 8o1, 8o2, 8o4, 8o5, 852, 862, 864, 872, 873, 889, 936, 95o, 997, 1o34, 1188, 1253, 1256, 1259, 1364, 1378, 1384, 1411, 1443, 1456, 1457, 15o4, 1524, 1552, 1554, 1646-1658, 1661, 1712, 1713, 1716, 1779, 183o, 1872, 1874, 1895-1897, 1951, 1965, 2oo3, 2o25, 2o38, 2o4o, 2o61, 21o8, 211o, 213o, 214o, 2141, 2145-2147, 2227, 2238, 2277, 2279, 2386, 2412, 2428, 2429, 25o9-2511, 2524, 2558, 26o1-26o3, 26o6, 2678, 2695, 274o, 2833, 2865, 2873, 2874, 2921, 2925, 2946, 2967, 2981, 3ooo, 3oo7, 3o16, 3o33, 3o62, 3o64, 3115, 314o, 3183, 3192, 3221, 3242-3244, 3249-3254, 33o3, 3314, 3323, 3324, 3335, 3336, 3347, 3356, 3357, 3366, 3367, 3383, 3391, 341o, 3496, 3519, 3575, 3594, 36o5, 3613, 3679

— Agricultural imports and exports 9o, 691, 78o, 795, 8o4, 1445, 1611, 1691, 1695, 1841, 1857, 193o, 1956, 2153, 2231, 2347-2353, 2895, 2911, 323o

— Agroindustry   253, 258, 361, 78o, 897, 911, 1388, 16o4, 1611,
     1618, 1619, 1638, 1639, 1841, 1925, 2o61, 2o72, 243o, 2692,
     355o, 3569, 3589, 36o5
Airport construction, airlines and air traffic   6o, 61, 3o5, 54o,
     7o2, 7o3, 755, 758, 777, 778, 817, 877, 1o31, 1o34, 1119,
     175o, 1873, 1874, 2o18, 2136, 2157, 2341, 25o3, 2779, 279o,
     3o68, 3112, 3211, 3411
Anthropological characteristics   759, 97o, 972, 11o1, 1894, 2o16,
     236o, 27o3, 333o, 3538
Arab Revolt: antecedents and course of events   196, 336, 426, 455,
     512, 515, 516, 628, 648, 784-786, 836, 91o, 931, 1o13, 1o28,
     1o29, 1o99, 1132, 1187, 1247, 1248, 1338, 1381, 144o, 1449,
     1471, 1477, 1519-1521, 1767, 1778, 1821, 1868-1871, 1921,
     1933, 1967, 2166, 2175, 2177, 2187, 2188, 2193, 22o9, 2465,
     2641, 2677, 2693, 286o, 2897, 3o11, 3o7o, 3o78, 32o2, 3225,
     3322, 337o, 3376, 3378, 3472
Armed forces, security policy and armament   75, 124, 142, 162, 274,
     282, 283, 42o, 453, 5o7, 5o8, 51o, 511, 684, 688, 7o5, 712,
     756, 781, 783, 87o, 888, 952, 961, 976, 991, 1oo5, 1o45,
     1o49, 111o, 1111, 1196, 1197, 1199, 1316, 1317, 1356, 138o,
     1393, 1394, 141o, 1462, 1534, 1537, 1544, 1545, 1559, 1759,
     1762, 1791, 1834, 1942, 2o14, 2o43, 2o44, 21o8, 211o, 2229,
     2234, 2245, 2258, 2288, 2292, 2316, 2356, 2366, 2446, 2452,
     2469, 2491, 2492, 25o7, 2516, 2518, 2532, 2634, 2635, 2713,
     2736, 2784, 279o, 2827, 2831, 2894, 2914, 2917-2919, 2938,
     2962, 2992, 2993, 3o27, 3o34, 31o6-31o8, 3237, 3296, 3352,
     3362, 3415, 3433, 3442, 3473, 3474, 3492, 3531, 3533, 3562,
     3585, 3598, 3611, 3615, 3621
Arts   146, 413, 44o, 1o72, 113o, 1389, 2oo6, 2746, 28o8, 3o85,
     3523, 3549, 3571, 364o, 3645
Banking and insurance   1o8, 2o6, 223, 347, 387, 515, 52o, 521, 57o,
     615, 633, 73o, 8o8, 884, 1o43, 1181, 1283, 164o, 1691, 1695,
     1749-1752, 1754-1756, 195o, 1985, 2156, 2251, 2471, 2495,
     2627, 2756, 2935, 3422, 3494
Biographies
— Muhamad ibn Abd al-Wahhab (17o3/o4-92)   86, 663, 729, 829, 1249,
     1343, 1856, 2248, 2418, 254o, 2551, 2572, 3299, 35o3, 3612
— King Abd al-Aziz (Ibn Sa'ud) (188o-1953)   149, 278, 415, 427,
     438, 533, 651, 843, 1o47, 1113, 1362, 1373, 1415, 1443, 1553,
     211o, 2111, 2131-2133, 2345, 239o, 2413, 2416, 2422, 2589,

— King Abd al-Aziz (Ibn Sa'ud)(1880-1953) (continuation)   2590, 2597, 2619, 2621, 2632, 2868, 3192, 3304, 3312, 3355, 3412, 3428
— King Sa'ud (1902-69)   416, 2740, 2775
— King Faisal (1905/06-75)   418, 482, 924, 1010, 1054, 1390, 1541, 1577, 1862, 2219, 2226, 2283, 2775, 2943, 2944, 3429, 3523, 3659
— King Khalid (1912-82)   441, 1390, 1542, 1892, 2629, 2746, 3670
— King Fahd (1922/23-)   441, 898, 1390, 1542, 2629, 2746, 3674
— Grand Sharif and King Husain ibn Ali (1853/54-1931)   336, 455, 1128, 1332, 2175, 2593, 2636
— Other statesmen and elite members   142, 149, 441, 455, 693, 866, 1132, 1235, 1236, 1390, 1559, 1892, 2175, 2512, 2629, 2746, 2945, 3070
— Well-known Western travellers in Arabia   145, 149, 384, 405, 424, 444, 474, 513, 531, 532, 623, 642, 643, 657, 685, 882, 901, 981, 1000, 1159, 1212, 1254, 1326, 1331, 1341, 1360, 1368, 1534, 1778, 1790, 1811, 1911, 1928, 2099, 2106, 2109, 2149, 2163, 2177, 2187, 2243, 2271, 2281, 2369, 2385, 2411, 2420, 2427, 2431, 2526, 2531, 2674, 2676, 2700, 2829, 2893, 2980, 3031, 3040, 3100, 3113, 3162, 3307, 3360, 3365, 3419, 3420, 3435-3437, 3445, 3451, 3488

Boundary disputes and settlements   98, 106, 111, 125, 194, 274, 287, 312, 468, 523, 534, 693, 822, 1080, 1084, 1096, 1097, 1124, 1133, 1140, 1142, 1203, 1282, 1347, 1361, 1381, 1401, 1449, 1527-1529, 1535, 1545, 1580, 1741, 1805, 1826, 1853, 1916, 1929, 1960, 1961, 1970, 2024, 2084-2087, 2289, 2391, 2395, 2476, 2559, 2571, 2590, 2611, 2667, 2671, 2689, 2704, 2772, 2802, 2839, 2860, 2902, 3001, 3053, 3134, 3137, 3189, 3199-3201, 3293, 3379, 3435, 3475-3477, 3480, 3481, 3497-3499

Building trade (s. also agriculture including animal production: water resources development and water usage; airport construction, airlines and air traffic; development of the material infrastructure: synopsis; manufacturing industry and industrial policy; oil and gas industry: production and transport as well as refining and shipment; port descriptions, construction and traffic; public health: sanitary and medical facilities and personnel; railroad construction and traffic; road construction and traffic; seawater desalination plants; towns: planning and development, architecture and furnishings as well as water resources development and water usage; transportation: synopsis; villages: architecture and furnishings as well as water resources development and water usage) 11o, 279, 488, 557, 582, 588, 598, 669, 725, 73o, 897, 986, 1o3o, 1o31, 1149, 1171, 124o, 1473, 1493, 16o7-161o, 1616, 1617, 1624, 1628-1632, 1751, 1757, 1939, 1973, 2145-2147, 2194, 2229, 2432-2442, 2477, 2489, 25o9-2511, 2622, 2722, 2723, 2757, 277o, 3229, 3252, 3254, 329o, 3359, 341o, 3559

Business guides and practices 81, 82, 94, 182, 198, 388, 526, 527, 551, 557, 56o, 562, 565, 566, 568, 57o, 574, 578, 581, 582, 584, 587-591, 593-612, 614, 66o, 714, 718, 719, 73o, 737, 78o, 789, 8o8, 8o9, 929, 966, 989, 99o, 1o91, 1176, 12o1, 1318, 14o7, 148o, 1487, 1488, 15o6, 1586, 16o3, 1613, 1663, 18o9, 1852, 2o37, 2o52, 21o4, 2119, 2127, 2128, 2184, 2217, 2218, 2247, 2255, 228o, 231o, 2317, 2318, 2371, 2453, 2474, 2481, 2487, 2489, 2495, 2675, 2737, 2746, 2755, 2764, 2778, 29o3, 2916, 2928, 2929, 2955, 2956, 3o36, 3o44, 3o69, 3177, 3271, 3272, 3284, 3288, 33o6, 3315, 3397, 34o1, 3457, 3471, 3495, 3516, 3542, 3545, 3548, 36o5, 3643, 3661

Cooperatives 129, 1129, 1691, 1695, 2685

Crafts and craftsmen 117, 617, 759, 85o, 852, 897, 915, 1157, 1235, 1285, 1291, 15o8, 15o9, 16o5, 2o17, 2o96, 2213, 2341, 2428, 2582, 31.49, 3192, 3323, 3324

Cultural policy 9o5, 1681, 2oo6

Development aid: bi- and multilateral

— Development aid received 35o, 463, 535, 546, 614, 74o, 777, 81o, 877, 934, 978, 1o51, 1157, 117o, 1256, 1262, 1345, 1377, 1538, 1587, 176o, 1789, 1878, 1997, 21o8, 211o, 212o, 2129, 213o, 2137, 2229, 2461, 2516, 265o, 2651, 274o, 2911, 2972, 3o71, 3o87, 3258, 3281, 3362, 3417, 3458, 3459, 3491, 3552, 3585

— Development aid granted 14o, 387, 463, 514, 668, 783, 813, 841, 842, 917, 1oo8, 1o38, 1422, 1754, 1794, 1867, 1932, 1964, 1989-1991, 2o12, 2158, 2258, 2314, 2438, 2472, 269o, 2746, 2984, 31o6, 34o5, 3413, 3448, 3556

Development of the material infrastructure: synopsis 279, 477, 48o, 563, 564, 766, 815, 843, 986, 1o38, 1162, 1278, 1446, 15o4, 1554, 1559, 16o3, 1748, 1957, 1997, 21o4, 21o5, 2114, 2118, 213o, 2228, 23oo, 24o2, 24o9, 2413, 2416, 2418, 242o,

Development of the material infrastructure: synopsis (continuation)
    2489, 2516, 2545, 2563, 2622, 2632, 2739, 2752, 2753, 2774,
    2777, 2837, 2854, 2863, 2911, 2938, 2968, 3o39, 3o49, 3o59,
    3o77, 3o92, 3197, 3231, 3257, 3259, 3287, 3421, 3494, 3495,
    35o8, 3518, 36o7, 3643, 3665, 3677, 368o
Domestic policy   418, 443, 5o7, 5o8, 614, 655, 673, 782, 818, 821,
    822, 9o6, 9o9, 975, 1oo5, 1o17, 1o96, 1113, 1169, 1282,
    1358, 1359, 1415, 1491, 1526, 17o8, 1773, 1776, 18o5, 1833,
    1836, 1862, 1866, 19o3, 1958, 1994, 2o57, 2118, 2139, 2262,
    2345, 2492, 2548, 2564, 2571, 2716, 2719, 2721, 2758, 2845,
    2917-292o, 2995, 3o25, 3154, 32o1, 3352, 3355
Economic development and structure   1o9, 11o, 116, 122, 13o, 174,
    175, 181, 198, 2o5, 2o6, 217, 219-221, 229, 232-237, 239,
    281, 282, 291, 299, 313, 318, 328, 344, 353, 358, 363, 368,
    4oo, 414, 425, 45o, 452, 463, 464, 483, 485, 552, 563, 564,
    569, 579, 58o, 586, 614, 619, 632, 659, 667, 672, 675, 676,
    697, 71o, 711, 725, 734, 764, 792, 798, 8o1, 8o5, 8o8, 8o9,
    815, 842, 864, 878, 884, 885, 897, 9o6, 9o7, 921, 952, 963,
    966, 973, 975, 989-991, 1oo5, 1o18, 1o19, 1o27, 1o38, 11o8,
    111o, 1111, 1121, 116o, 1171, 1173, 1182, 1257, 1278, 13o9,
    1315, 1334, 1362, 14o8, 14o9, 1427, 1461, 1482, 1495, 1589,
    159o, 16o1, 16o3, 1642, 1688, 1695, 1697, 1711, 1727, 1732,
    1751, 1752, 1756, 1772, 1774-1776, 1791, 1794, 18o4, 1822,
    1832, 1836, 1859, 1887, 1899, 19o1, 193o, 1931, 1937, 1938,
    1942, 1949, 195o, 1958, 1969, 1988, 199o, 1991, 1995, 1997,
    2o31, 2o37, 2o54, 2o56, 21o4, 211o, 2116, 2118, 2155, 2156,
    2223, 2227, 2242, 2251, 2256, 2259, 2276, 2285, 2292, 23oo,
    2318, 2323, 2334, 2335, 24o9, 2416, 2454, 2469, 247o, 2585,
    2622, 2675, 2737, 2746, 2751-2753, 2756, 2768, 2769, 2774,
    2777, 2778, 28oo, 281o, 2813, 2836, 2837, 2855, 2868, 2873,
    2876, 292o, 2933, 2937, 2946, 2951, 2959-2963, 2996, 3o22,
    3o42, 3o49, 3o95, 3147, 3154, 3192, 3231, 3256-3259, 3262,
    3266, 3269-3271, 3285, 3287, 3352, 3368, 34o1, 3421, 3446,
    3457, 3471, 3494, 3495, 3518, 3522, 3526, 3533, 3537, 3556,
    36o5, 36o7, 361o, 3621, 3643, 3646, 3661
Economic policy and development planning   54, 1o9, 138, 152, 193,
    2o7, 21o, 212, 22o, 221, 237, 239, 273, 318, 328, 4oo, 414,
    422, 442, 463, 484, 485, 5o2, 52o, 548, 576, 585, 614, 633,
    638, 646, 672, 675, 697, 7o9-711, 737-739, 745, 764, 78o,
    783, 792, 815, 878, 884, 885, 911, 93o, 952, 953, 963, 975,

Economic policy and development planning (continuation) 1oo8, 1o41,
    1127, 1162, 1171, 1176, 1214, 1257, 1315, 1318, 1391, 14o9,
    1422, 1438, 1442, 1462, 147o, 1473, 1482, 15o5, 1551, 1561,
    1586, 1589, 159o, 16o1, 16o3, 1627, 166o, 1661, 1682, 17o6,
    1715, 1722, 1728, 1738-174o, 1751, 1774, 1775, 18oo, 1822,
    1836, 1838, 1858, 186o, 1874, 1889, 1899, 19o1, 1937, 1949,
    195o, 1972, 1976, 199o, 1991, 2ooo, 2o12, 2o31, 2o43, 2o5o,
    2156, 22o6, 2216, 2227, 2235, 2256, 2258, 2318, 2322, 2323,
    2335, 2381, 247o, 2482, 2488, 2495, 2513, 2533, 2675, 2684,
    2737, 2745, 2754, 2755, 2759, 276o, 2787, 2788, 2791, 28oo,
    2822, 2855, 2873, 2933, 2936, 2938, 2955, 3o49, 3o79, 3o92,
    31o3, 313o, 3196, 3255, 3264-3266, 3297, 3311, 3318, 3352,
    3366, 3388, 34o1, 34o4, 3446, 3494, 3495, 3526, 3537, 3597,
    36o2, 36o7, 3643
Education and activities of children 139, 821, 1o23, 1516, 1518,
    2879
Educational system
— Synopsis 56, 88, 143, 144, 171, 333, 4o2, 413, 463, 478, 524,
    673, 8o1, 864, 879, 883, 892, 9o6, 937, 1o51, 1o74, 1o77,
    1177-118o, 1182, 14o4, 1534, 1543, 1559, 159o, 167o-1674,
    1679, 1681, 1683-1686, 1691, 1695, 171o, 1751, 1957, 2oo6,
    2oo9, 2o54, 21oo, 2216, 2223, 2242, 2329, 2333, 2455, 2493,
    2494, 25o7, 2713, 2733, 2741, 2794, 2836, 2863, 2894, 292o,
    2933, 3o82, 3o92, 31o2, 3139, 3223, 3274-3277, 328o, 3492,
    3523
— Development history 54, 57, 87, 115, 123, 2o1, 37o, 65o, 839,
    84o, 1o98, 1113, 1214, 1448, 22o8, 2216, 2223, 273o, 2773,
    2927, 2969, 3o11-3o13, 3355, 3483
— Primary education 57, 87, 95, 117, 18o, 2o1, 37o, 65o, 928,
    1o79, 128o, 1448, 1883, 2o15, 2252, 2332, 2647, 2866, 3o13,
    3o2o, 3166, 3167, 3275, 3324, 3483, 3539, 3577
— Secondary education 5o, 57, 76, 87, 115, 131, 153, 2o1, 3oo,
    3o1, 322, 325, 333, 385, 65o, 767, 788, 793, 861, 926, 938,
    1o77, 1o79, 1168, 13o5, 1423, 1424, 1436, 1448, 1492, 1982,
    1992, 2o15, 2o48, 2o49, 2192, 25o5, 2647, 27o7, 2834, 2866,
    2942, 2953, 2964, 3o13, 3o2o, 3o96, 3166, 3167, 3276, 3373,
    34o4, 3479, 3483, 352o, 3527-3529, 3535, 3568, 36o8
— Higher education 54, 57, 67, 1o5, 114, 151, 161, 167, 191, 2o1,
    296, 3o3, 3o4, 326, 327, 33o, 348, 369, 64o, 844, 9o5, 92o,
    952, 998, 1o39, 1185, 119o, 12o6, 1214, 1228, 127o, 1276,

350

— Higher education (continuation) 1304, 1305, 1308, 1319, 1356, 1366, 1376, 1413, 1419, 1447, 1448, 1483, 1501, 1515, 1539, 1557, 1558, 1563, 1599, 1645, 1700, 1701, 1719, 1747, 1992, 2005, 2023, 2029, 2090-2093, 2181, 2190, 2203, 2223, 2249, 2324, 2455, 2493, 2503, 2504, 2583, 2655, 2702, 2745, 2773, 2834, 2856, 2864, 2909, 2927, 2952, 3028, 3030, 3090, 3103, 3163, 3166, 3167, 3173, 3175, 3277, 3283, 3409, 3453, 3479, 3483, 3485, 3529, 3532, 3567, 3574, 3582, 3633, 3652, 3657
— Special education 57, 201, 1205, 1524, 1676, 2224, 2696, 3166, 3167, 3278, 3279
— Vocational education and training (without public adult education) 57, 121, 141, 184, 201, 250, 265, 331, 421, 422, 497, 619, 645, 694, 758, 793, 802, 810, 913, 982, 984, 985, 1051, 1070, 1105, 1211, 1224, 1225, 1399, 1444, 1448, 1538, 1644, 1680, 1728, 1748, 1864, 1881, 1883, 1972, 2038, 2050, 2182, 2185, 2216, 2227, 2229, 2290, 2295, 2432, 2697, 2698, 2861, 2886, 2935, 2965, 2966, 3013, 3086, 3112, 3124, 3166, 3167, 3192, 3204, 3206, 3238, 3404, 3485, 3489, 3521, 3574, 3608, 3644, 3657, 3673
— Girls' education 57, 201, 370, 454, 881, 1195, 1448, 1563, 1591-1593, 1709, 1719, 2078, 2263, 2355, 3166, 3167, 3483, 3489, 3525, 3526, 3652
— Public adult education 57, 164, 201, 332, 1165, 1209, 1448, 1669, 1677, 1678, 1687, 1883, 2078, 3166, 3167, 3447
— Teaching and administrative system and personnel 76, 78, 87, 95, 105, 153, 201, 300, 301, 322, 325-327, 370, 650, 767, 788, 793, 861, 880, 928, 938, 1076, 1079, 1107, 1165, 1168, 1190, 1205, 1206, 1209, 1227, 1270, 1276, 1305, 1419, 1423, 1424, 1436, 1448, 1492, 1501, 1549, 1563, 1982, 1992, 2015, 2049, 2103, 2181, 2185, 2199, 2249, 2252, 2324, 2329, 2332, 2490, 2679, 2701, 2702, 2707, 2730, 2732, 2909, 2942, 2953, 2964, 3028, 3090, 3096, 3122, 3166, 3167, 3175, 3193, 3280, 3350, 3373, 3447, 3453, 3479, 3483, 3485, 3520, 3527-3529, 3532, 3535, 3568, 3577, 3658
— Educational policy 54, 57, 168, 201, 332, 478, 650, 905, 1448, 1483, 1675, 1682, 2490, 2730, 2927, 3013, 3166, 3280, 3283, 3483, 3658, 3673

Elites
— Royal family 75, 124, 142, 194, 256, 280, 441, 549, 629, 712, 773, 781, 822, 870, 902, 961, 976, 1111, 1317, 1352, 1359,

— Royal family (continuation)   1362, 1526, 1534, 1542, 1759,
    1791, 1834, 1892, 1942, 2o14, 2143, 2282, 2283, 2292, 2315,
    2316, 2492, 2516, 2532, 2533, 2548, 256o, 2629, 2715, 2746,
    2747, 2917-2919, 2931, 2933, 2938, 3154, 3216, 3425, 3546,
    3674
— Remaining political elite   256, 441, 2265, 2316, 2492, 2533,
    2655, 2746, 3393
— Religious authorities   256, 271, 452, 533, 549, 744, 877, 9o2,
    9o5, 1113, 2265, 2292, 2315, 2492, 2533, 2931, 2933, 2938,
    3o11, 3o12
— Economic elite   139, 158, 159, 268, 441, 452, 461, 549, 632,
    659, 725, 945, 973, 1559, 2142, 2246, 2265, 2533, 2736, 2746,
    2938
— Other elites   452, 2466, 2746, 2938
Energy economy and policy   215, 463, 592, 884, 897, 1o31, 1o38,
    1691, 1695, 1751, 1756, 1827, 1874, 1917, 2o89, 2374, 277o,
    2835, 2895, 3146, 3343, 3654
Environmental pollution and protection   778, 3534
Fishery including pearl fishery   1o7, 2o4, 491, 492, 863, 9o8,
    1235, 1889, 2296, 2457, 2553, 2867, 32o9, 334o, 3458, 3459,
    36o5
Food preparation and diet   1o4, 253, 272, 289, 617, 661, 8o1, 821,
    13o2, 1451, 1512, 16o4, 1638, 1832, 1841, 1924, 2o96, 2121,
    2213, 243o, 2497, 2941, 2985, 3oo8, 3o57, 3487, 3515, 3631
Foreign policy
— Synopsis   192, 418, 443, 462, 483, 5o7, 543, 634, 646, 782, 819,
    87o, 975, 1oo5, 1o1o, 1o51, 1183, 1196, 1197, 1481, 1542,
    1759, 1793, 18o4, 18o5, 1836, 1862, 1946, 1957, 1958, 2o12,
    2o28, 2o44, 2143, 224o, 2283, 2319, 2396, 2513, 2721, 2845,
    2868, 2934, 3o22, 3o25, 3232, 3246
— Relations with Arab and Islamic states, people and organizations
    75, 97, 99, 113, 137, 149, 194, 283, 3o6, 3o8, 319, 341, 371,
    399, 416, 425, 437, 485, 5o3, 5o5, 5o8-511, 514, 544, 547,
    548, 673, 684, 733, 745, 774, 781, 783, 82o, 822, 841, 842,
    893, 898, 9o9, 961, 962, 1o2o, 1o45, 1o95, 11o9-1111, 112o,
    13o3, 1334, 1349, 1356, 14oo-14o3, 1415, 1452, 1453, 1472,
    15oo, 1522, 1525, 1533, 1534, 1544, 1545, 1578, 17o8, 1758,
    176o, 1773, 1776, 1777, 1791, 1794, 1798, 1799, 1818, 182o,
    1822, 1834, 1866, 19o3, 19o7, 1912, 192o, 1964, 1971, 1994,
    2o11, 2111, 2118, 2162, 2219, 2226, 2316, 2366, 2375, 2376,

— Relations with Arab and Islamic states, people and organizations (continuation)   2492, 2536, 2537, 2547, 2559, 2588, 2628, 2634, 2641, 2690, 2714, 2716, 2718, 2720, 2752, 2753, 2806, 2828, 2917-2919, 2975, 2989, 3024, 3029, 3058, 3093, 3106, 3154, 3199, 3201, 3217, 3219, 3226, 3234, 3273, 3313, 3351, 3353, 3369, 3392, 3424, 3442, 3461, 3465, 3480, 3481, 3498, 3523, 3543, 3556, 3558, 3562, 3598, 3606, 3621, 3662
— Relations with other developing countries   1579, 3024, 3154, 3556
— Relations with Western industrial states   111, 137, 155, 185, 186, 193, 283, 337, 340, 341, 350, 371, 416, 422, 423, 425, 427, 442, 482, 508-511, 517, 546, 548, 627, 628, 647, 651, 684, 688, 723, 774, 777, 781, 799, 834, 875-877, 893, 909, 978, 1083-1085, 1096, 1108-1112, 1119, 1128, 1144, 1150, 1194, 1199, 1203, 1226, 1262, 1282, 1295, 1303, 1314, 1316, 1325, 1334, 1352, 1353, 1362, 1377, 1381, 1415, 1437, 1452, 1453, 1462, 1519, 1521, 1522, 1533, 1534, 1538, 1545, 1568, 1580, 1708, 1760, 1762, 1789, 1791, 1799, 1818-1820, 1822, 1827, 1861, 1863, 1879, 1885, 1904, 1914, 1915, 1918-1920, 1963, 1970, 1996, 1997, 2014, 2024, 2030, 2083, 2085, 2094, 2110, 2111, 2137, 2170, 2229, 2232, 2235, 2287, 2289, 2306, 2308, 2316, 2346, 2363, 2365, 2366, 2376, 2406, 2465, 2491, 2492, 2516, 2549, 2588, 2590, 2591, 2634, 2641, 2650, 2651, 2661, 2671, 2694, 2714, 2740, 2860, 2902, 2911, 2917-2919, 2975-2979, 2988, 3001, 3024, 3027, 3029, 3053, 3054, 3107, 3154, 3189, 3199, 3201, 3218, 3225, 3226, 3234, 3237, 3298, 3316, 3351, 3355, 3362, 3369, 3461, 3523, 3552, 3556, 3562, 3572, 3585, 3611, 3621, 3642, 3678
— Relations with Eastern Bloc states   283, 510, 511, 774, 781, 909, 1110, 1111, 1194, 1414, 1452, 1760, 1920, 2336, 2375, 2491, 2492, 3467

Forestry   70, 77, 100, 112, 391, 3063

Geography: in depth   260, 298, 649, 1102, 1135, 1137-1139, 1266, 1720, 1882, 1884, 1908, 1923, 2237, 2291, 2556, 3032, 3184, 3245, 3282, 3361, 3384, 3493

Geography: synopsis (partially illustrated)   54, 110, 203, 290, 324, 351, 366, 370, 394, 417, 428, 437, 464, 497, 519, 558, 560, 562, 634, 635, 667, 670, 698, 730, 746, 764, 775, 792, 796, 801, 805, 807, 890, 899, 992, 993, 1035, 1108, 1117, 1174, 1214, 1257, 1260, 1261, 1269, 1312, 1317, 1323, 1348,

Geography: synopsis (partially illustrated) (continuation)   1350,
    1372, 1374, 1419, 1493, 1523, 1546, 1551, 1565, 1706, 1711,
    1715, 1718, 1746, 1774, 1782, 1783, 1824, 1855, 1860, 1890,
    1898, 1934, 1935, 1945, 1970, 2036, 2054, 2060, 2104, 2130,
    2184, 2235, 2239, 2285, 2370, 2378, 2481, 2487, 2507, 2541,
    2599, 2604, 2607, 2625, 2646, 2657, 2709, 2710, 2734, 2742,
    2750, 2756, 2761, 2775, 2778, 2779, 2807, 2808, 2852, 2853,
    2872, 2873, 2888, 2895, 2955, 2959-2963, 2985, 2987, 3025,
    3043, 3048, 3049, 3136, 3144, 3284, 3292, 3320, 3338, 3349,
    3362, 3385, 3391, 3400, 3401, 3430, 3439, 3443, 3444, 3450,
    3456, 3457, 3466, 3495, 3497, 3506, 3526, 3545, 3556, 3578,
    3593, 3609, 3638
Guest workers: composition and problems   275, 299, 320, 449, 451,
    452, 466, 472, 614, 620, 730, 939, 1297, 1311, 1356, 1393,
    1394, 1534, 1763, 1800, 2026, 2054, 2103, 2216, 2223, 2334,
    2507, 2713, 2874, 2875, 2919, 2933, 2938, 3364, 3646
Guidebooks and tourism   554, 1035, 1103, 1104, 1454, 1565, 2604,
    2605, 2654, 3397, 3466, 3593
History of the Western exploration of (Saudi) Arabia in the 18th
    to 20th centuries   170, 314, 384, 433, 500, 518, 624, 625,
    658, 678, 914, 1007, 1021, 1050, 1071, 1073, 1131, 1159,
    1213, 1322, 1324, 1325, 1327, 1455, 1479, 1532, 1564, 1570,
    1839, 1951, 1963, 2034, 2035, 2125, 2163, 2254, 2369, 2379,
    2392, 2408, 2410, 2413, 2424, 2425, 2428, 2448, 2449, 2464,
    2508, 2531, 2587, 2600, 2604, 2740, 2862, 2898, 3039, 3101,
    3126, 3179, 3220, 3540, 3604
Labor market and employment policy   54, 67, 89, 110, 191, 225, 275,
    299, 323, 331, 333, 449-452, 520, 548, 619, 633, 690, 694,
    730, 756, 808, 809, 884, 897, 904, 939, 940, 984, 985, 1053,
    1074, 1185, 1214, 1394, 1399, 1444, 1456, 1495, 1497, 1538,
    1585, 1601, 1642, 1680, 1691, 1695, 1727, 1730, 1763, 1800,
    1950, 1972, 1991, 2013, 2176, 2185, 2216, 2223, 2275, 2295,
    2300, 2334, 2360, 2361, 2713, 2875, 2933, 2938, 2951, 2970,
    3013, 3052, 3103, 3120, 3145, 3154, 3230, 3264, 3521, 3526,
    3672
Language modernization   3003, 3500
Legal and judicial system   83, 110, 135, 136, 168, 187, 195, 214,
    230, 274, 275, 355-357, 359, 401, 530, 578, 584, 589, 593,
    594, 601, 605, 633, 655, 673, 679, 680, 690, 708, 713, 730,
    808, 809, 830, 902, 903, 916, 952, 1091, 1093, 1113, 1125,

Legal and judicial system (continuation)   1176, 12oo, 12o2, 1234,
   1241, 1279, 1292, 1293, 1367, 1379, 1457, 148o, 15o6, 1559,
   156o, 1585, 1596, 16o3, 1663, 1691, 1695, 1733, 1734, 1753,
   1759, 1768, 18o1, 1812, 1813, 1815, 1817, 1835, 1852, 1891,
   1941, 196o, 1961, 197o, 1973, 2128, 2176, 2218, 2225, 2247,
   2261, 2369, 245o, 2451, 2476, 2484, 25o7, 2514, 2544, 2596,
   2724, 2737, 2764, 2774, 2781-2783, 2856, 2869-2871, 2883,
   2894, 2916, 2928, 2929, 2931, 2933, 3o11, 3o12, 3o18, 3o19,
   3o41, 3o51, 3o69, 3128, 3131, 3182, 3271, 3316, 34o2, 34o3,
   3475-3477, 3482, 3492, 3495, 3643, 3653
Leisure time occupations and facilities including sport   146, 172,
   173, 38o, 777, 1147, 1156, 1386, 1538, 1781, 2o81, 2432,
   2436, 2437, 2441, 26o5, 2818, 29o5, 3o17, 3478
Libraries and documentation   51, 296, 41o, 683, 1239, 1252, 1419,
   1562, 1691, 2oo6, 2o41, 223o, 2581, 2743, 3o35, 3123, 3554,
   3567, 3592
Living and working conditions of Westerners   11o, 127, 139, 172,
   173, 394, 494, 559, 684, 719, 73o, 749, 769, 81o, 834, 1oo6,
   1o32, 1128, 1566, 1855, 189o, 1934, 2o18, 2o52, 2o6o, 2184,
   23o8, 237o, 2489, 2622, 2671, 2689, 279o, 28o9, 2911, 2955,
   2957, 3399, 358o
Manufacturing industry and industrial policy   138, 147, 165, 2o6,
   2o8, 222, 225, 227, 231, 252, 268, 279, 291, 3o2, 32o, 323,
   398, 414, 463, 477, 48o, 548, 567, 571, 572, 578, 583, 59o,
   591, 595, 596, 598-6oo, 6o2-6o4, 6o6-6o9, 611-613, 714, 777,
   8o5, 8o9, 864, 884, 897, 9o6, 977, 1o18, 1o31, 1127, 1157,
   116o, 1162, 1163, 1167, 1173, 1186, 12o4, 124o, 1446, 1473,
   1474, 1561, 1586, 1597, 1598, 16oo-16o3, 16o6, 16o7, 1612-
   1617, 162o-1623, 1625-1637, 1639-1643, 1691, 1695, 17o2-
   17o4, 1748, 1751, 1755-1757, 1784, 1795, 1829, 185o, 1851,
   1858, 1874, 195o, 1969, 1975, 1977, 199o, 1991, 2o12, 2o13,
   2o54, 2156, 216o, 2167, 2227, 2228, 2258, 2299, 23oo, 2316,
   2322, 2326, 2335, 2473, 2482, 261o, 277o, 2771, 2796, 2811,
   2816, 2817, 2819-2821, 283o, 2855, 2895, 2948, 2958, 2968,
   31o4, 314o, 319o, 3195, 3196, 3212, 3229, 3231, 3233, 3235,
   3236, 3261, 3271, 3382, 3423, 3494, 3519, 3522, 3561, 3588,
   36o2, 3643, 3651
Money and monetary policy   1o8, 223, 347, 776, 1o38, 1o69, 1145,
   1146, 1181, 1218-122o, 1283, 1383, 1691, 1751, 1774, 1849,
   1985, 199o, 1991, 1999, 2134, 2155, 2156, 2276, 228o, 231o,
   2499, 2531, 2786, 2941, 3324, 3469, 347o

Nomads including partially settled tribes
— Economic bases   119, 246, 285, 3o9, 316, 471, 618, 656, 658, 7o4-711, 724, 726, 821, 822, 848, 849, 852, 853, 873, 914, 915, 918, 919, 951, 967, 1oo1, 1o93, 1o96, 1126, 1175, 1235, 125o, 1274, 128o, 1288, 13o2, 1392, 1451, 1484, 15o7, 158o, 1844-1846, 1951, 2o42, 2165, 2196, 221o-2213, 2215, 23o7, 2327, 2373, 24oo, 252o, 2521, 2552, 2558, 257o, 26o3, 2635, 2665, 2695, 2874, 288o, 2881, 2891, 2921, 2923, 3o57, 3o64, 3o97, 3o99, 3157-3161, 3168-3172, 3192, 3331-3333, 336o, 3389, 339o, 3435, 3438, 3524, 3541, 362o, 3631, 3637
— Material and immaterial culture   246, 285, 479, 5o1, 618, 624, 626, 68o, 7o8, 726, 821, 822, 845, 849, 852, 853, 914, 915, 951, 969, 971, 1oo1, 1o26, 1o93, 1o96, 1o97, 1125, 1175, 1235, 125o, 1273, 1286, 1288-129o, 1292-1294, 13o2, 132o, 1357, 1392, 1435, 1451, 1484, 15o2-15o4, 15o7, 1517, 1518, 1845, 1846, 1951, 1955, 2o42, 2o76, 2165, 2196, 221o-2213, 2215, 2274, 2327, 2373, 2377, 24oo, 2521, 2522, 2531, 2544, 2552, 2577, 2637, 2638, 2695, 288o, 2881, 2885, 2891, 3oo8, 3o57, 3o97-3o99, 3157-3161, 3168-3172, 3192, 3331-3333, 336o, 3377, 3389, 339o, 3416, 3511, 3541, 3565, 3631, 3666
— Social structure and social organization   119, 246, 293-295, 413, 5o1, 665, 7o4, 7o5, 7o8, 71o, 711, 726, 821, 845, 848, 849, 852, 853, 873, 951, 967, 1oo1, 1126, 1175, 1235, 125o, 1288, 1289, 13o2, 1392, 1451, 15o2-15o4, 15o7, 1529, 158o, 1845, 1846, 1986, 1991, 2164, 2165, 2196, 221o-2213, 2215, 2274, 2327, 2388, 252o, 2544, 2552, 2554, 256o, 2565, 2566, 2569, 257o, 2635, 2695, 27o5, 2874, 288o, 3o38, 3o57, 3o97, 3o99, 3161, 3172, 33o2, 3325, 3327, 3333, 336o, 3393, 3416, 3435, 362o, 3631
— Induced and spontaneous sedentarization   118, 119, 127, 176, 285, 316, 471, 517, 651, 7o4-7o7, 71o, 721, 8o3, 8o5, 821, 822, 869, 873, 879, 892, 918, 919, 951, 967, 979, 1o67, 1o68, 1o96, 1113, 1123, 1169, 1175, 125o, 1282, 13oo, 1392, 1443, 1491, 15o7, 151o, 1655, 1766, 1788, 1986, 2o77, 2212, 2345, 2373, 2377, 24o3, 2443, 2548, 2554, 2565, 2571, 26o2, 26o3, 2695, 27o5, 2785, 2874, 2894, 2921-2924, 2946, 3oo9, 3o33, 3o57, 325o, 3251, 3254, 3313, 3326, 3331, 3332, 3355, 3438, 3492, 3524, 3575, 3591
— Characteristics of pariah tribes   686, 821, 845, 846, 849, 852, 853, 1o14, 1o94, 1285, 1287, 1291, 1875, 1955, 221o, 2211, 2213, 2327, 2369, 2444, 2445, 257o, 2635, 2851, 2921

Oil and gas industry
— Synopsis 11o, 24o-244, 248, 254, 26o, 264, 266, 425, 653, 667, 676, 734, 9oo, 932, 968, 985, 1o36-1o38, 1216, 1278, 1295, 1296, 1334, 1346, 1425, 1445, 1466, 1467, 15o9, 1538, 1559, 1691, 1695, 1732, 1737, 1751, 176o, 1774, 181o, 1833, 1837, 186o, 1899, 19o2, 199o, 2o12, 2o24, 2o56, 2118, 2156, 2322, 2362, 26o7, 2632, 2694, 2754, 2817, 2837, 2863, 2865, 2868, 2873, 2888, 2911, 3o49, 3o6o, 3148, 3231, 3319, 3382, 3393, 3497, 365o
— Concessions and explorations 116, 149, 2o6, 213, 219, 226, 227, 31o, 337, 362, 372, 379, 381, 429, 463, 467, 468, 534, 652, 666, 681, 685, 725, 794, 799, 822, 884, 1o34, 115o, 1211, 1258, 1313, 1381, 1432, 1538, 1597, 1598, 1947, 1948, 197o, 2oo8, 2o44, 2o82, 2o84, 21o8, 213o, 2137, 2138, 2144, 2168, 2232, 2244, 2325, 2426, 2431, 2478, 2512, 259o, 2612, 2877, 2884, 29o4, 2972, 3o4o, 3o56, 3o59, 3o71, 3155, 318o, 3186, 3192, 33o8, 3362, 3365, 3386, 349o
— Production and transport 116, 12o, 166, 2o6, 213, 219, 251, 252, 262, 31o, 312, 343, 373, 375, 381, 429, 463, 467, 534, 681, 725, 751, 774, 835, 884, 987, 1211, 124o, 1422, 1431, 1495, 1497, 1597, 1598, 1736, 1874, 1948, 197o, 1983, 2oo7, 2o44, 2o83, 2o84, 21o8, 213o, 2137, 22o4, 2253, 2264, 2321, 2326, 2426, 2478, 2512, 2527, 2612, 2639, 2738, 274o, 2749, 2811, 2877, 2878, 2882, 2938, 2972, 3o56, 3o59, 3o71, 3118, 3152, 3181, 3192, 3362
— Refining and shipment 72, 2o6, 2o9, 213, 262, 263, 31o, 312, 376, 377, 429, 463, 467, 534, 835, 884, 1o86, 1211, 1495, 1581-1584, 1597, 1598, 1736, 1874, 1948, 197o, 2o44, 213o, 22o4, 2478, 2512, 2519, 2687, 274o, 2817, 2878, 2972, 3o56, 3o59, 3141, 3192, 32o8, 3233, 3235
— Characteristics of Aramco, its two predecessors and Trans-Arabian Pipe Line Company (TAP Line) including their personnel 116, 148, 155, 176, 185, 186, 191, 197, 2o2, 241, 242, 244, 245, 247, 248, 25o, 254, 257-26o, 265, 266, 279, 331, 349, 35o, 37o, 372-379, 381, 41o, 425, 429, 463, 635, 636, 653, 681, 699, 725, 75o, 866, 932, 945, 946, 961, 984, 985, 1o27, 1o51, 1157, 1211, 1252, 1253, 1258, 1295, 1296, 13o9, 1313, 1333, 1365, 1377, 1398, 14o4, 1478, 1524, 1534, 1538, 1554, 1559, 1566, 1581-1584, 176o, 1789, 1872, 1881, 1883, 19o2, 1924, 1926, 1948, 197o, 1996, 1997, 2o41, 2o44, 2o83, 21o8,

— Characteristics of Aramco, its two predecessors and Trans-Arabian Pipe Line Company (TAP Line) including their personnel (continuation) 2110, 2130, 2137, 2138, 2160, 2170, 2201, 2202, 2253, 2264, 2265, 2320, 2333, 2360-2362, 2426, 2478, 2512, 2548, 2612, 2632, 2655, 2662, 2705, 2735, 2739, 2740, 2790, 2877, 2878, 2911, 2954, 2972, 3003, 3013, 3052, 3056, 3059, 3060, 3071, 3118, 3119, 3145, 3181, 3192, 3203-3206, 3208, 3211, 3231, 3246, 3358, 3362, 3386, 3393, 3490, 3552, 3680
— Characteristics of other oil companies including their personnel 227, 270, 463, 534, 1081, 1167, 1597, 1598, 2044, 2160, 2265, 2516, 2972, 3104, 3105, 3119, 3231
— Governmental price and production policy 116, 120, 128, 137, 186, 283, 319, 422, 437, 462, 463, 506, 548, 638, 783, 866, 884, 961, 975, 1108, 1121, 1144, 1162, 1199, 1215, 1216, 1334, 1351, 1410, 1422, 1460, 1494-1497, 1534, 1535, 1736, 1762, 1777, 1791, 1798, 1799, 1822, 1833, 1836, 1851, 1859, 1904, 1910, 1944, 1974, 1983, 1991, 2014, 2031, 2043, 2044, 2143, 2150, 2169, 2170, 2198, 2258, 2283, 2325, 2346, 2372, 2491, 2492, 2735, 2945, 2972, 2973, 3029, 3058, 3066, 3232, 3234, 3320, 3338, 3386, 3464, 3494, 3562, 3585, 3610, 3621, 3635, 3678

Opposition groups and activities 75, 124, 127, 162, 281, 282, 319, 466, 507-509, 549, 681, 893, 905, 961, 976, 1005, 1110, 1196-1198, 1334, 1352, 1354, 1359, 1394, 1533-1535, 1542, 1569,1759, 1803, 1833, 1836, 1866, 1900, 1914, 1959, 1964, 2014, 2242, 2265, 2309, 2492, 2546-2548, 2826, 2868, 2917-2920, 2938, 3026, 3201, 3530, 3621

Ore deposits and metal mining industry 227, 291, 463, 625, 626, 661, 1038, 1049, 1051, 1509, 1559, 1597, 1733-1735, 2012, 2174, 2426, 2612, 2740, 3183, 3246, 3247, 3414, 3432

Peasants
— Material and immaterial culture 66, 117, 413, 741, 801, 855, 947, 1235, 1504, 2122, 2935, 3323, 3324, 3502, 3536, 3629
— Social structure and social organization 117, 741, 801, 1235, 1504, 1650, 3324, 3502, 3536, 3629

Pilgrimage
— Pilgrimage and pilgrim rites 52, 53, 68, 69,178, 354, 412, 481, 617, 623, 624, 682, 700, 701, 766, 832, 865, 1046, 1058, 1066, 1071, 1089, 1122, 1161, 1200, 1210, 1339, 1395, 1420,

— Pilgrimage and pilgrim rites (continuation)   1465, 1486, 1490, 1514, 1516-1518, 1547, 1559, 1792, 1809, 1943, 1981, 1998, 2067, 2254, 2342, 2409, 2447, 2508, 2636, 2664, 2670, 2744, 2832, 2892, 2912, 2913, 2947, 2974, 3011, 3037, 3055, 3067, 3094, 3227, 3228, 3344, 3371, 3394, 3396, 3407, 3512, 3603, 3677
— Pilgrim traffic and pilgrimage administration   52, 68, 69, 97, 154, 169, 177, 178, 189, 309, 314, 336, 389, 445-448, 456-459, 463, 465, 470, 475, 481, 495, 555, 617, 749, 765, 766, 777, 827, 828, 832, 834, 851-854, 865, 868, 895, 941, 1026, 1046, 1057, 1058, 1066, 1122, 1128, 1131, 1161, 1166, 1218, 1219, 1265, 1331, 1339, 1340, 1370, 1417, 1420, 1465, 1472, 1486, 1514, 1516-1518, 1547, 1576, 1691, 1695, 1698, 1699, 1751, 1792, 1842, 1873, 1906, 1943, 1980, 1981, 1998, 2012, 2067, 2069, 2070, 2105, 2107, 2110, 2112, 2113, 2254, 2280, 2307, 2327, 2343, 2344, 2379, 2396, 2403, 2404, 2409, 2447, 2508, 2515, 2517, 2523, 2525, 2527, 2623, 2626, 2645, 2663, 2664, 2670, 2671, 2744, 2860, 2862, 2896, 2915, 2947, 2999, 3010, 3011, 3021, 3037, 3045, 3142, 3199, 3222, 3227, 3241, 3309, 3344, 3348, 3371, 3378, 3389, 3390, 3394, 3396, 3407, 3507, 3509, 3590, 3603, 3607, 3616, 3636, 3660, 3677

Political history
— Synopsis   286, 425, 692, 766, 805, 921, 991, 1050, 1071, 1074, 1115, 1203, 1232, 1246, 1317, 1322, 1328, 1344, 1432, 1443, 1476, 1481, 1542, 1559, 1761, 1765, 1776, 1787, 1797, 1804, 1828, 1856, 1942, 1945, 1971, 1972, 1995, 1997, 2012, 2024, 2071, 2111, 2118, 2166, 2208, 2214, 2221, 2272, 2327, 2338, 2380, 2402, 2443, 2533, 2566, 2576, 2580, 2591, 2604, 2621, 2631, 2654, 2705, 2719, 2731, 2746, 2846, 2857, 2860, 2863, 2868, 2900, 2902, 2920, 3002, 3089, 3246, 3354, 3444, 3492, 3506, 3556, 3611, 3612, 3639
— First Saudi state (1745-1818): synopsis   86, 162, 465, 542, 545, 617, 663, 882, 1244, 1415, 1452, 1534, 1550, 1805, 1808, 2178, 2216, 2265, 2369, 2390, 2398, 2611, 2657, 3503, 3662
— First Saudi state (1745-1818): details   53, 63, 73, 85, 382, 443, 618, 729, 980, 1249, 1527, 1529, 1532, 1580, 1951, 2097, 2135, 2172, 2179, 2209, 2212, 2418, 2428, 2485, 2535, 2540, 2551, 2561, 2572, 2600, 2642-2644, 2688, 2700, 2848, 2899, 2907, 2908, 3299, 3305, 3355, 3408, 3418, 3426, 3427, 3584, 3622, 3632

— Second Saudi state (1824-91): synopsis   64, 86, 111, 162, 315,
    443, 465, 542, 1o14, 1249, 1415, 1452, 1534, 1773, 18o5,
    2265, 239o, 2611, 3226, 35o3
— Second Saudi state (1824-91): details   382, 744, 822, 995,
    1527, 1529, 153o, 1532, 158o, 1819, 1951, 2135, 2172, 22o9,
    2212, 2337, 2339, 234o, 2369, 2398, 2418, 2428, 254o, 2572,
    27o8, 3355, 3418, 3426, 3427, 3584
— Third Saudi state (since 19o2): synopsis   111, 155, 162, 345,
    371, 438, 512, 542, 627, 63o, 651, 663, 673, 797, 818, 87o,
    879, 912, 1o2o, 1o47, 1196, 1197, 1281, 1329, 1359, 1545,
    177o, 1773, 18o1, 18o3, 1861, 1876, 1919, 1958, 2o5o, 2o55,
    2132, 2217, 24o3, 2413, 2422, 2513, 2564, 2572, 2595, 2619,
    2657, 274o, 2774, 286o, 2927, 2934, 3o22, 3o25, 3o53, 3192,
    3312, 3339, 3455, 3497, 3596
— Third Saudi state (since 19o2): details   149, 196, 278, 283,
    337, 34o, 341, 35o, 415, 418, 423, 426, 427, 43o, 443, 485,
    517, 533, 559, 628, 693, 771, 821, 822, 834, 843, 876, 1o58,
    1o8o, 1o84, 1o96, 1o97, 11o9, 1112, 1113, 1128, 1169, 1183,
    1187, 1229, 1235, 1244, 1249, 1282, 1334, 1362, 1363, 138o,
    1415, 1452, 1472, 1525, 1527, 1529, 1534, 1553, 158o, 176o,
    1786, 18o5, 18o8, 1818, 1819, 1833, 1836, 1914, 1918, 1951,
    2o83, 2o94, 211o, 2131, 2133, 2162, 2163, 218o, 22o2, 22o6,
    22o9, 2212, 2226, 2265, 2345, 2386, 2387, 2389, 239o, 2393-
    2398, 2411, 2416, 2418, 242o, 2428, 2457, 2515, 2516, 2547,
    2548, 2571, 2578, 2579, 259o, 2611, 2664, 2671, 2894, 2931,
    2933, 2944, 2975-2979, 3oo1, 3189, 3199-32o1, 3217-3219,
    3226, 3324, 3355, 3363, 3369, 3378, 3379, 3392, 3412, 3428,
    3435, 3436, 348o, 3481, 3498, 3585, 3662
— The Hijaz under Turkish and Egyptian sovereignty (up to 1916)
    73, 74, 93, 179, 196, 3o9, 314, 315, 336, 382, 431, 455, 515,
    516, 617, 624, 626, 786, 787, 825, 839, 852, 853, 98o, 1o24,
    1o53, 1132, 1187, 1243, 1247, 1281, 1332, 1451, 1465, 1532,
    1567, 1797, 1999, 2o3o, 2o97, 2126, 2175, 2193, 22o7, 22o9-
    2211, 23o2, 23o3, 23o5-23o8, 2392, 2418, 2467, 2578, 2611,
    2618, 2688, 27oo, 3oo8, 3o1o, 3o12, 3o7o, 3114, 3164, 3222,
    33o5, 3355, 3371, 3395, 34o8, 3565, 3637
— Hashimite rule in the Hijaz (1916-25) (s. also Arab Revolt: an-
    tecedents and course of events)   336, 426, 455, 515, 559,
    628, 787, 839, 1o28, 1o29, 1o53, 1o82, 11o6, 1128, 1132,
    1187, 1249, 1329, 144o, 1449, 1468, 1519-1521, 1545, 1548,

— Hashimite rule in the Hijaz (1916-25) (continuation)   1778,
    1786, 1821, 1870, 2175, 2304, 2345, 2390, 2392, 2418, 2420,
    2465, 2515, 2578, 2593, 2611, 2664, 2860, 2976, 2994, 3053,
    3070, 3078, 3189, 3199, 3222, 3226, 3378, 3379, 3543
— Rashidi dynasty in Najd (1835-1921)   315, 406, 430, 443, 465,
    637, 851-853, 915, 1244, 1249, 1369, 1370, 1846, 1875, 1951,
    2164, 2171, 2209, 2211, 2212, 2281, 2339, 2340, 2389, 2418,
    2443, 2572, 2611, 2635, 2829, 3189, 3360, 3363, 3660
— Banu Khalid rule in al-Hasa (1680-1793/94 and 1819-30)   85,
    2079, 2418, 2428, 2551
— Al-Hasa under Turkish sovereignty (1871-1913)   430, 431, 616,
    627, 1112, 1235, 1244, 1532, 1796, 1819, 1951, 2126, 2212,
    2418, 2428, 3324, 3505, 3506
— Asir in the 18th and 19th centuries   297, 315, 337, 339, 431,
    732, 1272, 2030, 2260, 2357, 2415, 2547, 2880, 3114, 3501,
    3502
— Idrisi rule in Asir (1908-30)   337-341, 517, 628, 732, 742,
    743, 999, 1272, 1437, 2415, 2515, 2547, 2593, 2594, 2612,
    2880, 2988, 3189, 3199, 3201, 3378, 3392, 3498, 3502
Political system and individual political institutions   142, 149,
    162, 163, 255, 256, 281, 282, 318, 425, 432, 435, 438, 443,
    483, 485, 507, 549, 553, 601, 655, 673, 681, 712, 713, 730,
    774, 818, 879, 902, 905, 906, 909, 952, 1005, 1017, 1034,
    1038, 1074, 1108, 1113, 1229, 1278, 1333, 1358, 1359, 1379,
    1453, 1461, 1481, 1526, 1531, 1533, 1534, 1542, 1555, 1600,
    1721, 1759, 1793, 1801-1803, 1833, 1836, 1860, 1903, 1914,
    1942, 1943, 1945, 1958, 1964, 1970, 2050, 2057, 2105, 2180,
    2235, 2241, 2242, 2262, 2265, 2283, 2284, 2315, 2316, 2417,
    2423, 2451, 2492, 2513, 2516, 2533, 2539, 2548, 2564, 2580,
    2683, 2686, 2730, 2734, 2747, 2752, 2753, 2756, 2774, 2795,
    2845, 2856, 2883, 2894, 2917-2920, 2927, 2931-2934, 2938,
    2943, 2965, 3018, 3022, 3023, 3066, 3091, 3154, 3178, 3192,
    3216, 3313, 3351, 3352, 3355, 3393, 3425, 3492, 3517, 3556,
    3596, 3612, 3643
Population
— Methods and results of population estimates and census   89, 90,
    104, 189, 289, 450, 483, 631, 661, 732, 801, 806, 874, 940,
    1127, 1208, 1278, 1309, 1311, 1511, 1588, 1590, 1690, 1695,
    1696, 1800, 1832, 1950, 1951, 1968, 2012, 2042, 2054, 2220,
    2341, 2369, 2428, 2691, 2728, 2762, 2765, 2766, 2873-2875,

— Methods and results of population estimates and census (continuation)   2920, 2921, 2941, 2968, 3224, 3324, 3378, 3623, 3665
— Population policy   661, 874, 2885, 3268
Port descriptions, construction and traffic   269, 329, 617, 720, 811, 812, 814, 946, 1015, 1016, 1031, 1044, 1134, 1309, 1418, 1485, 1513, 1667, 1743-1746, 1874, 1888, 1951, 1965, 2021, 2022, 2148, 2296, 2297, 2310, 2379, 2390, 2434, 2483, 2542, 2543, 2660, 2763, 2787, 2788, 2797, 2798, 2801, 2822, 2858, 3005, 3207, 3294, 3305, 3390, 3457, 3563, 3566, 3583
Printing and press   18, 486, 487, 905, 942, 1072, 1268, 1450, 1691, 1759, 2006, 2242, 2466, 2538, 2656, 2659, 2745, 2756, 2814, 2825, 2920, 2926, 2965, 3010, 3123, 3130, 3345
Public administration system and personnel   134, 149, 151, 249, 318, 439, 614, 630, 673, 674, 979, 1034, 1050, 1075, 1207, 1439, 1551, 1644, 1691, 1722, 1759, 1972, 2047, 2050, 2110, 2151, 2206, 2216, 2257, 2292, 2305, 2330, 2331, 2392, 2415-2417, 2420, 2428, 2439, 2476, 2496, 2513, 2516, 2533, 2541, 2602, 2655, 2671, 2684, 2686, 2697, 2698, 2705, 2727, 2728, 2734, 2756, 2774, 2838, 2910, 2931-2933, 3018, 3103, 3124, 3125, 3145, 3324, 3492, 3517, 3556, 3597, 3627, 3665, 3666
Public finance and budgetary policy   108, 109, 120, 128, 133, 134, 175, 180, 206, 209, 210, 213, 216, 223, 237, 238, 284, 289, 291, 344, 363, 414, 463, 484, 521, 570, 617, 638, 676, 697, 730, 776, 841, 878, 884, 925, 1038, 1051, 1164, 1171, 1219, 1221, 1257, 1283, 1298, 1315, 1380, 1422, 1442, 1445, 1495, 1496, 1569, 1691, 1695, 1751, 1771, 1774, 1791, 1849, 1905, 1950, 1985, 1990, 1991, 2000, 2013, 2031, 2104, 2110, 2129, 2155, 2156, 2229, 2231, 2258, 2276, 2301, 2303, 2305, 2323, 2334, 2372, 2396, 2403, 2413, 2470, 2471, 2516, 2525, 2746, 2748, 2752, 2753, 2756, 2780, 2791, 2795, 2803, 2896, 2906, 2972, 3135, 3153, 3234, 3255, 3259, 3287, 3362, 3369, 3375, 3387, 3421, 3423, 3469, 3470, 3494, 3522, 3547, 3602, 3607, 3627
Public health
— Spread and control of sicknesses and diseases   55, 148, 261, 307, 314, 408, 475, 555, 617, 624, 639, 661, 687, 725, 760, 761, 765, 768, 770-773, 799, 821, 865, 941, 1059-1065, 1130, 1158, 1192, 1275, 1321, 1339, 1416, 1433, 1434, 1559, 1576, 1924, 1943, 1953, 1984, 2018, 2034, 2039, 2062, 2063, 2065,

— Spread and control of sicknesses and diseases (continuation)
2o66, 2154, 2222, 2267-227o, 2332, 2383, 2384, 2463, 2525, 2532, 2623, 2653, 2862, 2985, 3o14, 3o74-3o77, 3121, 3129, 3222, 3324, 3555, 3557, 3576, 359o, 3614, 3616, 3618, 3624, 363o, 3636, 3676

— Sanitary and medical facilities and personnel 1o2, 148, 2o1, 265, 314, 434, 463, 475, 555, 631, 661, 682, 687, 695, 696, 773, 8o1, 844, 865, 1o31, 1o58, 1o74, 1o88, 1158, 1192, 1228, 1231, 1256, 128o, 1321, 1339, 1433, 1434, 15o3, 1538, 1691, 1695, 1714, 1751, 1924, 1943, 1953, 1984, 2o18, 2o51, 2o54, 2o64, 2o75, 211o, 2112, 2113, 2154, 2275, 2332, 2432, 2433, 2435, 244o, 2441, 2463, 2476, 2532, 2533, 2545, 2622, 2623, 28o4, 2886, 2887, 2896, 2911, 2985, 3o1o, 3o21, 3o74, 3129, 3192, 346o, 3526, 3557, 3579, 3581, 359o, 3616, 3636, 3675

Radio and television 117, 267, 496-498, 715, 753, 8o1, 892, 9o5, 942, 1185, 145o, 1556, 1717, 1759, 2oo6, 2114, 2233, 2242, 2656, 2659, 2887, 292o, 2965, 3o15, 313o, 3345, 3644, 3647

Railroad construction and traffic 61, 224, 3o9, 311, 336, 392, 396, 431, 457, 459, 46o, 47o, 516, 658, 662, 695, 752, 826, 827, 896, 931, 946, 1o31, 1o9o, 1118, 1143, 1166, 1242, 1243, 1277, 1284, 1332, 1336, 1371, 1373, 1375, 1382, 14o6, 1449, 1451, 1471, 1576, 1764, 1847, 1848, 1886, 1888, 1913, 1967, 2o32, 2o45, 2o46, 2o59, 2177, 2187, 22o7, 221o, 2242, 23o1, 23o2, 23o4, 23o7, 2341, 24o1, 2421, 2459, 246o, 25o6, 25o8, 2518, 2532, 2545, 2584, 264o, 2677, 2823, 284o, 2896, 2982, 2983, 2999, 3143, 315o, 3192, 3194, 32o3, 3222, 33o9, 3371, 3434, 3472, 3486, 3517, 3637, 3664

Regional analyses and development 66, 84, 89, 169, 257, 258, 297, 397, 725, 732, 761, 762, 765, 858, 859, 874, 891, 954, 959, 96o, 974, 1o14, 1136, 1152, 1154, 1217, 1267, 1272, 13o6, 1332, 1391, 1412, 1429, 143o, 1456, 1457, 15o5, 1517, 1518, 1652, 1664, 1723, 1727, 1796, 1951, 2o8o, 2174, 225o, 2285, 24o4, 2541, 2555, 2557, 256o, 2562, 2567, 2568, 2573, 2574, 2576, 2578, 26o2, 2667, 2712, 2734, 2849, 2865, 288o, 2921, 2968, 3o38, 3o64, 3o99, 3192, 324o, 3324, 3327, 3328, 3336, 3337, 3398, 3449, 3663

Regional planning and policy 152, 858, 859, 959, 96o, 1391, 1412, 1429, 143o, 1483, 1652, 1665, 1716, 1724, 2285, 2854, 2954, 3342, 3449

Religious minorities: status and activities
— Muslims   87, 89, 549, 976, 1191, 1217, 1235, 1999, 2428, 2492, 2548, 2938, 3oo3, 3324, 3355
— Non-Muslims   87, 314, 559, 623, 624, 7oo, 749, 825, 834, 1o53, 1128, 1172, 1193, 1241, 1514, 1538, 1922, 194o, 1999, 2o3o, 211o, 2112, 2113, 2115, 2163, 23o6, 23o8, 2379, 2411, 26o8, 2617, 2623, 2624, 266o, 3114, 35o8
Research facilities and policy   258, 383, 78o, 8o2, 948-95o, 1188, 1319, 1384, 1411, 1895-1897, 1954, 2oo4, 2oo6, 2o41, 22o5, 2354, 236o, 2583, 2745, 31o3, 3127, 3132, 3579, 3649, 3654, 3657, 3677
Road construction and traffic   58, 59, 61, 62, 392, 466, 654, 748, 672, 946, 1o31, 1o34, 1o49, 1o5o, 1258, 1395, 1472, 1665-1668, 1814, 1874, 1978, 1979, 2112, 2113, 2124, 2222, 2242, 228o, 2328, 2341, 2515, 2525, 2545, 2584, 26o1, 26o2, 27o6, 2895, 299o, 3142, 3147, 3242, 3374, 3378, 3391, 3414, 3611, 3626, 3664, 3669, 3677
Seawater desalination plants   65, 181, 2o4, 2o6, 39o, 393, 463, 777, 922, 1713, 1874, 2oo3, 2o1o, 2895, 326o, 3289, 3454, 3668
Settlement types and pattern   66, 89, 413, 8o1, 2189, 2341, 2388, 2428, 2541, 26o1-26o3, 2873, 2874, 2923, 2954, 3251, 3324
Shipping   124, 2oo, 247, 489, 49o, 811, 812, 1o15, 1o16, 11oo, 1134, 1143, 137o, 1418, 1598, 1816, 1965, 1999, 2o21, 2o22, 226o, 2296, 2297, 23o6, 231o, 2379, 2458, 2542, 2543, 26o8, 2623, 2624, 266o, 2763, 28o1, 3oo6, 3o5o, 3294, 334o, 3341, 3389, 339o, 3423, 3566, 3667
Slave trade, treatment and liberation   317, 454, 5o1, 559, 617, 773, 9o5, 961, 1141, 1469, 1516, 1518, 1547, 1914, 2o3o, 2213, 23o8, 2388, 2392, 24o4, 2476, 2545, 2635, 2669, 2986, 3oo8, 3o88, 3169, 3551, 3599
Social expenditures and policy   136, 463, 633, 661, 8o8, 988, 1o74, 12o2, 1398, 1585, 1596, 1691, 1695, 17o5, 1729, 1751, 1924, 2o12, 2176, 22o6, 2227, 2258, 2286, 2533, 259o, 2596, 2648, 2812, 2931, 3128, 3192, 3264, 34o3, 3484, 3526, 36o5, 36o7
Tax system and collection   249, 346, 359, 52o, 522, 589, 633, 8o8, 8o9, 1o91, 1113, 1235, 1257, 1282, 148o, 1527, 1529, 158o, 1689, 1815, 1891, 195o, 197o, 2159, 2213, 2247, 25o1, 25o2, 2635, 2958, 3131, 3182, 3495, 3553, 36o5, 3671
Telecommunications and postal service   2o6, 463, 497, 715, 722, 757, 772, 1o34, 1134, 1268, 14o6, 145o, 1691, 1695, 1751,

Telecommunications and postal service (continuation)   1874, 1997, 2227, 2298, 2310, 2345, 2391, 2396, 2403, 2593, 2624, 3130, 3133, 3138, 3165, 3222, 3355, 3507, 3600, 3644

Total society (s. also nomads, peasants as well as townspeople)

— Material and immaterial culture   103, 132, 160, 349, 370, 502, 661, 816, 838, 847, 850, 1185, 1214, 1233, 1421, 1924, 1940, 1978, 2029, 2052, 2091, 2093, 2167, 2223, 2272, 2359, 2729, 2808, 2998, 3042, 3385, 3582, 3608, 3648, 3658, 3665

— Social structure and social organization   160, 162, 202, 271, 318, 328, 413, 509, 614, 664, 709, 712, 801, 805, 852, 866, 879, 979, 1191, 1237, 1282, 1299, 1310, 1321, 1353, 1355, 1448, 1470, 1958, 1964, 2055, 2078, 2090, 2242, 2259, 2265, 2274, 2284, 2292, 2334, 2359, 2513, 2533, 2563, 2575, 2655, 2705, 2713, 2808, 2874, 2880, 2933, 2938, 3003, 3022, 3099, 3338, 3339, 3351, 3352, 3385, 3393, 3447, 3665

— Cultural, social and socio-economic change   117, 124, 127, 132, 139, 162, 163, 183, 259, 334, 345, 348, 349, 413, 440, 461, 466, 471, 483, 485, 502, 614, 661, 667, 706-711, 721, 724, 731, 768, 783, 805, 816, 866, 905, 906, 944, 968, 1050, 1130, 1148, 1175, 1221, 1250, 1280, 1297, 1299, 1317, 1321, 1333, 1352, 1353, 1355, 1357, 1359, 1392-1394, 1421, 1432, 1453, 1504, 1507, 1509, 1510, 1533, 1539, 1558, 1776, 1802, 1803, 1836, 1844, 1845, 1900, 1903, 1919, 1924, 1936, 1942, 1957, 1958, 1987, 1991, 2023, 2042, 2053, 2055, 2092, 2093, 2108, 2110, 2114, 2115, 2151, 2194, 2242, 2259, 2262, 2284, 2323, 2359, 2362, 2417, 2420, 2425, 2507, 2513, 2546, 2563, 2564, 2604, 2655, 2659, 2689, 2719, 2730, 2832, 2847, 2923, 2933, 2934, 2938, 2941, 2943, 2969, 2970, 2998, 3004, 3043, 3057, 3064, 3073, 3077, 3192, 3213-3215, 3310, 3319, 3326, 3331-3333, 3338, 3339, 3362, 3382, 3385, 3393, 3517, 3610

Towns

— History and description   52, 53, 68, 90, 139, 156, 157, 172, 173, 189, 190, 288, 354, 386, 465, 469, 473, 481, 499, 556, 616, 617, 623, 624, 637, 679, 681, 708, 762, 763, 766, 807, 811, 812, 821, 825, 834, 852, 853, 857, 910, 915, 964, 1002-1004, 1014, 1051, 1056, 1057, 1071, 1073, 1100, 1103, 1104, 1128, 1134, 1151, 1153, 1155, 1161, 1215, 1230, 1245, 1251, 1264, 1265, 1271, 1339, 1342, 1369, 1396, 1454, 1475, 1485, 1514, 1547, 1559, 1664, 1694, 1843, 1876, 1927, 1951, 1965,

— History and description (continuation) 1979, 1987, 1998, 2oo1,
   2oo2, 2o21, 2o68, 2o69, 21o8, 211o, 2112, 2113, 2152, 2164,
   2184, 22oo, 2211, 2273, 2281, 2285, 2296, 2297, 23o6, 2338,
   234o, 2341, 2379, 2386, 2387, 239o, 2394, 24o1, 24o9, 2415,
   2421, 2425, 2428, 2467, 2468, 25o8, 2524, 2541-2543, 2545,
   255o, 2586, 2587, 259o, 26o4, 26o8, 2616, 2635, 2636, 2652-
   2654, 2664, 2668, 2711, 2779, 28o8, 285o, 286o, 2862, 2874,
   288o, 29o6, 2921, 3oo8, 3o1o, 3o21, 3o37, 3o42, 3o67, 3114,
   3188, 3192, 3241, 3294, 3324, 3329, 3334, 3344, 336o, 3363,
   3371, 3378, 338o, 3381, 3389, 339o, 3395, 35o6, 35o8, 3512,
   3566, 36o3, 36o4, 3677
— Planning and development 139, 189, 288, 292, 367, 393, 469,
   472, 473, 681, 727, 762, 763, 834, 86o, 871, 1127, 1221,
   1263, 1311, 1342, 1352, 1356, 1387, 1391, 1395, 1396, 1459,
   1467, 1478, 1483, 15o9, 1536, 1566, 1722, 1731, 1748, 1751,
   1785, 1823, 1874, 1919, 1979-1981, 1987, 2o54, 2o68, 2o69,
   2o88, 21o8, 2189, 2285, 2341, 2362, 2379, 24o9, 2413, 2425,
   2442, 2513, 2534, 2541, 2545, 26o2, 26o9, 2654, 274o, 2776,
   2789, 2792, 2793, 2838, 2842, 2844, 2854, 2855, 2874, 2911,
   2921, 2923, 2954, 2968, 3o42, 3o67, 3o81, 3118, 3176, 3185,
   3192, 3248, 326o, 3291, 3324, 3363, 3372, 3628, 3677
— Architecture and furnishings 53, 8o, 9o, 139, 149, 342, 469,
   473, 617, 727, 762, 763, 773, 777, 822, 823, 825, 832, 915,
   1o47, 1387, 1395, 1485, 1493, 1516, 1518, 1536, 157o-1575,
   1893, 2189, 2285, 2341, 2378, 2379, 239o, 2479, 2523, 2524,
   2531, 2532, 2541, 2587, 259o, 2596, 2622, 2723, 2843, 3oo8,
   3o1o, 3o42, 3o83, 3197, 3321, 3324, 36o4, 3628
— Water resources development and water usage (s. also seawater
   desalination plants) 9o, 188, 2o4, 365, 367, 39o, 393, 556,
   617, 762, 837, 914, 923, 1o31, 1o57, 1426, 1458, 15o4, 15o9,
   1516, 1518, 1646-1649, 1713, 1999, 2o1o, 21o8, 2141, 2227,
   2379, 2524, 2739, 2985, 2997, 3o1o, 3o21, 3244, 326o, 3324,
   3519, 3668

Townspeople
— Material and immaterial culture 53, 342, 617, 624, 773, 1o71,
   1965, 1998, 234o, 2394, 25o8, 2531, 2638, 3oo8, 3o1o, 3o11,
   3o37, 3321, 3322, 3324, 35o9, 3565
— Social structure and social organization 53, 74, 89, 472, 473,
   617, 624, 964, 1o56, 1235, 1342, 1516, 1518, 1951, 1987,

— Social structure and social organization (continuation) 1998, 2108, 2164, 2340, 2341, 2394, 2428, 2468, 2602, 2635, 2636, 2744, 3008, 3010, 3114, 3176, 3324, 3395, 3509

Trade and trade policy
— Foreign trade (s. also agricultural imports and exports) 96, 110, 206, 209, 211, 233, 235, 272, 289, 313, 463, 482, 517, 520, 525, 552, 565, 614, 617, 669, 728, 777, 809, 820, 875, 879, 884, 964, 977, 1012, 1038, 1083, 1091, 1134, 1138, 1148, 1163, 1184, 1240, 1397, 1408, 1414, 1426, 1445, 1472, 1497, 1513, 1532, 1540, 1546, 1567, 1691-1693, 1695, 1743-1745, 1751, 1765, 1807, 1816, 1840, 1880, 1888, 1893, 1899, 1901, 1930, 1938, 1947, 1951, 1952, 1966, 1990, 1991, 1999, 2037, 2104, 2127, 2129, 2134, 2156, 2227, 2229, 2231, 2260, 2265, 2280, 2306, 2310-2312, 2379, 2457, 2472, 2478, 2495, 2515, 2585, 2615-2617, 2649, 2752, 2753, 2756, 2760, 2763, 2787, 2810, 2815-2817, 2824, 2858, 2867, 2873, 2880, 2906, 2937, 3046-3049, 3147, 3187, 3218, 3224, 3229, 3259, 3263, 3269-3271, 3285-3288, 3305, 3390, 3431, 3435, 3440, 3441, 3457, 3463, 3494, 3516, 3518, 3560, 3578, 3634
— Domestic trade 110, 117, 169, 268, 361, 463, 561, 617, 637, 735, 759, 825, 852, 853, 884, 914, 1018, 1131, 1235, 1426, 1485, 1514, 1547, 1554, 1691, 1951, 1965, 2096, 2213, 2310, 2340, 2341, 2508, 2531, 2842, 2844, 2880, 2906, 3010, 3057, 3222, 3323, 3324, 3360, 3431, 3457

Transportation: synopsis (s. also airport construction, airlines and air traffic; port descriptions, construction and traffic; railroad construction and traffic; road construction and traffic as well as shipping) 59, 206, 207, 367, 463, 661, 716, 717, 792, 1074, 1295, 1296, 1298, 1309, 1576, 1691, 1695, 1751, 2108, 2227, 2254, 2392, 2712, 2744, 2756, 2855, 3072, 3382

Travel literature and personal accounts 53, 71, 127, 139, 145, 168, 170, 276, 280, 286, 309, 324, 354, 384, 386, 403, 405, 406, 409, 433, 435, 436, 454, 461, 465, 466, 481, 482, 512, 513, 516, 519, 529, 541, 550, 559, 616, 617, 623-626, 629, 635-637, 656-658, 677-679, 681, 684, 693, 695, 696, 700, 701, 722, 736, 748, 749, 770, 772-774, 821-825, 832, 834, 851-854, 856, 882, 886, 887, 889-891, 901, 914, 915, 927, 964, 980, 981, 999, 1000, 1002-1004, 1020, 1022, 1025, 1040, 1047, 1048, 1050, 1052, 1055, 1057, 1058, 1071, 1073, 1095-

Travel literature and personal accounts (continuation) 1o97, 11oo,
    1128, 115o, 1159, 1172, 1193, 1222, 1223, 123o, 1312, 132o,
    133o, 1331, 1337-1339, 1356, 1369, 137o, 1395, 1418, 1451,
    1478, 1489, 1516-1518, 1523, 1524, 1547, 1559, 1769, 18o9,
    1854, 1875, 1876, 1893, 1922, 1926, 1933, 1936, 194o, 1978,
    1993, 1998, 1999, 2o2o, 2o52, 2o81, 2o95, 21o1, 21o2, 21o6,
    21o8, 211o, 2112, 2113, 2161, 2173, 2183, 2193, 2194, 2195,
    22o7, 221o-2213, 2215, 2227, 2237, 2254, 2273, 2281, 2337-
    234o, 2358, 2367-2369, 2382-2384, 2386, 2387, 2389-2391,
    2394, 2399, 24oo, 24o5, 24o7-2412, 2414, 2415, 2419, 2421,
    2422, 2431, 2462, 2464, 2467, 2476, 2488, 2521, 2522, 2528-
    2532, 2545, 2549, 255o, 2589-2594, 2596, 26o8, 2614, 2616-
    2618, 2622, 2624, 2632, 2633, 2636, 2652, 2653, 266o, 2664,
    2667, 2668, 267o-2673, 2699, 27oo, 274o, 2881, 2888-289o,
    2892, 2897, 2911-2913, 2944, 2974, 3oo8, 3o1o-3o12, 3o21,
    3o43, 3o55, 3o56, 3o74, 3o76, 3o78, 3o8o, 3o87, 3o94, 3113,
    3114, 3126, 3156-3162, 3168-3172, 3174, 3186, 3188, 3197,
    3198, 322o, 3227, 3239, 3241, 3247, 33o5, 3317, 3321, 3358,
    336o, 3371, 3374, 3378, 3389-3391, 34o6, 3414, 3415, 3436,
    3437, 345o, 3462, 3468, 35o5-35o8, 351o, 358o, 3587, 36o1,
    3611, 3617, 3638, 366o, 3666

Villages
— History and description 79, 117, 617, 656, 658, 8o1, 852,
    853, 915, 1o42, 1451, 15o2-15o4, 1664, 1694, 1951, 2o21,
    2o22, 2191, 2197, 2211, 2212, 2394, 2421, 2541-2543, 26o1,
    2711, 2874, 288o, 2921, 2924, 3o42, 3324, 336o, 3629, 366o
— Planning and development 86o, 956-958, 1o42, 2191, 2197, 2428,
    2954, 3o42, 3324
— Architecture and furnishings 8o, 8o1, 85o, 856, 957, 958,
    1o42, 1493, 157o-1574, 2191, 2479, 2524, 2541, 2622, 2723,
    2843, 2874, 3o42, 3324, 3669
— Water resources development and water usage 914, 2985, 3324,
    3519
— Community development 332, 661, 1255, 1556, 17o5, 1726, 2186,
    225o, 268o, 292o, 2935, 2966, 3267

Wahhabiya: doctrine and practice 53, 87, 371, 411, 413, 485, 533,
    542, 641, 663, 729, 744, 783, 829, 879, 9o2, 9o5, 915, 952,
    955, 995, 1oo5, 1113, 1114, 1123, 1169, 12oo, 1249, 125o,
    1343, 1432, 1499, 1532, 1534, 1539, 155o, 18o5, 1856, 1936,

Wahhabiya: doctrine and practice (continuation)   1962, 1971, 2o19,
    2o71, 2212, 2214, 2236, 2248, 2272, 2292, 23o9, 2337, 2339,
    234o, 239o, 2443, 2451, 2475, 248o, 2485, 2486, 2546, 2551,
    2564, 2571, 2572, 2613, 262o, 2631, 2644, 2658, 2717, 2719,
    2841, 2927, 3oo2, 3o23, 3299, 3346, 3354, 336o, 3426, 3455,
    35o1, 35o3, 35o4, 35o6, 3591, 3612, 3632
Women: status and activities   127, 139, 158, 159, 277, 334, 37o,
    413, 454, 465, 655, 759, 773, 79o, 791, 821, 847, 852, 881,
    892, 944, 1o93, 113o, 1195, 1288, 13o2, 151o, 1516, 1518,
    1725, 1742, 178o, 1845, 1877, 1935, 2o27, 2o78, 2355, 2358,
    2468, 2484, 2486, 2532, 2544, 268o, 2938, 3o57, 3213-3215,
    33o2, 3452, 3526, 361o, 3653

II PERSÖNLICHE VERFASSER / AUTHORS

Abanami, A.A. 5o
Abbas, A.M. 3515
Abbas, H.A. 51
Abbasi, M.Y. 52
Abbassi, A. el- (pseud.) 53
Abbondante, P.J. 2156
Abdalwahed, A.M. al- 54
Abdel Azim, M. 55
Abd-el Wassie, A.W. 56, 57
Abdo, Albert N. 3516
Abdo, Assad S. 58-62
'Abdu, I. 3517
Abdul Bari, M. 63, 64
Abdul-Fattah, A.-R.A.-F. 65
Abdulfattah, K. 66
Abdulkader, A.A. al- 67
Abdullah, M.E. 3o28
Abdul Majid, H. 68
Abdul-Rauf, M. 69
Abdul-Samad, M. 3518
Abdulwahid, Y. 7o, 112
Abed, G.T. 1315
Abercrombie, T.J. 71
Abir, M. 73, 74
Abir, Mordechai 75, 3467
Abo Ali, S.A. 76
Abohassan, A.A. 77
Abo-Laban, M.A. 78
Abolkhair, Y.M.S. 79, 13o7

Abou el-Nasr, K. 8o
Abraham, N.A. 81
Abrahams, Anthony 82
Abrahams, A.M. 83
Abu-Bakr, A.S. 84
Abu-Hakima, A.M. 85, 86
Abu-Ihya, S. 87
Abu-Laban, B. 88
Abul-Ela, M.T. 89, 9o
Abul-Ezz, S. 91
Abul-Haggag, Y. 92
Abu-Manneh, B. 93
Abunabaa, A.M. 94
Abu Ras, A.S. 95
Abu-Rizaiza, O.S. 3519
Abu Sulaiman, A.W. 135
Achtnich, W. 1oo
Adelman, M.B. 1o3
Adham, M.H. 687
Adler, G. 1o4
Adwani, S.H. 352o
Afifi, M.H. 1224, 1225
Agarwal, D.P. 11o1
Ageel, H.A. 1o5
Ahmad, A. 1o8
Ahmad, Y.J. 1o9
Ahmed, S. 11o
Aiz, M. 112
Ajaji, A. 3521

Ajami, F.   113
Ajroush, H.A. al-   115
Akhdar, F.M.H.   116
Akkad, A.A.-H. al-   117
Akkad, H.A.-H. al-   118
Alabbadi, A.H.   119
Alageel, K.M.N.   120
Alaki, M.A.   121
Alam, M.S.   122
Alami, J.   123
Alan, R.   124
Albaharna, H.M.   125, 126
Albers, H.H.   127
Albokhair, Y.: s. Abolkhair, Y.M.S.
Albraikan, S.M.   128
Aldoasary, F.S.   3522
Alexander, L.T.   691
Algawad, M.A.   129
Alghamdi, A.A.S.B.   131
Alghofaily, I.F.   132
Algosaibi, G.A.   1120, 3523
Alhumaid, A.I.   133
Ali, A.M.   134
Ali, I. al-   3524
Ali, M.I.A.   135, 136
Ali, S.R.   137
Aliboni, R.   138
Alikattan, M.   3525
Alireza, M.   139
Alkazaz, A.   104, 140-142, 3657
Alkhowaiter, H.   143, 144
Allan, M.   145
Allen, M.J.S.   146
Allen, R.   147
Allen, T.E.   148
Almana, A.M.   3526
Almana, M.   149
Almas, H.M.   3527
Alnassar, S.N.   150

Alnimir, S.M.   151
Alohaly, M.N.   152
Alraegi, A.H.   153
Alrashid, S.A.   154
Alsaigh, M.N.H.   3528
Alstyne, R.W. van   155
Alter, H.W.   156, 157
Altorki, S.   158-160, 1780
Altwaijri, A.O.   3529
Alyahya, K.A.M.   161
Alyami, A.H.   162, 163
Alzamel, I.A.   164
Alzamil, A.   165
Amaldi, D.   3530
Ambah, S.   167
Ambroggi, R.   862
Amelunxen, C.   168
Amer, M.   169, 170
Amin, G.A.   174, 175
Amin, H.   176
Amin, M.   177, 178, 2254
Amr, S.M. al-   179
Amry, M.-A.Y.   180
Anani, F.M.   181
Anastos, D.   182
Anderer, K.   184
Anderson, G.W.   2985
Anderson, H.H.   3555
Anderson, I.H.   185, 186
Anderson, J.N.D.   187
André, M.   188
Ani, M.   1923
Ankary, K.M. al-   189
Annesley, G.: s. Valentia, G.
Ansari, 'A.-Q. al-   190
Ansary, A.O.T. el-   191
Anthony, J.D.   192-194
Antonius, G.   196
Apgar, M.   198
Appelman, H.   199

Arbose, J.   268-27o
Arcache, J.   3545
Arendonk, C. van   271, 272
Arfaj, N.A.   274
Arkadakshi, A.F.   275
Arle, M. d' (Pseud.)   276, 277
Armstrong, H.C.   278
Arndt, R.   279
Arno, A.   178o
Arnold, J.   28o
Arnon, Y.   281, 282
Asaad, M.M.A.   283
As'ad, I.A.   284
Asad, Mohammed A.   285
Asad, Muhammad: s. Weiss, L.
Asadallah, M.M.   287
Asadullah, M.   288
Asfour, E.Y.   289-291
Ashiry, H. el-   292
Ashkenazi, T.   293-295
Ashoor, M.-S.J.   296
Askari, H.G.   756
Assa, A.   298
Assaf, I.A.   299
Assaf, S.H.   3oo
Assaneea, A.A.   3o1
Ateque, H.I.   3o3, 3o4
Athubaity, M.M.   3532
Atiyah, A.M.   3o5
Attar, M.S. el-   3o6
Aulas, M.-C.   3o8
Auler, K.   3o9
Aurada, F.   31o-312
Avril, A. d'   314, 315
Awad, M.   316, 317
Awaji, I.M. al-   318
Ayoob, M.   319
Ayouti, Y. el-   32o
Ayubi, N.N.M.   321, 3533
Azouz, A.-A.H.   322

Azzi, R.   323, 324
Baâmer, S.I.   832
Babtain, A.-A.A.-W. al-   325
Babtain, I.A. al-   326
Backer, A.S.   327
Badia y Leblich, D.: s. Abbassi, A. el-
Badr, F.I.   328, 329
Badr, H.A. al-   33o
Badre, A.Y.   331
Bagader, A.A.   332
Bagais, M.O.   333
Baharna, H.M. al-: s. Albaharna, H.M.
Bahry, L.   334
Baker, A.J. el-   335
Baker, P.R.   336
Baker, T.D.   2886
Bakri, T.H.   3535
Baldry, J.   337-341
Baleela, M.M.   342
Ballantine, J.   343
Ballool, M.M.   344
Balta, P.   345
Bammate, N.O.-D.   1486
Bancal, J.-C.   346
Banyan, A.S. al-   348
Barakat, H.   349
Baram, P.J.   35o
Barcata, L.   351
Barham, P.R.   352
Barker, P.   353
Barny, F.J.   2o34
Barois, J.   354
Baroody, G.M.   355-357
Baroudi, E.   358
Barrak, I.A. al-   359
Barré, P.   36o
Barreveld, W.H.   361
Barrows, G.H.   362

Barry, Z.A.A.   363
Barth, H.K.   364-367
Bartholomew, P.   95o
Basabrain, A.A.A.A.   3536
Bashir, F.S. al-   368
Bassam, A.A.   3537
Bassam, I.A. al-   369
Bassam, N.A. al-   37o
Bassi, U.   371
Bastos, A.J.   3361
Bates, B.S.   372-381
Batrik, A.H.M. el-   382
Baumer, M.   383
Bawazeer, S.A.   385
Bawden, E.   386
Bayoumi, R.A.   27o3, 3538
Baz, F.   387, 388
Beaumont, A. de   389
Beaumont, R.   39o
Becher, R.   391
Becke, A.F.   931
Becker, H.   395
Becker, K.   396
Beckingham, C.F.   397
Bedore, J.   398-4oo, 3233-3236
Bédos, A.   182
Behrens, G.   4o1
Békri, C.   4o2
Belgrave, C.D.   4o3
Bell, G.L.   4o6
Bell, M.   4o7
Bell, S.D., Jr.   4o8, 2268,
  227o, 3o14, 363o, 3676
Bellotti, F.   4o9
Beltran, A.A.   41o
Ben Cheneb, M.   411
Ben Chérif   412
Benderly, B.L.   413, 2291
Ben Gabr, A.A.   414
Benkmann, H.G.   11o1

Benoist-Méchin, J.   415-418
Benton, G.   419
Berger, J.   42o
Bergmann, W.   421
Bergsten, C.F.   422
Bernasconi, P.   423
Bernleithner, E.   424
Berreby, J.-J.   425
Besson, Y.   426, 427
Bethmann, E.W.   428
Bianchini, M.   429
Bidwell, R.   433
Bienzle, U.   11o1
Bigelow, M.C.   1923
Bilainkin, G.   435, 436
Bilimatsis, J.S.   437
Bill, J.A.   438
Binder, F.   354o, 3541
Binsaleh, A.M.   439
Binzagr, S.   44o
Bird, K.   442
Birken, A.   443
Birket-Smith, K.   444
Birks, J.S.   445-452
Bishara, G.   453
Bitsch, J.   454
Björkman, W.   455
Blackwood, P.   456
Blake, G.H.   457, 458
Blanckenhorn, M.   459, 46o
Blandford, L.   461
Bleiber, F.   3543
Bligh, A.   462
Blume, H.   463, 464
Blunt, A.   465
Bobb, A.A.   2268-227o
Bochskandl, M.: s. Arle, M. d'
Bökemeier, R.   466
Boesch, H.   467, 468
Bokhari, A.Y.   469

Bonin, C.-E.   47o
Bonnenfant, P.   471-473
Bono, S.   474
Borel, F.   475
Boswinkle, E.   476
Bouaoula, A.   478
Boucheman, A. de   479
Boucher, B.P.   48o
Boulicaut, A. le   481
Boulter, V.M.   32o1
Bouteiller, G. de   482-485
Bouvat, L.   486, 487
Bowen, R.L., Jr.   489-492
Bowen-Jones, H.   493
Bowler, R.M.   494
Boxhall, P.   495
Boyd, D.A.   496-498
Boylan, F.T.   499
Bräunlich, E.   5oo, 5o1, 2327
Braibanti, R.   5o2, 5o3
Branscheid, V.   5o4
Braun, U.   5o5-511
Bray, N.N.E.   512, 513
Brébant, É.   3545
Bredi, D.   514
Breidenbach, K.   2861
Brémond, É.   515-517
Brent, P.   518
Bretholz, W.   519
Bricault, G.C.   52o, 93o
Bridge, J.N.   521
Bridi, G.S.   1984
Briner, E.K.   522
Brinton, J.Y.   523
Brittain, M.Z.   529
Brizard, S.   53o
Broadfoot, W.   531
Brockelmann, C.   11o2
Brodie, D.M.   532
Broucke, J.   533

Brown, E.H.   534
Brown, Glen F.   535
Brown, Grover F.   536
Brown, W.   537-539
Brownell, G.A.   54o
Bruce, J.   541
Bruyn, P. de   542, 543
Bruzonsky, M.A.   544
Brydges, H.J.   545
Bryson, T.A.   546
Buchan, J.   547-549, 3546
Buckingham, J.S.   55o
Buckner, R.G.   551
Buddenberg, J.   552
Büren, R.   553
Bürkner, F.C.   554
Buez, E.A.   555
Buhl, F.   556
Bullard, R.   559
Bunyan, J.   615
Burchardt, H.   616
Burckhardt, J.L.   617, 618
Burgoyne, J.A., Jr.   3548
Burki, S.J.   62o
Burleson, C.   2o52
Burningham, C.W.M.   621
Burrell, R.M.   622
Burton, I.   623
Burton, R.F.   624-626
Busch, B.C.   627, 628
Buschow, R.   629
Bushnak, M.'U.   63o
Bushra, S. el-   631
Bustani, E.   634
Bustani, N.   3555
Butler, G.C.   635, 636
Butler, S.S.   637
Buxton, J.   638
Buxton, P.A.   639
Cain, L.F.   64o

Calverley, Edwin E.   641, 642
Calverley, Eleanor T.   643
Campbell, C.P.   645
Campbell, J.C.   646, 647
Campbell, R.B.   3549
Candler, E.   648
Cantine, J.   3510
Capece Galeota Zuccoli, V.   649
Carami, M.S.G.   650
Carlson, T.   3550
Carmichael, J.   651
Carmichael, K.   652
Caroe, O.   653
Carré, O.   655
Carruthers, D.   656-658
Carter, J.R.L.   659, 660
Carter, L.N.   661, 2291
Carter, W.   662
Carver, J.L. (pseud.)   3551
Casillas, R.J.   3552
Caskel, W.   663-665, 2327
Caso, R.G.   3553
Cattan, H.   666
Cattin, J.   667
Cayre, G.   668
Celli, J.P.   3554
Chaglassian, H.T.   3555
Champenois, L.   673, 3022, 3024, 3025
Chang, R.S.-M.   408, 3576, 3630
Chapman, R.A.   674
Chatelus, M.   675, 676
Chatham, C.E., Jr.   3583
Chaudhuri, K.N.   272
Cheesman, R.E.   677-679
Chelhod, J.   680, 3302
Chenery, H.B.   2130
Cheney, M.S.   681
Chérif, A.   682
Childs, J.R.   683, 684

Chisholm, A.H.T.   685
Christian, V.   686
Chu, C.K.   687
Chubin, S.   688
Cipriani, L.   689
Clark, A.   1319, 1874, 3343
Clarke, J.   690
Clarke, S.H.   537-539
Clawson, M.   691
Clayton, G.F.   692, 693
Clements, F.   694
Clemow, F.G.   695, 696
Clerc, M.   3557
Cleron, J.P.   697
Clifford, M.L.   698
Clifford, R.L.   699
Cobbold, E.   700, 701
Colby, C.B.   702, 703
Cole, D.P.   704-711, 1392
Collins, M.   712
Conant, M.A.   1108
Coneybear, J.F.   716, 717
Cooke, H.V.   721
Cooksey, J.J.   722
Cooley, J.   723
Coon, C.S.   724-726
Cooper, W.W.   727
Coppock, J.D.   728
Corancez, L.A.O. de   729
Cornwall, P.B.   731
Cornwallis, K.   732
Costa, F.J.   2191
Cotran, E.   520
Cotton, D.   2104
Coulon, C.   733
Crane, C.R.   736
Crane, R.D.   737-739, 3561
Crary, D.D.   740, 741
Craufurd, C.E.V.   742, 743
Crawford, M.J.   744

Cressey, G.B. 746
Cresswell, E. 3o61
Crichton, A. 747
Crowe, P. 748
Cruz, D. da 75o-755
Cüceloglu, D. 2o91
Cumming-Bruce, N. 9oo
Cummings, J.T. 756
D. 757
Dabbagh, T.H. al- 758
Dabbagh, Z.M.A. al- 759
Daggy, R.H. 76o, 761
Daghistani, Abdal-Majeed I.
   762, 763
Daghistani, Abdulaziz I. 764
Daguillon, L. 765
Daham, A.A. 766
Daher, S.A. 3562
Daihan, M.A.-R. al- 767
Dajani, M.T. 1442
Dalenberg, C. 768
Dall, R.F. 769
Dame, L.P. 77o-772
Dame, Mrs. L.P. 773
Dammann, T. 774
Dana, L.P. 775
Daniels, R.J. 3563
Darrab, I.A. 3564
Darrat, A.F. 776
Davidson, B. 779
Davies, D.L. 237o
Davies, M.S. 237o
Davies, P. 78o
Dawisha, A.I. 781-783
Dawn, C.E. 784-787
Dayil, A.S.M. al- 788
Deane, C. 789
Deaver, S. 791
Decken, H. von der 792
Deeik, K.G. 793

DeGolyer, E. 794
Demeuse, P. 797
Denis de Rivoyre, B.L.: s.
   Rivoyre, D. de
DeNovo, J.A. 799
Dequin, H. 8oo-8o5
Descoudray, M.(?) 3565
Destrées, M.(?) 8o6
Desvergers, N. 8o7
Dhaher, A.J. 816
Dhanani, G. 818, 819
Dhohayan, A.I. 3567
Diab, M.A. 82o
Dickson, H.R.P. 821, 822
Dickson, V. 823, 824
Didier, C.E. 825
Dieckmann, P. 826, 3664
Dietrich, M. 11o1
Dietvorst, A.G.J. 828
Diffelen, R.W. van 829
Dilger, K. 83o
Dimock, W.C. 831
Dinet, A.É. 832
Dingelstedt, V. 833
Dingemans, H.H. 834
Dixon, H. 835
Djazzar, S.K. 687
Djemal, A. 836
Dodson, R.H.T. 837
Dörflinger, R. 184
Dohaish, A.A. 838-84o
Donini, P.G. 841, 842
Donkan, R.: s. Zischka, A.
Donovan, M. 52o
Dorozynski, A. 844
Dossary, S. al- 1785
Dossary, Saleh J. 3568
Dostal, W. 845-85o
Doughty, C.M. 851-854
Dowson, V.H.W. 855, 856, 3569

Draz, O. 357o
Drees, I.A. al- 861
Drouhin, G. 862
Drucker, J. 863
Duguet, F. 865
Duguid, S. 866
Duheash, O.A. 867
Dunbar, G.S. 868
Duncan, P.D. 869
Dyer, G. 87o
Earle, M.W. 871
Ebert, C.H.V. 872
Ebrahim, M.H.S. 873
Eddy, W.A. 876, 877
Edens, D.G. 878, 879
Edwards, F.M. 882
Efrat, M. 282
Egbert, R. 883
Eglin, D.R. 884, 2291
Ehlers, E. 885
Ehrenberg, C.G. 886
Ehrenberg, E. 888
Eigeland, T. 889-891
Eilts, H.F. 892-894
Eisele, F.R. 3361
Eisenberger, J. 895
Elefteriadès, E. 896
Elham, M. 898
Eliseit, H. 899
Ellwood, W. 9oo
Elmgren, S.G. 9o1
Elwan, O. 9o2
Emilia, A.D. 9o3
Ende, W. 18, 9o5
Entelis, J.P. 9o6
Erb, R.D. 9o7
Erdman, D.S. 9o8
Erris, T.S. el- 9o9
Esin, E. 91o
Ess, J. van 912

Etibi, M.H. al- 3571
Euting, J. 914, 915
Fabietti, U. 918, 919
Fadil, I. el- 27o3
Faheem, M.E. 92o
Faisal, K. al- 921
Faisal, M. al- 922, 923
Fakieh, O.A. 925
Faleh, N.A. al- 926
Falk, A. 927
Fallatah, I.M. 928
Fallon, N. 929, 93o
Falls, C.B. 931, 1967
Fanning, L.M. 932
Farag, W. 937
Faraj, A.H. 938
Fargues, P. 939, 94o
Farid, M.A. 941
Faris, A. el- 942
Faris, B. 943
Farmer, L. 944
Farmer, R.N. 945, 946
Farnworth, J. 537-539, 947-95o
Farouk-Sluglett, M. 3oo1
Farouky, S.T. 2o38
Farra, T.O.M. el- 951
Farsy, F.A.-S. al- 5o2, 952, 953
Faruki, M.T. el- 954
Faruqi, Z.-H. 955
Fathy, H. 956-958
Fayez, M. al- 1968
Felemban, A.A.H. (auch: A.H.S.) 959, 96o
Fellmann, W. 961
Feng, M. 3576
Feoktistov, A. 962
Ferguson, N. 963
Ferret, P.V.A. 964, 1o4o
Ferris, H.J. 965
Fiander, W. 966

Fiar, M.H. al-   967
Fiedler, V.   968
Field, H.   969-972
Field, M.   973-976
Field, P.   977
Fields, J.H.   978
Filali, M.   979
Filemban, S.N.   3577
Finati, G.   980
Finch, E.   981
Fingar, P.   982
Finielz, C.I.   983
Finnie, D.H.   984, 985
Finnie, R.   986
Fisher, S.N.   991
Fisher, W.B.   992, 993
Fitchett, J.   994
Fleischer, H.L.   995
Flint, J.   320
Floyd, B.N.   2726
Fogel, M.M.   997
Foley, J.A.   998
Forbes(-McGrath), R.   999, 1000
Forde, C.D.   1001
Forder, A.   1002-1004
Forslund, B.   3578
Forsythe, D.W.   1006
Foss, M.   1007
Fouad, M.H.   1008
Fougerouse, D.   3579
Frade, F.   1009-1011
Fraga, R.   1319
Franck, P.G.   725
Frankel, G.S.   1017
Franzmathes, F.   1018, 1019
Freeth, Z.   1020-1023, 3435
Fresnel, F.   1024, 1025
Fridolin   1026
Friedemann, J.   1027
Friedman, I.   1028, 1029

Frith, D.E.   1030, 1031
Frood, A.M.   1033, 1034
Fürstenmühl, R. von   1035
G.: s. Gobée, E.
Gabriel, E.   1036, 1037
Gälli, A.   1038
Gahtani, T.M.S. al-   1039
Gal, Y.   1110
Galinier, J.G.   964, 1040
Gamie, M.N.   1041, 3665
Ganoubi, A.I.   1042
Garabedian, G.A.   2039
Garbe, C.W.   1044
Gaspard, J.   1045
Gaudefroy-Demombynes, M.   1046
Gaulis, B.-G.   1047
Gaury, G. de   1048-1055
Gautier, É.-F.   1056, 1057
Gawawi, H.   1058
Gelpi, A.P.   1059-1065
George, A.R.   1067, 1068
Gerakis, A.S.   1069
Gérard, C.   3580
Gérard, I.   3580
Gérard, J.   3580
Gerken   1070
Germanus, J.   1071, 1072
Gervais-Courtellemont, J.-C.   1073
Ghaith, A.   1074
Ghamdi, Abdulaziz S. al-   3581
Ghamdi, Abdullah A. al-   1075
Ghamdi, Abdulrahim M. al-   1076
Ghamdi, Mohammed A.H.   1077
Ghamdi, Mohammed S.D. al-   3582
Ghanayem, M.A.   1078
Ghandoura, A.H.   1079
Ghannam, M.A. el-   402, 1224, 1225
Ghoneimy, M.T. el-   523, 1080
Ghorban, N.   1081
Giannini, A.   1082-1085

Gil Benumeya, R. 1o87
Gilbert, A.L. 1o88
Giles, M.L. 3583
Gillen, F. 1923
Gillen, S.C. 1923
Gingrich, A. 85o
Ginsberg, A. 2o91
Girvin, E. 1o89
Gismann, A. 55
Glick, L.A. 1o91
Glubb, F. 1o92
Glubb, J.B. 1o93-1o97
Gobée, E. 1o98, 1o99
Gobineau, J.A. de 11oo
Goedde, H.W. 11o1
Goeje, M.J. de 11o2
Goellner, W.A. 11o3, 11o4
Götz, W. 11o5
Goglia, L. 11o6
Gohaidan, M.S.S. 11o7
Gold, F.R. 11o8
Goldberg, D. 11o9
Goldberg, J. 111o-1112
Goldrup, L.P. 1113
Goldziher, I. 1114
Gómez Aparicio, P. 1115
Gondrecourt, A. de 1116
Gordon, E. 1117
Gordon, N. 1118
Gormly, J.L. 1119
Gosaibi, G.A. al-: s. Algo-
    saibi, G.A.
Gottheil, F.M. 1121
Gouilly, A. 1122
Gouin, É. 3584
Gouldrup, L. 1123
Goy, R. 1124
Gräf, E. 1125, 1126
Gräfen, R. 1127
Grafftey-Smith, L. 1128

Grandguillaume, G. 1129, 113o
Grant, C.P. 1131
Grayson, B.L. 3585
Greenidge, C.W.W. 1141
Greenip, W.E. 1142
Griessbauer, L. 1143
Griffith, W.E. 1144
Griffith-Jones, S. 1145, 1146
Griggs, L. 1147
Grimaldi, F. 1148
Grindley, W. 1149
Grobba, F. 115o
Grohmann, A. 1151-1155
Grunwald, K. 1157
Guaiz, S.A. al- 1158
Guarmani, C. 1159
Güldner, W. 116o
Guellouz, E. 1161
Günthardt, W. 1162
Gurashi, H.D. al- 1165
Guthe, H. 1166
H., A.S. 1167
Haass, R. 3586
Haast, A. 1498
Habeeb, M.M.S. al- 1168
Habib, J.S. 1169
Hablützel, H. 117o
Hablützel, R. 1171
Habshush, H. 1172
Hacker, B. 3587
Haddad, G.A. 1173, 3588
Haddad, M.S. al- 1175
Haddad, N.A. 2267-227o, 3o14
Haddadeen, M.S. 1176
Hadie, M.A. 1177-118o
Hadj, E. el- 1181
Hafidh, N. al- 1182
Hafiz, F.A. 1183
Hafiz, O.Z. 1184
Hafiz, T.K. 1185

Hagra, H.H.   3589
Hague, B.C.   1186
Haidar, M.   1187
Haj, F.M. al-   1188
Hajrah, H.H.   1189
Hakim, M.H.   1190
Hakima, A.M.: s. Abu-Hakima, A.M.
Hakken, B.D.   1191
Halawani, A.W. el-   1192
Halévy, J.   1193
Halfpenny, A.F.   947
Hallam, H.M.   1194
Hallawani, E.A.-R.   1195
Halliday, F.   1196-1199
Halm, H.   1200
Halmos, E.E., Jr.   1201
Hamad, H.S. al-   1202
Hamadi, A.M.   1203
Hambleton, H.G.   1204
Hamdan, A.I.   1205
Hamdan, M.Z.   1206
Hamdan, S.A. el-   2645
Hamdan, Y.A. al-   1207
Hamid Daoud, S.A. el-   1208
Hamidi, A.S.   1209
Hamídulláh, M.   1210
Hamilton, C.W.   1211
Hamilton, J.A.   1212
Hamilton, J.G.   3243
Hamilton, P.   1213
Hammad, M.A.   1214
Hammoudeh, S.M.   1215, 1216
Hamouda, M.   3590
Hamza, F.   1217
Hanf, T.   1321
Hans, J.   1218-1220
Hansen, H.   1221
Hansen, T.   1222
Hansen, W.G.   1223

Harby, M.K.   1224, 1225
Hare, R.A.   1226
Hariri, H.B.   1227
Harrell, G.T.   1228
Harrington, C.W.   1229
Harrison, P.W.   1230-1237, 3591
Harsaghy, F.J., Jr.   1238, 1239, 3592
Harsham, P.   1240
Hart, P.T.   1241
Hartmann, M.   1242-1244
Hartmann, R.   1245-1251
Hartzell, M.E.   1252
Hasan, K.A.   1823
Hashagen, J.   1254
Hashe, A.M.   1255
Hassan, H.M.   1256
Hassanain, M.A.   1257
Hasson, R.C.   1258
Hatem, M.A.-K.   1259
Haupert, J.S.   1260, 1261
Hayes, S.D.   1262, 1263
Hayit, B.   1265
Hazard, H.W.   1266-1269
Hazzam Dawsari, F.S. al-   1270
Headley, R.L.   1271-1273
Heady, H.F.   1274
Hebshi, H.B. el-   1276
Heck, G.   3593
Hecker, M.   1277
Hecklau, H.   1278
Hejailan, S.   1279
Helaissi, A.S.   1280
Helmensdorfer, E.   1281
Helms, C.M.   1282
Henin, C.   1283
Hennig, R.   1284
Henninger, J.   1285-1294
Henry, J.C.   1295, 1296
Hentig, W.O. von   1297-1299

Herrick, A.B.   3361
Herzog, R.   13oo
Hess, J.J.   13o1, 13o2
Heykal, M.H.   13o3
Hibon, A.   1832
Hibshy, M.A.   13o4, 13o5
Hickson, L.   178o
Hidore, J.J.   13o7
Hilaissi, N. el-   13o9
Hilālī, T.D. al-   131o
Hill, A.G.   1311
Hirashima, H.Y.   1312
Hirst, D.   1313
Hirszowicz, Ł.   1314
Hirth, L.   11o1
Hitti, S.H.   1315
Hoagland, J.   1316
Hobday, P.   1317, 1318
Hobson, R.   1319
Hoel, L.   132o
Hörder, M.-H.   1321
Hötzl, H.   3594
Hoffman, I.E.   1111
Hogarth, D.G.   1322-1332
Holden, D.   1333, 1334
Holm, H.M.   1335
Holman, J.K.   1336
Holt, A.L.   1337
Holzhausen, R.   1338
Homeyer, B.   1oo
Hommel, F.   11o2
Hoog, P.H. van der   1339, 134o
Hope, W.E.S.   1341
Hoppe, H.H.   11o1
Hopper, H.   1342
Hopwood, D.   1343
Horack, M.M.   2985
Hoskins, H.L.   1344-1346
Hosni, S.M.   1347
Hottinger, A.   1348-1359

Hourani, A.   136o
Houri, I.S. el-   27o3
Howard, B., Jr.   1361
Howarth, D.A.   1362, 1363
Hoye, P.F.   1364, 1365
Hoyt, C.K.   1366
Hoyt, M.P.   1367
Huber, C.   1369, 137o
Hubert-Rodier, L.   3595
Hudson, M.C.   3596
Hüber, R.   1371-1373
Hübner, G.   1374
Hughes, H.   1375
Hulais, H.Y.   1376
Hull, C.   1377
Humaidan, S.H.   1378
Humphreys, R.S.   1379
Hurewitz, J.C.   138o, 1381
Hurren, B.J.   1382
Husain, Z.A. al-   3597
Hussain, M.K.   1383
Hussain, Z.   1384, 1385
Hussaini, A.   1386
Hussayen, M.A. al-   1387
Hussein, F.   1388
Hussein, T.   1389
Ibrahim, A.A. al-   1391
Ibrahim, S.E.   1392-1394
Idries Shah, H.   1395
Ikin, E.W.   2o16
Ilam, H.M.D.M.   1396
Ileri, M.   1397
Inayatullah, S.   14o5
Inshaullah, M.   14o6
Ioffe, A.Y.   1414
Iqbal, M.   1415
Irving, T.B.   1417
Irwin, E.   1418
Isa, A.S.   1419
Isaac, E.   142o

Iseman, P.A.   1421
Iskandar, M.M.   1422
Ismaeel, A.U. al-   1423
Ismail, S.   4o2
Issa-Fullata, M.M.   1424
Issawi, C.P.   1425, 1427
Issawi, H.F. el-   1428
Ives, G.O.   1431
Iwanowitsch, J.   184
Izzard, M.   1432
Izzedine, C.   1433, 1434
Jabber, P.   2356, 3598
Jabbur, J.   1435
Jabr, S.M.   1436
Jacob, H.F.   1437
Jacobs, A.   1438
Jadallah, S.M.   1439
Jäschke, G.   144o
Jafary, A.A. al-   1441
Jakubiak, H.E.   1442
Jalabert, L.   1443
Jallal, A.A. al-   1444
Jamjoom, M.A.   1445
Jamjoum, A.S.   1446
Jammaz, S.I.   1447
Jansen, M.E.   2254
Jarrar, G.   1448
Jarvis, C.S.   1449
Jasir, A.S. al-   145o
Jaussen, J.A.   1451
Jazairi, M.Z. al-   1452
Jeandet, N.   1453
Jeffery, A.   1455
Jerash, M.A. al-   1456, 1457
Jiabajee, N.A.   1458
Job, C.   3594
Johany, A.D.   146o
John, H.J.   3361
Johns, R.   1334, 1461
Jomard, E.F.   1463, 1464

Jomier, J.   1465, 3396
Jones, K.W.   1466, 1467
Jong, F. de   1468
Jong, G.E. de   1469
Jordan, A.A., Jr.   147o
Jouin, Y.   1471
Jovelet, L.   1472
Jum'ah, 'A.S.   1475
Jung, E.   1476, 1477
Jung, H.   1478
Jurji, E.J.   3245
Juynboll, G.H.A.   1479
Juynboll, T.W.   3599
Jwaideh, Z.E.   148o
Kabbaa, A.S.   1481
Kaddoura, M.   1482
Kadhim, M.   1989
Kadi, M.A. al-   1483
Käselau, A.   1484
Kahlenberg, C.   1485
Kaïdi, H.   1486
Kaikati, J.G.   1487, 1488
Kamal, Ahmad   1489, 149o
Kamal, A.H.   1491
Kamerbeek, G.J.   36oo
Kamookh, A.A. al-   1492
Kampffmeyer, G.   11o2
Kanoo, A.L.   1493
Kanovsky, E.   1494-1497
Kapoor, A.   1498
Karabuda, B.   36o1
Karabuda, G.   36o1
Karout, Z.I.   1499
Kashmeeri, B.O.   15oo
Kashmeeri, M.O.   15o1
Katakura, M.   15o2-15o4
Katanani, A.K.   15o5
Kattan, R.A.S.   36o2
Kay, E.   15o6
Kay, S.   15o7-151o, 3669

Kaylani, H.M. al- 289, 1511, 1512
Kayser, M. 1513
Kazem Zadeh, H. 1514, 3603
Kazemzadeh, I.: s. Kazem Zadeh, H.
Kazmi, Z.A. al- 1515
Keane, J.F. 1516-1518
Kebell, M. 2471
Kedourie, E. 1519-1522
Keiser, H. 1523, 1524
Kelidar, A.R. 1525, 1526
Kelly, J.B. 1527-1535
Kelly, K. 1536
Kennedy, E.M. 1537, 2356
Kershaw, R.M. 1539
Keun de Hoogerwoerd, R.C. 1540
Khadduri, M. 1541, 1542
Khadra, O.A. 1543
Khalaf, A.A. 1544
Khalil, M. 1545
Khammas, M. 1546
Khan, G.A. 1547
Khan, M.A.A. 1548
Khan, M.A. Saleem: s. Saleem Khan, M.A.
Khan, Mohammad S. 1549, 2252
Khan, Mu'īnuddīn A. 1550
Khashoggi, H.Y. 1551
Khatib, Abdel Basset el- 1552
Khatib, Abd al-Hamid al- 1553
Khatib, Abdul Rahman 1554
Khatib, M.F. el- 1555
Khattab, M.K. 1556
Khayat, A.A. 1557
Khedaire, K.S. al- 1558
Kheirallah, G. 1559, 1560
Khudr, A.S. 1561
Khurshid, Z. 1562
Khuthaila, H.M. 1563

Kiernan, R.H. 1564
Kimball, S.T. 1566
Kimche, D. 1567
Kimche, J. 1568, 1569
King, G.R.D. 1570-1575, 3604
King, M.C. 1062
King, R. 457, 1576
Kirazian, H. 1758
Kirchner, R.A. 1759, 2291
Kirk 1030
Kirk, George E. 1760
Kirk, Grayson L. 1761
Kisnawi, M.M. 3608
Klare, M.T. 1762
Klebnikoff, S. de 1763
Kleihauer, E. 1101
Klein, P. 3254
Kleist, H. von 1764, 1765
Klemme, M. 1766
Kley, H.D. 3609
Klingmüller, E. 1767, 1768
Klippel, E. 1769
Klopp vom Hofe, P. 1770
Knauerhase, R. 1171-1777,3610,3672
Knightly, P. 1778
Kobori, I. 1779
Koch, K.-F. 1780
Koester, D. 1785
Kohn, H. 1786-1788
Kohne, E. 1101
Kolko, G. 1789
Koner, W. 1790
Konzelmann, G. 1791, 1792
Kopf, W. 3611
Korany, B. 1793, 1794
Kordi, K.A.K. 1795
Kornrumpf, H.-J. 1796, 1797
Koszinowski, T. 141,1798-1805,3674
Kourouklis, S.D. 1807
Krajewski, L. 1808

Krause, W.W.   1809, 1810
Krenkow, F.   1811
Kroon, A.   862
Krüger, H.   1812, 1813
Krüger, K.   1814
Kruse, H.   1815
Kruyt, J.A.   1816
Kuhaimi, S.A.A. al-   1817
Kumar, R.   1818, 1819
Kuniholm, B.R.   1820
Kurban, A.K.   1984
Kurd, A.A. el-   1821
Kurdi, M.A.M.   1822
Kurdi, T.M.K.   1823
Kurian, G.T.   1824
Kussmaul, F.   726
Kutschera, C.   1825
Kuwaiz, A.I. el-   1827
L.   1828
Labaki, B.   1829
Labban, S.A.   1830, 1831
Labonne, M.   1832
Lacey, R.   1833, 1834
Lacher, J.   1835
Lackner, H.   1836
Lamare, P.   1839
Lambsdorff, O.   1840
Lamer, M.   1841
Lammens, H.   1842, 1843, 2999
Lancaster, W.O.   1844-1846
Landau, J.M.   1847, 1848
Landsberg, H.H.   691
Landsberg, M.   1849
Lange, W.   1850, 1851
Langefeld-Wirth, K.   1852
Langella, V.   1853
Langer, S.   1854
Lanier, A.R.   1855
Laoust, H.   1856, 3612
Larson, T.J.   1923

Lasky, H.   1861
Lateef, A.   1862, 1863
Lateef, N.A.   1864, 1865
Laurent, F.   1866
Law, J.   1867
Lawrence, T.E.   1868-1871
Lawton, J.   1319, 1872-1874, 3613
Leachman, G.E.   1875, 1876
Leaman, O.   1877
Lean, O.B.   1878
Leatherdale, C.A.   1879
Lebkicher, R.   1881-1884
Lebling, B.   1885
Lederer, A.   1888
Lee, E.   1890
Lee, J.F.K.   520, 1891
Lees, B.   1892
Leeson, H.S.   3614
Lefebvre, T.   1893
Lehmann, H.   1894, 2016
Leiden, C.   438
Leipold, L.E.   1898
Leithead, C.S.   1192
Lemaud, C.   1900
Lenczowski, G.   1902-1904
Lesch, W.   1908
Leveau, R.   1910
Levi Della Vida, G.   1911
Levy, P.   1913
Lewis, B.   3396
Lewis, C.C.   1914
Lewis, W.H.W.   1915
Liebesny, H.J.   1916
Linabury, G.O.   1918, 1919
Linde, G.   1920, 3615
Lippens, P.   1922
Lipsky, G.A.   1923, 1924
Lipsmeier   1070
Little, A.D.   1925
Littman, J.   1926

Littmann, E.   1927, 1928
Litwak, R.   1929
Lloyd, E.M.H.   1930
Lochner, R.K.   1933
Loewenthal, N.P.   1934
Loir, R.   1935, 1936
Lonchampt, J.   1937
Londres, A.   1940
Long, D.E.   1941-1946, 2938, 3616
Longrigg, S.H.   1947, 1948
Looney, R.E.   1949, 1950
Lorimer, J.G.   1951
Loustaunau, C.A.   1952
Loutfi, Z.I.   1953
Love, M.   1319
Lowth, G.T.   3617
Lucas, I.A.M.   1954
Lundbaek, T.   1955
Lunde, P.   891, 2254
Lupien, J.P.   1956
Lustig, M.W.   103
Lyautey, P.   1957
Lynn, R.   2091
Maaskola, P.   1958, 1959
MacDonald, C.G.   1960, 1961
Macdonald, D.B.   1962
MacDonald, W.N.   1963
MacIntyre, R.R.   1964
Mackie, J.B.   1965
Maclean, H.W.   1966
MacMunn, G.F.   1967
Madani, A.R. al-   1968
Madani, G.O.   1969
Madani, M.O.   1970
Madani, N.O.   1971
Madi, M.A.F.   1972
Mahassni, H.   1973
Mahmoud, A.   1974
Mahnke, H.-J.   1975

Makin, W.J.   1978
Makari, J.G.   3618
Makki, M.S.   1979
Makky, G.A.W.   1980, 1981
Makoshi, A.A.-R.   1982
Malaika, Y.   1983
Malak, J.A.   1984
Malik, A.A.   1985
Malik, S.A.   1986, 1987
Mallakh, R. el-   1988-1990, 3368
Malleess, S.M. al-   1992
Mallmann, W.   888
Malmignati, D.   1993
Malone, J.J.   1994-1997
Maltzan, H. von   1998, 1999
Mamméri, H.   2000
Mandaville, James   2001-2003, 2565
Mandaville, Jon   2004
Mandily, O.A.   2189
Mangat-Rai, C.R.   2005
Mani, M.A. al-   2006
Mankour, N.   2009
Manners, I.R.   2010
Manning, R.   3461
Mansfield, P.   2011
Mansour, H.O.   2013
Mansur, A.K. (pseud.)   2014
Mantynen, H.   2733
Manuie, M.A.   2015
Maranjian, G.   1894, 2016
Mares, W.J.   2462
Marett, W.C.   2018
Margoliouth, D.S.   2019
Marimont, H. (Pseud.)   2020
Marks, M.M.   2023
Marlowe, J.   2024
Marotz, G.   2025
Marr, P.A.   1251
Marshall, D.   2026-2028

Marsouqi, H.A. al- 2029
Marston, T.E. 2030
Martan, S.S. 2031
Martin, H.E. 3619
Martin, L. 2032
Marx, E. 3620
Masia, M. 2033
Mason, A.D. 2034
Massignon, L. 2035, 2036
Mathieu, A.L. 2038
Matossian, R.M. 2039
Matouk, A. 2040
Matthews, A.T.J. 1923
Matthews, C.D. 2041, 2042
Maull, H.W. 2043, 2044, 3621
Maunsell, F.R. 2045, 2046
Maurizi, V. 3622
Mazroa, S.A. al- 2047
Mazroe, H.M.H. al- 2048
Mazyed, M.I. al- 2049
Mazyed, S.M. al- 2050
McCarthy, J. 3623
McComb, D.E. 408, 2267, 2269, 3014
McConahay, M.-J. 2051
McDonald, J. 2052
McDonald, W.N. 2053
McGregor, R. 2054
McHale, T.R. 2055-2057
McIrwin Abu-Laban, S. 88
McKenna, J. 2058
McLoughlin, A. 2059
McMaster, B. 2060
McMillan, W.M. 2061
McNiel, J.R. 2062-2066, 3624
Mead, W.R. 3360
Mecci, M.S. 2068-2070
Mech, P. 2071
Medawar, G.S. 289, 2072-2074, 3625

Meeker, M.E. 2076
Meglio, R. di 2077-2079
Meigs, P. 2080
Meinertzhagen, R. 2081
Mejcher, H. 2082, 2083
Melamid, A. 2084-2088
Melibary, A.R. 2089
Melikian, L.H. 2090-2093
Melka, R.L. 2094
Memon, A.F. 2095
Memun Abul Fadl, S. 2096
Mengin, F. 2097
Merkel, G. 3626
Merriam, J.L. 2098
Mertens, R. 2099
Mertz, R.A. 2100
Messer, E. 2101
Messerschmidt, E.A. 2102
Messiha, S.A. 2103
Metcalf, J.E. 989
Metta, V.B. 2105
Meulen, D. van der 2106-2115
Meyer, A. 2116
Meyer, W.C. 2117
Meyer-Ranke, P. 2118
Middleton, D. 2125
Midhat, A.H. 2126
Mihdar, A. al- 2128
Mikesell, R.F. 2129, 2130
Mikusch, D. von 2131-2133
Milburn, W. 2134
Miles, S.B. 2135
Miller, A.D. 2137
Milyani, H.M. 3627
Mineau, W. 2138
Mishari, H. 2140, 2141
Mishlawi, T. 2142, 2143
Mittelmann, G. 395, 2145-2148
Mittwoch, E. 2149
Mixon, J.W. 2150

Mizjaji, A.D. al-   2151
Moberg, A.   2152
Moe, L.E.   2153
Mofti, F.A.   3628
Mohammed, S.K. el-   2154
Moliver, D.M.   2155, 2156
Momberger, M.   2157
Moneef, A.A.   2159
Moneef, I.A. al-   2160
Monfreid, H. de   2161
Monheim, C.   2162
Monroe, E.   2163
Montada, W.   2146
Montagne, R.   2164, 2165
Monteil, V.   2036
Montgomery, J.A.   2166
Montgomery, P.A.   2167
Moore, F.L., Jr.   2168
Moran, T.H.   2169
Morano, L.   2170
Mordtmann, J.H.   2171, 2172
Moritz, B.   1102, 2173, 2174
Morris, J.   2175
Morrison, O.F.   2176
Morrison, W.D.   2928, 2929
Morsey, K.   2177
Morsi Abbas, A.   2178
Morsy Abdullah, M.   2179
Mortimer, E.   2180
Moshaikeh, M.S.H.   2181
Mostyn, T.   2183
Moulla, E.A. al-   2185
Mourad, F.   2186, 3629
Mourant, A.E.   1894, 2016
Mousa, S.   2187, 2188
Moussali, M.S.   2189, 2190
Moustapha, A.F.   2191
Muarik, S.A. al-   2192
Mücke, H. von   2193
Mueller, J.H.   2194, 2195

Müller, V.   2196
Mughram, A.A.   2197
Muir, J.D.   2198
Mulla, M.A.   2199
Mulligan, W.E.   1272, 2200-2202, 2555, 2559
Munif, H.M.   1554
Munro, J.   2203
Murphy, C.J.V.   2204
Murr, K.   2205
Murray, E.S.   408, 2267, 2268, 3014, 3630, 3676
Murshid, T.A.   2206
Musil, A.   2207-2215, 3631
Mussa, S.I.   2216
Mustafa, A.   1063
Mustafa, Z.   2217, 2218
Mutawia, H.H.   2219
Myers, H.E.   716, 717
Myers, R.J.   2220
Mylrea, C.S.G.   2221, 2222
N., A. de   3632
Nabti, F.G.   2223
Nader, A.   2224
Nafa, M.A.   2225
Nagel, T.   2226
Nahari, A.M. al-   2230
Naiem, A.M. el-   2231
Nairab, M.M.   2232
Najai, A.M.   2233
Nakhleh, E.A.   2234, 2235
Nallino, C.A.   2236, 2237
Nallino, M.   2238-2243
Narayanan, R.   2244, 2245
Naseer, A.A.   2246
Nasr, K.S.   2247
Nasri, H.Y.   2248
Nassar, F.M. al-   2249
Nasser, S.A.   2250
Natto, I.A.   2252

Nawwab, I.I.   2254
Nazer, I.S.   2255
Neaim, H.A. al-   2257
Neetix, H.W.   2258
Nehme, M.G.   2259
Neimans, R. von   2260
Nettelbeck, J.   3633
Neumann, R.G.   2262
Niblock, T.   2265
Nichols, R.L.   2267-2270, 3014, 3676
Niebuhr, B.G.   2271
Niebuhr, C.   2272, 2273
Nieuwenhuijze, C.A.O. van   2274
Nifay, A.M. al-   2275
Nimatallah, Y.A.-W.   2276
Nixon, R.W.   2277, 2278
Noaim, A.   2279
Noel-Brown, S.J.   2280
Nölker, H.   184
Nolde, E.   2281
Nollet, R.   2282
Nolte, R.H.   2283, 2284
Noris, J.   2285
Norton, M.   2286
Noty, A.A. al-   3589
Novik, N.   2288
Nunè, E.   2289
Nyrop, R.F.   2291-2293
Obaid, A.S. al-   2295
Obojski, R.   2298
Ochel, W.   2299, 2300
Ochsenwald, W.L.   2301-2310
Ocwieja, F.A.   2311
Odell, R.M.   2312
O'Donnell, P.D.   2313
Önder, Z.   2315, 2316
O'Hali, A.A.   2319
Olayan, S.S.   3635
Olsen, G.R.   2322
Omair, S.A.   2323

Omar, W.   3636
Omer, A.   3538
Omran, A.N. al-   2324
Oppenheim, M. von   2327
Orlowski, J.   1101
Os, J. van   2328
Osaimi, M.S.M.J. al-   2329
Osman (auch: Othman), O.A.   2330, 2331
Owens, M.V.   2332
Ownby, P.   2333
Ozoling, V.   2334, 2335
Page, S.   2336
Palgrave, W.G.   2337-2340
Pape, H.   2341
Paréja, F.   2342-2344
Paret, R.   2345
Park, T.W.   2346
Parker, J.B., Jr.   2347-2353
Parry, M.S.   2354
Parssinen, C.   2355
Pascal, A.   2356
Passama, J.   2357
Pastner, C.M.   2358
Patai, R.   2359
Pauling, N.G.   2360, 2361
Pechel, J.   2362
Peck, M.C.   2363-2366
Pelly, L.   2367-2369
Pendleton, M.   2370
Penrose, E.   2372
Peppelenbosch, P.G.N.   2373
Perera, J.   2374, 2375
Peretz, D.   2376
Pershits, A.I.   2377
Pesce, A.   2378, 2379
Pesenti, G.   2380
Petit-Laurent, J.   2381
Peucker, K.   2382
Peursem, G.D. van   2383-2385

Philby, H.St.J.B. 2386-2427
Philipp, H.-J. 2428-2431, 3637
Phillips, T.O. 3638
Phoenix (pseud.) 2443
Pickens, C.L. 3o45
Pieper, W. 2444, 2445
Pierre, A.J. 2446
Pilar Serrano de Lababidy, M. del: s. Serrano de Lababidy, M. del Pilar
Pinto, O. 2448
Pirenne, J. 2449
Piscatori, J.P. 245o, 2451
Pivka, O. von 2452
Planagan, A.J. 2453, 2454
Plant, M. 21o4
Plas, C.O. van der 2456
Plass, J.B. 2457
Plaut, S.E. 462
Plüddemann, M. 2458
Pönicke, H. 2459, 246o
Polk, W.R. 2462
Pollitzer, R. 2463
Pollog, C.H. 2464
Porath, Y. 2465
Prax 2467, 2468
Preece, R.M. 2469
Presley, J.R. 247o-2472
Prest, M. 2473
Prisse d'Avennes, E. 2475
Pritzke, H. 2476
Prochazka, T., Jr. 2479
Pröbster, E. 248o
Puin, G.-R. 2485, 2486
Putman, J.J. 2488
Puttendörfer 1o7o
Pye, B. 2489
Qadi, S.H. 249o
Quandt, W.B. 2491, 2492
Qubain, F.I. 2493

Quinlan, T. 2495
Quota, B.M.N. (oder: B.N.M.) 2496
Quotah, M.M.N. 2497, 2498
Quraishi, H.E.A. al- 2499
Raddady, M.M. al- 25oo
Rafei, N. 25o1, 25o2
Ragette, F. 25o3
Rahnama, S. 25o4
Raieky, M.I. al- 25o5
Rajehi, M.O.R. 25o7
Ralli, A. 25o8
Ramm, H. 25o9-2511
Rand, C.T. 2512
Raoof, A.H. 2513
Rasheed, M.S. al- 2514
Rashid, S.A.A. al- 2517
Rásky, L. 2518
Raswan, C.R. 252o-2522
Rathjens, C. 2523-2527
Raunkiaer, B. 2528-2531
Rautenbach, L. 2532
Rawaf, O.Y. al- 2533
Rayess, F. 2534
Raymond, J. 2535
Razvi, M. 2536, 2537
Reeser, R.H. 869
Reeves Palmer, M. 2538
Regli, E. 589
Rehatsek, E. 254o
Reichert, H. 2541
Reintjens, H. 2544
Reisch, M. 2545
Reissner, J. 2546-2548
Rendel, George 2549
Rendel, Geraldine 255o
Rentz, G. 1272, 1882, 1884, 2551-258o, 3639
Reynolds, B. 2581-2583
Ricci, C. 2584

Rice, A.G.   2585
Richards, J.M.   2586, 2587
Ridda, M.   2588
Riedl, H.   85o
Rifaï, T.   191o
Rihani, A.F.   2589-2598
Riley, C.L.   2599
Ritter, C.   26oo
Ritter, W.   26o1-26o7
Rivoyre, D. de   26o8
Robbins, R.R.   2611
Robinson, W.I.   947
Robson, J.   2613
Roches, L.   2614
Rochet d'Héricourt, C.E.X.
   2615-2618
Rock, A.   2619
Roeder, A.   262o
Rörig, H.   2621
Rösel, W.   2622
Roff, W.R.   2623
Rohlfs, G.   2624
Roloff, M.   2625
Ronall, J.O.   1157, 2626, 2627
Rondot, P.   2628-2631
Roosevelt, K.   2632
Rosen, G.   2633
Rosen, S.J.   2356, 2634
Rosenfeld, H.   2635
Rosenthal, E.   2636
Ross, H.C.   2637, 2638
Rossi, E.   2639-2641
Rousseau, J.B.L.J.   2642-2644
Rowley, G.   2645
Royce, C.H.   1923
Rubelli, L.   2649
Rubin, B.   265o, 2651
Rüppell, E.   2652, 2653
Rugh, W.A.   2654-2659
Russel, S.   266o

Rustow, D.A.   2661, 2662, 3678
Rutter, E.   2663-2669
Rutter, O.   267o
Ruwayha, F.A.   1554
Ryan, A.   2671
Ryckmans, G.   2672-2674
Ryder, W.   2675
Rypka, J.   2676
S., E.   2677
Saab, R.   2678
Saad, E.O. al-   2679
Sa'ad, N.M.   268o
Saade, R.F.   2681
Saadi, M. el-   1954
Saadûn, M.   2682
Saaty, M.A.   2683
Sabaa, H.M.A.   27o3, 3538
Sabab, A.A.A. al-   2684
Sabbagh, G.   2685
Sabban, A.A.S. al-   2686
Sabini, J.   2687, 2688
Sablier, É.   2689
Sabra, N.   269o
Sabri, M.M.   2691
Sabry, Z.I.   2692
Sachar, H.M.   2693, 2694
Sacher, R.   2695
Sadek, F.M.   2696
Sadhan, A. al-   2697
Sadik, F.A.   2698
Sadleir (auch: Sadlier), G.F.
   2699, 27oo
Safadi, A.I. al-   27o1
Saggaf, A.A.   27o2
Saha, N.   27o3, 3538
Sahwell, 'A.S.   27o4
Said, A.H.   27o5
Saif, J.A.   27o6
Saif, S.M. al-   27o7
Sakkar, S. al-   27o8

Salah, S.   2709, 2710
Salah, Y.S.   2711
Salam, E.M.A. el-   2835
Salam, Y.M.   2712
Salamé (auch: Salameh), G.
    2713-2718
Saleem Khan, M.A.   2719-2721
Saleh, F.S. al-   2722
Saleh, M. al-   3640
Saleh, M.A. eben   2723
Saleh, Nabil A.   2724
Saleh, Nasser O. al-   2725,
    2726
Saleh, Nassir A.   2727, 2728
Salem, F. al-   2729
Salem, M.S. al-   2730
Salibi, K.S.   2731
Salloom, H.I. al-   2732
Sammak, A.   2733
Samman, N.H.   2734
Sampson, A.   2735, 2736
Sams, T.A.   2737
Samuel, A.P.W.   2703, 3538
Sanger, R.H.   2739, 2740
Santamaria, R.   3641, 3642
Santani, A.   2741
Sardar, Z.   2743, 2745
Sarhan, S.   2746
Sarkis, R.   1181
Saud, M.A.T. al-   2747
Sauer, G.   2829
Sauer, H.D.   2830
Saunders, H.H.   2831
Saussey, E.   2832
Savignac, R.   1451
Saxen, A.   2833, 3496
Sayed, A.M.M. el-   2834
Sayed, M.E. el-   3028
Sayigh, A.A.M.   2835
Sayigh, Y.A.   2836, 2837

Sayrafi, Y.H.   2838
Sbit, A.-R.S. as-   2006
Scasso, C.   2840
Schaade, A.   1102
Schacht, J.   2841
Schamberger, R.   407
Scharabi, M.   2842-2844
Schechterman, B.   2845
Scheltema, J.F.   2846
Schemeil, Y.   2847
Schleifer, J.   2848-2851
Schliephake, K.   2852-2855
Schmid, P.   2856
Schmidt, W.   2857
Schmidt-Pathmann, W.   2858
Schmitz-Kairo, P.   2859, 2860
Schnabel, A.   2861
Schnadelbach, R.T.   1536
Schnepp, B.   2862
Schoedl, P.F.   2863
Scholtyssek, S.   104
Scholz, F.   2865
Schott, G.   2867
Schreiber, F.   2868
Schütze, R.A.   2869-2871
Schulz, G.   2853, 2872
Schulze, J.   2285
Schuster, W.   2873
Schwarz, E.   184
Schweizer, G.   2874, 2875
Scott, R.W.   2877, 2878
Seabrook, W.B.   2881
Seaman, B.W.   182, 2883
Sebai, Z.A.   2703, 2885-2887,
    3538
Seering, R.   2888
Seetzen, U.J.   2889-2893
Seflan, A.M. al-   2894
Seiz, G.   3644
Sékaly, A.   2896

Selow-Serman, K.E.   2897
Seoudi, M.A.   2898
Serageldin, I.   3646
Sergeant, R.B.   2899
Serrano de Lababidy, M. del Pilar   2900
Sertoli Salis, R.   2901
Seton-Williams, M.V.   2902
Severino, D.   2905
Sha'afy, M.S. al-   2906
Shaafy, M.S.M. el-   2907, 2908
Shadly, R.A.   2909
Shadukhi, S.M.   2910
Shaffer, R.   2911
Shah, I.A.   2912, 2913
Shah, S.A.   2914
Shair, I.M.   2915
Shaked, H.   282, 2634, 2917-2919
Shaker, Farid A.   2189
Shaker, Fatina A.   2920
Shamekh, A.A.   2921-2925
Shamekh, M.A.R. al-   2926
Shami, I.A. al-   2927
Shamikh, M.A.-H. (oder: M.A.-R.) el-   3645
Shamma, S.   2928, 2929
Shanneik, G.   2930-2933
Sharabi, H.B.   2934
Sharbiny, S.U.T.   2935
Sharshar, A.M.   2936, 2937
Shaw, J.A.   2938
Shawly, A.T.   2939
Shea, T.W.   2940, 2941
Shebl, A.H.   2942
Sheean, V.   2943, 2944
Sheehan, E.R.F.   2945
Shehata, M.H.   2887
Sheikh, Abdulrahman Abdulaziz A.H. al-   2946

Sheikh, Abdul Ghafur   2947
Sheikh, Abid M.S.   2948
Sheikh, F.S. el-   2703
Sherbini, A.A. el-   2949, 2950
Sherbiny, N.A.   2951, 3646
Sheshsha, J.A.   2953
Shiber, S.G.   2954
Shilling, N.A.   2955-2957
Shinawi, A.A.K.   2958
Shirreff, D.   900, 2959-2962
Shoaib, M.S.   2964
Shobaili, A.S.   2965, 3647
Shobokshi, S.   2966
Shomrany, S.A. al-   2967
Shuaiby, A.M. al-   2968
Shukri, A.I.   2969, 2970
Shurrab, S.   2971
Shwadran, B.   2972, 2973
Sikandar Begam   2974
Siksek, S.G.   331
Silverfarb, D.N.   2975-2979
Sim, K.   2980
Simansky, N.   2981
Simmersbach, B.   2982, 2983
Simmons, A.   2984
Simmons, J.S.   2985
Simon, K.   2986
Simpich, F.   2987
Simpson, C.   1778
Sinclair, C.A.   449-452
Sindi, A.M.   2989
Sindi, S.B.   2990
Singer, S.F.   2991
Singh, H.   480
Siradj, A.   2994
Sisley, T.   2995
Skilbeck, D.   716, 717
Skinner, M.   756
Skornia, V.   2997
Slaugh, F.S.   2998

Slemman, H. (pseud.): s. Lammens, H.
Sluglett, P.   3oo1
Smalley, W.F.   3oo2
Smallwood, T.O.   33o3
Smeaton, B.H.   3oo3
Smith, Adam   3oo4
Smith, Alan M.   3oo5, 3oo6
Smith, G.R.   146
Smith, J.P.   1316
Smith, J.T.   3oo7
Smith, T.R.   3576
Smith, W.R.   3oo8
Smithers, R.   3oo9
Snavely, W.P.   878
Snodgrass, F.O.   237o
Snouck Hurgronje, C.   3o1o-3o12
Snyder, H.R.   3o13
Snyder, J.C.   4o8, 3o14, 3576, 363o, 3676
Sobaihi, M.A.   3o15
Sobaihi, S.M. al-   3o16
Solaim, S.A.   3o18, 3o19
Soliman, T.M.A. al-   3o2o
Souami, R.   3228
Soubhy, S.   3o21
Soulié, G.J.-L.   673, 3o22-3o25
Sowaygh, I.A. al-   3o28
Sowayyegh, A.H. al-   3o29
Sowygh, H. ibn Z. el-   3o3o
Spärck, R.   3o31
Sparrow, G.   3o32
Sparroy, W.   1547
Speetzen, H.   3o33
Spiegel, S.L.   3o34
Spies, O.   3o35
Spinks, W.   1o3o, 3o36
Spiro, S.   3o37
Sprenger, A.   3o38
Sprengling, M.J.   3o39

Squire, C.B.   3o4o
Stachels, R.   3o41
Staehelin, W.   3o43
Stanton, H.U.W.   3o45
Stauffer, T.B.   3o52
Steffen, H.   3o53
Steffen, W.G.   3o54
Stegar, W.   3o55
Stegner, W.   3o56
Stein, L.   3o57
Steinbach, U.   3o58
Steineke, M.   1882, 1884, 3o59
Stevens, G.P., Jr.   3o6o
Stevens, J.H.   3o61-3o65
Stevens, P.   3o66
Stewart, D.S.   3o67
Stifani, E.I.   3o68
Stilgoe, E.   3o69
Stitt, G.M.S.   3o7o
Stocking, G.W.   3o71
Stöhr, J.   2285
Stokes, B.R.   3o72
Storm, I.P.   3o74
Storm, W.H.   3o75-3o77
Storrs, R.   3o78
Stresemann, E.   3o8o
Strika, V.   3o81-3o85
Strippelmann, W.-D.   3o86
Strohm, J.L.   3o87
Stross, L.   3o88
Stuhlmann, F.   3o89
Subait, A.S.   3o9o
Sudairy, M.A. el-   3o92
Sudais, M.A.-S. al-   3648
Sullivan, R.R.   3o93
Sultan Jahan Begam   3o94
Suraisry, J.E.   3o95
Surur, R.S.   3o96
Sweet, L.E.   3o97, 3o98
Sykes, P.M.   31oo, 31o1

Szyliowicz, J.S. 3102, 3103
Taher, A.H. 3104, 3105
Tahir-Kheli, S. 3106
Tahtinen, D.R. 3107, 3108
Takieddine, L. 289
Takroni, M.H. 3109
Talhouk, A.M.S. 3110, 3111, 3649
Tallon, P. 3112
Tallqvist, K. 3113
Tamimi, F.M. al- 727
Tamisier, M. 3114
Tannous, A.I. 3115-3117
Tarabzune, M.R. 3119
Tarbush, S. 3120
Tarizzo, M.L. 3121, 3630
Tarkanian, L.E.H. 2167
Tartir, M. 3490
Tashkandi, M.O. 3122
Tashkandy, A.-J. 3123
Taton, R. 3650
Tawail, M.A. al- 3124
Tawati, A.M. 3125
Taylor, B. 3126
Taylor, D.C. 3127
Taylor, F.W. 3128
Taylor, J.W. 3129
Tayyeb, M.A. al- 3130
Thaddeus, J.D. 1984, 2039
Thesiger, W.P. 3156-3162
Thomas, A. 3166, 3167
Thomas, B.S. 3168-3172
Thomas, K. 3173
Thomas, R.A. 2985
Thubaity, A.M. al- 3175
Thubaity, K.K.M. al- 3176
Thum, W. 3177
Tibi, B. 3178
Tidjani, H. 1486
Tidrick, K. 3179

Tietjen, W.V. 3180
Timm, W. 3182
Titi, C. 3183
Tomiche, F.-J. 3184
Tomkinson, M. 3185, 3186
Tončić von Sorinj, D. 3187
Tončić von Sorinj, M. 3188
Topf, E. 3189
Torki, S. al-: s. Altorki, S.
Tothill, J.D. 3191
Totten, D.E. 3192
Towagry, A.M. 3193
Townsend, J. 3195, 3196
Toy, B. 3197, 3198
Toynbee, A.J. 3199-3202
Tracy, W. 3173, 3204-3212
Traini, R. 3213-3215
Trautz, M. 3220, 3360
Treskow, C. von 3221
Tresse, R. 3222
Trial, G.T. 3223
Troeller, G. 3225, 3226
Trunec, H. 3227
Turki, Abdel M. 3228
Turki, Abdulaziz M.I. al- 3229
Turki, M.I. al- 3230
Turky, H. 3231
Turner, L. 398, 3232-3236
Turner, W.O., Jr. 3237
Turpin, R.E. 3238
Tuson, P. 3239
Tweedie, W. 3240
Tweedy, O. 3241
Twitchell, K.S. 3242-3246
Twitchell, N.G. 3247
Uhlig, D. 3249-3254
'Uthaymin, A.-A.S. al- 3299
Uvarov, B.P. 3300, 3301
Uzayzi, R.Z. 3302
Vajda, A.J. de 3303

Valancogne, A.   3304
Valentia, G.   3305
Vander Clute, N.R.   3306
Vander Werff, L.L.   3307
Vasco, G.   3309
Vasile, R.C.   3310
Veccia Vaglieri, L.   3312, 3313
Vesey-Fitzgerald, D.F.   3317
Vianney, J.J.   3318, 3319
Vicker, R.   3320
Vickery, C.E.   3321, 3322
Vidal, F.S.   3323-3337
Vieille, P.   3338, 3339
Villiers, A.J.   3340, 3341
Vincent-Barwood, A.   3342, 3343
Vitray-Meyerovitch, E. de   3344
Voll, J.   3346
Voss, B.   3347
Vredenbregt, J.   3348
Waagenaar, S.   3349
Wabli, S.M. al-   3350
Wagner, D.   3351-3353
Wahba, H.   3354, 3355
Wahby, M.   520
Walford, G.F.   3358
Wallace, D.M.   3359
Wallin, G.A.   3360
Walpole, N.C.   3361
Walt, J.W.   3362
Walters, G.J.   3397
Ward, M.D.   2346
Ward, P.   3363
Ward, R.J.   3364
Ward, T.E.   3365
Warren, P.   3366
Wathen, A.L.   3243
Waterbury, J.   3368
Watt, D.C.   3369
Watt, W.M.   726
Waugh, T.   3370

Wavell, A.J.B.   3371
Wedin, W.E.   3372
Weeks, A.E.   3373
Weidauer, R.   3374
Weintraub, Sidney   3375
Weir, S.   3377
Weisl, W. von   3378
Weiss, L.   286, 3379-3381
Weiss, W.   3382-3385
Weissman, P.   281
Wells, D.A.   3386-3388
Wellsted, J.R.   3389, 3390
Welzel, C.S.   3391
Wenner, M.W.   3392, 3393
Wensinck, A.J.   3394-3396
Westlake, H.J.H.   3397
Wetzstein, J.G.   3398
Whayne, T.F.   2985
Wheatcroft, A.   3400
Wickens, G.M.   2899
Wiedensohler, G.   3402, 3403
Wieland, E.   3404
Wieland, T.K.   3361
Wien, J.   3405
Wiese, E.   3406
Wiese, J.   3407
Wiet, G.   3408
Wilkins, C.E.   3409
Willert, L.   3410
Williams, Keith   3411
Williams, Kenneth   3412
Williams, Maurice J.   3413
Williams, Maynard O.   3414, 3415
Willson-Pepper, C.R.   3416
Wilmington, M.W.   3417
Wilson, A.T.   3418
Wilson, J.C.   3419, 3420
Wilson, Rodney   3421, 3422, 3651
Wilson, Rodney J.A.   3423
Wilton, J.   3424, 3425

Winder, R.B.   3223, 3245, 3426-3429
Windfuhr, V.   3430
Windsor, G.W., Jr.   3431, 3432
Wingerter, R.B.   3433
Winkler, E.   3434
Winstone, H.V.F.   1o21, 1o22, 3435-3437
Wirth, E.   3438, 3439
Wissa-Wassef, C.   3442
Wissmann, H. von   726, 2106, 2524, 3443-3445
Wizarat, S.   3446
Wöbcke, M.   3593
Wohaibi, M.N.   3447
Wohlers-Scharf, T.   3448
Wohlfahrt, E.   3449, 345o
Wolff, G.   3451
Wong, S.S.   3453
Wood, J.G.   3454
Wood, Junius B.   3455
Woods, G.   3456
Wozelko, H.   3457
Wray, T.   3458, 3459
Wren, G.R.   346o
Wright, C.   3461
Wright, J.K.   3462
Wylde, A.B.   3463
Yackel, M.P.   3o59
Yamani, A.Z.   3464
Yaseen, N.H.   3652
Yeganeh, M.   1425
Yegnes, T.   2917-2919, 3465
Yodfat, A.   3467
Young, A.   3468
Young, Arthur N.   3469, 347o
Young, G.B.   3471
Young, H.W.   3472
Young, M.J.L.   838, 84o
Young, P.L.   3473, 3474

Young, R.   3475-3477
Younger, S.   3478
Zafer, M.I.   3479
Zahlan, R.S.   348o, 3481
Zaid, A.M.   3482
Zaid, Abdulla Mohamed al-   3483
Zaid, I.A. al-   3484
Zaidi, A.A. al-   3485
Zaidi, Z.H.   3486
Zaidy, A.K. al-   3487
Zakka, N.M.   3488
Zalatimo, Y.N.   3489
Zamel, A.   349o
Zawadzky, K.   3491
Zedan, F.M.   3492
Zehme, A.   3493
Zein, Y.T. el-   3494
Zillmann, G.   3495
Zimmermann, F.   3496
Zischka, A.   843, 3497
Zötl, J.   3594
Zoli, C.   3498, 3499
Zughoul, M.R.   35oo
Zulfa, M. al-   35o1, 35o2
Zwemer, S.M.   35o3-3512

NACHTRÄGE / ADDITIONS

Abd al-Hay, A.A. al-   3653
Abulfaraj, W.H.   3654
Abul-Nasr, S.E.   3655
Abu Yaman, I.K.   3656
Baltow, A.M.   3658
Beling, W.A.   3659
Blunt, W.S.   366o
Crecelius, D.   3662
Cuinet, V.   3663
Hanna, A.T.   3676
Heim, A.   3666

Heuglin, M.T. von 3667
Momenah, A.A. 3675
Rasch, B. 3677
Rustow, D.A. 3678
Smith, J. 3679
Todd, W.F. 3680

III PERSÖNLICHE HERAUSGEBER / EDITORS

Adams, M.   290
Ahmad, M.   86
Aitchison, C.U.   111
Alexander, S.S.   291
Alexander, Y.   2366
Amirsadeghi, H.   3586
Antonius, S.   167
Arberry, A.J.   2564
Ayoob, M.   319
Badawi, Z.   2744
Bakr, M.A.   2895
Bankes, W.J.   980
Barth, H.K.   2726, 2875
Barton, J.L.   3504
Baumhauer, O.   384
Beck, C.E.   937
Beling, W.A.   193, 357, 404, 503, 953, 1946, 2355, 2365, 2580, 2659, 2697, 2989, 3029, 3336, 3659
Bell, F.E.E.   405
Bettelheim, A.   2149
Bidwell, R.   430-432
Black, J.S.   3008
Blume, H.   364, 443, 463, 1200, 2874, 3544
Böckstiegel, K.-H.   1768, 1813, 3316

Boni, S.   842
Bowen-Jones, H.   2728
Bricault, G.C.   1977
Brown, L.C.   2093
Bullard, R.   558
Carruthers, D.   1159
Chamieh, J.   3556
Chrystal, G.   3008
Chubin, S.   1311, 1358
Churchill, C.W.   3323
Churchill, R.   3477
Clarke, J.I.   2054, 2728
Collins, R.O.   693
Cooper, C.A.   291
Corcoran, K.R.   730
Crowe, T.   749
Dearden, A.   790
Deighton, L.C.   2100
Dietzel, K.H.   3444
Dovifat, E.   2825
Ehrenberg, C.G.   887
Esposito, J.L.   2451
Field, P.   1181
Fischel, W.J.   2553
Fisher, S.N.   724, 741, 3052
Fisher, W.B.   2054
Fishburne, P.B., III   2217, 2255, 3548, 3553

Franck, P.G. 725
Gabrieli, F. 848
Garnett, E. 853, 854
Ghonemy, M.R. el- 118
Giertz, G. 1998
Goitein, S.D. 1172
Gonzalez, N.L. 3335
Good, D. 3o59
Gosciniak, H.-T. 852
Gräf, E. 1293
Graves, P.P. 1132
Grothe, H. 2625
Gruber, G. 26o3
Grün, E. 2273
Grün, R. 2273
Haddad, H.S. 1174
Haddad, W.W. 23o5
Hoenerbach, W. 144o
Hofmeier, R. 18o1
Hopwood, D. 1525, 2372, 2572, 2836, 2941
Hunter, D.E. 3333
Ismael, T.Y. 2513
Issawi, C.P. 1426, 2377, 2553
Itzkowitz, N. 2o93
Jacques, T. 64, 1847
Janzen, J. 918
Keegan, J. 87o
Kennedy, W.J. 1538
Kerr, M.H. 3598, 3646
Kettani, M.A. 2895
Kilmarx, R.A. 2366
Kilner, P. 1565
Klute, F. 3443
Knowles, A.S. 2773
Kosiński, L.A. 445
Koszinowski, T. 367, 511, 9o2, 1448, 18o5, 18o6, 2o44, 2548, 2855, 2933
Kritzeck, J. 785, 1268, 1435

Kruse, F. 2893
Lamping, H. 26o3
Larson, R. 64, 1847
Lawrence, A.W. 1871
Lechleitner, H. 26o2
Leeds, A. 3o97
Lerg, W.B. 2814
Long, D.E. 1945
Lutfiyyah, A.M. 3323
Lutz, W. 26o3
Maged, M. 33o6
Mallakh, D. el- 1991
Mallakh, R. el- 1991
Mansfield, P. 2o12
Meckelein, W. 1oo, 26o6
Meek, M.E. 64, 1847
Meinecke, M. 292, 2844
Milmo, S. 1565
Milson, M. 362o
Moore, A. 1181
Mostyn, T. 2184
Müller, D.H. 1854
Müller, R. 26o3
Nader, C. 167
Nallino, M. 2237
Nelson, C. 7o5
Nelson, R.M. 83, 1891
Niblock, T. 194, 452, 549, 71o, 919, 1199, 12o4, 1343, 151o, 1997, 2265, 2266, 3oo1, 3o66, 3422, 3481
Nijim, B.K. 1174
Nötzold, G. 1374
Nohlen, D. 1278
Nuscheler, F. 1278
Ochsenwald, W.L. 23o5
Prakke, H. 2814
Pratt, W.E. 3o59
Prothero, R.M. 445
Purdy, A. 2487

Rajan, M.S.   2245
Rashid, I. al-   2515, 2516
Reich, B.   1945
Reischauer, R.D.   2563
Rödiger, E.   3390
Robertson, N.   2612
Ryan, J.   3521
Ryder, W.   2963
Saad, A.T.   3521
Sardar, Z.   2744
Schiavone, G.   2371
Schmieder, O.   3444
Schmitthenner, H.   3444
Schmolke, M.   2814
Schönborn, M.   1801
Scholz, F.   918
Scoville, S.A.   2880
Seifert, W.W.   2895
Sherbiny, N.A.   906, 2951
Sherwood, M.A.   2952
Shilling, N.A.   1643, 3561
Shiloh, A.   1924
Shirreff, D.   2963
Silvert, K.H.   2284
Simmonds, K.R.   3477
Sinclair, R.W.   2988
Sinor, D.   64, 1847
Steinbach, U.   1801, 2043
Stone, R.A.   3073
Sweet, L.E.   3098, 3099, 3341
Tachau, F.   3393
Taylor, A.   1261
Teaf, H.M.   725
Tessler, M.A.   906
Thomas, R.H.   3174
Thompson, J.H.   2563
Tramontana, A.   842
Trietsch, D.   3224
Udovitch, A.L.   1533
Vayda, A.P.   3097

Voigt, W.   2486
Vorlaufer, K.   2603
Wallace, J.   1565
Weintraub, R.   3376
Weintraub, Stanley   3376
Welch, J.   3477
Whelan, J.   147, 488, 3401
Wherry, E.M.   3504
Whitten, P.   3333
Wickersham, W.G.   2217, 2255, 3548, 3553
Wilhelmy, H.   2726, 2875
Willoughby-Osborne   2974
Winkler, E.   2602
Winder, R.B.   785, 1268, 1435
Wright, K.M.   2215
Yassin, S.   3598, 3646
Young, T.C.   2333
Zahlan, A.B.   167
Ziock, H.   2133
Ziwar-Daftari, M.   493, 3423
Zwemer, S.M.   3504

NACHTRAG / ADDITION

Hunt, S.B.   3678

IV KORPORATIVE VERFASSER UND HERAUSGEBER / CORPORATE AUTHORS AND EDITORS

Agricultural Science Faculty, American University of Beirut   2692
Algemene Bank Nederland   13o
American Association of Collegiate Registrars and Admissions Officers   3166, 3167
American Friends of the Middle East   171, 428, 876, 3166
Arab Information Center   2o1
Arab Support Committee   3o26
ARAMCO 24o-264, 355, 831, 997, 183o, 1831, 1884, 2354, 2941, 311o, 3326
Army Military Personnel Center   331o
Arthur D. Little International Inc.   282o
Australian Department of Overseas Trade   313
Bechtel (Power) Corporation 392-394, 986, 1934
Botschaft der Bundesrepublik Deutschland, Dschidda   477
British Bank of the Middle East   3661
British Council   524
British National Export Council   525
British Overseas Trade Board   526, 527
Building Management and Marketing Consultants, Ltd.   557
Bundesstelle für Aussenhandelsinformation   56o-612, 792
Bundesverband der Deutschen Industrie e.V.   613, 614
Business International S.A.   632
Central Office of Information   67o
Centre d'Études et de Recherches sur l'Orient Arabe Contemporain   473, 655, 668, 671, 673, 676, 94o, 113o, 1763, 1937, 2631, 2847, 3448
Centre for Arab Gulf Studies, University of Basrah   2334, 2721

Centre for Economic, Financial and Social Research and Documentation S.A.L.   672
Chase World Information Corporation   11o, 1867
Committee for Middle East Trade   714
Committee for the World Atlas of Agriculture   8o4
Confederation of British Industry   718, 719
Council for Middle East Trade   356o
Council of British Manufacturers of Petroleum Equipment (CBMPE)   734
Delegation der Liga der Arabischen Staaten   796
Deutsche Bank   8o8, 8o9
Deutsche Gesellschaft für Agrar- und Ernährungshilfe in Entwicklungsländern e.V.   5o4
Deutsche Stiftung für Entwicklungsländer   116o, 1438
Deutsches Hydrographisches Institut   811, 812, 3566
Doxiadis Associates   857-86o
Economic Commission for Western Asia (ECWA)   874, 2949
Economic Research Institute, American University of Beirut   289, 331, 728, 82o, 1511, 1512, 2o73, 3625
Elektrizitäts-Actien-Gesellschaft, vorm. W. Lahmeyer & Co., Consulting Engineers   897
España, Ministerio de Asuntos Exteriores   911
ETCO Consulting Engineers   1664, 1665
European Consortium for Political Research   2322
FAO   118, 335, 361, 476, 621, 862, 933-936, 965, 983, 117o, 1274, 1428, 1766, 1864, 1865, 1878, 2o38, 2o61, 2456, 2981, 33o3, 3515, 3573, 3574
Farm Machinery Corporation   3575
First National City Bank   989, 99o
Ford Foundation   3oo9
Foreign Operations Administration   2278
Forschungsinstitut der Friedrich-Ebert-Stiftung   1oo5
France, Ministère de la Guerre   1o13
——, Ministère des Affaires Étrangères   1o14
——, Service Hydrographique   1o15, 1o16
G. Candilis Metra International Consultants   1722, 1723
Gesellschaft zur Beförderung der Entdeckung des Innern von Africa   617, 618

Great Britain   1133
——, Admiralty   1134-1137
——, Board of Trade   1138
——, Foreign Office   1139
——, Parliament   1140
GTZ   104, 184, 421, 810, 1070, 1105, 2861, 3404
Hofuf Agricultural Research Centre   1896
ILO   451, 1398, 1399
Ingenieurgemeinschaft Lässer-Feizlmayr Consulting Engineers   3668
Institut für Agrarsoziologie, landwirtschaftliche Beratung und angewandte Psychologie der Universität Hohenheim   1897
Institut für Publizistik der Freien Universität Berlin   2825
Institut für Publizistik der Universität Münster   2814
Institut National de la Recherche Agronomique   1832
Institute of Management Sciences   727
Institute of Practitioners in Advertising   1407
International Bank for Reconstruction and Development (IBRD)   1409
International Institute for Strategic Studies (IISS)   783, 1410
International Iron and Steel Institute   165
International Land Development Consultants B.V. (ILACO)   1411, 1412
Istituto Affari Internazionale   138, 1802, 2690
ITALCONSULT   1429, 1430, 1651, 1652
Kingdom of Saudi Arabia   1580-1586
——, Agency for Technical Cooperation Administration   1587
——, Central Planning Office (Organization)   943, 1588-1590
——, General Directorate for Girls' Education (of Girls' Schools)   1591-1593
——, General Directorate for the National Guard   1594
——, General Directorate for Youth Sponsorship   1595
——, General Organisation for Social Insurance   1596
——, General Petroleum and Mineral Organization (PETROMIN)   1597, 1598
——, Imam Mohammad Ibn Saud Islamic University   1599
——, Industrial Studies and Development Centre   1600-1643
——, Institute of Public Administration   1644
——, King Abdulaziz University   1645
——, Ministry of Agriculture and Water   70, 112, 954, 1188, 1388, 1552, 1646-1662, 2098, 3524, 3569, 3570, 3589, 3605
——, Ministry of Commerce and Industry   1663
——, Ministry of Communications   1664-1668

Kingdom of Saudi Arabia, Ministry of Defence and Aviation 1669
——, Ministry of Education 1670-1687
——, Ministry of Finance and National Economy 1688-1698, 3671
——, Ministry of Hajj 1699
——, Ministry of Higher Education 1700, 1701
——, Ministry of Industry (and Electricity) 1702-1704
——, Ministry of Information 1705-1721, 3606
——, Ministry of the Interior 1722
——, Ministry of the Interior for Municipal Affairs 1723, 1724
——, Ministry of Labour and Social Affairs 1725-1730
——, Ministry of Municipal and Rural Affairs 1731
——, Ministry of Petroleum and Mineral Resources 1732-1737
——, Ministry of Planning 1738-1740, 3607
——, Permanent Delegation (Mission) to the UN 1741, 1742
——, Ports Authority 1743-1746
——, Riyad University 1747
——, Royal Commission for Jubail and Yanbu 1748
——, Saudi Arabian Agricultural Bank 1749
——, Saudi Arabian Airlines 1750, 2778, 2779
——, Saudi Arabian Monetary Agency 1751-1753
——, Saudi Fund for Development 1754
——, Saudi Industrial Development Fund 1755-1757
Klöckner-Humboldt-Deutz AG 3374
Königreich Saudi-Arabien, Ministerium für Arbeit und Soziale Angelegenheiten 1781
——, Ministerium für Erziehung 3673
——, Ministerium für Information 1782-1784
Leichtweiss-Institut für Wasserbau der Technischen Universität Braunschweig 1895-1897
Lloyds Bank 1931
London Chamber of Commerce 1938, 1939
Marineleitung 2021, 2022
Mecca Cultural Club 2219
Metra Consulting Group 789, 2104
Middle East Airlines Airliban 2119
Middle East Supply Centre 2121-2124
Midland Bank 2127
Munzinger-Archiv 1390
Nah- und Mittelost-Verein e.V. 1019, 2227-2229

National Commercial Bank   2251
Nationale Schweizerische UNESCO-Kommission   1162
Oberkommando der Kriegsmarine   2296, 2297
OECD   1o9, 2314
Österreich, Bundeskammer der Gewerblichen Wirtschaft   3634
Office Belge du Commerce Extérieur   2317, 2318
OPEC   2325
Philipp Holzmann AG   2432-2442, 25o9
Plunkett Foundation for Co-operative Studies   2685
Price Waterhouse & Company   2474
Professional Business Reports on Market Development N.V.   2481
Reichs-Marine-Amt   2542, 2543
Riyadh Chamber of Commerce and Industry   261o
Royal Institute of International Affairs   2646
Royaume d'Arabie Séoudite, Ministère de l'Information   2647
——, Ministère du Travail et des Affaires Sociales   2648
Sarabex Ltd.   52o
Saudi Consulting House   282o, 3643
Saudi Plastics Products Company Ltd.   2821
Schweizerische Bankgesellschaft   2876
Seven Arabian Markets, Ltd.   29o3
Shair Management Services   2916
Shiloah Center of Middle Eastern and African Studies   1494
SIPRI. Stockholm International Peace Research Institute   2992, 2993
Society of Petroleum Engineers of AIME, Saudi Arabia Section   2556, 2561
South Asian Student Association   3o26
SRI International   1149
Städtebauliches Institut im Fachbereich ORL der Universität Stuttgart   3o42
Stanford Research Institute   869, 3o44
Statistisches Bundesamt   3o46-3o49
The Associate Consulting Engineers   253
The Brookings Institution   2492
UNO   317, 619, 956, 1255, 2966, 31o5, 3256-3268
UNCTAD/GATT   3269-3272
UNESCO   4o2, 1182, 1224, 1549, 2696, 3274-328o
UNIDO   2835
UNITER/State of California Conference   923

Universität Tübingen, Sonderforschungsbereich 19, Tübinger Atlas
 des Vorderen Orients 2547
Université Catholique de Louvain, Institut des Pays en Développe-
 ment, Centre de Recherches sur le Monde Arabe Contemporain
 (C.R.M.A.C.) 3282
University Securities Ltd. 3284
US Department of Agriculture 1335, 2153
US Department of Commerce 1952, 2453, 2454, 3285-3290, 3516
US Department of Health, Education, and Welfare 640, 883
US Department of Housing and Urban Development 3291
US Department of the Interior 3292
US Department of State 3293
US Hydrographic Office 3294
US Information Agency 3295
US-Saudi Arabian Joint Commission on Economic Cooperation 3297,
 3298
Verein Deutscher Maschinenbau-Anstalten (VDMA) e.V. 3315
WAKUTI Karl Erich Gall KG (WAKUTI GmbH, Consulting Engineers)
 704, 1654-1656, 1658, 3253, 3356, 3357, 3496
WHO 687, 2463

Bibliothèque nationale / CCOE

CATALOGUE COLLECTIF DES OUVRAGES EN LANGUE ARABE ENTRES ACQUIS PAR BIBLIOTHEQUES FRANCAISES DE 1952 – 1983

Union catalogue of Arabic books
in French Libraries 1952–1983

Herausgegeben von Mohamed Said und Georges Haddad
unter Mitwirkung des Instituts du Monde Arabe

1984. 2667 Seiten. Linson. DM 800,—
ISBN 3-598-10510-X

1952 hat die Bibliothèque nationale mit dem Catalogue Collectif des Ouvrages Etrangers (CCOE), einem Verzeichnis ausländischer Bücher in französischen Bibliotheken begonnen. 1970 wurden die arabischen Bücher abgetrennt und ein eigenständiger Bereich speziell für arabische Literatur eingerichtet.

Der Bestand beläuft sich auf 3.500.000 Karten mit Büchern in arabischer Schrift und Bücher von arabischen Autoren in lateinischer Schrift geschrieben.

Durch die Teilnahme von großen Bibliotheken mit umfangreichen arabischen Beständen (Bibliothèque nationale, Institut des langues et civilisations orientales, Bibliothèque nationale et universitaire de Strasbourg) wird mit deren Karteikarten der CCOE regelmäßig erweitert und stellt ein vollständiges Bestandsverzeichnis dar.

1978 regte die Mission de la recherche au Ministere des Universités an, die Karteikarten für die Öffentlichkeit in Katalogform zugänglich zu machen.

1983 wurde der Katalog mit der Hilfe des Institut du Monde Arabe zusammengestellt und jetzt vom K. G. Saur Verlag veröffentlicht.

K·G·Saur München · New York · London · Paris

K·G·Saur Verlag KG · Postfach 71 10 09 · 8000 München 71 · Tel. (089) 79 89 01
K·G·Saur Inc. · 175 Fifth Avenue · New York, N.Y. 10010 · Tel. 212-9821302
K·G·Saur Ltd. · Shropshire House · 2-20 Capper Street · London WC1E 6JA · Tel. 01-637-1571
K·G·Saur, Editeur SARL. · 6, rue de la Sorbonne · 75005 Paris · Téléphone 354 47 57

Bibliographien zur Regionalen Geographie und Landeskunde / Bibliography on Regional Geography and Area Studies
Hrsg. v. Walter Sperling / Lothar Zögner

Walter Sperling
Bd. 5: **Landeskunde DDR**
Eine kommentierte Auswahlbibliographie. Ergänzungsband 1978–1983
1984. ca. 600 Seiten. Linson DM 56,–
ISBN 3-598-21135-X

Dieser Band ist eine Fortführung und die ideale Ergänzung zum 1978 vorgelegten Landeskunde DDR-Band. Die einzelnen Titel sind jetzt alphabetisch nach Autoren geordnet. Annotationen werden kapitelweise gegeben und zwar so, daß der Benutzer zunächst zu Einführungen und Standardwerken hingeführt wird und erst im Anschluß daran an speziellere Literatur. Personen- und Sachregister (kumuliert mit dem geographischen Register sowie ein Titelverzeichnis) sichern raschen Zugang zu der gewünschten Information.

Bereits erschienen:

Bd. 1: **Landeskunde DDR**
Eine annotierte Auswahlbibliographie
Bearb. u. komm. v. Walter Sperling
1978. XXII, 456 S. Lin. DM 48,–
ISBN 3-7940-7038-0

Bd. 2: Eckart Ehlers
Iran
Ein bibliographischer Forschungsbericht / A Bibliographic Research Survey. Mit Kommentaren und Annotationen / With Comments and Annotations.
1980. XIII, 441 S. Lin. DM 76,–
ISBN 3-598-21132-5

Bd. 3: Armin Hetzer / Viorel S. Roman
Albanien / Albania
Ein bibliographischer Forschungsbericht / A Bibliographic Research Survey. Mit Titelübersetzungen und Standortnachweisen / With Location Codes.
1983. 653 S. Lin. DM 110,–
ISBN 3-598-21133-3

United Arab Emirates
Business Directory 1984/85 Edition
Published by ATCO Araba Business International
1984. 721 Seiten. Kunststoffeinband.
DM 398,–
ISBN 3-598-07506-5

Der Band verzeichnet über 1200 wichtige Firmen, ihre Tochtergesellschaften und die großen Importeure. Ein mehrsprachiger Index bietet Zugang in Englisch, Französisch, Deutsch, Spanisch und Italienisch.

K·G·Saur München · New York · London · Paris
K·G·Saur Verlag KG · Postfach 71 10 09 · 8000 München 71 · Tel. (089) 79 89 01
K·G·Saur Inc. · 175 Fifth Avenue · New York, N.Y. 10010 · Tel. 212-9821302
K·G·Saur Ltd. · Shropshire House · 2-20 Capper Street · London WC1E 6JA · Tel. 01-637-1571
K·G·Saur, Editeur SARL · 6, rue de la Sorbonne · 75005 Paris · Téléphone 354 47 57

Ref Z 3026 .P48 1984
Philipp, Hans-Jurgen.
Saudi Arabia

MAR 3 1 1986